Microwave and Radio Frequency Applications

Proceedings of the Third World Congress on Microwave and Radio Frequency Applications

"Bridging Science, Technology, and Applications"

Related titles published by The American Ceramic Society:

Microwaves: Theory and Application in Materials Processing V (Ceramic Transactions Volume 111)
Edited by David E. Clark, John G.P. Binner, and David A. Lewis
©2001, ISBN 1-57498-103-X

Microwaves: Theory and Application in Materials Processing IV (Ceramic Transactions Volume 80)
Edited by David E. Clark, W.H. Sutton, and David A. Lewis
©1997, ISBN 1-57498-025-4

Microwaves: Theory and Application in Materials Processing III (Ceramic Transactions Volume 59)
Edited by David E. Clark, Diane C. Folz, Steven J. Oda, and Richard Silberglitt
©1995, ISBN 1-57498-002-5

Other books of interest:

Introduction to Drying of Ceramics
Denis A. Brosnan and Gilbert C. Robinson
© 2003, ISBN 1-57498-046-7

Introduction to Ceramic Engineering Design
Edited by David E. Clark, Diane C. Folz, and Thomas D. McGee
© 2002, ISBN 1-57498-131-5

Innovative Processing and Synthesis of Ceramics, Glasses, and Composites VI (Ceramic Transactions Volume 135)
Edited by Narottam P. Bansal and J.P. Singh
© 2002, ISBN 1-57498-150-1

Innovative Processing and Synthesis of Ceramics, Glasses, and Composites V (Ceramic Transactions Volume 129)
Edited by Narottam P. Bansal and J.P. Singh
© 2002, ISBN 1-57498-137-4

The Magic of Ceramics
David W. Richerson
© 2000, ISBN 1-57498-050-5

For information on ordering titles published by The American Ceramic Society, or to request a publications catalog, please contact our Customer Service Department at 614-794-5890 (phone), 614-794-5892 (fax), customersrvc@acers.org (e-mail), or write to Customer Service Department, 735 Ceramic Place, Westerville, OH 43081, USA.

Visit our on-line book catalog at www.ceramics.org.

Microwave and Radio Frequency Applications

Proceedings of the Third World Congress on Microwave and Radio Frequency Applications, September 2002, in Sydney, Australia

Edited by

Diane C. Folz
Virginia Polytechnic Institute and State University

John H. Booske
University of Wisconsin-Madison

David E. Clark
Virginia Polytechnic Institute and State University

John F. Gerling
Gerling Applied Engineering, Inc.

Published by

The American Ceramic Society
735 Ceramic Place
Westerville, Ohio 43081
www.ceramics.org

Proceedings of the Third World Congress on Microwave and Radio Frequency Applications, September 2002, in Sydney, Australia

Cover Illustration (Box): Microwave energy can easily be transferred to metal shapes using a susceptor crucible courtesy of E.B. Ripley, T.L. White and P.A. Eggleston, and appears as figure 1 in their paper "Direct Microwave Coupling to Metals at Elevated Temperatures," which begins on page 241.

For information on ordering titles published by The American Ceramic Society, or to request a publications catalog, please call 614-794-5890.

4 3 2 1–06 05 04 03

ISBN 978-1-57498-158-2

Contents

Microwave Synthesis and Reactions

Microwave Processing of Inorganic Materials

Systems and Diagnostics

Agricultural Products

Applications

Electronic Materials

Waste Treatment and Recycling

Biomedical Research and Health Issues

Preface

In 1988, the Materials Research Society held its first symposium on microwave processing of materials and over the next decade the Materials Research Society and The American Ceramic Society held symposia on the subject on an annual basis, with the symposia alternating between the two societies. Other professional organizations, such as the International Microwave Power Institute, the American Chemical Society, the Association for Microwave Power in Europe for Research and Education (AMPERE) amongst others, also have sponsored symposia in microwave processing. The First World Congress on Microwave and RF Processing was held in Orlando, Florida in January 1997. Its success resulted in the planning of a series of Congresses to be held every two to three years.

In April 2000, the Second World Congress on Microwave and Radio Frequency Processing was held in Orlando, Florida. This Congress was unique in that it provided both interdisciplinary and cross-disciplinary approaches to the use of microwave energy for materials processing. There were over 250 attendees from 25 countries, including chemists, physicists, polymer scientists, electrical engineers, materials scientists, engineers, microwave experts, equipment manufacturers, utility representatives and end-users (i.e., those already using microwave energy as well as those considering its use). Over 150 technical papers were presented at the Congress covering a wide range of radio frequency (RF) and microwave applications in materials processing, including sintering, joining, polymer curing, rubber vulcanization, chemical synthesis, waste remediation, oil separation, medical diagnostics, microelectronics and more. A number of papers were presented that covered the fundamentals of microwave-materials interactions, non-thermal effects, and process modeling.

This Third World Congress on Microwave and Radio Frequency Applications was the first Congress to be held outside the U.S., making the meeting a truly international event. While it was not clear in the early planning stages, the meeting in Sydney, Australia was a tremendous success, especially given the troublesome global economic and political environment.

As it was with the first two Congresses, it is clear that we need more participation of the users (and potential users). The researchers and equipment manufacturers have demonstrated their ability and willingness to address and solve technical issues that will allow implementation. What is needed is more input from the users who know their markets and can help identify processes where microwave energy can provide technological and/or economic benefits. For the Fourth World Congress on Microwave and Radio Frequency Applications, scheduled for Austin, Texas in November 2004, we will again reach out to the industrial community in hopes they will share their successes and their needs for the future.

Editors:

Diane C. Folz — David E. Clark

John H. Booske — John F. Gerling

Note: The papers in this book are peer-reviewed. Each paper underwent at least two reviews; one review by a member of the microwave processing research community and a second review by one of the editors of the book.

Acknowledgements

The editors are grateful for the support provided by various professional societies, government agencies and companies. We also thank the staff of the Tour Hosts Pty Limited for managing the Congress as well as Peter May and the Institution of Engineers-Australia for serving as the hosts of the Congress.

Of course, the most important constituents of any symposium are the speakers, authors, session chairs and manuscript reviewers. Thanks to the creativity and dedication of these individuals, the Congress was high quality and smoothly run, resulting in an important and timely publication.

Supporting Organizations

Advanced Manufacturing Technologies, Australia (exhibitor)

Australian Nuclear Science & Technology Organisation (ANSTO), Australia (exhibitor)

CoberMuegge LLC, U.S.A. (exhibitor)

Communications & Power Industries, Inc., Australia (exhibitor)

EM Software & Systems (SA) PTY LTD, South Africa (exhibitor)

Fiso Technologies, Canada (exhibitor)

Gerling Applied Engineering, Inc.

Institution of Engineers – Australia

Microdry: The Nemeth Engineering Group, U.S.A

PSC, Inc. – A Litzler Company, U.S.A. (exhibitor)

R.F. Schiffmann Associates, Inc.

University of Wisconsin - Madison

Virginia Polytechnic Institute and State University, U.S.A.

Endorsed by

Association of Microwave Power in Europe for Research and Education

Institute of Electromagnetic Wave Application – Japan

Materials Research Society

National Institute of Ceramic Engineers

The American Ceramic Society

Virginia Polytechnic Institute and State University – Microwave Processing Research Facility

Congress Chairs

Luke Nadj, Chair, Advanced Manufacturing Technologies, Inc., Australia
Robert F. Schiffmann, Vice Chair, R.F. Schiffmann Associates, Inc., U.S.A.

Technical and Industrial Program Chairs

John Booske, Chair, University of Wisconsin – Madison, U.S.A.

Diane C. Folz, Vice Chair, Virginia Polytechnic Institute and State University, U.S.A.

John F. Gerling, Industrial Program Chair, Gerling Applied Engineering, Inc., U.S.A.

Organized by the Microwave Working Group – 3rd World Congress Leadership

Bernard Krieger, Cober Electronics, U.S.A.
Jon Binner, Loughborough University, U.K
John Booske, University of Wisconsin-Madison, U.S.A.
David Clark, Virginia Polytechnic Institute and State University, U.S.A.
Diane Folz, Virginia Polytechnic Institute and State University, U.S.A.
John Gerling, Gerling Applied Engineering, Inc., U.S.A.
David Lewis, SOLA International Research Centre, Australia
Robert F. Schiffmann, R.F. Schiffmann Associates, Inc., U.S.A.

International Technical Program Committee

Duangduen Atong
Technology Centre (MTEC)
Thailand

Jon Binner
Loughborough University
United Kingdom

Gary Bond
University of Central Lancashire
United Kingdom

Rod Boswell
Australian National University
Australia

Diane Folz
Virginia Polytechnic Institute & State University
U.S.A.

Camelia Gabriel
Microwave Consultants Ltd.
United Kingdom

John Gerling
Gerling Applied Engineering, Inc.
U.S.A.

Susan Hagness
University of Wisconsin-Madison
U.S.A.

Milan Hajeck
Academy of Sciences of the Czech Republic
Czech Republic

Ruth Kiminami
Universidada Federal de Sao Carlos
Brazil

David Lewis
SOLA International Research Centre

Christina Leonelli
Universita' degli Studi di Modena e Reggio Emilia
Italy

Anton Mavretic
Advanced Energy Corporation
U.S.A.

Kirill Rybakov
Institute of Applied Physics
Russian Academy of Sciences
Russia

Bob Schiffmann
R.F. Schiffmann Associates, Inc.
U.S.A.

Rebecca Schulz
Corning, Inc.
U.S.A.

Miyake Shoji
Osaka University
Japan

Maria Stuchly
University of Victoria
Canada

Nguyen Tran
Industrial Research Institute of Swinburne
Australia

Ian Turner
Queensland University of Technology
Australia

Introduction

The Microwave Working Group: An Unusual Team with an Important Mission

Bernard Krieger, President

Robert Schiffmann, Vice President (and 3rd World Congress Co-Chair)

The Microwave Working Group is the organizer of the Third World Congress on Microwave and RF Processing whose proceedings are documented by the papers selected by the editors for this book. The Congress took place from the 22nd though the 26th of September of 2002 at the Sydney Convention and Exhibition Center in Sydney, Australia under the sponsorship and implementation by the Institution of Engineers Australia.

We are often asked why we have such an unusual name as the "Microwave Working Group," who is this group and what is its purpose? It is easier to explain by saying what we are not. We are not a society, although we are incorporated, and we don't collect dues. Our team is small and fluid in its makeup, but its accomplishments are very big. Our purpose is to bridge science, technology and applications in the area of microwave and RF processing. We coordinate with societies, governments, utilities, and universities on a worldwide basis wherever there is involvement with microwave or RF technology. In conjunction with these constituencies, we produce world congresses to present the best and latest in science together with applications oriented programs designed to attract potential users. Our congresses then document selected papers in hard bound volumes, such as this book, so that all interested parties may have important reference material.

The Third World Congress was attended by delegates from 19 countries and contained well over 120 technical papers in 3 _ days of oral and poster sessions. These papers covered a broad range of topics including: Industrial Microwave and RF Processing; Microwaves and Plasmas; Systems and Diagnostics; Inorganic Materials; Chemistry & Organic Materials; Dielectric Properties; Biological Effects and Standards for Electromagnetic Exposure; and Modeling.

The Technical Program was preceded by three Workshops: 1. Human Exposure to Electromagnetic Fields; 2. Electromagnetic Modeling for Microwave and RF Processing; and 3. Hands-on Course on High-Power Microwave Processing. All were well attended by an enthusiastic audience. Our thanks go out to Drs. Camelia Gabriel, Vadim Yakovlev and Nguyen Tran for their hard work in organizing these workshops and to all their speakers.

The Technical Program Committee took the bold step of devoting a single day to a single topic: Health Effects of RF and Microwave Radiation covering the issues of Sources of Exposure, Dosimetry, Research Methodologies and Standards. The sessions were not only well attended, but there was lively discussion between presenters and the audience. We are grateful to Dr. Camelia Gabriel for all her hard work organizing these sessions and to all the excellent speakers.

The organization of the Microwave Working Group is extremely flexible and unique. Professionals in the field of microwave and RF technology volunteer their efforts on a dynamic basis. The core of the organization for the Third World Congress, in addition to the authors of this article, include: Dr. John Booske of the University of Wisconsin who carried the heavy load of Technical Program Chair. John designed the sessions, picked the session chairs, and solicited and judged the papers together with his team of session chairs. Diane Folz of Virginia Tech, Vice Program Chair, worked closely with Dr. Booske and assisted him. Our organizational scheme is that Diane will be the Technical Program Chair for the Fourth World Congress. Dr. David Clark of Virginia Polytechnic Institute and State University, Symposium Chairman for the Second World Congress in Orlando in 2000, gave us the benefit of his experience on an executive level, as did Dr. David Lewis of SOLA, the Technical Program Chair of the First World Congress in 1997 in Orlando.

John Gerling, Gerling Applied Engineering, organized the industrial applications focus of the Third World Congress. John is also the Treasurer of the Microwave Working Group, an unusual position since we do not have money in our Treasury. However, we must be heavily involved in financial planning to assist the host organizations that provide the resources and overall arrangements. Dr. Jon Binner, University of Loughborough, was the Technical Program Chair for the Second World Congress. He provides an important technical and scientific credential as well as close liaison to an important endorsing society, the AMPERE association, which is the microwave society of Europe. Jon is chairman of their 9th International Conference on Microwave and High Frequency Heating, September 2003.

The host for the Third World Congress was the Institution of Engineers, Australia. For the Second World Congress it was the American Ceramic Society and for the First World Congress, the Electric Power Research Institute.

Our mission of bridging science, technology and applications makes us unique; particularly since we realize that for applications to be successful, they must be economically practical and readily understood by the user community. Further, the fact that we are not a society and have the flexibility to link with different groups permits us to work with different endorsers and societies focused on different disciplines and who are user oriented and can benefit from our technology.

We are already planning the Fourth World Congress on Microwave and Radio Frequency Applications, to be held in 2004, and we plan for that event to link with a large process oriented society in North America and have our congress at the same time and at the same place as their Annual Technical Meeting so that large groups of attendees will be able to benefit from our sessions.

Our hosts for the Third World Congress in Sydney were wonderful and of the highest quality. We linked together prior to September 11, 2001 and that event put heavy burden on Luke Nadj, Chairman of the Third World Congress, and Peter May, Executive Director of the Institution of Engineers Australia. We also thank Fiona Driver and her excellent staff at Tour Hosts who handled the day-to-day organization of the Congress. We appreciate all their efforts and the high quality of the event they produced.

Fundamentals And Modeling

ABOUT TRAPPING OF MODES

Jean-Louis Mousson, Christophe Lohr, Pierre Pribetich and Didier Stuerga

Groupe d'**E**tudes et de **R**echerches sur les **M**icroondes
GERM/LRRS, UMR 5613 CNRS-Université de Bourgogne,
BP 47870, 21078 Dijon cedex, France

INTRODUCTION

Efficient microwave processing have imposed new requirements in applicator design. Hence, interests in using numerical analysis to extract a more complete understanding of the capabilities of the electromagnetic devices has dramatically increased within the last several years. In response to increased interest in energy efficiency and control of processing, the authors have designed new trends in applicator design centered on improving heating uniformity, increasing applicator coupling efficiency and the development of electrical feedback control of microwave processing systems. Most of these developments have been possible by the use of numerical methods. However, this works would show the interest of full waves analysis based on the residue theories[1-3]. This approach is necessary to forecast and analyze the influence, or more precisely, the contribution of each eigen functions or modes to the electromagnetic fields distribution obtained.

Due to the dielectric loss, cut-off frequency vanishes and all modes can propagate along the structure. In fact, inhomogeneous lossy loadings remove the monomodal character of the waveguide. The conventional modal analysis requires both tedious and completely mathematical manipulations and numerical root searching. Due to the high level of losses characteristic equation admits poles and root searching in complex plane with perturbation methods fails. Hence, due to the high level of losses, we cannot use a classical perturbation theory as is usually done. Furthermore, an iterative research procedure on zero will not be suitable because it is not possible to find a good initial value to start the iterative procedure. This is the reason why we have chosen to perform the resolution in the complex plane, with a method based on the residue theory.

MATHEMATICAL FORMULATION

Principles

Many electromagnetic problems relating to opened or closed structures lead to Equation (1) :

$$F(z) = 0 \tag{1}$$

where F(z) is an analytic function having polar singularities and z is a complex variable. Our method calculates the zeros of this kind of analytic function in particular areas of the complex plane, specially inside the region where the function is holomorphic and single valued.
Our method is based on the well-known formula given by Equation (2).

$$M_n = \frac{1}{2\pi i}\oint_D \frac{F'(z)}{F(z)} z^n dz = \sum_i^n m_i z_i^n - \sum_i^n m_i p_i^n \tag{2}$$

where F(z) is the complex function, z_i are the zeros, and p_i are the poles[4]. This Equation gives the nth-order moments M_n of the n zero of a holomorphic function within a closed region. The integral has therefore to be calculated as the sum of the integrals along the circumferences constituting the contour of D. Each integrals has been numerically calculated by means of the well known formula of Delves and Lyness[5]. Once the moments M_n are known, the zero can be calculated by solving an algebraic equation of degree n.

Main steps

During the first step, the examined functions are evaluated along the circumferences in the points chosen for the circumference. It is possible that some zeros may be previously known and given among the input variables. In any case, the number n of unknown zeros must be less than 5. During the first steps, calculus could be interrupted if the variation of the function is too rapid between consecutive points along the circumference. In such case it becomes impossible to determine the number of the zeros in D and it is needed to increase number of points for calculating the integral. The calculus could be interrupted if either a zero or a pole is recognized along one of the circumference.

At the second step, the integral which gives the moments is evaluated. Finally, the zeros are calculated.

Accuracy and limits

The precision which can be obtained is essentially influenced by the variation rate of the examined function along the contour of region D and consequently by the number of points taken on the contour. It is therefore clear that the a zero or a pole close to the contour results in decreased accuracy. In such circumstance there are two ways out : either to increase the number of points or to modify the region D. In any case, it should be borne in mind that on increasing the number of points, the accuracy reaches a maximum. The number of points for which the maximum is obtained depends upon the function and upon the region position.

SOME RESULTS

Applicator design is complex because efficient energy transfer depends of many parameters which changes as heating material takes place. During heating, the material complex dielectric constant may exhibit several change with temperature (increase or decrease). Changes can also result of chemical reactions induced by heating.

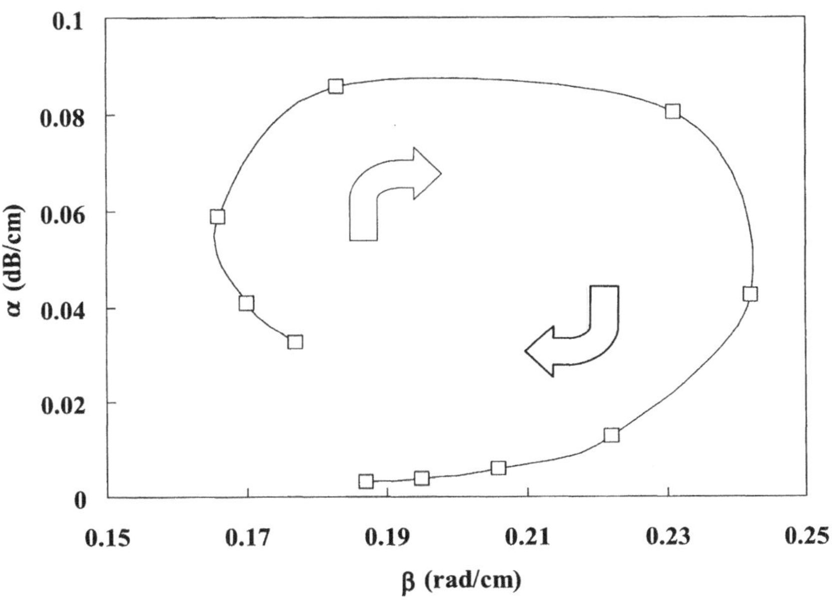

Fig. 1 : Change of propagation constants during heating (from 10°C to 100°C by 10°C steps)

Hence, during microwave treatment both the real and imaginary parts of the material complex dielectric constant may dramatically change.
These changes of materials dielectric characteristics results in a nonlinear, variable equivalent admittance for the applicator. Numerical results shown correspond to a cylindrical waveguide (diameter equals to 180 mm) centrally loaded with a water pipe (diameter equal to 40 mm). The frequency is equal to 2.45 GHz (microwave heating frequency, ISM frequencies). Fig. 1 describes the change of the phase constant and attenuation constant of the first TE^z modes. In complex plane, the complex propagation constant covers a loop during heating.

CONCLUSION

Numerical tools for trapping eigen modes of high lossy industrial applicators have been presented. This trapping methods is based upon the residue theories. Modes established for lossless structures have been extended to high-loss systems and limits of classical perturbations approaches can be completely avoided.

A key goal for the design of optimized microwave industrial applicators for microwave heating is taking into account dielectric losses. The information made available by this way is a valuable contribution to a more accurate understanding of electromagnetic distribution within the applicator. Moreover, our approach should make predictive control and design of optimized applicators. It is a viable alternative to the trial and error methods currently used for designing microwave applicator for industrial heating applications.

REFERENCES

[1]A. Calmels, D. Stuerga and P. Pribetich, "Modeling propagation in high-power microwave devices," *Microwaves and Opt. Lett.* **21** [6] 477-82 (1999).

[2] D. Stuerga, A. Calmels and P. Pribetich, "Modeling propagation for high power cylindrical microwave applicators," *Microwaves and Opt. Lett.* **30** [3] 193-95, (2001).

[3]D.J. Hill, C.D. Rudd, and M.S. Johnson, "Design and application of a cylindrical mode applicator for the in-line microwave preheating of liquid thermoset," *J. of Microwave Power and Electromagnetic Energy*, **33**, [4] 216-230. (1998).

[4]P. Lampariello and R. Sorrentino, "The ZEPLS program for solving characteristic equations of electromagnetic structure," *IEEE Trans. MTT*, **23** 457-458, (1975).

[5]L.M. Delves and J.N. Lyness, "A numerical methods for locating the zeros of an analytic function," *Math. Comput.* **21** 543-560 (1967).

MICROWAVE PROCESSING OF TEMPERATURE-DEPENDENT AND ROTATING OBJECTS: DEVELOPMENT AND EXPERIMENTAL VERIFICATION OF FDTD ALGORITHMS

P. Kopyt, M.Celuch-Marcysiak, W.K.Gwarek
Institute of Radioelectronics, Warsaw University of Technology, Warsaw, Poland

ABSTRACT

The FDTD (Finite Difference Time Domain), which has been previously proven to be effective in modeling high-frequency electromagnetic problems in the telecommunications industry, has also recently been applied in microwave power engineering. Experimental validation of FDTD predictions for temperature-varying media parameters is presented in this work. The paper further describes a simple algorithm that makes it possible to take into account object rotation – important in simulations of domestic microwave ovens.

INTRODUCTION

The modelling of microwave interactions with objects that are heated in microwave devices has been successfully accomplished by the microwave power research community in a number of cases, e.g.[1][2][3]. Recently several commercial electromagnetic software packages, originally developed for the telecommunications industry, have also been applied to microwave heating problems. Their ranking in terms of suitability and effectiveness conducted in [4] points to QuickWave-3D [5], a conformal FDTD electromagnetic simulator based on the method of [6], as a promising tool for microwave processing.

Two important features have received limited attention in both academic and commercial software to date. First of all, the physical parameters of heated media tend to vary, often substantially, with the increase of temperature. This converts the electromagnetic problem into a non-linear one, more difficult to handle than typical linear electromagnetic problems in telecommunications. Secondly, in real-life heating devices the greater uniformity of power distribution within the heated objects is often achieved by moving the object inside the cavity. In industrial ovens, the foodstuffs are passing through a cavity on a conveyor belt, while in small-scale household devices the object is placed on a rotating shelf.

Accurate modelling of microwave heating scenarios requires that thermal dependence of media parameters as well as movement of the objects be taken into account. This first issue has been considered in a recent paper [7], where a special regime for operating the electromagnetic software with variable media has been mathematically developed and implemented. The aim of the present paper is twofold: to experimentally validate the methodology of [7] and to extend the software system by taking load movement into account. To this end, we shall first describe the development of a new Microwave Power Laboratory allowing software validation under controlled frequency and power conditions, and with a future possibility of temperature measurements in rotating loads. Experimental results obtained in this laboratory will be compared to the computational predictions of the software system. Then a simple but effective method to computationally model an object heated and rotated inside a domestic microwave oven will be proposed. In all computations, the electromagnetic analysis is conducted by the existing conformal FDTD solver [5], media parameters are controlled as proposed in [7] while object movement is simulated through a set of external scripts in the Matlab [8] environment.

DEVELOPMENT OF MICROWAVE POWER LABORATORY

The accuracy of the simulation software employed to model microwave heating processes needs to be verified. The measurement equipment recently acquired by the Institute of Radioelectronics has been employed in the task and presented in Fig. 1a. It consists of a microwave oven (by Plazmatronika, www.plazmatronika.pl) with adjustable power level controlled with a PC, and a signal conditioner with a set of eight thermometers (by Fiso Technologies, www.fiso.com) that can register temperature changes simultaneously.

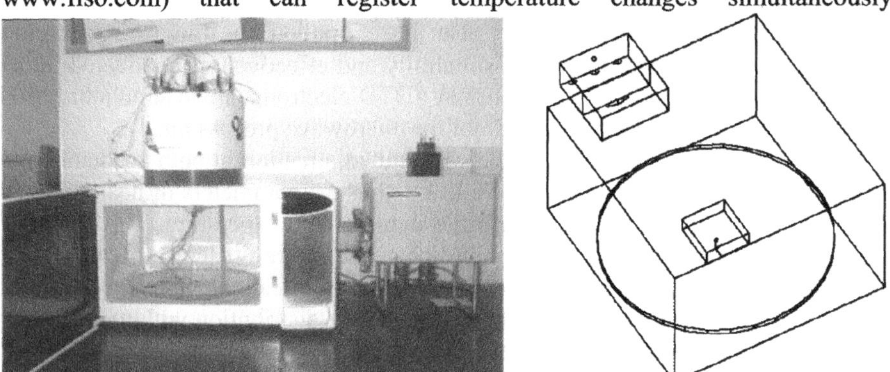

Fig. 1. Measurement system (a); model of the cavity and the sample (b)

During the simulations we assume that the oven is fed by an ideal magnetron tuned to 2.45 GHz. This assumption has been verified experimentally by measuring the magnetron spectrum for two different loads placed inside the oven cavity. One was a glass of water and the other — a sample of bread 60(w) × 60(d) × 20(h) mm. Both were placed centrally on the oven turn-table. The measurements have been performed with a spectrum analyser connected to a half-wavelength dipole. Fig. 2. shows the results. It is clear that the spectrum is narrow (about 5MHz) and the change of the load has caused only a small shift of the magnetron operating frequency (by 10MHz).

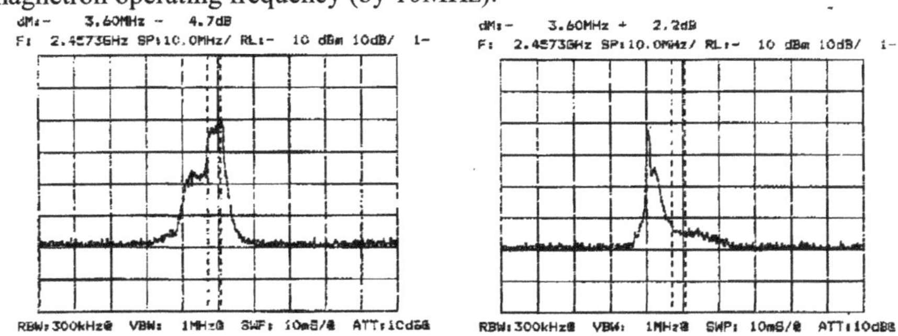

Fig. 2. Microwave oven magnetron spectrum: a) for water; b) for bread

However, what also needs to be checked is the frequency characteristics of reflections. If for a given load a strong resonance occurs in the vicinity of 2.45 GHz, then even a small shift of magnetron operating frequency may abruptly change the working conditions. While we cannot easily perform wide-band measurements in the physical set-up, one FDTD simulation with pulse excitation and Fourier-transform processing quickly produces the results shown in Fig. 3. The reflections are constant in the neighbourhood of 2.45 GHz, which confirms that we can further assume an ideal magnetron working at this frequency.

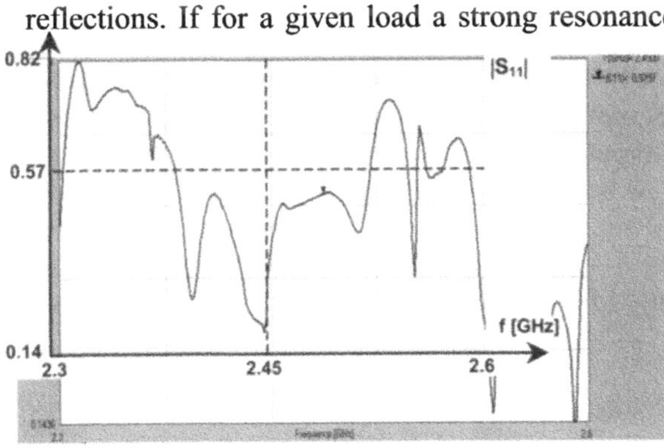

Fig. 3. Simulated reflection coefficient of the microwave oven with bread sample

EXPERIMENTAL VALIDATION OF THE ACCURACY OF SIMULATION WITH VARIABLE MEDIA PARAMETERS

In the simulations an additional module – called BHM (*Basic Heating Module*) [7] – has been used that takes into account the temperature-induced changes of the media parameters. The changes are based on the data stored in an external file containing enthalpy and the corresponding media parameters: electrical permittivity, conductivity (losses) as well as temperature. Example data further used for bread are graphically presented in Fig. 4.

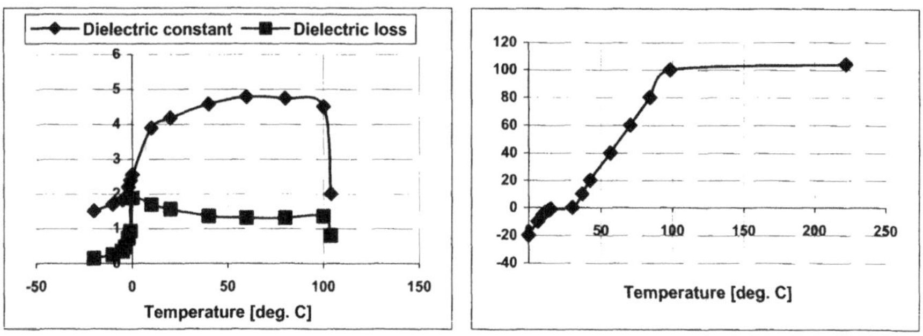

Fig. 4. Data contained in the file describing the media: a) dependence of the electrical permitivitty and conductivity on the temperature; b) dependence of the temperature in a cell on the enthalpy in this cell.

The BHM module repetitively modifies the media parameters according to the dissipated power and data provided in the file. The total heating time has to be divided into smaller timesteps in order to accurately model the gradual changes of the media parameters. The operations performed in each step are illustrated in Fig. 5. First, the steady-state needs to be reached in order to calculate the dissipated power envelope in the simulated circuit. Then the lossy materials are being heated for the time equal to the thermal timestep length. Therefrom enthalpy is obtained and the media parameters are modified according to the file, and temperature in each cell is changed.

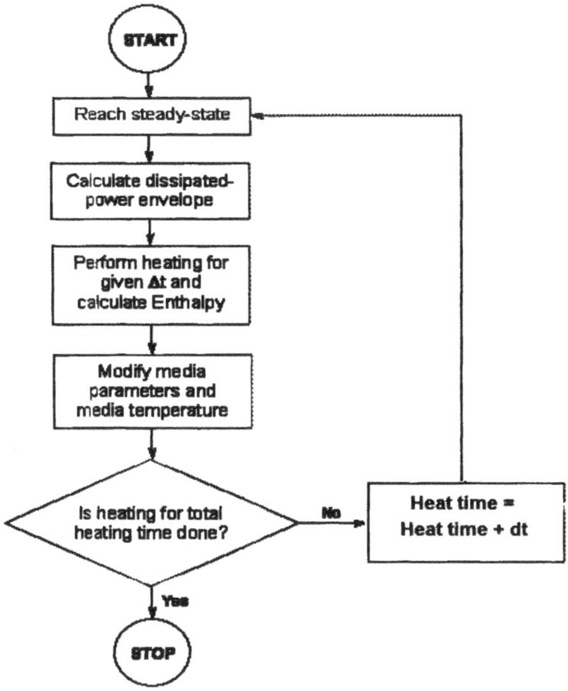

Fig. 5. Operation of the BHM module

We have performed simulations and then measurements with the sample of bread 60(w) × 60(d) × 20(h) mm placed centrally on the oven turntable. The power level has been set to 100W and the heating lasted around 5 minutes. Temperature has been registered at 8 locations indicated in Fig. 6b, in the middle layer of the sample. Since the main goal of the experiment has been to check the accuracy of the software, we have performed the heating without sample rotation. It has made the computation time much shorter as there has been no need for repeated simulations of the oven with sample at various angular positions.

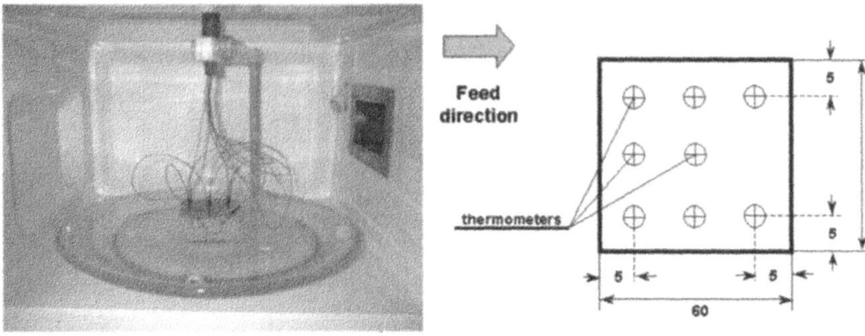

Fig. 6. The oven cavity with bread sample (a); placement of thermometers in sample (b)

The experimental data and the simulation results are presented in Fig. 7a. The comparison has been made in the hot spot, which is just off the sample center. The curves are close to each other with the biggest discrepancy occuring an the beginning of the heating process. This can be ascribed to the inertia of the thermometer.

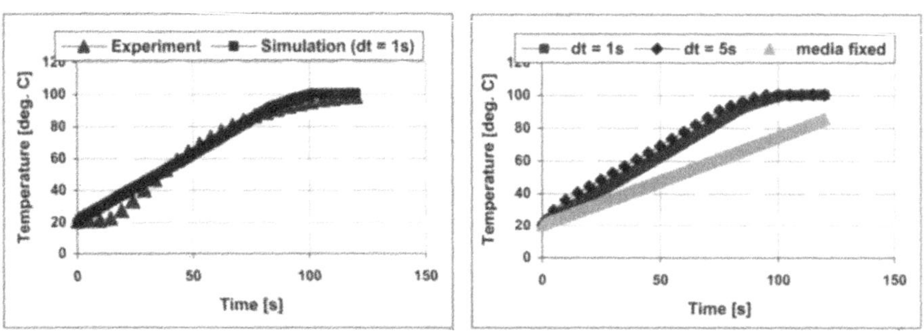

Fig. 7. a) measurements and simulations (in the hot spot); b) simulation results (two different heating timesteps – 1s and 5s – and one case of fixed media parameters)

The inertia of the thermometers used in the measurements has been checked experimentally for water and for bread. The thermometer has been placed first in a glass filled with water at room temperature (25°C) and then transferred to a glass filled with much hotter water. The output has been exhibited in Fig. 8a. A similar experiment has been conducted for bread. It seems that in case of water — where contact resistance between the thermometer and the medium is low — time of temperature rise is around 4s., while in bread the resistance is much higher and the time is as high as 20s. The results obtained for bread have been presented in Fig. 8b.

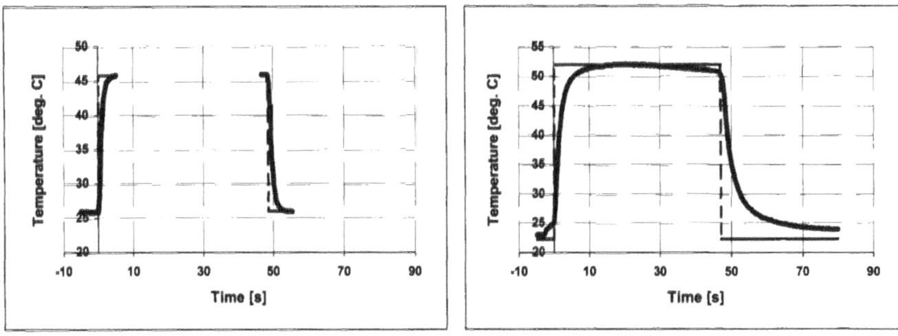

Fig. 8. Response of the thermometer used in experiment: a) for water; b) for bread

Finally, let us note that the simulation data in Fig.7a have been obtained for a heating time step set to 1s. We have later checked that a time-step as big as 5s could be used in this case; the differences shown in Fig.7b are small since the heating power is relatively low. When heating with higher power levels, the timestep should be shorter in order to avoid instabilities in the simulations. Comparing BHM results to standard FDTD analysis with constant media parameters (Fig.7b), we find that even slow changes of parameters, such as those in Fig.4a, can affect the accuracy of the simulation.

NEW METHOD OF SIMULATION OF ROTATING LOADS

Further simulations have been conducted with a model of an example microwave oven. The whole geometry of the model together with excitation and the sample object being heated have been prepared with a single macro. By modifying its parameters one can place the sample object at any angle inside the oven. Next, using an external routine written in Matlab [8], one can trigger QuickWave-3D simulation several times, each time for a different object position. The calculated dissipated power pattern in the object cross-section will be different for each angle.

The collected information comes in the form of snapshots of dissipated power in the object's cross-sections. If the object is rotated, the snapshot is rotated as well. Obviously direct summing of the data for different rotation angles is not possible. This task requires that the snapshots of dissipated power be first brought back to the original position — 0 degrees. Since in each case the angle of rotation is known, a standard formula for coordinate system rotation can be applied to each cell.

$$x' = x\cos(\alpha) + y\sin(\alpha), \qquad (1a) \qquad\qquad y' = -x\sin(\alpha) + y\cos(\alpha) \qquad (1b)$$

where x, y are the coordinates of the cell-centers of a rotated snapshot, while x' and y' are the coordinates of the cell-centers after the rotation back to the reference position.

After the rotation to the reference position, it is possible to average the dissipated power for all the angular positions of the object.

VERIFICATION OF THE ROTATING LOAD MODEL

Fig. 9 presents three images that have been rotated back and are ready for averaging. During all the simulations the object (a parallelepiped 40x40x20mm) is placed centrally on the shelf and is rotated around its center. The object is made of meat modelled as a material of relative electric permitivity $\varepsilon_r = 50 - j20$.

The accuracy of the averaging procedure for a different number of distinct positions has been tested. As a reference, we have used the averaged pattern $\mathbf{P^5}$ obtained by summing up power snapshots taken every 5 degrees. Table 1 contains the results of comparison between the reference pattern and patterns based on a smaller number of angular positions. The relative error has been defined as:

$$e(a) = \frac{\sum_{i=1}^{N} \sum_{j=1}^{M} \left(P_{ij}^{5} - P_{ij}^{a}\right)^2}{NM}$$

where, P^a is a pattern being verified against the pattern P^5, N and M are the dimensions of the patterns (given in cells).

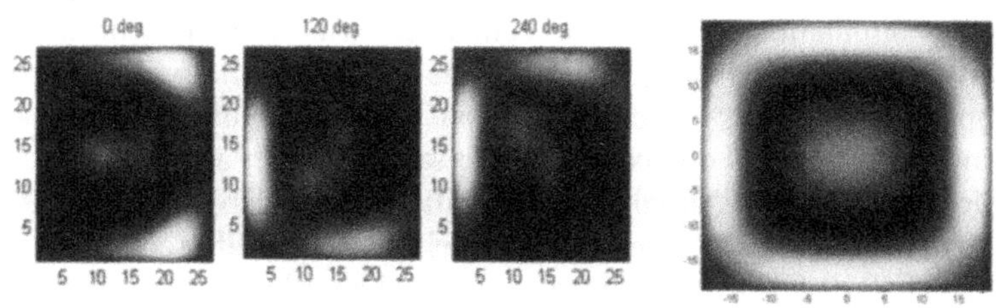

Fig. 9. Dissipated power pattern for 3 angular positions of the object (from 0 to 240 with the step of 120 degrees) and the averaged heating pattern.

Table I. Relative error of averaging (for sets 7, 8 the mirroring has been used)

Set number	1	2	3	4	5	6	**7**	**8**
Step [°]	5	10	20	30	45	60	[1]	[2]
No. angles	72	36	18	12	8	6	**6**	**7**
Effective no. angles	72	36	18	12	8	6	**10**	**12**
Error [%]	0	0.09	0.30	1.16	4.00	5.14	**1.90**	**1.13**
[1] angles in set no. 7: 0°, 36°, 72°, 180°, 216°, 252°; [2] angles in set no. 8: 0°, 30°, 60°, 90°, 180°, 210°, 240°								

The error grows as the number of modelled angular positions decreases. It is clear that due to the symmetry of the problem one can find pairs of snapshots that are mirrored copies of one another. It is sufficient to simulate one result and sum it up twice with and without mirroring. Sets 7 and 8 contain angles only from the first and third quadrants of the coordinate system, and mirrored. Their error rate is low despite the relatively small number of electromagnetic simulations performed.

CONCLUSION

We have presented the experimental verification of the simulation tool used in modelling heating processes. It has been shown that the accuracy of the computations is high enough to ensure a good agreement with measurements. Next, the software has been used to model problems with load rotation, which is an important issue in microwave power applications and has not been previously addressed. An effective and simple approach has been proposed. It is a post-processing method that is easy to implement with standard mathematical routines found in e.g. the Matlab package. This is important in design of microwave ovens and microwaveable food packages.

REFERENCES

[1] P.O. Risman and M. Celuch-Marcysiak, "Electromagnetic modelling for microwave heating applications," *13th Intl. Microwave Conf. MIKON-2000*, Wroclaw, Poland, May 2000, pp. 167-182.

[2] M. Sundberg, P. Kildal and T. Ohlsson, "Moment method analysis of a microwave tunnel oven," *Journal of Microwave Power and Electromagnetic Energy*, **33** [1], 36–48.

[3] B. Wäppling-Raaholt and P.O. Risman, "FDTD simulation of a microwave heating process: effects of oven parameters on heating uniformity," *3rd Intl. Conf. on Predictive Modelling*, Leuven, September 2000, pp.271–273.

[4] V. V. Yakovlev, "Examination of contemporary electromagnetic software capable of modelling problems of microwave heating," *Proc. 8th Intern. Conference on Microwave and HF Heating*, Sept. 2001, Bayreuth, Germany,

pp. 19-20. Complete paper in Trends in Microwave and HF Heating, Springer Verlag, 2001 (to be published).

[5] W. K. Gwarek, M. Celuch-Marcysiak, M. Sypniewski, A. Wieckowski, *"QuickWave-3D Software Manual,"* QWED. Poland 1999.

[6] W. K. Gwarek, "Analysis of an arbitrarily-shaped two-dimensional microwave circuits by FD-TD method," *IEEE Trans. on Microwave Theory and Techniques*, **36** [4], pp. 738-744, April 1988.

[7] M.Celuch-Marcysiak, W.K.Gwarek, M.Sypniewski, "A novel FDTD system for microwave heating and thawing analysis with automatic time-variation of enthalpy-dependent media parameters", *8th International Conference on Microwave and High Frequency Heating*, Bayreuth, Sept.2001.

[8] Matlab 5.1, Mathworks – university licence.

A COMPARISON OF SEMI-ANALYTICAL AND NUMERICAL SOLUTIONS FOR THE MICROWAVE HEATING OF A LOSSY MATERIAL IN A THREE-DIMENSIONAL WAVEGUIDE

B. Liu and T.R. Marchant,
School of Mathematics and Applied Statistics,
University of Wollongong,
Wollongong, 2522, N.S.W., Australia.

I.W. Turner and V. Vegh,
School of Mathematical Sciences,
Queensland University of Technology,
GPO Box 2434, Brisbane, Q4001, Australia.

ABSTRACT

The microwave heating of a three-dimensional block in an infinitely long rectangular waveguide propagating the TE_{10} mode is considered. The electrical conductivity (the dielectric loss) is assumed to be a function of temperature, and modelled by the Arrhenius law. A coupled set of equations is obtained that describes the electromagnetic fields and the temperature distribution in the block.

The numerical solutions of this problem are obtained by two methods, the well known FD-TD scheme and a frequency domain method which makes the further assumption that a single TE_{10} mode exists in the waveguide and material. The results show that an S-shaped temperature versus power response curve is possible, with thermal runaway occurring when the solution jumps from the lower (cool) branch to the upper (hot) branch of this curve. The results confirm the validity of the single-mode assumption and show a good comparison between the numerical and semi-analytical solutions.

INTRODUCTION

Industrial microwave heating has a history dating back to the 1940s (see [1]), and there are many heating, drying and thawing applications in industry. One key benefit of microwave heating is its speed of processing compared to conventional convective heating, which arises primarily due to the volumetric heating phenomenon associated with microwave processing. This phenomenon, which sees the bulk of the energy delivered directly to the core of the sample, can be used to eliminate the case-hardening of products that are heated and dried by classical convective methods. Other advantages of microwave heating are the more compact heating systems offered by the technology, the superior final product quality, and cleaner heating and drying environments.

In general, controlled uniform heating is difficult to achieve with microwave processing, because rapid overheating (thermal runaway) can occur in some materials for a small increase in incident power. The phenomenon of thermal runaway is analogous to the combustion of an exothermic chemical reactant. The rate at which a chemical reaction proceeds increases with temperature, hence if the reaction is exothermic then the heat generated will result in positive feedback that can cause an explosion. The rate at which microwave radiation is absorbed by a material is also temperature-dependent. Hence, thermal runaway in this context is the result of an analogous feedback process.

A study of microwave heating involves solving both Maxwell's equations and the forced heat equation. Maxwell's equations describes the propagation of the microwave radiation in the material and the microwave waveguide or cavity, while the forced heat equation describes the heat absorption and diffusion in the block. Due to the temperature-dependent material properties, the governing equations are nonlinearly coupled and unable to be solved analytically. Numerical and semi-analytical solution methods have been used with success however, and are briefly reviewed below.

[2] considered the steady-state microwave heating of a three-dimensional block by the transverse magnetic waveguide mode TM_{11}, in a long rectangular waveguide. The electrical conductivity and the thermal absorptivity were assumed to be temperature dependent while both the electrical permittivity and magnetic permeability were assumed constant. Both a quadratic relation and an Arrhenius-type law were used for the temperature-dependency. Semi-analytical solutions, valid for small thermal absorptivity, were found for the steady-state temperature and the electric-field amplitude using the Galerkin method. At the steady-state, the temperature versus power relationship was found to be multivalued; at the critical power level, thermal runaway occurs when the temperature jumps from the lower (cool) temperature branch to the upper (hot) temperature branch of the solution. Examples were presented

in the limits of small and large heat-loss with an excellent comparison with numerical solutions obtained.

[3] considered the small Biot-number heating of a ceramic slab in a TE_{103} waveguide applicator including an iris with a variable aperture. By considering the small Biot-number limit, which allows the heat equation to be averaged, an expression is obtained relating the temperature and incident power. This relation was shown to be bistable, or S-shaped. The microwave cavity can be tuned by varying the aperture width because this changes the S-shaped response curve. Hence, there exists an optimal aperture width that sinters the ceramic with minimum power. Moreover, varying the aperture width (and hence the S-shaped curve) allows the sintering temperature on the upper branch to be reached without feedback control.

The finite-difference time-domain (FD-TD) method is commonly used to solve Maxwell's equations numerically (see [4,5,6,7]). [7] applied the FD-TD method to the microwave heating of a lossy material in a three-dimensional cavity, where a rectangular waveguide propagating the TE_{10} mode excites a cavity. It was shown that the assumption of the existence of a single TE_{10} mode within the waveguide provided a computed electric-field that did not satisfy the divergence condition. The authors developed a full-mode numerical model that included improved numerical implementations of the appropriate boundary conditions at corners and at dielectric interfaces to improve the accuracy of the FD-TD field approximations. The numerical solutions corresponded closely to experimental results from case studies involving the heating of plastic blocks and food on a tray.

[4] considered a hybrid analytical and numerical method for the heating of a ceramic material in a single mode cavity containing an iris. For low-loss ceramics loaded within high-Q cavities, the usual FD-TD numerical scheme was shown to be inefficient. To overcome this inefficiency the authors split the problem by solving numerically with no iris, and then by analytically matching the numerical solution to theoretical results obtained for the electromagnetic field near the iris. The full solution is then obtained in a computationally efficient manner, with some six fold computation time improvements over the FD-TD numerical scheme alone.

The aim of this paper is to compare semi-analytical and numerical solutions for the heating of a three-dimensional dielectric block in a rectangular waveguide by the TE_{10} mode. Numerical solutions will be obtained by time-domain (using the usual FD-TD scheme) and frequency domain methods. A key assumption of the frequency domain method is that of a single waveguide mode, the validity of which will be examined. The FD-TD method for solving Maxwell's equations in three-dimensions typically require excessive amounts of computer resources, so comparison of the CPU usage of each scheme will also be made. As the geometry of the problem is regular, semi-analytical solutions developed using the method of [2] are also presented, and shown to be reasonably accurate. It is envisaged that the techniques developed here will be useful for the development of efficient hybrid analytical and numerical schemes for the heating of blocks in arbitrarily shaped microwave cavities and thus forms the foundations of future research to be conducted by the authors.

In §2 the governing equations are presented, while §3 contains brief details of the numerical schemes. A comparison of the numerical and semi-analytical solutions is presented in §4. The results highlight the good agreement between the numerical and semi-analytical solutions for the computed steady-state power versus temperature S-shaped curves.

GOVERNING EQUATIONS

Maxwell's equations of electromagnetism are given by

$$\begin{array}{ll} \nabla \cdot \mathbf{D} = \nabla \cdot (\epsilon \mathbf{E}) = \rho, & \nabla \cdot \mathbf{B} = \nabla \cdot (\mu \mathbf{H}) = 0, \\ \nabla \times \mathbf{E} = -\frac{\partial}{\partial t}(\mu \mathbf{H}), & \nabla \times \mathbf{H} = \frac{\partial}{\partial t}(\epsilon \mathbf{E}) + \sigma \mathbf{E}, \end{array} \tag{1}$$

based on the assumption that the material is homogeneous, isotropic and ohmic, so that the current $\mathbf{J}$ and the displacement current $\mathbf{D}$ are both proportional to the electric field $\mathbf{E}$ and the magnetic field strength $\mathbf{H}$ is proportional to the magnetic flux density $\mathbf{B}$. Here σ is the electrical conductivity, ϵ is the electrical permittivity and μ is the magnetic permeability. In general all material properties are temperature dependent.

Maxwell's equations are coupled with the forced heat equation

$$\rho C T_t = \nu \nabla^2 T + \frac{1}{2} \omega \epsilon \tan \delta \mathbf{E} \cdot \overline{\mathbf{E}}, \tag{2}$$

where ν is the thermal conductivity, ρ is the density, C is the specific heat of the material, ω is the frequency of the radiation and $\tan \delta$ is the loss tangent.

Typically the loss tangent is temperature dependent; power-law and exponential temperature dependencies have been widely used in modelling microwave heating phenomena. Arrhenius-type temperature dependencies, which are bounded at large temperatures, have also been used. This law is motivated from statistical mechanics and models the physical saturation of the conductivity at high temperatures. See [8] for a discussion of the physics of an Arrhenius-type absorption law and its application to the sintering of ceramics. In this paper, an Arrhenius-type temperature dependency, of the following form is used,

$$\tan \delta = \frac{\sigma}{\omega \epsilon} = \alpha (1 + 20 e^{-\frac{T_a}{(T - T_a)}}), \tag{3}$$

where T_a is the ambient temperature and α is a constant. Equation (3) is equal to unity at the ambient temperature of T_a, while as the temperature becomes large, (3) is bounded ($\tan\delta \to 21$).

The boundary conditions for the electric-field at the interfaces with free space are

$$(\mathbf{E} - \mathbf{E}_f) \times \mathbf{n} = 0, \quad (\mathbf{H} - \mathbf{H}_f) \times \mathbf{n} = 0, \quad (\epsilon\mathbf{E} - \epsilon_0\mathbf{E}_f) \cdot \mathbf{n} = 0, \tag{4}$$

where $\mathbf{E}_f$ and $\mathbf{H}_f$ are the electric and magnetic fields in the free space incident upon the material. On the surface of the waveguide $\mathbf{E} \times \mathbf{n} = 0$, as there is no electric-field in a conductor. For simplicity, this work assumes a fixed temperature boundary condition, $T = T_a$, is applied on each of the slab's surfaces.

The frequency domain

In their most general form the electric and magnetic fields can be written as follows:

$$\mathbf{E} = E_j\mathbf{e}_j = U_j(x,y,z)e^{-i\omega t}\mathbf{e}_j, \mathbf{H} = H_j\mathbf{e}_j = V_j(x,y,z)e^{-i\omega t}\mathbf{e}_j, j = 1,2,3, \tag{5}$$

where $\mathbf{e}_1$, $\mathbf{e}_2$ and $\mathbf{e}_3$ are unit vectors in the x, y and z directions respectively. The electric and magnetic fields are written as steady-state amplitudes modulated by the frequency. As the time scale for electromagnetic propagation is much smaller than the time scale for thermal diffusion the time derivatives of the electric and magnetic field amplitudes, U_j and V_j, and of the electrical conductivity, σ, can be ignored. Substituting (5) into (1) gives

$$\nabla^2 U_j - \tfrac{\partial}{\partial x_j}\nabla \cdot \mathbf{U} + {k_1}^2(1 + i\tfrac{\sigma}{\omega\epsilon})U_j = 0, \quad j = 1,2,3, \tag{6}$$

as the governing equation for the steady-state components of the electric-field amplitude, where the wavenumber $k_1 = \omega/c$ and c is the velocity of the radiation in the block.

A single-mode assumption, that the electric field in the block maintains the form of the TE_{10} mode, is made so that $U_1 = U_3 = H_2 = 0$, and there is only one electric-field amplitude component, $U_2 = U_2(x,z)$. The governing equation (6) then becomes

$$U_{2xx} + U_{2zz} + {k_1}^2(1 + i\frac{\sigma}{\omega\epsilon})U_2 = 0. \tag{7}$$

The steady-state amplitude equation (7) and the forced heat equation (2) are non-dimensionalised by the scalings

$$t' = \frac{t\nu}{\rho C l^2}, \quad x' = \frac{x}{l}, \quad y' = \frac{y}{l}, \quad z' = \frac{z}{l}, \quad U_j' = \frac{U_j}{U_{inc}}, \quad T' = \frac{T}{T_a} - 1, \tag{8}$$

where $2l$ is the length of the block, U_{inc} is the incident amplitude of one of the components of the electric-field, ν is the thermal conductivity and T_a is the ambient temperature. The non-dimensionalisation results in the scaled frequency, power, electrical conductivity and wavenumbers having the forms

$$\omega' = \frac{\omega l^2 \rho C}{\nu}, \quad \beta' = \frac{l^2 \rho C U_{inc}^2 \omega \epsilon}{2\nu T_a}, \quad \sigma' = \frac{\sigma}{\omega \epsilon}, \quad k_1' = k_1 l. \tag{9}$$

A finite three-dimensional block of non-dimensional length 2, width $2n$ and height $2m$ is considered. It is assumed that the waveguide has width $2n$ and height $2m$ so that the block completely fills the cross-section of the waveguide, which is infinitely long. The waveguide is aligned to ensure that the waveguide mode propagates in the z-direction. Here, the fundamental transverse electric waveguide (TE_{10}) mode is considered. The TE_{10} waveguide mode, which is the usual choice for single-mode commercial waveguides, is

$$E_2 = \cos(\frac{\pi x}{2n})e^{i\theta}, \quad H_1 = -\frac{k_r}{\mu\omega}\cos(\frac{\pi x}{2n})e^{i\theta}, \quad H_3 = -\frac{i\pi}{2\mu\omega n}\sin(\frac{\pi x}{2n})e^{i\theta}, \tag{10}$$

where $\theta = k_r z - \omega t$, $k_r^2 = k^2 - \frac{\pi^2}{4n^2}$, and k is the wave number in free space. Note that for the microwaves to propagate in the waveguide $k > \frac{\pi}{2n}$, which is the cut-off wavenumber.

In nondimensional form the equations are

$$\begin{aligned} &U_{2xx} + U_{2zz} + {k_1}^2(1 + i\tan\delta)U_2 = 0, \\ &T_t = T_{xx} + T_{yy} + T_{zz} + \beta\tan\delta|U_2|^2 = 0, \end{aligned} \tag{11}$$

For the TE_{10} mode the boundary conditions, from (4), are

$$\begin{aligned} &U_{2z} + ik_r U_2 = 2ik_r\cos(\frac{\pi x}{2n}), \quad z = -1, \\ &U_{2z} - ik_r U_2 = 0, \quad z = 1, \\ &U_2 = 0, \quad x = \pm n, \\ &T = 0, \quad x = \pm n, \quad y = \pm m, \quad z = \pm 1. \end{aligned} \tag{12}$$

THE NUMERICAL SCHEMES

The Frequency Domain Solution Method

The frequency domain solution is found by resolving (11) with boundary conditions (12). The heat equation is solved using the explicit DuFort-Frankel method, which gives second-order accuracy at each time step. The Helmholtz equation for the electric-field amplitude is discretised using central differences with the discretisation giving rise to the following matrix

$$A\mathbf{U}' = \mathbf{b}, \tag{13}$$

where the vector $\mathbf{U}'$ represents the components of electric-field amplitude. The system matrix $A \in \mathbb{C}^{n\times n}$ had dimension $n = 5000$ and (13) is solved iteratively via $GMRES$, which is a Krylov subspace method. It was found that the subspace $\mathcal{K}_\ell(\mathbf{A}, \mathbf{b})$ required dimension in the range $\ell \in [100, 200]$ to reduce the system residual to 1×10^{-6}. The use of a preconditioner to improve the performance of the iterative solver will be the subject of future work.

The Time Domain Solution Method

The Maxwell's equations (1) were discretised using the FD-TD staggered in space and time methodology [1,5,6] over a given Cartesian Yee lattice. As discussed in §2, under the TE_{10} assumption the following field components are obtained from (1):

$$\frac{\partial \mathbf{H}_1}{\partial t} = \frac{1}{\mu}\frac{\partial \mathbf{E}_2}{\partial z}, \frac{\partial \mathbf{H}_3}{\partial t} = -\frac{1}{\mu}\frac{\partial \mathbf{E}_2}{\partial x}, \frac{\partial \mathbf{E}_2}{\partial t} + \frac{\sigma}{\epsilon}\mathbf{E}_2 = \frac{1}{\epsilon}\left(\frac{\partial \mathbf{H}_1}{\partial z} - \frac{\partial \mathbf{H}_3}{\partial x}\right). \tag{14}$$

To increase the spatial stability of the numerical solver [7], the electric and magnetic field components have been computed on cell edges and faces, respectively. By applying the leapfrog time discretisation [5] to (14) and rearranging the equation for H_1 the following discrete analogue of the x-component of the magnetic field is obtained:

$$H^{n+\frac{1}{2}}_{1(i,j+\frac{1}{2},k+\frac{1}{2})} = H^{n-\frac{1}{2}}_{1(i,j+\frac{1}{2},k+\frac{1}{2})} + \frac{\delta_t}{\mu}\left(\frac{E^n_{2(i,j+\frac{1}{2},k+1)} - E^n_{2(i,j+\frac{1}{2},k)}}{\delta_z}\right). \tag{15}$$

Similar equations for H_3 and E_2 can also be obtained. Once the electric and magnetic fields have been computed for a sufficiently large number of periods, the power distribution can be calculated from the electric field alone:

$$P_{abs} = \frac{1}{q\lambda_T}\int_{p\lambda_T}^{(p+q)\lambda_T} \sigma \mathbf{E}^2(t)dt, \tag{16}$$

where p is the number of periods before the absorbed power (P_{abs}) is computed, q is the number of periods for which the electric field has been time-averaged to obtain the power and λ_T is the period of the electromagnetic wave inside the waveguide. Typically $p >> q$ and $q = 1$ is chosen so that the electric and magnetic fields have reached a so-called steady state inside the waveguide. In the numerical simulation performed here, $p+q=10$ was used and the power is computed according the following discrete form of (16):

$$P^n_{(i,j,k)} = \frac{1}{16}\sigma\left(E^n_{2(i,j+\frac{1}{2},k)} + E^n_{2(i+1,j+\frac{1}{2},k)} + E^n_{2(i,j+\frac{1}{2},k+1)} + E^n_{2(i+1,j+\frac{1}{2},k+1)}\right)^2. \tag{17}$$

The power distribution obtained from (17) is used in the forced heat equation (2) to compute the heat distribution inside the load. This is achieved through a forward time and centred space discretisation of equation (2) with the time step chosen to satisfy the stability requirement for this method.

The time step associated with the FD-TD algorithm is $O(10^{-12})$, however, for the heat equation the time step of this scheme is much larger $O(10^{-2})$. Consequently, the power distribution can be obtained in an instance of the time scale of the heating. Therefore, the power is recomputed only when the properties of the material due to heating have substantially changed. Typically, the constraint on the recomputation of the power is triggered when

$$e = \left|\frac{\varepsilon''_{n+1}(T_{\max}) - \varepsilon''_n(T_{\max})}{\varepsilon''_n(T_{\max})}\right| < 5 \times 10^{-t}, \tag{18}$$

where, e is the measure of the error in the dielectric properties due to heating and t is a positive integer. In (18), n refers to the levels at which the power has been computed. It was found that varying the value of t in (18) slightly did not affect the quality of the final solution considerably. However larger values of t does impact substantially on the overall computational time. Equation (18) merely determines how long the heat equation is iterated before the power is recomputed and for the results reported here the value of $t=2$ was used.

RESULTS AND COMPARISONS

The following physical properties were used for all simulation results given in this section:

$$\nu = 37.7\,kg\,m\,s^{-3}K^{-1},\ C = 940\,J\,kg^{-1}\,K^{-1},\ \rho = 3.58\times 10^3\,kg\,m^{-3}. \tag{19}$$

The amplitude of the electric field is related to the average input power (P_{av_in}) by

$$E_0 = \sqrt{\frac{2P_{av\_in}\mu_0\omega}{\gamma}},\quad \gamma = \sqrt{\omega^2\mu_0\varepsilon_0 - \left(\frac{\pi}{a}\right)^2}. \tag{20}$$

A cube of dimension $a = 0.1m$ is considered where the cube fills the cross-section of the waveguide. The results are presented in non-dimensional form, with the FD-TD numerical solution non-dimensionalised for comparison purposes. As a cube is considered $m = n = 1$ in (12). Other parameters in non-dimensional form are $\alpha = 0.005$, $k = k_1 = 2.57$ and $k_r = 2.44$.

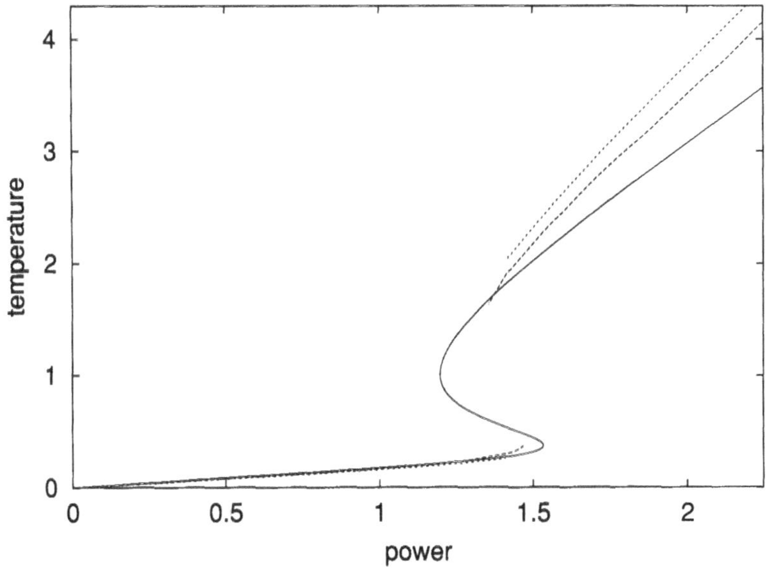

Figure 1: The steady-state power versus temperature curve. Shown are the semi-analytical solution, the frequency domain (large dashes) and the FD-TD (small dashes) numerical solutions.

Figure 1 shows the steady-state temperature versus power curve. Presented are the nondimensional maximum temperature in the block and the nondimensional power β. Shown are the semi-analytical solution and the frequency domain and FD-TD numerical solutions. The semi-analytical solution is found using the Galerkin method. Trial functions for the temperature and the electric-field amplitude are assumed and expressions for the decay of the electric-field and the power versus temperature curve are found by averaging over the slab, see [2] for more details of the technique.

The figure shows that a S-shaped response occurs. Thermal runaway arises at a critical β with the solution jumping from the lower (cool) branch to the upper (hot) branch. On the lower (cool) branch the comparison between the three solutions is excellent. On the upper (hot) branch the comparison is still very good with a difference of 8%, at $\beta = 2$. Moreover, the semi-analytical solution is very accurate, even for large β. The difference between the semi-analytical and frequency domain solutions at $\beta = 2$ is 10%. No numerical results are available for the second solution branch, as it is unstable.

For the chosen value of $k = 2.57$ both the TE_{10} and the TM_{11} modes can propagate in the waveguide as all higher modes are cut-off. However, examination of the FD-TD numerical solution shows that no other mode, besides the incident TE_{10} mode, is present in the waveguide or the heated material. Hence, the single-mode assumption, made in the frequency domain solution, is reasonable.

The frequency domain solutions are found using a spatial discretisation of $\Delta z = 0.125$, while the FD-TD solutions are found with a discretisation of $\Delta z = 0.067$. Hence the difference between the numerical solutions on the upper branch is most likely due to discretisation error. Both the numerical solutions were found using a single Silcon Graphics processor capable of 800 MFlops. Each steady-state solution point takes a similar amount of CPU; 7 minutes for the frequency domain method and 10 minutes for the FD-TD method.

CONCLUSIONS

A comparison of the simulation results for a homogeneously loaded infinitely long waveguide using a semi-analytical method and both frequency and time domain numerical methods has been presented in this paper. The results highlight that all three methods show good agreement for the S-shaped curve for the ceramic like material. Furthermore, the assumption of a single mode within the waveguide is suitable for use in the frequency domain method.

Further work will concentrate on efficient numerical solutions for other waveguide scenarios such as a short-circuited guide and a cavity with an iris. The validity of the single mode assumption for other waveguide modes, such as the TM_{11} mode and in cavities will also be investigated. These investigations will lead to the authors developing efficient hybrid analytical and numerical schemes for microwave heating in waveguides and cavities.

REFERENCES

[1] A.C. Metaxas and R.J. Meredith, "Industrial Microwave Heating", Peter Peregrinus Ltd. London, UK. (1983)

[2] B. Liu and T. R. Marchant, "The microwave heating of three-dimensional blocks: semi-analytical solutions", *IMA J. Appl. Maths.* **67** 145-175 (2002).

[3] G.A. Kriegsmann, "Cavity effects in microwave heating of ceramics" *SIAM J. Appl. Math.* **57**, 382-400 (1997)

[4] C. V. Hile and G. A. Kriegsmann, "A hybrid numerical method for loaded highly resonant single mode cavities". *J. Comput. Phys.* **142**, 506–520 (1998).

[5] H. Zhao and I.W. Turner, "An analysis of the finite-difference time-domain method for modelling the microwave heating of dielectric materials within a three-dimensional cavity system" *J. Micro. Power* **31**, 199-214

(1996).

[6] K. S. Yee, "Numerical solution of initial boundary value problem involving Maxwell's equations in isotropic media" *IEEE Trans. Antennas Propagat.* **14**, 302-307 (1966).

[7] Y. Liu, "Fourier analysis of numerical algorithms for the Maxwell equations, *J. Comput. Phys.* **124**, 396-416 (1996).

[8] V. M. Kenkre, L. Skala, M. W. Weiser and J. D. Katz, "Theory of microwave effects on atomic diffusion in sintering: basic considerations of the phenomenon of thermal runaway"; pp. 179–183 in *Microwave processing of materials II* Edited by W. B. Snyder, W. H. Sutton, M. F. Iskander and D. L. Johnson. Material Research Society (1991).

DETERMINATION OF VOLUME FRACTION AND PARTICLE SIZE IN DISPERSED SYSTEMS BY DIELECTRIC SPECTROSCOPY AND PRINCIPAL COMPONENT ANALYSIS

Marc Regier and Helmar Schubert
Institute of Food Process Engineering
University of Karlsruhe
Kaiserstr. 12
76131 Karlsruhe
Germany

ABSTRACT

In this work a new method for the determination of volume fraction and particle size in dispersed systems is presented. The method is based on dielectric spectroscopy by an open ended coaxial probe combined with principal component analysis. After a short introduction in the theory of principal component analysis, results of the volume fraction determination of oil-in-water emulsions and of the particle size determination of polystyrene suspensions are presented. It is shown, that the new method is well suited for both tasks even at highly concentrated dispersions, making it possibly interesting for an in- or online quality control during continuous dispersion production.

INTRODUCTION

Important parameters, that govern many physical properties of fluid dispersed systems like emulsions or suspensions, are the volume fraction and the particle size distribution of the dispersed phase. For example, these properties have strong influences on the rheology [1], the colour or on sensorial properties.

In the case of a continuous production of dispersions an exact and fast control of the volume fraction and the particle size distribution is therefore essential to ensure a constant product quality. That is why online methods are desired, that overcome the problems of the conventional offline-quality control.

In the case of volume fraction determination, established methods are sample drying, weighing and chemical analysis or density measurements. Whereas the first is by far too slow for an online control the latter only works for pure two-component-systems.

Even more problematic is the online particle size measurement [2]. Laser diffraction systems only work in dispersions with low volume fraction, so that highly concentrated dispersions have to be diluted, what may have undesired influences on the particle size distribution. The disadvantage of the alternative method of ultrasound spectroscopy is the necessary knowledge of the dispersion's properties that are sometimes difficult to obtain.

For the purpose of both, the volume fraction- and the particle size-determination dielectric spectroscopy (frequency range between 100 kHz and 6 GHz) in combination with principal component analysis is studied in this work. Studied dispersed systems, presented here, are oil-in-water-emulsions and supensions of polystyrene in water.

THEORY OF THE PRINCIPAL COMPONENT ANALYSIS

Practical fluid dispersions like emulsions or suspensions often consist of more than two chemical species. Therefore a simple calibration procedure, which relates the dielectric constant at one frequency with the interesting property, as the volume fraction of the dispersed phase or the mean particle size will not succeed, for the evaluation a frequency range is required, instead. The consequence is a large dataset. One possible method for its analysis is the principal factor analysis, which is a mathematical-statistical technique. It was first used in the behavioral sciences and has been applied succesfully to physical and chemical problems for more than 30 years, now [3]. The aims of the principal component analysis are to determine from a large dataset of sample spectra the main factors and the weights (also named loadings) of the factors in each sample spectrum.

This can be mathematically described by a rotation of the axes or a decomposition of a datamatrix [D] into two matrices, the principal component matrix [R] and the matrix of weights [C]:

$$[D] = [R] * [C] \tag{1}$$

The theoretical mathematical and statistical background can be found in detail in [3], only the main steps should be treated here at the example of dielectric spectroscopy data:

The measured spectrum (ε' or ε'' over frequency), which is already discretisized by the measurement procedure itself, can be written as a column vector D_i, where the values of ε' or ε'' respectively are listed in their natural order of increasing frequency. n is the the total number of frequency steps. All m spectra yield then the data matrix

$$[D]=\begin{pmatrix} \varepsilon(\varphi_1,f_1) & \varepsilon(\varphi_2,f_1) & \ldots & \varepsilon(\varphi_m,f_1) \\ \varepsilon(\varphi_1,f_2) & \varepsilon(\varphi_2,f_2) & \ldots & \ldots \\ \ldots & \ldots & \ldots & \ldots \\ \varepsilon(\varphi_1,f_n) & \ldots & \ldots & \varepsilon(\varphi_m,f_n) \end{pmatrix} \tag{2}$$

, where $\varepsilon(\varphi_i,f_j)$ is the value of ε with varied property φ_i (e.g. dispersed phase fraction or particle size) and at the frequency, f_j.

In the next step the covariance matrix [Z] is formed by the multiplication of the data matrix with its transposed matrix:

$$[Z]=[D]^T *[D] \tag{3}$$

It can be shown [3] that the principal factors can be determined by a multiplication of the data matrix [D] by a matrix [Q], where [Q] is the sorted eigenvector matrix of the covariance matrix. The loadings matrix (also called weight matrix) of the principal components are then calculable by the transposed of the eigenvector matrix.

The individual loadings of the principal components should show the main effects of the varying interesting properties and can be correlated by a polynomial function with them.

MATERIALS AND METHODS

Oil-in-Water-Emulsions

The base materials were demineralized water together with laurylethylenoxide (LEO-10) as an emulsifier and vegetable oil. The concentration of the emulsifier was kept constant at a level of 5g/l(water), which is two to three times the critical micell concentration, so that stable emulsions could be produced. The oil-in-water-emulsions are produced in a batch ultrasonic process, which is described in detail for example in [4]. The used batch emulsifying process (20 min, $P/V=4*10^4$ W/m^3) yielded a mean droplet size (Sauterdiameter) of approximately 5µm, which was determined by a laser scattering device (Coulter LS 230). In order to get various oil volume fractions of the same droplet size distribution (which also could influence the dielectric properties), all emulsions were produced by diluting a highly concentrated, stabilised emulsion (70%).

Monodispersed Polystyrene-Suspensions

The polystyrene-dispersions are produced by microparticles GmbH (Germany) as monodispersed aqueous suspensions with a dispersed phase fraction of 10 mass%. The particle sizes range from approximately 200 nm up to 1100 nm.

Measurement Of Dielectric Properties

The complex dielectric constants of the pure materials and of the produced emulsions were characterized by a commercial open ended coaxial line reflectory dielectric probe (HP85070B) together with a network analyzer (HP8753D). With this device it was possible to measure the dielectric properties in a frequency range between 200 MHz and 6GHz. By using a modified probe, according to [5], it was possible to widen the frequency range down to 50 kHz, what is important for the particle size dependency of the dielectric properties.

The system is normally used for homogeneous fluids or solids with a smooth surface. Due to the large penetration depth (mm-range) in comparison with the small particle size (sub-µm-range) it should be well suited for more heterogenous materials like emulsions and suspensions, as well.

Principal Component Analysis Software

The statistical method of factor analysis is implemented in various mathematical software packages. Nevertheless we used a self-written code [6], since it can be flexibly changed to our needs and give helpful insights into intermediate results.

RESULTS AND DISCUSSION

Oil-in-Water-Emulsions

The measured dielectric loss spectra between 200 MHz and 6 GHz of the emulsions are shown in figure 1. It can be clearly seen that the dielectric properties are mainly influenced by the dipolar relaxation behaviour of the water molecules with their relaxation frequency outside of the measured frequency range (nearby 20 GHz). The displayed error bars represent the standard deviation of three measurements. They are sufficiently small to distinguish the different oil fractions varied in 10 % steps.

The measured spectra of ε'' is used as starting matrix of the principal component analysis. The principal component analysis software yields the principal factors that are shown in figure 2.

The first principal component shows the typical dependency, which is known from the dipolar relaxation and can be assigned to the water fraction. The other factors are more than one decade smaller in absolute values. Therefore we tried a reconstruction of the measured spectra with only one basic vector: the first principal component with its corresponding weights. This reconstruction can be seen in figure 3.

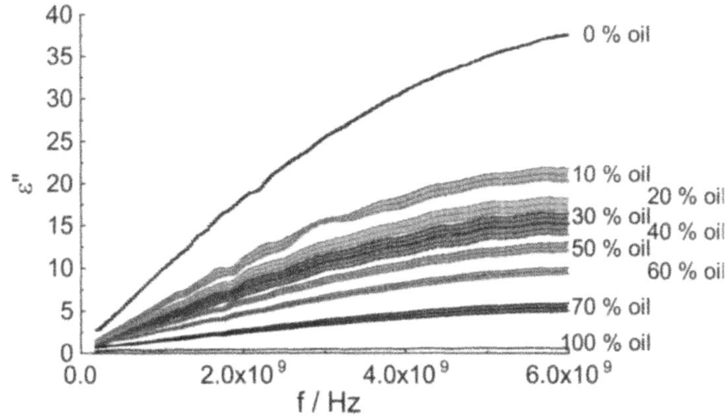

Fig. 1: Measured values (with standard deviations as error bars) of the imaginary part of the dielectric constant over the frequency.

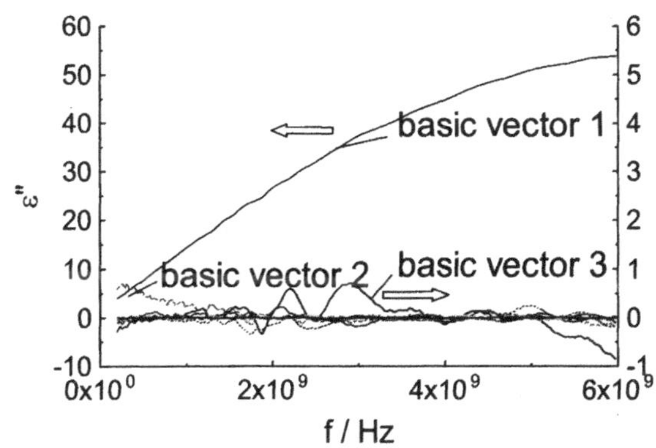

Fig. 2: The principal components of ε". Already the typical form, but also the absolute value of only the first basic vector lead to the total number of principal components of 1.

Except for the dielectric spectrum of the pure components, the reconstructed spectra lies completely within the error bars of the measurements. Nevertheless only one principal component is used for the volume fraction calibration. This is done in figure 4, where the oil volume fraction is shown in dependency of the weight of the first principal component together with a parabolic function, fitted through the discrete data points.

In order to further check the results of the factor analysis another measurement spectrum (of 5 % oil volume fraction) should be reproduced by a target testing procedure. Within this procedure it can be tested if the determined principal components are sufficient for reproducing the measurement spectrum or if another one has to be added. Figure 5 shows the result: again the reproduced spectrum lies within the error bars of the measured spectrum, another basic vector is not needed. The resulting weight of the first principal component is 0.51, which fits excellently to the interpolation of already known weights, shown in figure 4.

The calibration-validation procedure can also be done for the other emulsions: the result is a so called internal cross-correlation diagram, which is shown in figure 6. Again, a good correlation between the determined and the „real" oil volume fractions can be recognized.

These results show, that dielectric spectroscopy in combination with principal component analysis is able to determine dispersed phase fractions of oil-in-water emulsions. Further studies showed, that even in the case of variations in a third ingredient, like a salt content, this analysis can yield reliable values of the dispersed phase fraction.

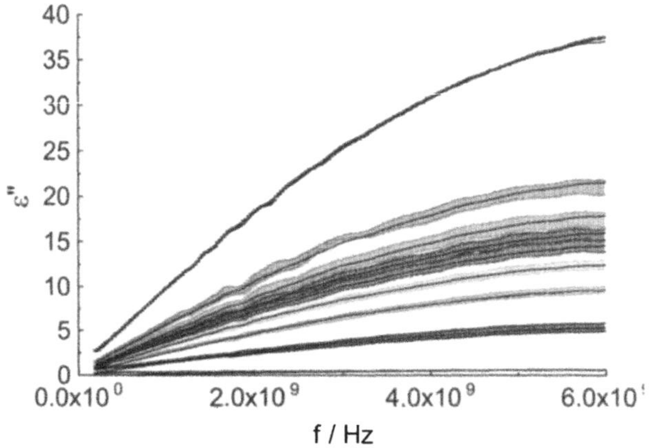

Fig. 3: The reconstruction (dark lines) of the measured spectra with only the first principal vector succeeds within the measurement error bars.

Monodispersed Polystyrene-Suspensions

In the case of the monodispersed polystyrene-suspensions the dispersed phase content was kept constant but the particle size varied. It should be studied, if the dielectric spectra of the supensions are influenced by the particle size, and if it is then possible to determine the particle sizes from the spectra.

In figure 7 the dielectric spectra of the polystyrene-supsensions are presented. Clearly, a dependency of the particle size on the dielectric spectra can be observed, especially at low frequencies. This dependency makes it possible to apply the principal component analysis successfully. In this case two principal components can be identified and their weights can be used for a correlation with the particle size. The internal cross correlation in figure 8 shows the good agreement between particle size determination by dielectric spectroscopy and the "real" values.

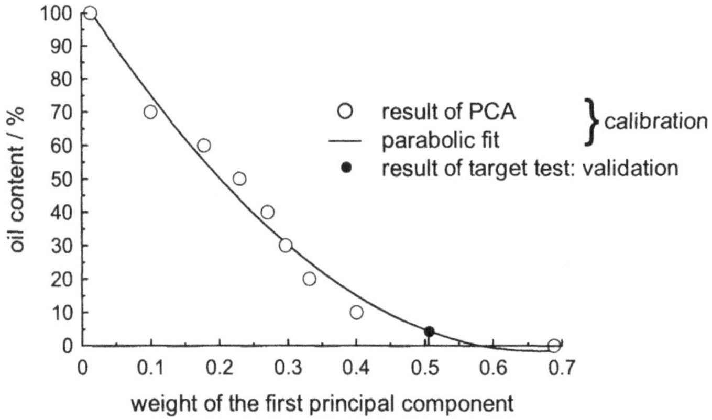

Fig. 4: Correlation of the oil volume fraction with the weight of the first principal component. The unfilled points are fitted by a parabolic function. The filled point shows the result of a target test procedure of an emulsion of 5 % oil content.

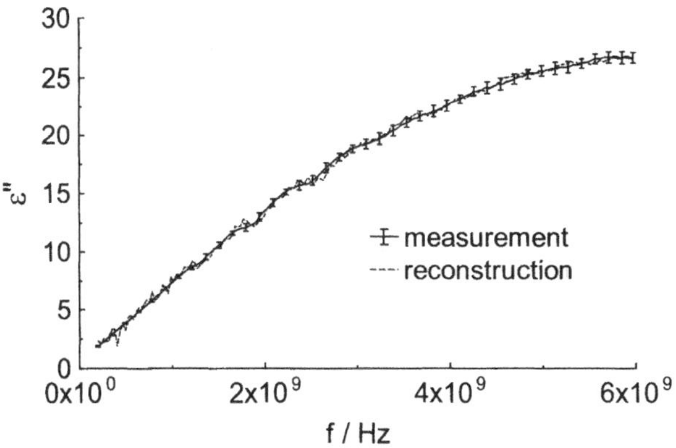

Fig. 5: The reproduction of a measured spectrum of 5 % oil content by the basic components succeeds within experimental error bars.

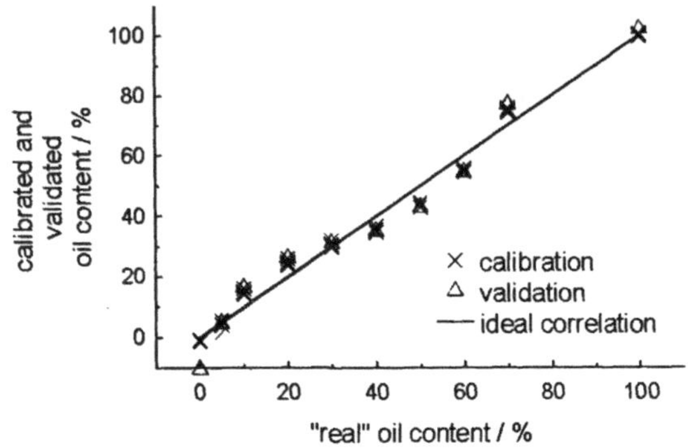

Fig. 6: Correlation of the determined oil volume content with the "real" oil volume content.

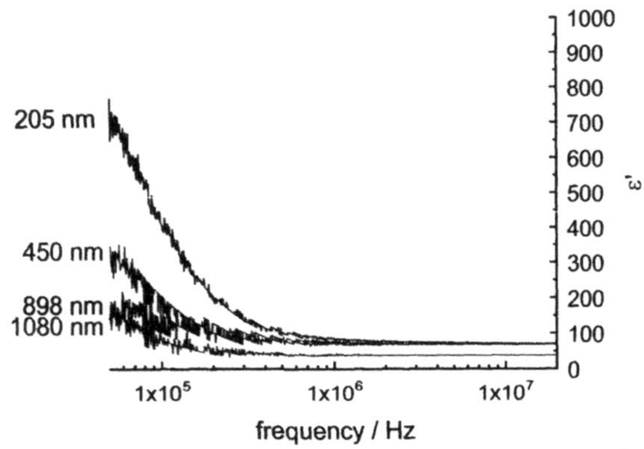

Fig. 7: Dielectric spectra of monodispersed polystyrene-suspensions.

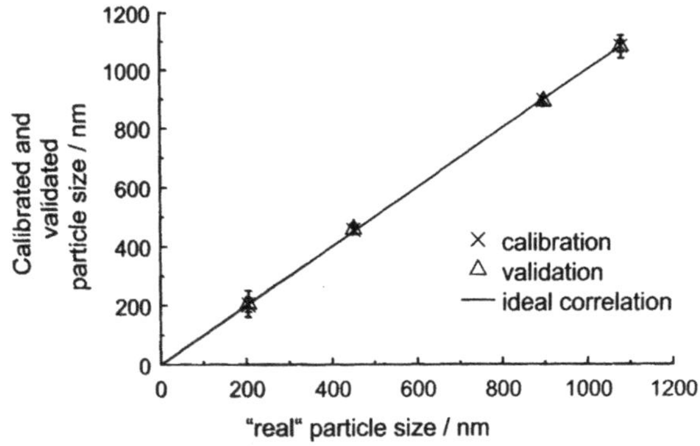

Fig. 8: Correlation of the determined particle size with the "real" size.

CONCLUSIONS AND OUTLOOK

The presented results of the dielectric spectroscopy in combination with principal component analysis shows that this is a new method for dispersed phase fraction- and particle size- determination. Due to its suitability also for highly concentrated systems and its speed, it should be interesting for an in- and online quality control during production.

REFERENCES

[1] C. Gerhards, “Rheological characteristics of emulsions”, Food Ingredients Europe Conference Proceedings 1995, 175-177, (1995).

[2] E. Kissa, *Dispersions: Characterization,Testing, and Measurement*, Dekker, (1999).

[3] E.R. Malinowski, D.G. Howery, *Factor Analysis in Chemsitry*, Wiley, (1980).

[4] O. Behrend, K. Ax, H. Schubert, “Influence of Continuous Phase Viscosity on Emulsification by Ultrasound“, Ultrasonics – Sonochemistry 7, 77-85, (2000).

[5] S.S. Stuchly, G. Gregory, L. Anderson, A. Kraszewski, “A New Sensor for Dielectric Measurements“, IEEE Trans. Instrumentation & Measurement, IM-35, 139-141, (1986).

[6] M. Regier, *Über dielektrische und Magnetresonanz-Methoden zur Charakterisierung disperser Systeme*, filed Dissertation, University of Karlsruhe, (2002).

COMPUTATIONAL ANALYSIS OF A TAPERED SLOTTED WAVEGUIDE AS A RADIATING ELEMENT

Vyacheslav V. Komarov
Saratov State Technical University
Politekhnicheskaya St., 77
Saratov 410054, Russia

Vadim V. Yakovlev
Worcester Polytechnic Institute
100 Institute Rd.
Worcester, MA 01609

ABSTRACT

An irregular (tapered) waveguide with a non-linear longitudinal slot suggested in paper[6] as a radiator providing uniform field distribution of the surface of an elongated dielectric slab the near-field zone is analyzed with the use of the 3D conformal FDTD simulator *QuickWave-3D*. Advanced computational study reveals the structure's limited potential for microwave high power applications because of its low radiation efficiency and uneven radiated electric field. Valuable capabilities of the used software in analysis and design of microwave applicators are also demonstrated.

INTRODUCTION

The recent review of electromagnetic (EM) software[1,2] has shown that there are several simulators on the market, which are suitable for modeling of processes and systems of microwave (MW) heating. *QuickWave-3D* (*QW3D*) (http://www.qwed.com.pl), the 3D full-wave conformal FDTD simulator has been identified as a particularly useful tool due to a number of implemented specific extensions and helpful functions (field envelopes, energy coupling, basic heating module, and others)[2]. The heating intensity in the food loads in a MW oven computed with *QW3D* was presented in[3]. This modeling package was used to characterize single- and multi-mode systems, which are typical in MW chemistry[4]. In paper[5], *QW3D* was utilized to find characteristics of a 915 MHz water load and especially those ones which cannot be measured. This experimentally verified computational analysis has suggested that many existing MW systems and elements developed on the basis of extensive experimentations and simplified calculations could be accurately studied with the use of this computational tool in order to characterize EM processes in these devices more precisely and to get the ideas on possible ways of their modification and optimization.

The present paper pursues such a goal with respect to a special radiating element (RE)[6], which reportedly provides relatively uniform field distribution on the surface of an irradiated dielectric slab in the near-field zone. Rigorous analysis of this structure appears to be extremely difficult, so it has been synthesized by its authors from an admittance integral equation and with the use of the perturbation theory under a number of assumptions. A RE with the claimed capability would be of high value for many MW heating applications

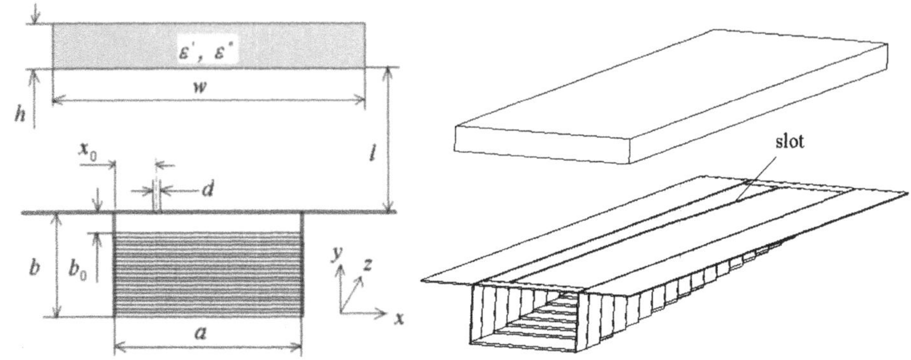

Fig. 1. Vertical profile and 3D view of the applicator with the irradiated dielectric layer

(particularly, in wood processing). However, the method used for its design doesn't allow engineers to figure out important characteristics of the system in detail, and experimental options here are either limited, or expensive. We show that valuable and reliable information about this RE can be generated by accurate *QW3D* simulation and with the use of modest computational resources.

APPLICATOR

The irregular (i.e., tapered) rectangular waveguide with non-linear longitudinal radiating slot cut in its broad wall (Fig. 1) has been proposed in[5] for heating of sheet and band materials. The radiation from the RE is supposed to be uniform when the amplitude of the transverse current induced by the TE_{10} mode is constant with respect to the longitudinal z-coordinate. The RE's authors suggest that this amplitude primarily depends on the waveguide narrow wall, the distance from this wall to the slot, and the fundamental mode's attenuation constant. From this assumption, the slot was found to be of curvilinear profile.

Paper[6] also describes an experimental verification of the synthesized RE at 2.45 GHz, but does not provide complete information about the experiment: material parameters of the irradiated plate are not specified, presence of other objects near the RE is not mentioned, etc. Therefore, direct comparison between the experimental data and simulations is not possible. For this reason, our goal has been to create a parameterized computer model wherein all structure dimensions, materials parameters and excitation can easily be changed, and independently validate with this model the principle of the RE operation and compute its characteristics inaccessible for the simplified analysis preformed by the RE's authors.

MODEL

The scenario has been modeled by *QW3D* as an open structure. The slotted waveguide and the wood plate were considered surrounded by the surfaces imitating the near-to-far field transformation and the absorbing boundary conditions. The slot profile $x_0(z)$ has been approximated by two straight-line segments. Following[6], in our computations, the length of the slotted waveguide was 1,000 mm; $a \times b = 90 \times 45$ mm and $a \times b_0 = 90 \times 9$ mm were the input

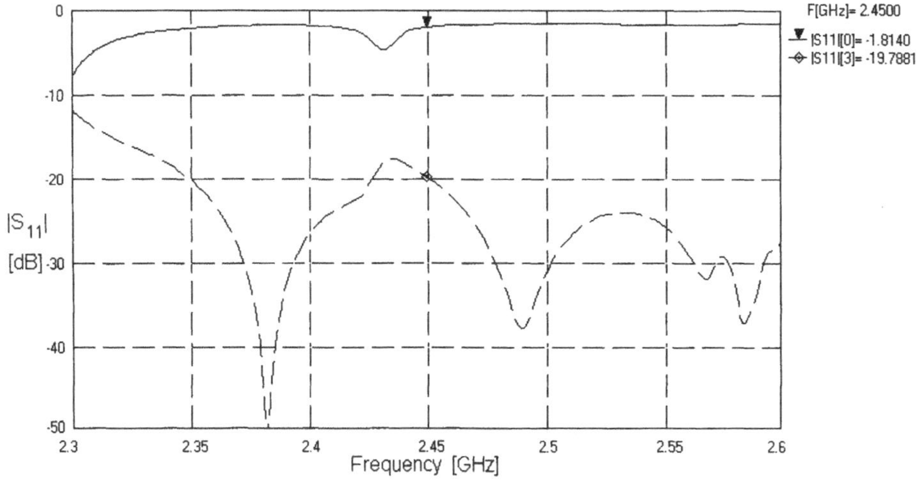

Fig. 2. Return loss with (dotted curve) and without (solid curve) a matched load in the tapered end of the waveguide.

and output waveguide cross-sections respectively; the width of the slot d was constant (2 mm). The rectangular wood plate of 1,000 mm length (w = 150 mm, h = 20 mm) was placed in parallel to the waveguide (with l = 100 mm). Dielectric constant ε' and loss factor ε'' of the wood plate unknown in the experiment[6] were taken equal to 3.8 and 0.87 respectively.

Simulation with a wide-band pulse excitation has been used to compute reflections in the slotted waveguide and to calculate the radiation pattern of the RE. Sinusoidal modal excitation at 2.45 GHz (TE_{10} mode) has been applied then to determine distribution of the field and dissipated power.

The space within the coordinate planes with the absorbing boundary conditions has been discretized with the use of a non-uniform mesh consisting of the cells varying in size from 2×4×10 mm to 10×8×10 mm: the larger cells are in air while the smaller ones are placed within the wood plate and around the narrow slot to guarantee its good approximation. These cell sizes are smaller than typically recommended for a conventional FDTD discretization (10 cells per wavelength in a medium), thus the accuracy of computation is supposed to be satisfactorily high. To make sure that this is the case, the results have been tested on a denser mesh, and no noticeable change have been found.

The chosen non-uniform mesh (total about 1 million cells in the model) has requested about 90 MB of RAM. It took a Pentium III 850 MHz PC near 3 hours to get steady state results. The developed model is quite adequate and flexible, so it can be used for detailed studying of the structure and finding its specific characteristics as well as exploring the ways of improvement of its performance.

RESULTS

Reflection

In the experimental validation[6], the field radiated from the RE was measured in the presence of the matched load in the tapered end of the waveguide. Our simulations show that

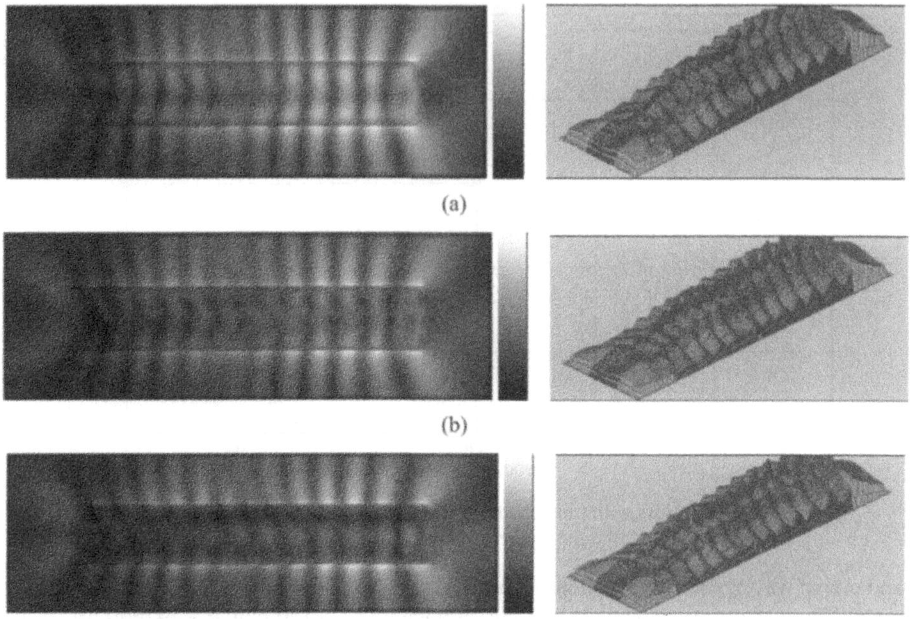

Fig. 3. Thermal (left) and hilltop (right) presentations of the electric field (envelope) in the horizontal planes through the wood plate: top (a), centre (b), and bottom layers (c).

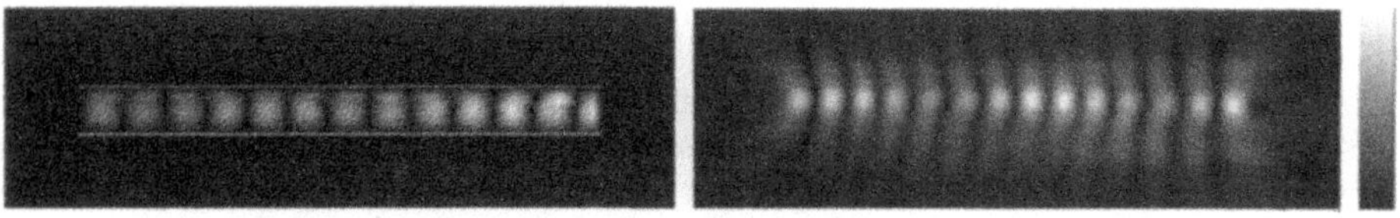

Fig. 4. Electric field in the horizontal planes: instantaneous field inside the waveguide (7 mm from the wall with the slot (left)) and a field envelope outside the waveguide (7 mm from this wall (right)).

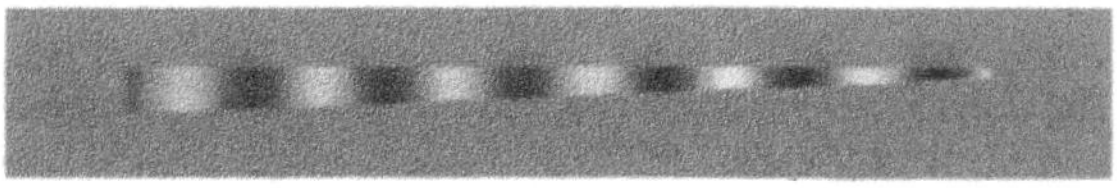

Fig. 5. Instantaneous electric field in the vertical plane through the center of the waveguide

in this case the return loss at 2.45 GHz is -19.8 dB, i.e., only about 1% of power is reflected back to the source (Fig. 2). However, without the waveguide load, the return loss is -1.8 dB, which means that as little as 34% of generated EM energy leaves the waveguide for the outer space. This suggests that the construction under consideration does not appear to be an efficient radiator: the low level of reflections determined in[5] in fact means that a larger part of the magnetron power was absorbed by the terminal dummy load rather than radiated.

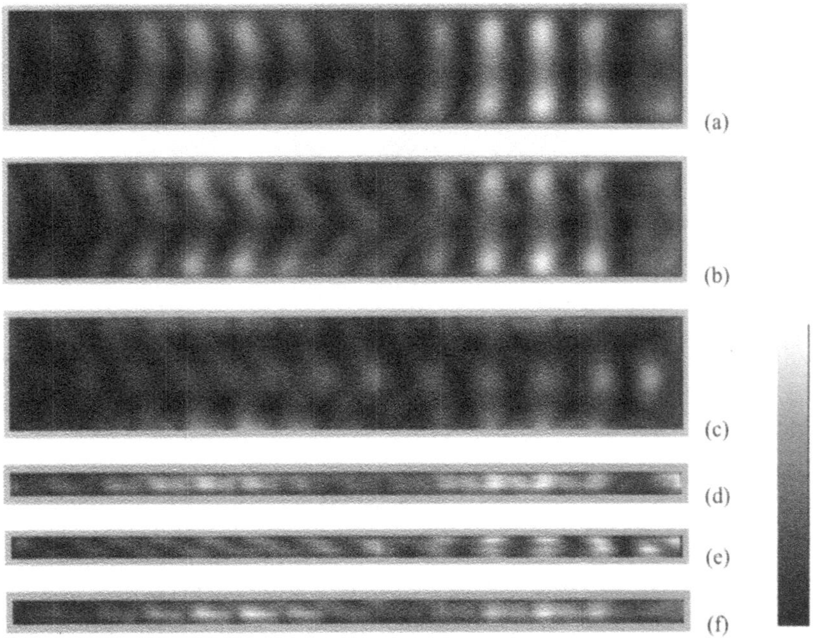

Fig. 6. Patterns of dissipated power in the wood plate: in the horizontal planes through the top (a), central (b), and bottom layers (c); in the vertical planes through the layers on the left (d) and right (f) edges and through the center (e).

Electric Field and Dissipated Power

In contrast to the suggested uniformity[6], the patterns of the electric field in the horizontal planes through the wood plate (Fig. 3) reveal increasing field magnitude towards the plate's edges. Moreover, in the longitudinal direction, the field distribution is characterized by alternating maxima and minima. This standing wave pattern depends on material parameters, so we conclude that the experiment[6] likely dealt with a different (unspecified) wood material. Fig. 4 shows that the pattern of the full field in the waveguide near the top wall is slightly different from the TE_{10} mode, which is explained by contributions by others than the vertical field components emerged due to the longitudinal slot. It is also seen that the trend in the patterns of the radiated field revealed in the planes through the wood plate (alternating maxima and minima along the *z*-axis shown in Fig. 3) is formed in outer space right near the slotted wall.

The pattern in the *yz*-plane (Fig. 5) suggests that the field magnitude inside and outside the waveguide is very different; numerical comparison gives the difference by at least the order of two. This confirms that the considered slotted structure is characterized by low radiation efficiency. It appears that the use of this element in high power applications without the internal matched load is not possible, but employing such a load would diminish the value of the structure as an RE.

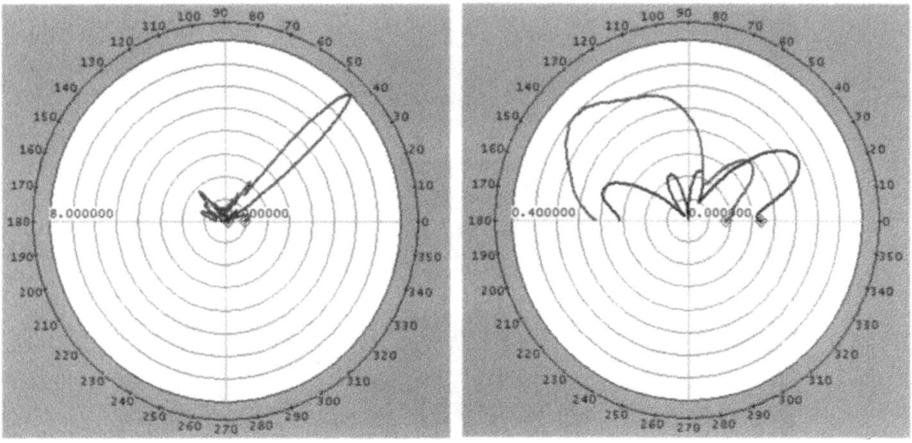

Fig. 7. Two spherical components of the radiation patterns in the longitudinal (left) and transverse (right) vertical planes.

Computations also show that the EM power is dissipated in the wood also non-uniformly forming the "hot" and "cold" spots generally following the alternating maxima and minima of the electric field (Fig. 6).

Radiation Pattern

As shown in Fig. 7, in the *yz*-plane, the radiation pattern of the considered RE is characterized by a strong and narrow main lobe rotated at approximately 45° from the positive direction of the *z*-axis whereas in the *xy*-plane the radiation is more uniform though much weaker: the peak of the transversal distribution is 23 times smaller that the longitudinal one. These angle distributions seem to be well matched with the horizontal field patterns in Fig. 3 showing a higher concentration of the field along the plate's edges and on the part of the plate located above the tapered end of the slotted waveguide. It is also worth mentioning that strong directivity of the radiation pattern can increase the possibility that the device would violate the safety standards.

CONCLUSION

In this paper, we have presented specific characteristics of the tapered slotted waveguide[5] to allow engineers to evaluate its applicability in microwave heating applications as a radiating element. Our computational analysis has shown its low radiation efficiency, high directivity of the main narrow lobe, and alternating maxima/minima in the magnitude of the radiated electric field. The depicted results generated by the *QW3D* model and containing the data on the internal reflections, field distributions and heating patterns in the wood plate provide comprehensive description of the key EM processes in the applicator that can be used for further numerical study of the system towards the improvement of heating uniformity, radiation efficiency, and safety of the device.

REFERENCES

[1] V.V. Yakovlev, "Commercial EM codes suitable for modeling of microwave heating – a comparative review," In: U. van Reinen, M. Gunther, D. Hecht, Eds., Scientific Computing in Electrical Engineering, *Lecture Notes in Computational Sciences and Engineering*, vol. 18, Springer Verlag, pp. 87-96, 2001.

[2] V.V. Yakovlev, "Examination of contemporary electromagnetic software capable of modelling problems of microwave heating," *Proc. 8th Conf. Microwave and HF Heating, Bayreuth, Germany, September 2001*, pp. 19-20.

[3] P.O. Risman, "A microwave oven model: examples of microwave heating computations," *Microwave World*, vol. 19, No 1, pp. 20-23, 1998.

[4] P.O. Risman, "Differences between multi-mode and single mode systems for microwave chemistry," *Proc. 37th Microwave Power Symp., Atlantic City, NJ, July 2002*, pp. 5-8.

[5] G. Eves, and V.V. Yakovlev, "Analysis of operational regimes of a high power water load," *J. Microwave Power & Electromagn. Energy*, vol. 37, No 3, pp. 127-144, 2002.

[6] M.D. Davidovich, and V.P. Meschanov, "Synthesis of configuration of uniformly radiating longitudinal slots in the sections of nonregular rectangular waveguides," *IEEE Trans.*, vol. MTT-46, N 2, pp. 188-191, 1998.

MICROWAVE HEATING OF SILICON WAFERS IN CYLINDRICAL TM CAVITIES: A MATHEMATICAL MODEL

Gregory A. Kriegsmann
Department of Mathematical Sciences
Center for Applied Mathematics and Statistics
New Jersey Institute of Technology
University Heights
Newark, NJ 07102

ABSTRACT

A mathematical model is presented which describes microwave heating of silicon wafers in TM_{101} and TM_{111} cylindrical cavities. The underpinnings of this model are based upon the assumptions that the thickness of the wafer is much less than its radius, and that the modal pattern and amplitude are not affected, to leading order, by the sample's presence. The model takes the form of a two-dimensional nonlinear heat equation which is decoupled from Maxwell's equations. This heat equation is solved numerically and yields spatial and temporal approximations to the heating process; these are in qualitative agreement with experiments.

INTRODUCTION

Recent experiments by Booske and his colleagues [1-2] have shown that silicon wafers can be rapidly heated, with moderate power, in cylindrical TM_{111} and TM_{101} cavities. The implications of these experiments suggest that microwave heating may an efficient mechanism for two important materials applications; annealing single wafers and bonding several wafers together.

The mathematical problem that describe these heating processes is an initial, boundary value problem involving the heat and Maxwell's equations. These equations are nonlinearly coupled through the temperature dependence of the effective electrical conductivity. This feature coupled with the complexity of the cavity yields a formidable mathematical and physical problem which must be resolved by numerical methods, such as, finite difference and finite elements [3]. Even with modern high powered computers this can be a very time consuming and difficult task.

As a first step in understanding the heating process a simple mathematical model is proposed. It is a two dimensional diffusion equation that describes the evolution of the wafer temperature. This equation contains both radiative and convective loss terms which describe heat transfer on the wafer surfaces, and a source term which takes into account the modal pattern of the cavity and the detuning of the cavity. This model is based upon the assumptions that the thickness of the wafer is much less than its radius, that the modal pattern is not affected, to leading order, by the sample's presence, and that the cavity detuning can be expressed by a simple functional which incorporates an average of the temperature [4].

The source term in the heat equation has no angular dependence for the case of TM_{101} excitation. This makes the equation one-dimensional and easily solved by a number of numerical techniques. In particular, the steady state temperature can be very efficiently found by using a shooting method, along with a continuation technique [4]. This yields temperature distributions and a temperature-power relationship which agrees well with the experimental results of Reference 2. On the other hand, if TM_{111} excitation is used, then the source term has strong angular dependence and the heat equation is two-dimensional. An alternating-direction finite difference algorithm has been developed to produce accurate approximate answers to this equation. The results of numerical simulations are in qualitative agreement with experiments. Specifically, rapid heating is described on the time scales seen experimentally, the functional dependence of the maximum wafer temperature on the input power is in agreement with experiments, and temperature variations of the silicon wafer, driven by this mode are specifically obtained.

FORMULATION

A thin silicon wafer of thickness L and radius R_0 is placed nearly at the bottom of a cylindrical cavity of height H and radius R_1. The wafer rests upon three quartz pins, each of height H_P, which act like the legs of a stool. In the heating application reported in Reference 2, $L \sim 360\mu m$ and $R_0 \sim 36mm$, so that the fineness ratio $\delta = L/R_0 << 1$. In addition, R_1 is slightly larger than R_0 and $H_P \sim 1mm$.

The temperature in the wafer satisfies the heat equation

$$\rho C_p \frac{\partial}{\partial t} T = K\{\frac{\partial^2}{\partial X^2} + \frac{\partial^2}{\partial Y^2} + \frac{\partial^2}{\partial Z^2}\}T + \frac{1}{2}\sigma(T)|\mathbf{E}|^2, \quad \mathbf{X} \in D, \qquad (1a)$$

where D denotes the region occupied by the wafer, i.e., $D = \{(X, Y, Z) | 0 < R = \sqrt{X^2 + Y^2} < R_0, \quad -\frac{L}{2} - Z_0 < Z < \frac{L}{2} - Z_0\}$, and $Z = -Z_0$ is the location of the center-plane of the wafer. In this equation the density, thermal capacity, thermal conductivity, and electrical conductivity of silicon are denoted by ρ, C_p, K, and $\sigma(T)$, respectively, and $\mathbf{E}$ is the electric field. Although the thermal parameters are temperature sensitive their effects are second order compared to those caused by $\sigma(T)$; consequently they are neglected here. The boundary conditions take into account convective and radiative losses and are given by

$$K \frac{\partial}{\partial Z} T + h(T - T_A) + se(T^4 - T_A^4) = 0, \quad Z = \frac{L}{2} - Z_0 \tag{1b}$$

$$K \frac{\partial}{\partial Z} T - h(T - T_A) - se(T^4 - T_A^4) = 0, \quad Z = -\frac{L}{2} - Z_0 \tag{1c}$$

where h is the effective heat transfer coefficient, s is Boltzman's constant, e the emissivity of silicon, and T_A the ambient temperature of the surrounding environment. Implicit in these conditions are the assumptions that the environment, including the cavity walls, are held at the fixed temperature T_A, and there is no heat lost through the stool that supports the sample. In addition the thinness of the wafer leads to the boundary condition

$$\frac{\partial}{\partial R} T = 0, \quad R = R_0, \quad -\frac{L}{2} - Z_0 < Z < \frac{L}{2} - Z_0 \tag{1d}$$

which assumes there is no heat loss through its edge. Finally, the initial condition of the wafer is the ambient temperature,

$$T(X, Y, Z, 0) = T_A. \tag{1e}$$

The electric and magnetic fields within the wafer satisfy the time harmonic Maxwell's equations

$$\nabla \times \mathbf{E} = i\omega\mu_0 \mathbf{H}, \quad \nabla \times \mathbf{H} = \{-i\omega\epsilon + \sigma(T)\}\mathbf{E} \tag{2}$$

where ϵ is the permittivity of silicon, $\sigma(T)$ is again its effective electrical conductivity, ω the frequency of the microwave source, and μ_0 the permeability of free space. Inside the cavity and exterior to the wafer the same equations hold with $\sigma = 0$ and $\epsilon = \epsilon_0$, the permittivity of free space. The boundary conditions require the vanishing of the tangential electric field on the walls of the cavity, and the continuity of the tangential magnetic and electric fields across the surface of the wafer.

MODELING

The solution of (1-2) taking into account the excitation of the cavity by an external source is a formidable mathematical problem. The difficulties in obtaining an accurate approximation of the electromagnetic fields and temperature distribution in the wafer stem from two roots. The first is the nonlinear coupling caused by the temperature dependence of the effective electrical conductivity, $\sigma(T)$. The second is the complicated geometry and concomitant physics, such as cavity detuning, induced by the feeder waveguide which connects the source to the cavity. Numerical methods such as finite differences and finite elements are, in general, the only avenue to obtain accurate approximate answers to this problem [3]. However, certain simplifications can be made which significantly reduce the computational burden and which yield results that contain considerable physical insight into the heating process. These simplifications will now be described.

As mentioned in the previous section the fineness ratio of the silicon wafer $\delta = L/R_0 << 1$ is a small parameter. Using a systematic perturbation scheme, as in Reference, 4 it is relatively easy to show that the temperature is given by

$$T = T_0(X, Y, t)\{1 + O(\delta)\} \tag{3}$$

where T_0 is the leading term in an asymptotic expansion and the $O(\delta)$ term represents corrections which are the same order as δ. It is important to observe here that the leading order temperature is independent of Z in this limit. The correction term in this expansion will not be pursued here.

By integrating equation (1a) over the thickness of the wafer L, using the divergence theorem, and applying the boundary condition (1b-c), it is found that the function T_0 satisfies

$$\rho C_p \frac{\partial}{\partial t} T_0 = K\nabla_{\perp}^2 T_0 - 2\mathcal{L}(T_0) + \frac{1}{2}\sigma(T_0)\frac{1}{L}\int_{-Z_0-L/2}^{-Z_0+L/2} |\mathbf{E}|^2 \, dZ \tag{4a}$$

$$\mathcal{L}(T_0) = \frac{1}{L}\{h(T - T_A) + se(T^4 - T_A)\} \tag{4b}$$

where $\nabla_{\perp}^2$ is the Laplacian operator in X and Y and the term $\mathcal{L}(T_0)$ takes into account the heat loss on the two faces of the wafer. The boundary condition (1d) now becomes

$$\frac{\partial}{\partial R} T_0 = 0 \qquad R = R_0 \tag{4c}$$

and the initial condition (1e) remains the same.

All that is needed now to complete the model is an approximation of the electric field within the sample. Once this is obtained, equation (4) yields a two-dimensional nonlinear heat equation which models the microwave heating of the wafer.

As a first approximation, the presence of the thin wafer does not perturb the electromagnetic field distribution. Assuming the height of the cavity H is adjusted so that either a TM_{101} or a TM_{111} mode is excited [2], the magnetic and electric fields are given by

$$\mathbf{H} = H_0\, R_0 \{ -\frac{\partial}{\partial R}\psi\, \hat{\theta} + \frac{1}{R}\frac{\partial}{\partial \theta}\psi\, \hat{R} \} \tag{5a}$$

$$\mathbf{E} = i\frac{H_0\, R_0}{\omega\epsilon} \{ \frac{\partial}{\partial Z}[\frac{\partial}{\partial R}\psi\, \hat{R} - \frac{1}{R}\frac{\partial}{\partial \theta}\psi\, \hat{\theta}] - \lambda\psi\hat{Z} \} \tag{5b}$$

where $\hat{R}, \hat{\theta}, and \hat{Z}$ are unit vectors in their corresponding directions. The scalar function ψ is given by

$$\psi_{111} = \cos\theta\, J_1(z_1^1 R/R_0) \cos(2\pi Z/H) \tag{5c}$$

for the TM_{111} mode and

$$\psi_{101} = J_0(z_1^0 R/R_0) \cos(2\pi Z/H) \tag{5d}$$

for the TM_{101} mode. Here J_n is the Bessel function of order n, z_1^n is its first zero, $\lambda \equiv \lambda_n = (z_1^n/R_0)^2$ in (5b), with $n = 0, 1$ for the TM_{101} and TM_{111} modes, respectively. Finally, the parameter H_0 is the strength of the mode which is not known without properly taking into account the coupling of the cavity and source.

In the experiments described in Reference 2 the wafer is placed near the bottom of the cavity and thus $Z_0 \sim -H/2$. Since the wafer is very thin, $Z \sim Z_0$ and this forces the first two terms in (5b) to be very small. Consequently the electric field is approximated in the sample by

$$\mathbf{E} = H_0\, R_0 \frac{i\lambda_n}{\omega\epsilon} \cos(n\theta) J_n(z_1^n R/R_0)\hat{Z} \tag{6}$$

for the TM_{1n1} mode, $n = 0, 1$. Inserting this expression into (4a) yields

$$\rho C_p \frac{\partial}{\partial t} T_0 = K \nabla_{\perp}^2 T_0 - 2\mathcal{L}(T_0) + \frac{E_0^2}{2} \mathcal{S} \tag{7a}$$

$$\mathcal{S} = \sigma(T_0) (\frac{z_1^n}{N_1})^4 \cos^2(n\theta) \, J_n^2(z_1^n R/R_0) \tag{7b}$$

where $N_1 = \sqrt{\epsilon_1/\epsilon_0}$ is the index of refraction of silicon and $E_0 = H_0/\omega\epsilon_0 R_0$, the electric field strength of the exciting mode.

Before proceeding further there are two interesting points to be made about the source term in (7). First, for a fixed E_0 it is proportional to $(z_1^n)^4$. Since $z_1^1 = 3.831 > z_1^0 = 2.405$ the source term is larger for the TM_{111} mode than the TM_{101}. This fact was observed experimentally in Reference 2. Secondly, there is more spatial variation, namely in the θ direction, for the TM_{111} mode and this causes nonuniform heating. Thus, there is a trade off in using the more efficient mode.

At this juncture the relationship between the electric field strength of the exciting mode E_0 and the power supplied by the microwave source P_S must be determined. Once this is done, equation (7) along with (4c) and (1e) will determine the temperature in the wafer.

In the simplest approximation it is assumed that the sample has no effect on the resonant mode. Then E_0^2 is given by

$$E_0^2 = \frac{Z_0 P_S}{\pi R_0^2}, \qquad Z_0 = \sqrt{\mu_0/\epsilon_0} \tag{8a}$$

where Z_0 is the impedance of free space. This assumption precludes the effect of cavity detuning where the strength of the mode E_0 is reduced by the increase in the effective electrical conductivity of the material as it is heated [4]. That is, E_0^2 in (8) is no longer a constant proportional to P_S, but also depends upon, in some global fashion, the temperature of the wafer. This dependency has been explicitly calculated for the related problem of heating a thin ceramic slab in a TE_{103} cavity [4] and is given by

$$E_0^2 = \frac{Z_0 P_S}{\pi R_0^2} \frac{1}{(1 + \chi \iint_\Omega \mathcal{S} \, dx \, dy)^2} \tag{8b}$$

where Ω represents the surface of the slab and χ is proportional to the Q of the cavity. Although the analogous relationship between E_0 and P_S has not yet been derived for the present cavity configuration, (8b) should still be qualitatively correct and will be used in the forthcoming numerical simulations. Finally, if the cavity has a low Q, then χ is small and (8b) reduces to (8a).

THE NUMERICAL METHODS

The mathematical model of the heating process is now described by equations (7-8), boundary condition (4c), and initial condition (1e). This problem is nonlinear and its solution must be found by an accurate numerical method. If the excitation is from a TE_{101} mode, then there is no angular dependence in the source and the problem becomes one-dimensional. This is solved by a straight forward Crank Nicolson method [5]. The results show that the temperature increases with time and approaches a steady state value which depends upon the power. This occurs on the order of 15-25 seconds. These results are not shown here. Here, only the steady state temperature are considered and consequently, a shooting method coupled with a continuation technique is used to rapidly obtain accurate approximations [4]. On the other hand, if the excitation is from a TM_{111} mode, then the problem is two-dimensional. The method chosen to solve this problem is an alternating direction finite difference scheme [5] applied to the equation in its polar form. In each of the r and θ sweeps a Crank Nicolson method is used to insure stability and to allow larger time steps in the simulations. The method is first order accurate in Δt and second order in the spatial step sizes. The loss of second order accuracy in Δt is caused by the splitting method, but is very worthwhile as only simple tridiagonal matrices are inverted. Finally, once T_0 is approximated on the grid in the $r - \theta$ plane, interpolation is used to approximate T_0 in terms of x and y.

SIMULATIONS

The thermal and material parameters of silicon are used in the following examples, $h = 38W/m^{2o}K$, $R_0 = 36\,mm$, $L = 360\mu\,m$, $\delta = L/R_0 = 0.01$, the height H of the cavity is chosen to excite either the TM_{101} or the TM_{111} mode at 2.45 Ghz., and the ambient temperature is taken to be 300^oK. The effective electrical conductivity is modeled by

$$\sigma = \sigma_A 10^{f(T)} \tag{9a}$$

$$f(T) = c_1[1.0 - e^{-c_2(T-T_A)/T_A}] \tag{9b}$$

where $\sigma_A = 2/\Omega m$ is the value of the conductivity at T_A, $c_1 = 6.882$, and $c_2 = 0.257$. Equation (9) fits the experimental conductivity data presented in Reference 2 for a particular doping concentration, except a lower temperatures where there is an initial decrease on the conductivity. Finally, the index of refraction was taken to be $N_1^2 = 3.42$.

Figure 1 shows the steady state temperature at the center of a wafer, for TM_{101} excitation, as a function of the microwave power. The value of $\chi = 0.1$ was chosen so that the numerical and experimental results agreed for $P_S = 800$ watts. The remainder of the numerical and experimental results [2] are in excellent agreement. If this parameter were set to zero, then there would be no steady state temperature above sixty watts because a thermal runaway would occur. That is, the power being absorbed by the wafer is greater than the ability of its surface to disipate thermal power. Finally, for each power level shown in Figure 1 the temperature decreased monotonically from the center of the sample to its edge, reflecting the structure of the modal pattern of the exciting mode.

Figure 1

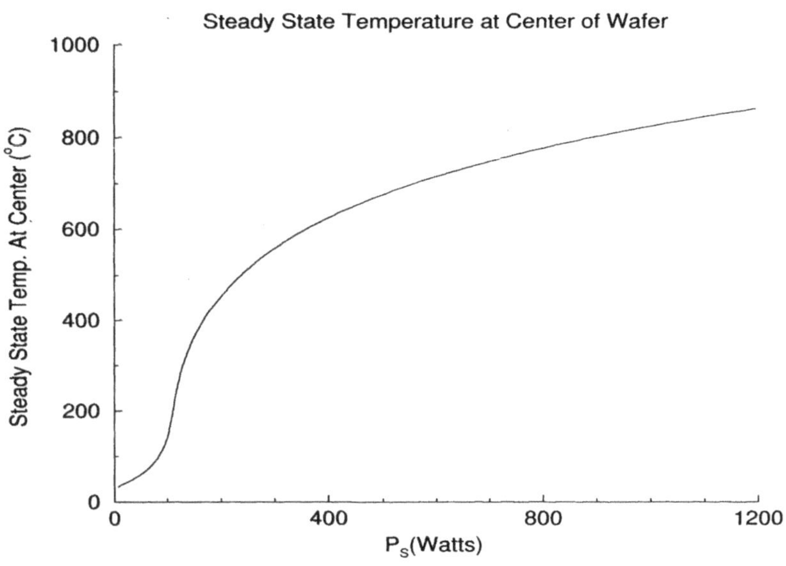

Figure 2 shows a two dimensional representation of the temperature for TM_{111} excitation at 500 watts. It clearly shows the modal pattern being hotter near $\theta = 0, \pi$ and near the maximum of J_1. It took approximately 20 seconds for the temperature to reach steady state with a maximum value of 900^oC. These are both in good qualitative agreement with experiments [2].

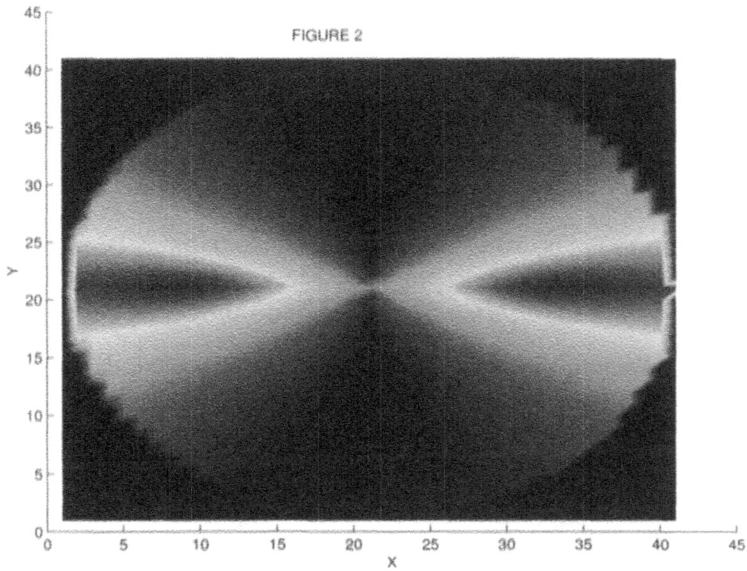

CONCLUSIONS

A simple model of microwave heating of a silicon wafer in a cylindrical TM_{1n1} cavity was presented for $n = 0, 1$. It was based upon three assumptions. The first required the thickness of the wafer to be much smaller than its radius; this is valid for the types of samples considered in Reference 2. The second assumed that wafer did not affect the modal structure of the electric field. The validity of this reasonable assertion is the topic of ongoing research. If a correction to the field is found, then it will just modify the source term (7b). The third modeled cavity detuning by a simple functional of the temperature. This functional was derived for the related problem of heating a thin slab in a rectangular cavity; its validity is also the subject of ongoing research. If its form is different than the one used in this paper, then (8b) must be modified. In any case the resulting model will still be a two-dimensional heat equation with the same qualitative features. Finally, in spite of these assumptions the numerical results obtained from the model are in good qualitative, and in some cases excellent quantitative, agreement with experiments.

ACKNOWLEDGMENTS

This work was supported by the Department of Energy Under Grant No. DE-FG02-94ER25196, and the National Science Foundation Under

Grant No. DMS 9305828.

REFERENCES

1. Thompson, K., Booske, J.H., Gianchandani, Y.B., and Cooper, R.F., "Electromagnetic Annealing for the 100 nm Technology Node", *IEEE Electronic Device Letters*, Vol. 23, pp. 127-129 (2002).

2. Thompson, K., Booske, J.H., Gianchandani, Y.B., and Cooper, R.F., "Direct Si-Si Bonding by Electromagnetic Induction Heating", *Journal of MicroElectroMechanical Systems*, in press.

3. Fu, W. and Metaxas, A., "Numerical Prediction of Three Dimensional Power Density Distributions in a Multi-Mode Cavity", *Journal of Microwave Power and Electromagnetic Energy*, Vol. 29, No. 2, pp.67-75 (1995).

4. Kriegsmann, G. A., "Pattern Formation in Microwave Heated Ceramic Cylinders and Slabs", **IMA Journal on Applied Mathematics**, Vol. 66, No. 1, pp. 1-32 (2001).

5. Mitchell, A.R. and Griffiths, D.F., The Finite Difference Method in Partial Differential Equations, John Wiley and Sons, New York City, 1980.

THE DIELECTRIC PROPERTIES OF FRUIT, VEGETABLES AND MEAT

Sheryl A. Barringer, Oya Sipahioglu and Cavit Bircan
Department of Food Science & Technology
The Ohio State University
2015 Fyffe Road
Columbus, OH 43120-1007 USA

ABSTRACT

Predictive equations for the dielectric properties at sterilization temperatures are necessary to determine heating patterns during industrial microwave processing and during reheating of ready to eat microwaveable foods. However, the temperature dependence of the dielectric properties of meats, vegetables and fruits that are commonly thermally processed are not readily available in the literature. The objective of this study was to generate predictive equations for the dielectric properties at 2450 MHz of vegetables, fruits and meats at 0-130°C, as a function of temperature, moisture, ash content and water activity.

For most of the samples, the dielectric constant decreased with increasing temperature, increased with increasing moisture content and decreased with increasing ash. The dielectric loss factor either decreased then increased, or simply increased with increasing temperature. The inflection point was dependent on the ash content. As ash content decreased, the inflection point occurred at higher temperatures. The dielectric loss factor increased with increasing ash content, but showed no clear correlation to moisture content.

INTRODUCTION

Currently, microwave pasteurized and sterilized meals are available in Belgium and Japan. The major advantage of microwave sterilization over retorting is that it is a high temperature short time (HTST) treatment that better preserves the flavor and quality characteristics of the product[1]. Microwave sterilization is three to five times faster than retorting depending on the product[2]. To predict the quality of the food that results from a microwave process, equations capable of predicting the temperature distribution within the food have been developed. This involves simultaneously solving Maxwell's equations of electromagnetism and the heat equation where all thermal and electrical properties of the food are nonlinearly dependent on temperature[3,4]. A common feature of most equations is the assumption of constant dielectric properties during heating. The dielectric properties of most materials vary with temperature and composition, in some cases dramatically, therefore it would be valuable to develop temperature dependent equations that predict dielectric properties of foods to increase the accuracy of the present temperature distribution models.

While effects of individual food components have been investigated, it is not yet possible to predict the effect of the composition of the food on the dielectric properties. Literature reporting the dielectric properties of many foods collected at different temperatures and frequencies is available[5,6]. Others[7] have developed equations predicting dielectric properties as a function of salt, moisture, fat content, density, temperature and frequency for different food groups and have also created predictive equations for foods and various subsets[8].

The objective of this study was to determine the effect of temperature, moisture and ash on the dielectric properties of turkey meat, ham, other meats, vegetables and fruits at 2450MHz and 0-135°C.

MATERIALS AND METHODS

Meats included beef, ham, turkey, chicken breast and thigh, cod, perch and salmon. Vegetables included carrot, parsley, spinach, radish, turnip, yam, potato, garlic, parsnip, and broccoli. Fruits included cucumber, apple, pear, banana, and corn. The samples covered a moisture range of 95.89-57.30% and an ash range of 0.26-1.56%. Nineteen ham samples were prepared at different moisture (38.2-68.9%) and ash contents (1.78-6.80%)[9]. Turkey samples were brined in 0.5% sodium tripolyphosphate, 1.75% sodium chloride, and 0-3% sodium lactate and collected both raw and reduced moisture (0.84-0.90 a_w)[10].

The dielectric constant and loss factor were measured using an open ended coaxial probe and a network analyzer (85070B and 8752C, Hewlett-Packard, Palo Alto, CA) interfaced with a computer. The probe lies facing upwards inside a cylindrical jacketed stainless steel sample holder (2.5cm inner diameter, 3cm height), which is connected to an oil bath (RTE 140, Neslab Co., Newington, NH) that adjusts the temperature of the sample. The sample holder was placed between two steel plates and sealed using bolts (Fig 1). O-rings were placed between the sample holder and the plates to prevent moisture loss. A temperature probe connected to a temperature acquisition unit (HP 34970A, Hewlett-Packard, Palo Alto, CA) was fit permanently into the upper plate and reaches into the center of the sample. The samples were placed underneath a weak spring which held the sample against the probe if the samples shrunk, and prevented liquid accumulation. This ensured that the sample, and not the expelled liquid, was being measured. The sample holder was sealed to prevent any moisture loss as steam. A thin metal sheet was placed between the sample and the spring to prevent the spring digging into the sample.

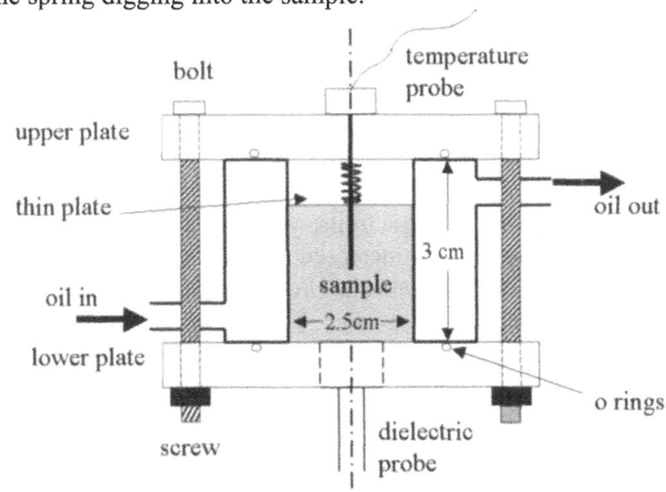

Figure 1. Sectional view of stainless steel sample holder for dielectric property measurements.

The sample (a cylinder 2.5cm x 2.5cm) was initially refrigerated to 5°C and the oil in the jacketed sample holder cooled the temperature probe to 5°C. The first measurement was made when the temperature reading was stable. The oil bath was then increased to 5°C above the temperature of measurement. When the sample reached the desired temperature the dielectric properties were measured and the oil temperature increased another 5°C. Thus, the temperature gradient in the sample is from the stated temperature to 5°C warmer. A temperature range of 5 to 135°C was measured in 5°C intervals at 2450MHz. All measurements were performed in triplicate, with fresh samples.

Calibration was performed using a short, air, and water before every test. The sample holder was stabilized so that the probe could not be moved after calibration. The dielectric constant and loss factor were calculated from the phase shift and the magnitude of the reflected signal by the software.

Ash content of the samples was determined by dry ashing at 500°C in a muffle oven (Thermolyne, Dubuque, IA) overnight. The moisture content of the samples was determined by drying in a vacuum oven (National Appliance Co., Skokie, IL) at 75°C for 6 hr. Water activity (a_w) was determined using a water activity analyzer (CX-2, Decagon Devices Inc., Pullman, WA).

The response variables, dielectric constant and loss factor, were fit by using multiple regression from combinations of temperature, moisture content, ash, and water activity. The fitting principle was least squares, which is the most appropriate method to employ when the data are expected to exhibit significant random errors[7]. The models and all predictors included in the models had a significance of p#0.001. Quality of fit was assessed from adjusted coefficient of determination (R^2_{adj}) of the equation. The measurements were done in triplicate. The data were analyzed using JMPin software (SAS Institute Inc., Cary, NC).

RESULTS AND DISCUSSION

Dielectric constant

Effect of Temperature: In most fruits, vegetables and meats, the dielectric constant decreased as temperature increased (Fig 2, Table 1). However, there were some exceptions to this general rule. Potatoes showed a sudden increase in the dielectric constant between 65 and 70°C (Fig 3). The gelatinization temperature of potato starch is 62-68°C[11]. Gelatinized starch has a higher dielectric constant than ungelatinized starch[12]. Therefore it is reasonable that the dielectric constant would increase during gelatinization. Why this occurs is most likely due to hydrogen bond stabilization that facilitates polarization of water molecules by the electromagnetic field[12].

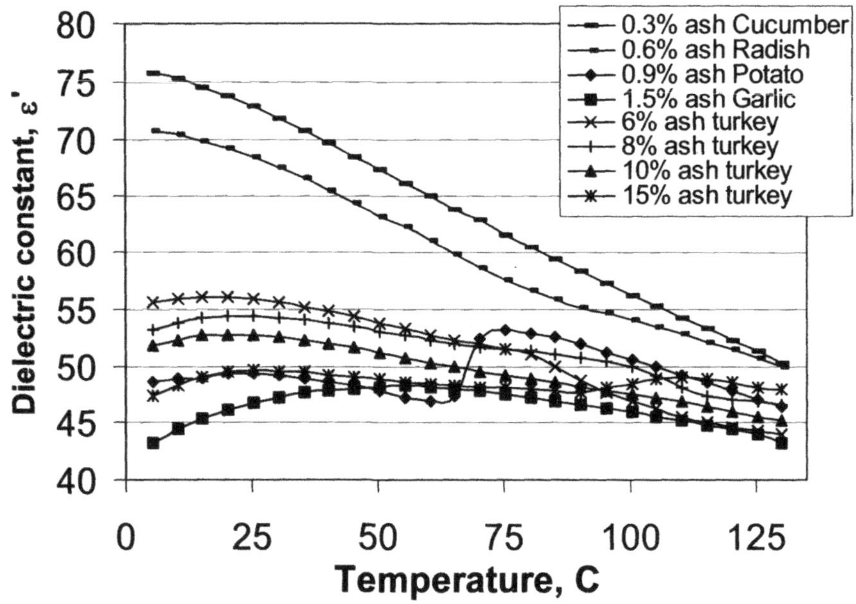

Figure 2. The dielectric constant at 2450MHz decreases with increasing temperature and ash, with some exceptions.

Another exception to the general rule was garlic (Fig 4). The dielectric constant first increased to 55°C then decreased in a smooth curve. Looking closely, this same increase and decrease occurred for most samples between

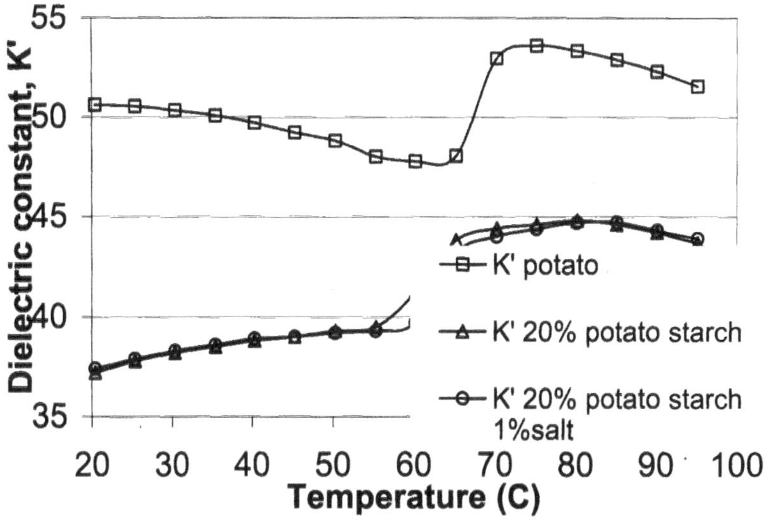

Figure 3. The dielectric constant of potato increases at the same temperature as a solution of potato starch, with or without salt.

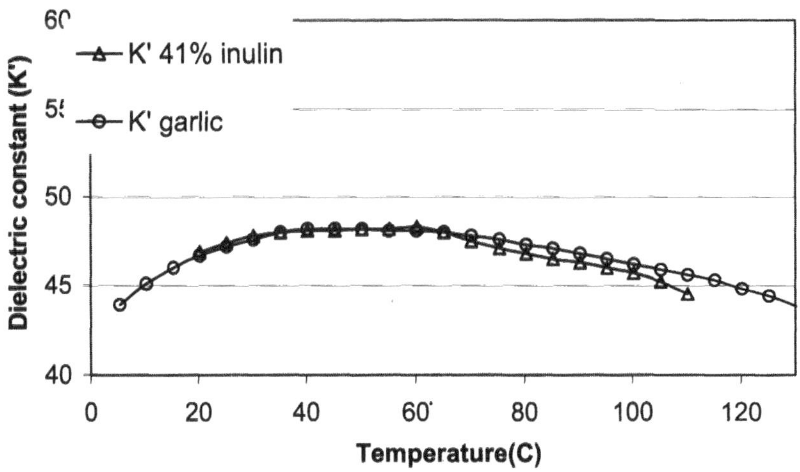

Figure 4. The dielectric constant of garlic and inulin show similar trends

refrigeration and room temperature. This slightly unusual temperature dependence of the dielectric constant occurs because it is a combination of the dielectric constant of bound and free water. The dielectric constant of bound water increases as temperature increases, while the dielectric constant of free water decreases[7]. The combination of these two produce the measured dielectric constant.

The other exception occurred for low moisture samples, such as partially dried turkey and ham (Fig 5). The dielectric constant of these samples, with 36-51% moisture, increased with temperature instead of decreasing. This same behaviour has been observed in corn[13] and salted ham[14]. The increase with temperature occurs as the bound water component of the dielectric constant dominates over the free water component.

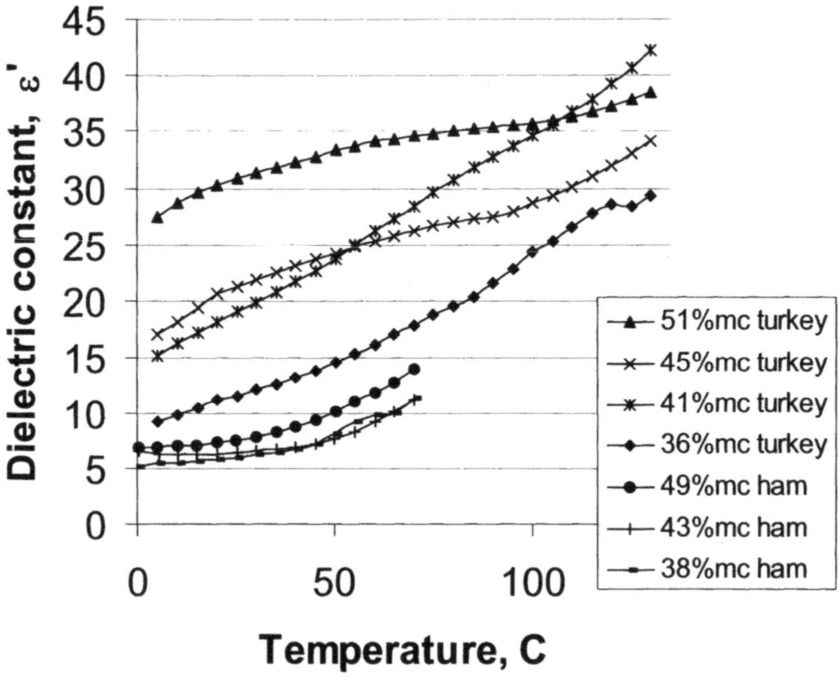

Figure 5. The dielectric constant at 2450MHz increases with temperature when moisture is tightly bound.

Effect of Moisture: The dielectric constant increases as the moisture content increases, for all samples (Fig 5, Table 1). This has frequently been observed, since water is the major food component that contributes to the dielectric constant.

Effect of Ash: The dielectric constant of food items decreases as the ash content increases (Fig 2, Table 1). This occurs because the ash binds the water, decreasing its ability to move in response to the electromagnetic field. In some samples, such as among fruits or unsalted meat samples, no correlation to ash was found. This is likely because both the total amount and range of ash within those samples is small.

A possible exception to this trend occurred at high temperatures among the turkey samples (Fig 2). Above 100°C, the samples were in the reverse order according to ash content. The dielectric constant of the highest ash sample, with 15% ash, was fairly independent of temperature. It varied from 47 to 49 for the entire temperature range measured. The lower ash samples decreased with increasing temperature, so that above 100°C their dielectric constant was lower than samples with greater ash content. Thus the trend reverses and at high temperatures, the dielectric constant increases as the ash content increases.

Dielectric loss factor

Effect of Temperature: The dielectric loss factor either decreased then increased, or simply increased, with increasing temperature (Fig 6, 8, Table 1). This occurs because the dielectric loss factor is composed of two components, dipole loss and ionic loss. The dipole loss, mainly from the water, decreases with temperature. The ionic loss, from dissolved salts, increases with temperature. The combination of the two factors causes an initial decrease, then increase with temperature. The transition temperature, or temperature at which the dielectric loss factor vs temperature curve changes from decreasing to increasing, is dependent on the ash content (Fig 5). The higher the ash content, the lower the transition temperature.

Effect of Ash: The dielectric loss factor increases with increasing ash content (Fig 6, 8, Table 1). This behaviour has been observed by many researchers, and occurs because the ionic loss component increases strongly with temperature.

Effect of Moisture: There is no clear effect of moisture content on the dielectric loss factor. Looking at subgroupings of samples, the dielectric loss factor increased with moisture content for turkey, increased then decreased for other meats, and decreased then increased for vegetables and fruits. Other authors have found similarly conflicting results. The dielectric loss factor peaks at a moisture content between 44-67% at 2800 MHz[15], decreases linearly with moisture content at 20, 300, and 2450 MHz for grains[16], decreases with increasing moisture content (35 to 98%) at 2450 MHz for maple syrup[17], increases with increasing moisture content until 20-30%, levels off and may decrease for higher moisture contents for low moisture content foods[18], and increases with increasing moisture content at 3000 MHz for high moisture levels[14].

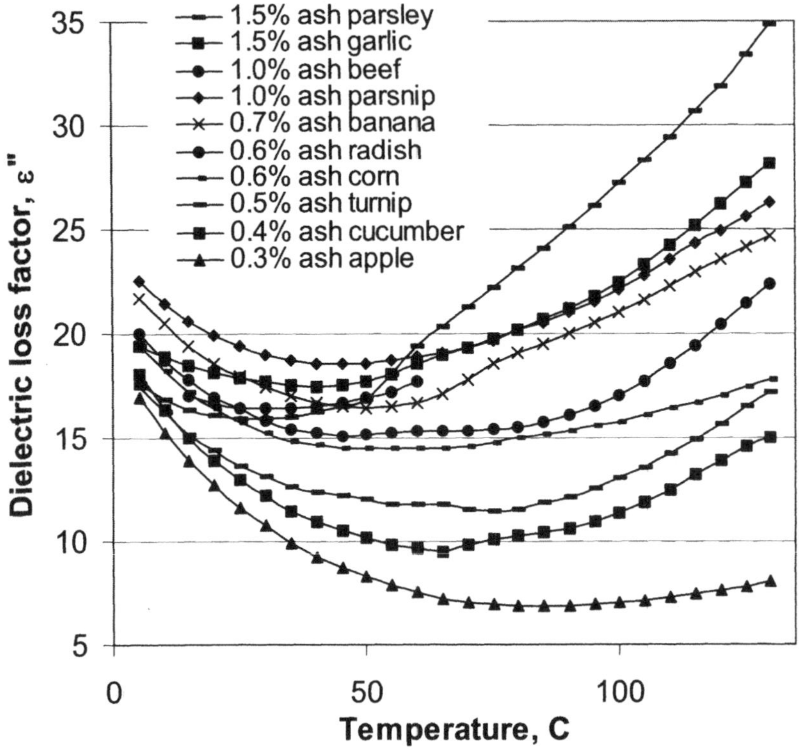

Figure 6. The dielectric loss factor at 2450MHz increases then decreases with temperature, and increases with ash content.

Equations

Equations that predict the dielectric constant of subgroups of vegetables, fruit, ham, turkey and other meats as a function of temperature, moisture or water activity, and ash were produced (Table 1). These equations can be used to calculate the temperature dependent dielectric properties of these products, if the composition and temperature are in the same range of those in this study.

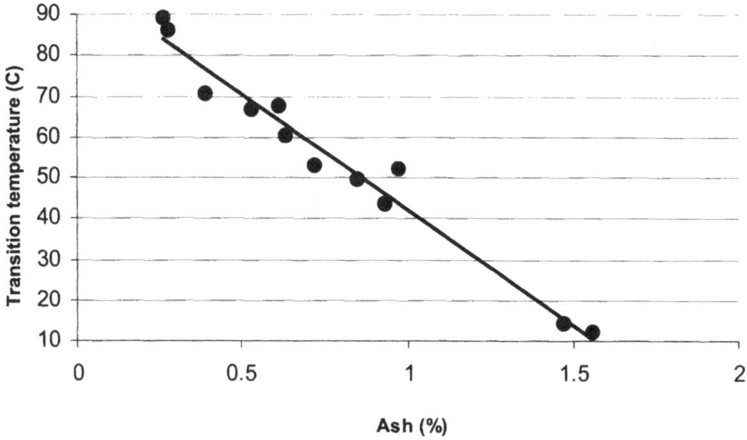

Figure 7. The transition temperature of the dielectric loss factor at 2450MHz is a function of ash content.

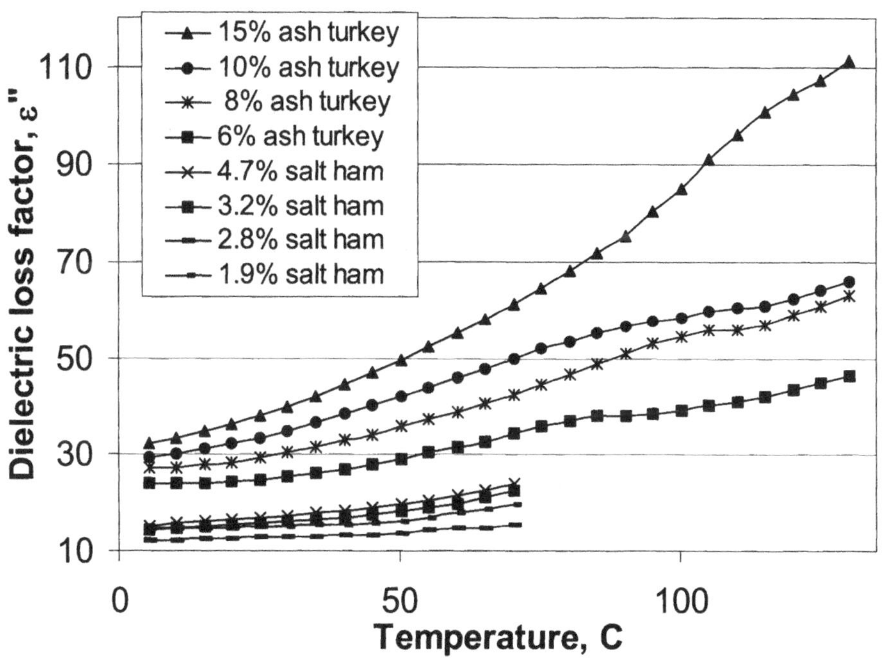

Figure 8. The dielectric loss factor at 2450MHz increases with ash content.

Conclusions

For most of the samples, the dielectric constant decreased with increasing temperature. Some exceptions were potato, garlic and low moisture ham and turkey. The dielectric constant increased with increasing moisture content. The dielectric constant decreased with increasing ash. Some exceptions were groups of samples that contained little ash, and a reversal of the trend at high temperatures.

The dielectric loss factor either decreased then increased, or simply increased with increasing temperature. The inflection point was dependent on the ash content. As ash content decreased, the inflection point occurred at higher temperatures. The dielectric loss factor increased with increasing ash content, but showed no clear correlation to moisture content.

Table 1. Predictive equations for the dielectric constant and the loss factor as a function of temperature, moisture and ash at 2450MHz.

Dielectric constant		R^2_{adj}
Vegetables	$\varepsilon' = -243.6 + 1.342\cdot T + 4.593\cdot M - 426.9\cdot A + 376.5\cdot A^2 - 0.01415\cdot M\cdot T - 0.3151\cdot A\cdot T$	0.856
Fruits	$\varepsilon' = 22.12 + 0.2379\cdot T + 0.5532\cdot M - 0.0005134\cdot T^2 - 0.003866\cdot M\cdot T$	0.828
Ham	$\varepsilon' = -25.49 + 1.063\cdot M - 1.041\cdot A + 0.03452\cdot T$	0.817
Turkey	$\varepsilon' = -352.2 + 1.888\cdot T + 0.5095\cdot Ad + 420.4\cdot a_w - 2.042\cdot a_w\cdot T$	0.840
Other meats	$\varepsilon' = 6.49 + 0.689\cdot M - 0.0748\cdot T$	0.820
Dielectric loss factor		
Vegetables	$\varepsilon'' = -100.02 - 0.1611\cdot T + 0.001415\cdot T^2 + 2.429\cdot M - 378.9\cdot A + 316.2\cdot A^2$	0.855
Fruits	$\varepsilon'' = 33.41 - 0.4415\cdot T + 0.001400\cdot T^2 - 0.1746\cdot M + 1.438\cdot A + 0.001578\cdot M\cdot T + 0.2289\cdot A\cdot T$	0.887
Ham	$\varepsilon'' = -150.2 + 5.243\cdot M + 6.220\cdot A - 0.2845\cdot T - 0.04322\cdot M^2 - 0.4732\cdot A^2 + 0.002245\cdot T^2 + 0.1090\cdot A\cdot T$	0.852
Turkey	$\varepsilon'' = -197.8 - 0.3192\cdot T + 0.6737\cdot Ad + 0.0017\cdot T^2 + 230.1\cdot a_w + 0.0450\cdot Ad\cdot T$	0.867
Other meats	$\varepsilon'' = -608 - 0.107\cdot T + 0.00177\cdot T^2 + 12.6\cdot M - 0.0839\cdot M^2 + 279\cdot A - 122\cdot A^2$	0.890

T= temperature (°C), M= moisture (%), A= wet basis ash (%), Ad=dry basis ash (%), a_w= water activity.

REFERENCES

[1]T. Ohlsson, "Microwave processing in the food industry," *European Food and Drink Rev* **7** 9-11 (1991).

[2]C.E. Morris, "Breakthrough: world's first commercial microwave-sterilization system," *J Food Eng* **63** [3] 98, 101-102, 106 (1991).

[3]H. Zhang and A.K. Datta, "Coupled electromagnetic and thermal modeling of microwave oven heating of foods," *J Microwave Power and EE* **35** [2] 71-85 (2000).

[4]J.M. Hill and T.R. Marchant, "Modeling microwave heating," *Appl Math Modelling* **20** [1] 3-15 (1996).

[5]M. Kent, *Electrical and dielectric properties of food materials* London: Science and Technology Publishers. 35-135p (1987).

[6]W.R. Tinga and S.O. Nelson SO, "Dielectric properties of materials for microwave processing – tabulated," *J Microwave Power EE* **8** [1] 23-65 (1973).

[7]R.K. Calay, M. Newborough, D. Probert D and P.S. Calay, "Predictive equations for the dielectric properties of foods," *Int J Food Sci Tech* **29** 699-713.(1995).

[8]E. Sun, A. Datta and S. Lobo, "Composition-based predictions of dielectric properties of foods," *J Microwave Power EE* **30** [4] 205-212 (1995).

[9]O. Sipahioglu, S. A. Barringer, I. Taub and A. Prakash. "Modeling the dielectric properties of ham as a function of temperature and composition." *J Food Sci* in press (2003)

[10]O. Sipahioglu, S. A. Barringer, I. Taub and A. P. P. Yang, "Characterization and Modeling of Dielectric Properties of Turkey Meat" *J Food Sci* in press (2003)

[11]R.L. Whistler and J.R. Daniel, "Carbohydrates". In: Fennema OR, *Food Chemistry*. New York: Marcel Dekker, Inc. 114p (1985).

[12]B.D. Roebuck, S.A. Goldblith and W.B. Westphal, "Dielectric properties of carbohydrate-water mixtures at microwave frequencies," *J Food Sci* **37** 199-204 (1972).

[13]S.O. Nelson, "Frequency and moisture dependence of the dielectric properties of high-moisture corn" *J Microwave Power* **13** [2] 213-218 (1978).

[14]N.E. Bengtsson and P.O. Risman, "Dielectric properties of foods at 3GHz as determined by a cavity perturbation technique. II. Measurements on food materials" *J Microwave Power* **6** 107-123 (1971).

[15]T. Funebo and T. Ohlsson, "Dielectric properties of fruits and vegetables as a function of temperature and moisture content," *J Microwave Power EE* **34** [1] 42-54 (1991).

[16]S.O. Nelson, "Frequency, moisture, and density dependence of the dielectric properties of small grains," *Trans. ASAE* **35** [2] 625-629 (1987).

[17]D. Favreau, V. Sosle and G.S.V. Raghavan, "Dielectric properties of maple sap and syrup at 2.45 GHz," *J Microwave Power EE* **32** [2] 96-100 (1997).

[18]R.F. Schiffman, "Microwave and dielectric drying" In: Mujumdar AS. *Handbook of Industrial Drying*. Marcel Decker Inc.: New York. 352p (1995).

Microwave Synthesis and Reactions

SYNTHESIS OF NANO-SIZED INORGANIC COMPOUNDS INDUCED BY MICROWAVE IRRADIATION

Yuji Wada and Shozo Yanagida
Material and Life Science, Graduate School of Engineering, Osaka University, Yamada-oka, Suita, Osaka 565-0871, Japan
ywada@mls.eng.osaka-u.ac.jp and yanagida@mls.eng.osaka-u.ac.jp

ABSTRACT

We succeeded in synthesizing nano-sized inorganic compounds under precise control of size and its distribution. Two examples are demonstrated in this paper.

Nano-sized nickel metal particles having narrow size-distribution were prepared through reduction of $Ni(OH)_2$ with ethylene glycol under microwave irradiation. The average particle size was controlled between 5-8 nm by changing the irradiation time and irradiation power.

CdS nanocrystallites are controlled in size and size-distribution by microwave-assisted reaction of cadmium acetate with thiourea in N,N-dimethylformamide (DMF). The absorption onset of CdS nanocrystallites is shifted to longer wavelength with increase of MW irradiation time, indicating the particle size growth under prolonged irradiation. However, when the microwave irradiation of the solution is punctually and intermittently repeated under keeping the solution at ambient temperature before each irradiation, the onset is pinned at the same wavelength and only the optical density of the absorption band is increased, suggesting that the particle growth occurs only during the continuous irradiation and stops once when the system is cooled down.

INTRODUCTION

The size-control and the size-distribution of semiconductor nanocrystallites are of great importance from the viewpoint of application such as photocatalysts and materials for optics and electronics.[1-3] Endeavors have been made to develop the methods to control size and size-distribution in the preparation of nanoparticles, e.g., using stabilizers (terminators) such as thiol, selenol, phosphate, and phosphine oxide; regulating the reaction space by using matrices such as zeolite, glass, polymer, reverse micelle, vesicle, LB film, multilayer film, xerogel, and silica.[4]

MW heating has recently begun to attract the attention of chemists as a new method not only for synthesizing organic compounds but also for preparing inorganic ones[5-8] such as nano-sized inorganic particles,[9-14] molecular sieve films,[15] and self-assembled monolayers.[16] Among the nano-sized inorganic materials reported so far, one can find nickel nanocrystallites,[17] platinum nanocrystallites,[9] iron oxides, Ni/NiO composite,[10] aluminum nitride,[11] iron oxide nanoparticles,[12,13] and alumina-ceria nano-sized composites.[14]

Although the formation mechanism of monodispersed nanoparticles under MW irradiation has not been entirely understood, we can point out several factors important for precise control of the particle size and its distribution: the purity of starting materials, effect of a reactor wall, temperature change, heating rate, homogeneity of a solution, and a stabilizing reagent. We describe in this paper two examples, achieving precise control of their sizes and size distribution.[15,16]

EXPERIMENTAL

A MW oven (Sanyo EM-650T, 2.45 GHz, max. power 650 W) was modified by Shikoku Keisoku Kogyo Co. so that the temperatures of samples inside the oven can be monitored *in situ* under irradiation. A thermo fiber sensor (Anritsu Meter Co., LTD., 10117-*S) was inserted through a hole made in the ceiling of the oven. This hole was covered with a stainless sleeve to avoid MW leakage from the oven, which was monitored by a MW detector (Narda Corp., Model 8201) for safeguarding.

A dispersion solution of nickel hydroxide (nacalitesque, light-green powder, 3.0×10^{-4} mol) in ethylene glycol (2 ml) containing polyvinylpyrrolidone (PVP, average molecular weight: 40000, 0.133 g) and hydrogen tetrachloroplatinate(II) (1.6×10^{-6} mol) was placed in a glass vial.

An N,N-dimethylformamide (DMF) solution (10 ml) of $Cd(OAc)_2 \cdot 2H_2O$ (2.5×10^{-5} mol) and thiourea (2.5×10^{-5} mol) in a Pyrex glass vial was placed in a MW oven and irradiated with MW continuously or intermittently after cooling down the irradiated solution to ambient temperatures before each irradiation. The temperature of the solution was found raised to 343 K by 15-s irradiation and to 403 K by 41-s irradiation as examples.

RESULTS AND DISCUSSION

Ni NANOCRYSTALLITES

Microwave irradiation was used to prepare nano-sized particles of Ni metal through the reduction of nickel hydroxide in ethylene glycol in the presence of a platinum salt. Nano-sized nickel metal particles much smaller than those obtainable through the conventional heating method were successfully obtained.

Hydrogen tetrachloroplatinate(II) and PVP should act as a precursor of Pt catalyst for nucleation of Ni metal and a stabilizer of small Ni metal particles, respectively, as suggested in the literature.[17] When $Ni(OH)_2$ was introduced into the solution and stirred, a light-green powder of $Ni(OH)_2$ was dispersed therein. When the dispersion solution was irradiated by microwave (2.45 GHz) at 200 W, the temperature of the solution was raised to 423 K in 5 min and the color of the solution changed to gray. Droplets of the solution were placed onto a carbon-coated copper mesh and dried at about 373 K under a vacuum for observation via transmission electron microscopy (TEM). The TEM image (Fig. 1) showed the formation of aggregated cubic nano-sized particles. The primary size distribution determined via TEM was centered at 7 nm and ranged between 4 and 12 nm (Fig. 2). Analysis of the diffraction ring pattern indicated that the particles consisted of metal nickel and traces of remaining $Ni(OH)_2$. Observation of the solution sample after irradiation via dynamic laser scattering revealed a size and distribution almost identical to those obtained in the TEM observation, indicating that the particles were mono-dispersed in the solution.

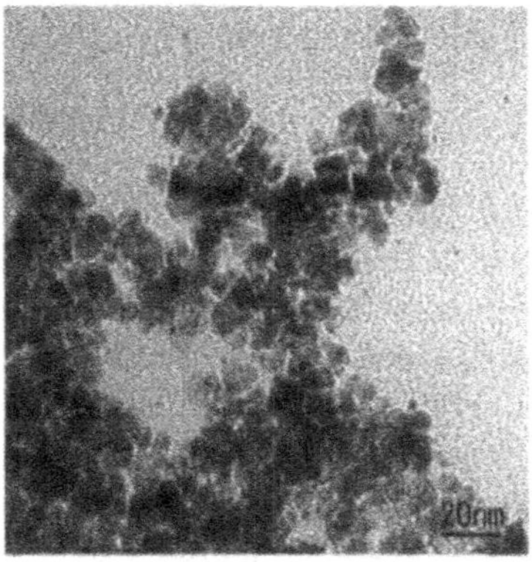

Fig.1 TEM picture of Ni nanocrystallites synthesized under MW irradiation.

The preparation of Ni metal particles using ethylene glycol and polyvinylpyrroridone was performed in the presence of hydrogen tetrachloroplatinate(II) previously under conventional heating by Hegde et al.[17] The particle size of the Ni nanoparticles prepared in the present work was ten times smaller than those reported by Hegde. The Ni particles formed via the microwave method were cube-shaped, whereas Hegde reported that the Ni particles prepared by the conventional heating method were spherical. Thus, the growth process of Ni particles for the microwave-irradiating method would be different from that for the conventional method. The growth of Ni particles can be controlled by the irradiation power and time.

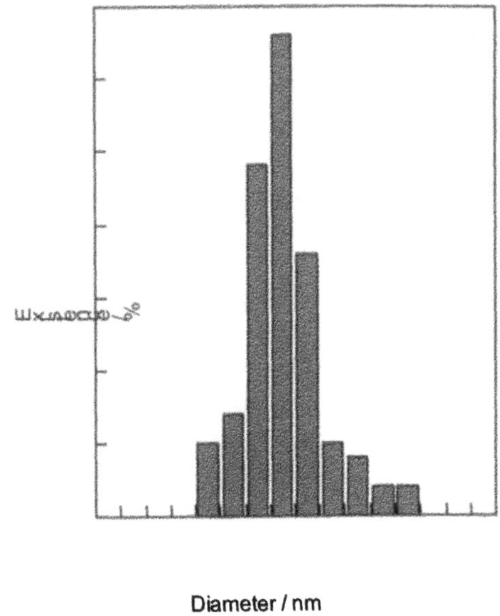

Fig. 2 Size distribution determined by TEM.

CdS NANOCRYSTALLITES

Reaction Conditions for MW Irradiation

We successfully attempted the preparation of colloidal dispersion of CdS nanocrystallites through the reaction of cadmium salt with sulfide anion (S^{2-}) under MW irradiation. Thiourea was selected as a source of S^{2-} because it is known to generate S^{2-} through thermal decomposition under assistance of a base as a catalyst (see Scheme 1).[18-21] We selected cadmium acetate as an appropriate salt for examining the preparation method from the examined several salts, because photoluminescence was observed for the yellowish colloidal dispersion prepared only from the acetate salt.

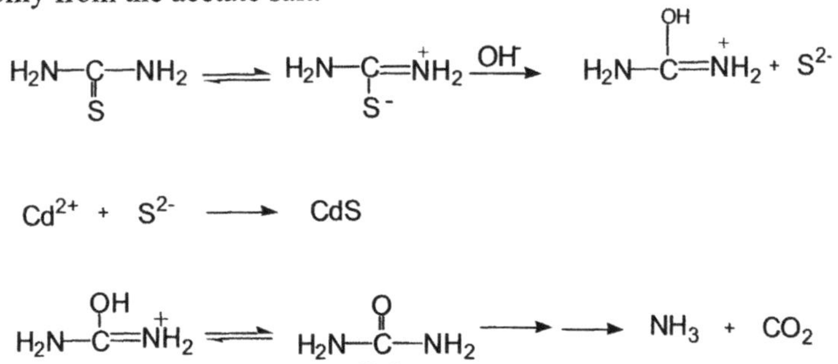

Scheme 1

When the DMF solution containing $Cd(OAc)_2 \cdot 2H_2O$ and thiourea was irradiated by MW, the reaction mixture became transparently yellowish and its color was gradually strengthened by further irradiation, indicating the formation of colloidal CdS in the system. The temperature of the solution was found raised up to 353 K after the irradiation for 18 s. A TEM image of the solution after the 18-s irradiation revealed the formation of nanocrystallites of CdS (Fig. 3). The electron diffraction pattern showed that the CdS nanocrystallites possessed the hexagonal crystal structure. The size distribution of CdS nanocrystallites was centered at 2.6 nm and ranged between 1.5 and 4 nm (Fig. 4). No CdS nanocrystallites were observable for the sample solution irradiated for 15 s, suggesting the presence of the threshold in the MW power for the release of S^{2-} from thiourea.

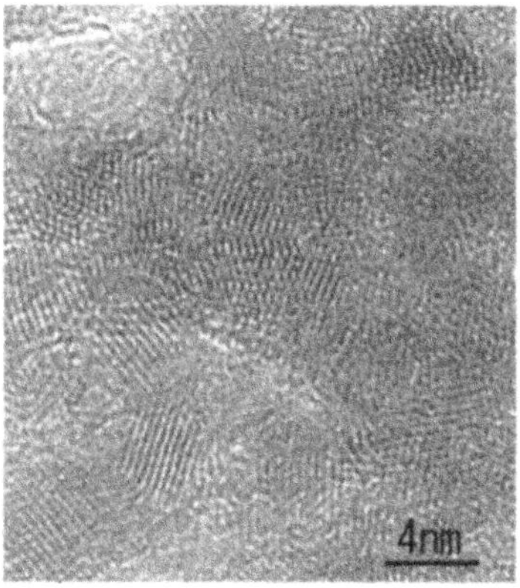

Fig. 3 Size distribution determined by TEM.

Water included in the Cd salt as water of crystallization and also in DMF should give hydroxide anion as a base in the presence of DMF. MW-assisted hydrolysis of thiourea catalyzed by hydroxide leads to release of S^{2-} in the system (see Scheme 1). Furthermore, DMF molecules will interact weakly with the surface of the fresh CdS nanocrystallites by coordination to exposed Cd cations, resulting in stabilizing the nanocrystallites without coagulation.[22]

For comparison of the MW heating method with the conventional heating, the same starting solution was heated up to 353 K in a hot oil bath and kept for 30 s at the temperature. The color of the solution was changed to yellow but precipitation was visually recognized.

In general, the synthesis of nanoparticles from their precursors in solutions consists of two main processes: nucleation and particle growth. In addition, termination of the particle growth is also important. Each step can be controlled by changing preparation conditions such as starting materials, solvent, additives as stabilizer, and reaction time and temperature. When the synthesis is carried out under conventional heating, temperature distribution of the solution will be inhomogeneous, because the reaction system is heated by thermal conduction through a container wall, convection, and agitation. The above two processes, nucleation and particle growth, should depend on the local temperature distribution of the solution, and then the inhomegeneity in the temperature distribution should cause broad distribution in the particle size. The MW heating should lead to homogeneous, i.e., molecular-level heating of the polar solution system, achieving automatic control of the above two processes indispensable for the selective preparation of nano particles.[20]

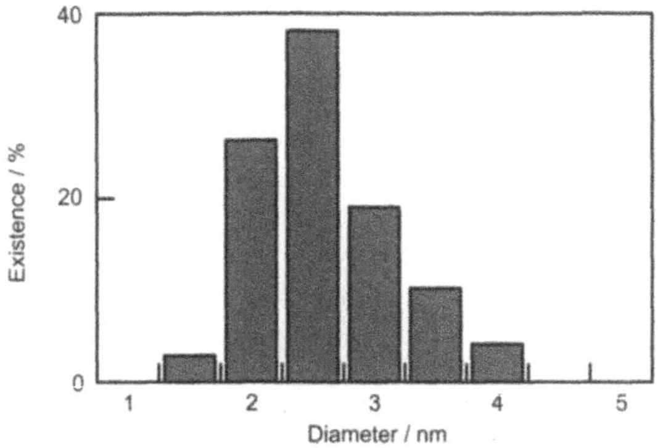

Fig. 4 Size distribution determined by TEM.

Pinning of the Nanocrystallite Growth

For the solution prepared by 18-s irradiation, the absorption onset was observed at 422 nm, being in agreement with that estimated by the dependence of the bandgap energy on the particle size observed by TEM (2.6 nm).[23-25] The absorption spectra of the colloidal solution prepared by MW irradiation moved to long wavelength in their onsets when the irradiation time was prolonged. By increasing the irradiation time up to 27 s, the sample was heated up to 373 K and gave the absorption onset at 445 nm. This shift of the absorption onsets indicated size-growth of the CdS nanocrystallites induced by the continued irradiation. The growth of the particle size could be estimated to be 0.2-0.4 nm from the shift of the absorption onset observed for the two samples.[23] The integrated absorption intensities were also increased with increasing the irradiation time, showing that the concentration of CdS nanocrystallites was also increased with an increase of the size of the nanocrystallites.

The colloidal CdS solution was repeatedly and intermittently irradiated several times keeping the time of irradiation constant. Every time after the solution was irradiated, the reaction mixture was cooled down to ambient temperature. Under such MW heating the absorption onset was pinned at the same wavelength and only the optical density of the absorption band of the CdS solutions was increased. (Fig. 5). By intermittent irradiation of the solution for 18 s, 21 s, or 27 s, the absorption onsets were pinned at 430 , 440 , or 460 nm, respectively. These results showed that the punctual and intermittent MW irradiation increased the number of the CdS nanocrystallites without growth in the particle size, leading to the size-controlled production of CdS nanocrystallites.

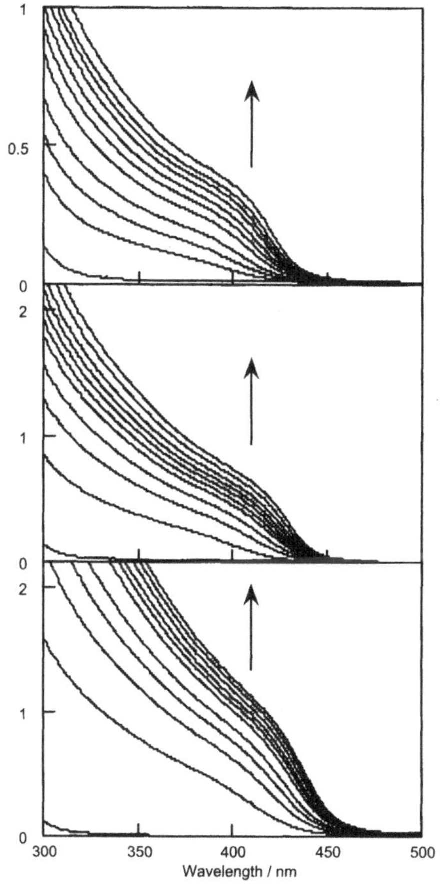

Fig.5 Absorption spectra of the colloidal CdS solutions prepared by the intermittently repeated MW irradiation. The solutions were intermittently irradiated for the constant times, 18 s (upper), 21 s (middle), and 27 s (lower), respectively. The solutions were cooled to room temperature after each irradiation.

SUMMARY

We showed the striking advantages in applying MW irradiation for synthesizing nanomaterials, which should be very attractive not only for industrial application but also for a new area of "Nano-tech". We still need to continue endeavor for understanding special characters of MW chemistry.

REFERENCES

1 A. Henglein, *Top. Curr. Chem.*, 1988, **143**, 113.

2 M. L. Steigerwald and L. E. Brus, *Acc. Chem. Res.*, 1990, **23**, 183.

3 H. Weller, *Angew. Chem. Int. Ed. Engl.*, 1993, **32**, 41.

4 K. E. Gonsalves, S. P. Rangarajan and J. Wang, in *Handbook of Nanostructured Materials and Nanotechnology*, ed. H. S. Nalwa, Academic Press, San Diego, 2000, Vol. 1; pp 1.

5 *Microwave-enhanced Chemistry*, ed. H. M. Kingston and S. J. Haswell, American Chemical Society, 1997.

6 C. R. Strauss and R. W. Trainor, *Aust. J. Chem.*, 1995, **48**, 1665.

7 A. Zlotorzynski, *Critical Reviews in Analytical Chemistry,* 1995, **25**, 43.

8 D. E. Clark and W. H. Sutton, *Annu. Rev. Mater. Sci.*, 1996, **26**, 299.

9 W. Yu, W. Tu and H. Liu, *Langmuir,* 1999, **15**, 6.

10 O. Plachik, S. Avivi, D. Pinkert and A. Gedanken, *Nano Structured Materials,* 1999, **11**, 415.

11 P. D. Ramesh and K. J. Rao, *Adv. Mater.*, 1995, **7**, 177.

12 P. Rigneau, K. Bellon, I. Zahreddine and D. Stuerga, *Eur. Phys. J. App. Phys.*, 1999, **7**, 41.

13 Q. Li and Y. Wei, *Materials Research Bulletin,* 1998, **33**, 779.

14 B. Djuric'ic', S. Pickering, D. McGarry, P. Tambuyser and P. Thomas, *J. Mater. Sci.*, 1999, **34**, 1911.

15 Y. Wada, H. Kuramoto, J. Anand, T. Kitamura, T. Sakata, H. Mori, and S. Yanagida, *J. Mater. Chem.* **11**, 1936-1940 (2001).

16 Y. Wada, H. Kuramoto, T. Sakata, H. Mori, T. Sumida, T. Kitamura, and S. Yanagida, *Chemistry Letters* , 607-608 (1999).

17 M. S. Hegde, D. Larcher, L. Dupont, B. Beaudoin, K. Tekaia-Elhsissen, and J. Tarascon, *Solid State Ionics*, **93**, 33 (1997).

18 S. C. Sahu and S. N. Sahu, *Thin Solid Films,* 1993, **235**, 17.

19 M. Froment and D. Lincot, *Electrochim. Acta,* 1995, **1995**, 1293.

20 P. J. George, A. Sáchez and L. Huang, *J. Cry. Growth,* 1996, **158**, 53.

21 P. O'Brien and T. Saeed, *J. Cry. Growth,* 1996, **158**, 497.

22 H. Hosokawa, H. Fujiwa, K. Murakoshi, Y. Wada, S. Yanagida and M. Satoh, *J. Phys. Chem.,* 1996, **100**, 6649.

23 K. Murakoshi, H. Hosokawa and S. Yanagida, *Jpn. J. Appl. Phys.,* 1999, **38**, 522.

24 Y. Wang and N. Herron, *Phys. Rev. B,* 1990, **42**, 7253.

25 T. Vossmeyer, L. Katskas, M. Giersig, I. G. Popovic, K. Diesner, A. Chemseddine, A. Eychmüler and H. Weller, *J. Phys. Chem.,* 1994, **98**, 7665.

SYNTHESIS OF CARBON NANOTUBES BY MICROWAVE PLASMA CHEMICAL VAPOR DEPOSITION AT LOW TEMPERATURE

Wang Jian-hua, Wang Sheng-gao
Province key laboratory of plasma chemistry & advanced materials,
Wuhan Institute of Chemical Technology,
Wuhan, Hubei 430073, People' s Republic of China

ABSTRACT

In this paper pure carbon nanotubes (CNTs), straight CNTs and aligned CNTs were synthesized by microwave plasma chemical vapor deposition (MWPCVD) at the low temperature of about 570 ℃. By decreasing the carbon precursor content or strengthening the etching ability of some specials, such as atomic hydrogen, atomic oxygen or OH radicals, the purity of CNTs can be improved. For straight CNTs, it can be produced under the catalytic effect of Fe-Co-Ni alloy particles due to its uniform catalytic activity around its circumference. Aligned CNTs on nickel substrate were also synthesized by MWPCVD. All these research results may be useful in the study of the properties of CNTs and for potential applications of CNTs.

INTRODUCTION

Carbon nanotubes (CNTs) are novel materials with unique electrical and mechanical properties. Since the discovery of CNTs, there have been numerous reports describing the synthesis, nature and applications of CNTs.[1,2] It is well known that most of the natures of CNTs depend on diameters and chiralities, but it is very difficult to control diameters and chiralities of CNTs at high temperature. Furthermore, the high temperature growth process districts some applications of nanotubes. For example, flat panel display is one of the most promising applications of CNTs in the future, but the synthesis of CNTs at high temperature is not suitable for this application because high temperature will damage substrate. For these reasons, synthesis of CNTs at low temperature has received a great deal of attention. Among all of the synthesis process of CNTs, chemical vapor deposition process, especially plasma chemical vapor deposition process (PCVD), can herald the synthesis of CNTs at low temperature. Plasma-enhanced hot filament chemical vapor deposition (PE-HF-CVD) was applied to the synthesis of CNTs on glass at temperature below 666℃. [3] Microwave plasma chemical vapor deposition (MWPCVD) was used to synthesize CNTs at temperature ranging from 520 to 700 ℃. [4] But some research results indicated that at very low growth temperature, a large amount of carbonaceous particles were formed together with CNTs and the CNTs are curly. Obviously, the existence of carbonaceous particles and the curly shape of CNTs are bound to have some bad effects on the property and applications of CNTs. So, it is very important to synthesize pure CNTs and straight CNTs at low temperature. Furthermore, CNT alignment is particularly important in fabricating cold-cathode flat panel displays. But high growth temperature is unsuitable for the fabrication of devices. [3] So, synthesis of carbon nanotubes at low temperature has become a challenging issue.

In our research, high pure CNTs, aligned CNTs and straight CNT films were synthesized by MWPCVD at low temperature of about 570 ℃. Some parameters, which affect the purity and morphology of CNTs, were discussed in the paper.

EXPERIMENT

Fig.1 shows a schematic diagram of the MWPCVD apparatus. The MWPCVD section consisted of four pieces of equipment: a quartz-glass deposition system, a 2450 GHz, 800 W microwave source, a waveguide unit and a pumping unit with a turbo molecular pump. A portable radiation thermometer (Ircon, Inc.) (UX 40P) is used to measure the exact temperature of the atalyst surface. In our experiment, three kinds of catalysts are used to synthesize CNTs. The catalysts and CNTs are prepared as follows.

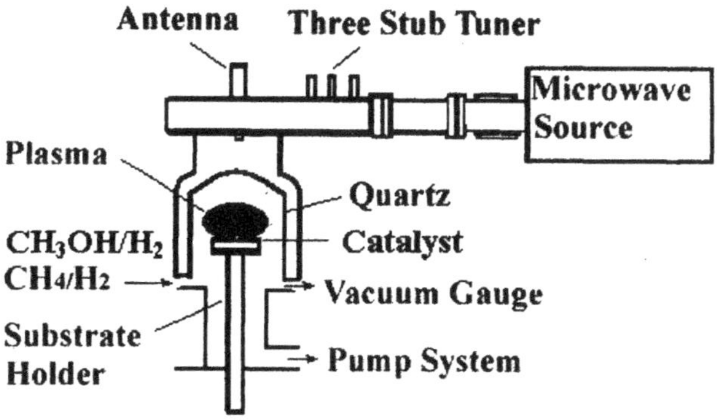

Fig. 1 The photograph of the MWPCVD apparatus

Growth of High Pure CNTs

Catalyst used in this experiment was prepared by sol-gel process. In a typical experiment, 0.1 mol cobalt nitrate was dissolved in 0.8 mol ethanol. The pH value of the solution was kept at 1.8 by adding appropriate amount of hydrochloric acid and 0.1 mol tetraethoxysilicane was added into the solution slowly. The mixture was stirred for about 1 h until it become clear, then kept at 60℃. After the gelation of the mixture, the gel was dried at 120℃ for 24 h so as to remove a greater part of water and ethanol. After being calcined at 550℃ for 2 h, the dried sample was put into the chamber of the MWPCVD apparatus and the cobalt oxide was reduced to cobalt. During the reduction process, the total pressure in the chamber was kept at 4 kPa, the microwave plasma input power was 400 W, and the flow rate of H_2 was 50 sccm (standard cubic centimeter per minute). CNTs were synthesized for 0.5 h at the same chamber after 2 hours' reduction process. During CNTs growth, the catalyst surface was in contact with the lower edge glow generated by microwave plasma. No heating source was used in the experiment. The temperature of about 570 ℃ was achieved only by microwave plasma power. In this process, the flow rates of H_2, the microwave plasma input power and the total pressure in the chamber were the same as those of reduction process. In order to analyze the effects of some parameters on the purity of CNTs, CH_4 and CH_3OH were used as carbon sources, respectively, and different flow rate of CH_4 were adopted.

Synthesis of Aligned CNTs

In this work, continuous Ni thin film, which was used as the precursor of catalyst for the synthesis of CNTs, was electrodeposited on Ni substrate by a two-probe direct current method in 5 wt. % $NiSO_4$ and 2 wt. % H_3BO_3 solution at 25 ℃. During the electrodeposition process, the voltage and the current were kept at 40V and 0.1 A/cm^2, respectively, and the electrodeposition time was about 20 s. Then the Ni films, with the nickel substrates, were putted into the deposition chamber of MWPCVD apparatus. Before the synthesis of CNTs, the continuous Ni film was etched by hydrogen plasma for 4 min. During the etching process and CNTs growth period, the flow rate of CH_3OH was 10 sccm, and the flow rate of H_2 was 50 sccm, the total pressure in the chamber was kept at 4 kPa, the microwave plasma input power was 400 W.

Synthesis of Straight CNT Film

The process of straight CNT film growth was roughly the same as that of aligned CNTs growth. The only difference was that the Fe-Co-Ni alloy thin film, instead of Ni thin film, which was electrodeposited on Ni substrate by a two-probe dc method in 2 wt. % $FeSO_4$, 2 wt. % $NiSO_4$, 2 wt. % $CoSO_4$ and 2 wt. % H_3BO_3 solution at 40 ℃, was used as the precursor of catalyst.

After the synthesis of CNTs, scanning electron microscopy (SEM) (JSM-5510LV) and transmission electron microscopy (TEM) (JEM-100CXII) were used to investigate the morphology and structure of the yields.

RESULTS AND DISCUSSION

During the CVD process, CNTs were synthesized under the catalytic effect of catalysts. The most effective catalysts are iron, nickel and cobalt. The catalytic effect of catalyst is thought to be related to a combination of factors such as catalytic decomposition of hydrocarbons and the possibility of the carbon to diffuse through and over the catalyst particles extremely rapidly. The latter factor allows CNTs to grow by a mechanism of diffusion and precipitation. There are two steps during the synthesis of CNTs. The first step is the nucleation of CNTs and the other is the growth of CNTs. [5] Under the catalytic effect of catalyst particles, the carbon precursor is deposited on the surface of these particles. At a suitable temperature, carbon atoms will diffuse into the particles. When the carbon is saturated in the carbon-catalyst system, the nucleation will take place on the surface of catalyst particle. Then, more carbon species are diffused into and through the catalyst particles to form CNTs. During the CVD process, the forms of yields are determined by reaction parameters, especially the partial pressure of carbon and the reaction temperature. At very low temperature, only a small amount of carbon atoms can diffuse into the catalyst particles to form CNTs. If there are more carbon precursors, other undesirable forms of carbon, such as amorphous carbon nanoparticles, will be co-produced. In our experiment when the flow rate of CH_4 was about 10 sccm, some other carbonaceous nanoparticles were also deposited and attached on the surface of CNTs (see Fig. 2(a)). With the decrease of the flow rate of CH_4, the amount of other carbonaceous nanoparticles decreased. Fig. 2(b) shows the TEM image of carbon yields synthesized by MWPCVD when the flow rate of CH_4 is 5 sccm. We can found that most of the yields were CNTs. Compared to Y. C. Choi's research result, [4] the key reason why pure CNTs could grow at low temperature in our experiment is the small flow rate of CH_4. Because of the existence of catalysts, the deposited carbon forms CNTs preferentially. Furthermore, even if some carbonaceous nanoparticles were deposited at the same time, most of them would be etched by H_2 plasma because of high ratio of H_2/CH_4. So when the flow rate of CH_4 is very small, CNTs instead of other carbonaceous nanoparticles are deposited. From the research result we can conclude that the purity of CNTs can be improved by strengthening the etching ability of H_2 and decreasing the carbon precursor content in the chamber.

To further discuss the effect of the etching ability of plasma on the purity of CNTs, CH_3OH was used as carbon source, and the flow rate of CH_3OH was kept at 10 sccm. Fig. 3 shows the SEM image of CNTs. Compared to Fig. 2(a), no carbonaceous nanoparticles was found attached on the CNTs. The research results show that CH_3OH, which substitute for CH_4, can improve the purity of CNTs.

When we used CH_3OH as carbon source, there were a lot of chemical specials with high internal energy, such as radicals and ions, produced in plasma. The carbon atoms contribute to high rate deposition of CNTs and carbonaceous nanoparticles. Some of the other specials, such as atomic hydrogen, atomic oxygen, OH radicals, contribute to etch some carbon yields, and the etching ability of atomic oxygen and OH radicals is stronger than that of atomic hydrogen. Because CNTs is more stable than carbonaceous particles, at an appropriate condition, carbonaceous nanoparticles will be etched and most of the yields are CNTs. Then in our experiment, atomic oxygen and other oxygen-containing radicals directly decomposed from CH_3OH plays an important role in the synthesis of pure CNTs.

Fig. 2 TEM images of CNTs synthesized by MWPCVD at different ratio of CH_4/H_2 (a: the ratio of CH_4/H_2 is 10/50; b: the ratio of CH_4/H_2 is 5/50)

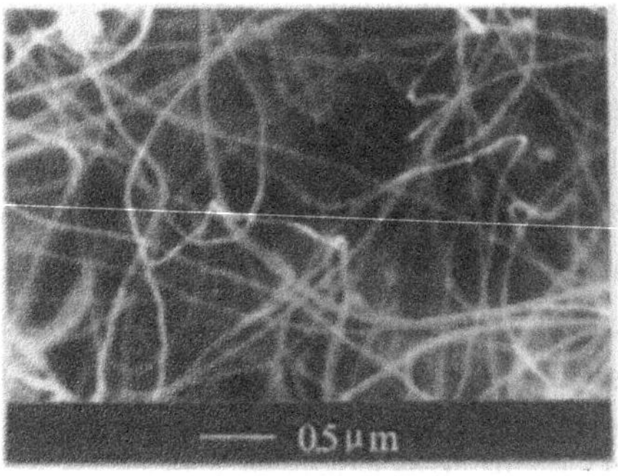

Fig. 3 SEM image of CNTs synthesized by MWPCVD using CH_3OH as carbon source

Fig. 4(a) shows the morphology of Ni film electodeposited on Ni substrate and Fig. 4(b) shows the morphology of the yields grown under the catalytic effects of Ni film. We can find that CNTs protrude from the surface of Ni substrate. Because of the existence of Ni particles attached on the end of each carbon nanotube, the image of CNTs is not clear. But the alignment of CNTs can be detected from the dull morphology of CNTs. It can be deduced that most as-synthesized CNTs are approximately perpendicular to the surface of the Ni substrate.

In order to examine the growth direction of the CNTs under the catalytic effects of Ni film, part of the yield-covered area was pressed with tweezers for 5 min before SEM observation (Fig. 5a). In this low-magnification image, because of the pressing of tweezers, the alignment of CNTs can be detected, as indicated by arrow A. The high-magnification SEM image correspond to arrow A clearly shows that the CNTs have good alignment (Fig. 5b). Ni particle can be found on the top end of each CNT. This is responsible for the observation result shown in Fig. 4(b). Because of the strong etching ability of oxygen decomposed from CH_3OH, no carbonaceous particles was found on the surface of CNTs.

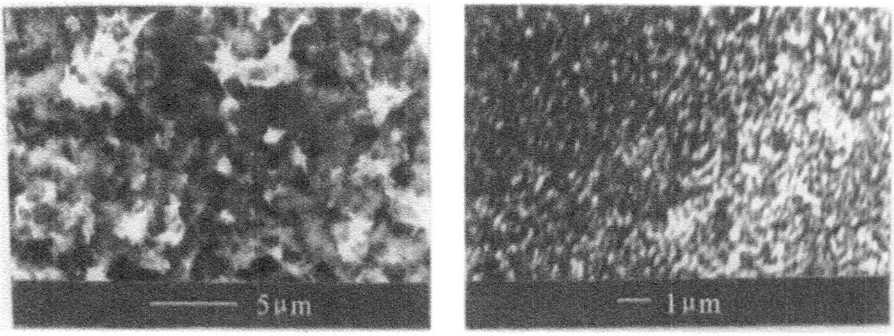

Fig. 4 SEM surface morphology of the nickel layers (a) and of aligned CNTs (b)

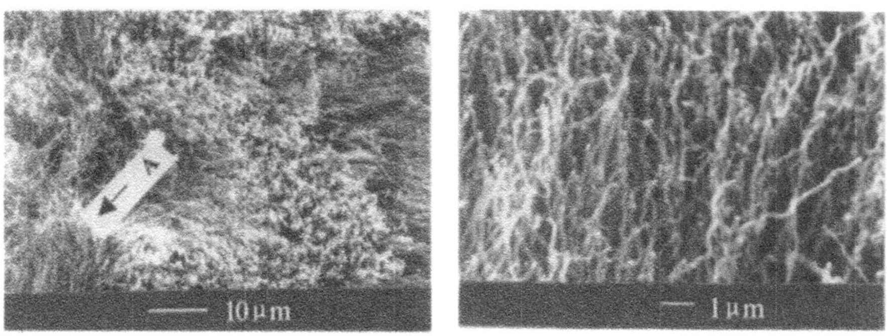

Fig. 5 Low-magnification (a) and high-magnification (b) SEM images of aligned CNTs grown on Ni substrate

From the above research, we found that most of CNTs synthesized at low temperature were curly. This shows that the CNTs have a great number of defects. Some researchers pointed out that straight CNTs couldn't be produced at low temperature because the too low growth temperature would result in structure defects. [4] But in our research, straight CNTs can be produced at low temperature about 570℃. Fig. 6 shows the low and high magnification SEM images of the straight CNT film grown on nickel substrate. From the SEM images, we found that most of the CNTs are straight and no CNTs tangled together. Even a few CNTs are not straight, the curvature radius is relatively big. It is well known that when a single crystal of metal is used as a catalyst, different crystal planes have different catalytic performance. L. F. Sun [6] pointed out that when just one transition metal is used as catalyst, due to different catalytic activity on the surface of the nanoparticles, the growth rate along the circumference is not uniform, and this causes the CNTs to curve towards the lower growth rate side. While for Fe-Co-Ni alloy nanoparticles, we can speculate that in our experiment condition the catalytic activity on the surface of the Fe-Co-Ni alloy is nearly uniform, and the uniform growth rate along the circumference of round alloy particles plays a very important role in the growth of straight CNT films.

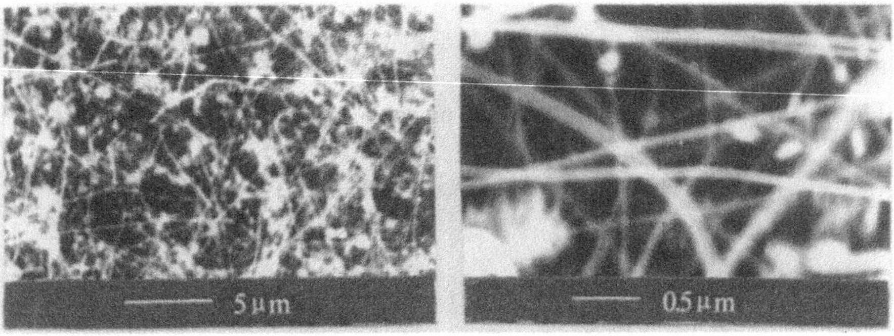

Fig. 6 Low-magnification (a) and high-magnification (b) SEM images of straight CNT films

CONCLUSIONS

CNTs were synthesized by MWPCVD at low temperature. The purity of CNTs can be improved by strengthening the etching ability of H_2 and decreasing the carbon precursor content in the chamber. And when CH_3OH was used as carbon source, atomic oxygen and other oxygen-containing radicals directly decomposed from CH_3OH plays an important role in the synthesis of pure CNTs. Straight CNTs can be synthesized at low temperature under the uniform catalytic activity of Fe-Co-Ni alloy particles around their circumference. Furthermore, aligned CNTs were synthesized at the low temperature of about 570 ℃

REFERENCES

[1] Bradley. S. Files, Brian. M. Mayeaux, Carbon nanotubes, Adv. Mater. Proc., 156, 4), 47-49, 1999

[2] P. Chen, J. Lin, K. L. Tan, Carbon nanotubes: a future material of life, IUBMB life, 49, 2, 105-108, 2000

[3] Z. F. Ren, Z. P. Huang, J. W. Xu, J. H. Wang, P. Bush, M. P. Siegal, P. N. Provencio, Synthesis of large arrays of well-aligned carbon nanotubes on glass, Science, 282, 5391, 1105-1107, 1998

[4] Y. C. Choi, D. J. Bae, Y. H. Lee, B. S. Lee, Growth of carbon nanotubes by microwave plasma-enhanced chemical vapor deposition at low temperature, J. Vac. Sci. Technol. A 18, 4,1864-1868, 2000

[5] S. B. Sinnott, R.Andrews, D. Qian A. M. Rao, Z. Mao, E. C. Dickey, F. Derbyshire, Model of carbon nanotube growth through chemical vapor deposition, Chem. Phys. Lett., 315, 1-2, 25-30,1999

[6] L. F. Sun, J. M. Mao, Z. W. Pan, B. H. Chang, W. Y. Zhou, G. Wang, L. X. Qian, S. S. Xie, Growth of straight nanotubes with a cobalt-nickel catalyst by chemical vapor deposition, Appl. Phys. Lett., 74, 5, 644-646, 1999

SYNTHESIS OF COMPOSITE POWDERS AND COATING OF FIBRES BY COMBINED CVD-PECVD IN A MICROWAVE HEATED FLUIDISED BED REACTOR

R. Tap and M. Willert-Porada
Chair of Materials Processing
University of Bayreuth
95440 Bayreuth
Germany

ABSTRACT

The development of a new coating process for powdered materials is presented here. The combination of a thermal chemical vapour deposition (CVD) process and a plasma enhanced chemical vapour deposition (PECVD) process both in a fluidised bed is shown for the coating of carbon fibres with titanium.

INTRODUCTION

In materials processing coating of powders is a general practise for supported catalyst materials, like e.g., platinum metals deposited on a high specific surface area support as well as for homogenous distribution of additives, e.g. Y_2O_3 on Si_3N_4-ceramic powders to facilitate the formation of a liquid phase during sintering at high temperatures. In the last years coating of fibres has received much attention for composites materials specially in the automotive industry. Carbon fibre reinforced metal matrix composites offer many attractive properties such as low density, excellent strength and stiffness in fibre axis and a low coefficient of thermal expansion [1]. To achieve these properties a special focus has to be kept on the fibre/matrix interface. The poor wettability of liquid metals like magnesium on carbon, reaction between the fibre and matrix material as well a diffusion of matrix into the fibre make a coating of the fibres necessary [2]. Besides using a sacrificial material like pyrolised carbon the use of ceramic coatings is a major strategy. Coatings of nitrides, carbides, borides and oxides of titanium, zirconium, hafnium, silicon and aluminium can be taken into consideration; TiN and SiC are the most promising and most investigated candidates [3,4]. To preserve the good mechanical properties of the fibres only thin coatings of these materials should be deposited on the surface.

Several techniques and processes have been developed to obtain these thin and uniform coatings. For these requirements CVD (Chemical Vapour deposition) is the most suitable process [5,6] although other techniques like dip coating are reported [7]. The transport of reactant or removal of by-products by the gas phase gives access to the whole (inner and outer) surface of the fibres. An excellent dispersion of the powdery materials, e.g., within a fluidised bed, would be advantageous for the deposition of very thin and regular coatings.

Thermal CVD on porous substrates, fibres or powders, has natural limitations as well. Laminar flow and diffusion of molecules into small pores is limited by the size and geometry of pores. Therefore, some additional driving forces for material transport would significantly improve the applicability of CVD as an industrially viable coating method for porous and particulate materials. In order to retain a thin film structure of the coating or a small size of the catalyst particles as well as a high specific surface area of the materials, a low processing temperature is essential. Therefore, plasma enhancement to lower the temperatures or high frequency radiation to enhance transport of materials could be suitable for improved CVD on porous materials.

First reports on applications of microwave (MW) heated and plasma enhanced CVD reactors are:

1. MW-plasma heated circulating fluidised bed (ETH Zurich, Switzerland [8])
2. MW heated rotary kiln (University of Erlangen, Germany, [9])
3. MW heated "warm wall" CVD reactor [10]
4. MW heated stationary fluidised bed reactor [11].

The reported applications of microwave radiation to facilitate CVD on powders, fibres or porous materials have numerous disadvantages. The plasma heated circulating fluidised bed process is limited by the size of the MW-plasma zone generated inside the circulating fluidised bed reactor and shows low flexibility because a reduced operating pressure is needed.

With the rotary kiln applicator high temperatures can be achieved, however plasma options are not reported nor is an alternation between thermal CVD and plasma enhanced processing, PECVD possible. In the "warm wall" reactor powders can not be processed uniformly at satisfactory throughput rates. In the stationary fluidised bed reactor only particulate deposits, i.e. no thin coatings, were achieved up to now.

The largest interest in coated powders and fibres is to optimise the composition and the activity of the coating. Therefore new concepts for coating using thermal CVD as well as PECVD need to be developed. The potential of MW application to achieve these goals is by selective microwave heating for thermal CVD within the nucleation step followed by a microwave atmospheric plasma enhanced coating at lower temperatures. The laboratory scale apparatus for such a combined process will be presented along with results on TiN-coated carbon fibres.

EXPERIMENTAL

Substrate carbon fibres (Sigrafil C) from SGL Carbon were used. The dimension of a single fibre is 10 μm in diameter and 500 μm in length. BET-surface measurements (ASAP 2010, Micromeritics) gave a value of 0.98 m^2/g. The electric resistance of the fibres is 5 μΩm. Dielectric properties of the fibres are not available.

Titanium-(IV)-chloride (Merck-Schuchardt) was used as a titanium source for the deposition of titanium nitride. $TiCl_4$ is a liquid at room temperature with a boiling point of 136°C. The main reaction for the coating is:

$$2\,TiCl_4 + 4\,H_2 + N_2 \xrightleftharpoons{800°C} 2\,TiN + 8\,HCl\,, \qquad \Delta H_r\text{: } 82{,}2\ kJ/mol \quad (1.)$$

The fibres were coated in a two zone fluidised bed (Fig.1). The lower zone is a microwave heated fluidised bed for thermal CVD.

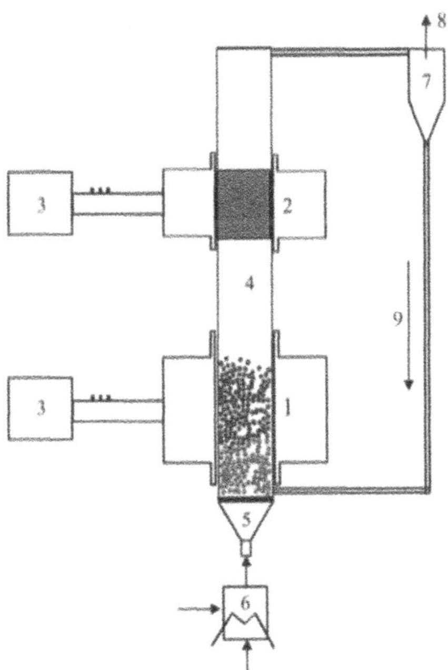

1. Zone of Thermal CVD
2. Zone of Plasma Enhanced CVD
3. Microwave source
4. Fluidised Bed Reactor
5. Inlet for Fluidisation Gas
6. Precursor Inlet with Vaporiser
7. Cyclone
8. Gas exhaust
9. Particle recirculation

Fig. 1 Sketch of CVD-PECVD fluidised bed reactor

The reactor tube is made of quartz glass and has an inner diameter of 40 mm and a height of 140 mm. The brass resonator (H x W x D: 140 x 140 x 170 mm) acts as a multimode applicator. A microwave source consisting of a 6 kW (2.45 GHz) magnetron head (Richardson) is attached by a waveguide. The electric field in the guide was perpendicular to the axis of the quartz tube. The forward and reflected power can be measured by an impedance analyser and optimised via an automatic stub-tuner.

The upper zone is a microwave generated plasma zone. In the centre of the cylindrical aluminium applicator (∅ 95 mm, H: 43 mm) a quartz glass tube with inner diameter of 40 mm contains the fluidised bed. The height of the plasma source is about 44 mm. For stable plasma conditions the dimensions of the applicator were designed only to allow the TM_{010}–mode to propagate. A microwave source consisting of a magnetron head (Muegge) with a maximum power of 2 kW at 2.45 GHz was used. The electric field in the waveguide as well as in the cavity was parallel to the axis of the quartz tube. A 3-stub tuner was used to adjust the electromagnetic field.

The cavity for the plasma zone was designed with the help of electromagnetic simulation using the software ANSYS Multiphysics. The diameter of the empty cavity was fitted that resonant conditions for the TM_{101}-mode were achieved at 2.45 GHz. Field measurements with thermo-sensitive paper were able to approve the calculated results.

The fibres were introduced onto the gas distributor plate at the bottom of the fluidised bed reactor. By adjusting the gas velocity the fluidised bed could be establish either in the thermal CVD applicator in a relative dense form or in the plasma zone where the fibres are much more loosely. The fluidisation gas consisting of a mixture of Argon with 5 volume-% Hydrogen during the coating process; also nitrogen was used.

The precursor $TiCl_4$ is filled into a vaporiser and fed with the nitrogen into the reactor.

Entrained fibres are separated in the cyclone from the gas and can be refilled into the fluidised bed. The exhaust passes a washing bottle to remove any unreacted precursor and the by-product HCl gas.

In a typical experiment the bed is fluidised to the desired height. In the case of thermal CVD the bed is heated up to its decomposition temperature in the range between 800 to 900 °C. In the case of PECVD the microwaves are switched on and the plasma is ignited. Then nitrogen containing the precursor is fed into the reactor and the reaction takes place. Due to unstable fluidisation conditions such as collapsing of the bed or refeeding of entrained particles the gas flow had to be increased respectively decreased for several seconds. After the decomposition time the nitrogen stream with precursor is switched off as well as the microwave power. To cool down the bed and to desorb remaining precursor the fluidisation kept on for several minutes.

The fibres are analysed on the BET-surface (ASAP 2010, Micromeritics), on their titanium content via combustion analysis and on the morphology and composition on the surface by FEM (LEO Gemini), REM (JEOL SM840) and EDX (Oxford).

RESULTS

In the following part the results on coating of carbon fibres with TiN is described. The fibres coated by thermal CVD, PECVD or combinations of both are compared with the untreated fibres. The aim is to achieve a uniform and thin coating of the particles to keep their good mechanical properties and make them suitable for fibre/matrix composites.

Although the simulation of the field distribution and the design was done for an unloaded cavity a stable and centred plasma could be observed during the experiments. The weak loading of the fluidisation gas with fibres seems to have not noticeable influence on the resonance conditions. Nevertheless high dielectric fluidised beds can have significant influences on the field distribution of the cavity and it must be taken into account during the design of a single mode cavity.

The table 1 gives an overview on the experiments and conditions.

Process	Process parameters			Results	
	u [m/s]	T [°C]	t [s]	Ti-content [wt%]	BET-surface [m²/g]
Uncoated fibres	/	/	/	/	0.98
PECVD	0.111		600	1,76	1,72
CVD	0.033	900	600	0.50	2.10
1) PECVD 2) CVD	1) 0.088 2) 0.033	1) 500 2) 900	1) 300 2) 600	0.82	1.75

Tab.1 Overview on the presented experiments

The untreated carbon shows a raw structure along the axis of the fibre which is a result of the fabrication process of these fibres. Polyacrylonitrile long fibres are pyrolysed to carbon and respectively milled to the desired length. The milling would also explain the slight impurities on the surface. The high magnification image shows a fine scaley surface structure in a dimension less then 100 nm.

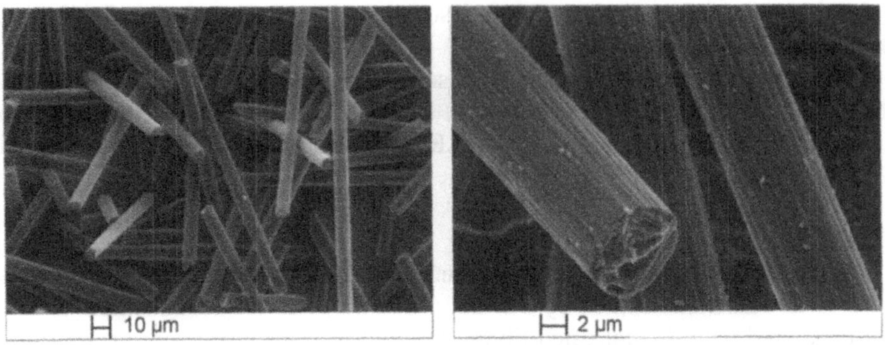

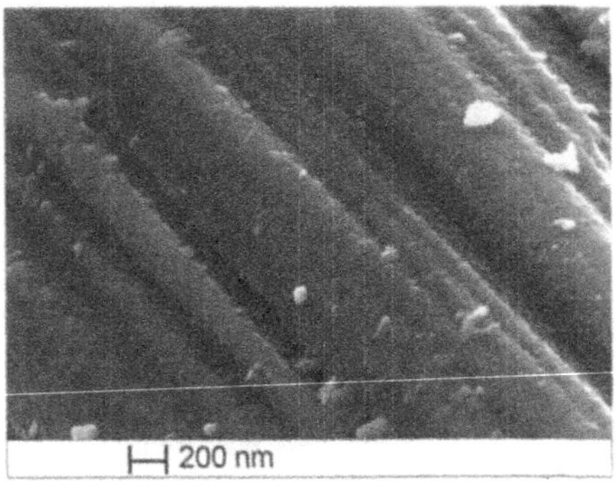

Fig. 2 FEM pictures of the original untreated carbon fibre (Sigrafil C; SGL Carbon).

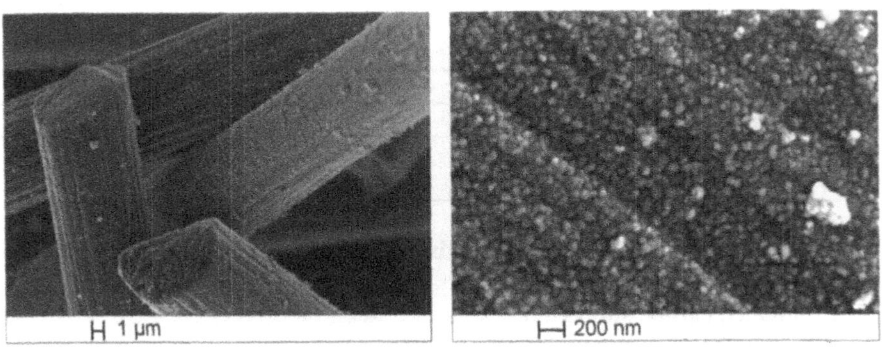

Fig.3 FEM pictures of carbon fibres coated in a PECVD fluidised bed

The coating via PECVD in the fluidised bed shows a relative uniform morphology consisting of small particles in the range of 50 nm. The raw structure of the fibres is still noticeable so that the thickness of this coating is small. Several larger particles can be seen due to the impurities of the original fibres or single grains which have grown faster than the other grains on the surface.

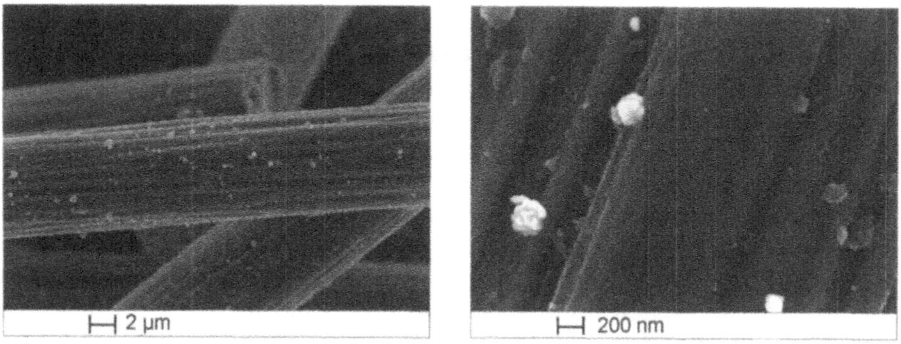

Fig.4Carbon fibres coated in a fluidised bed by thermal CVD

The fibres coated in a thermal CVD process show only a very weak deposition of particles. The particles that were formed are therefore much lager than in the PECVD process. In the pure thermal process the surface offers only a few places for crystallites to grow.

Fig.5 Carbon fibres first coated in a PECVD fluidised bed then in the CVD zone

The fibres coated in a combination of plasma enhanced and thermal CVD process show both coating types discussed in the single processes before. Due to the plasma process there is again a relatively uniform coating of particles in the size of 50 nm. This particle layer acts as seeds for the deposition in the thermal

CVD process. Therefore larger particles grow in this second process step. The size of these particles is in the range of 200 nm.

DISCUSSION

The results on the carbon fibres show that the coating in microwave heated fluidised beds is possible. Depending on the way the necessary reaction enthalpy is provided either as volume heating of the particles in a classical CVD process or by introducing a plasma the resulting coating is affected. The use of plasma has the effect that the surface of the substrate is activated. This results in a uniform small sized particle coating. The process is therefore an excellent technique for thin coating or surface etching applications. Limits for this step are solid throughputs and precursor concentration in the plasma zone. The thermal CVD process will not modify the surface and therefore deposition will only take place at already existing seed crystals. On the other hand the range for precursor concentration and particle density is greater than for plasma heating which leads to higher deposition rates and flow rate of particles.

With the combination of both processes an exact adjustment of the desired coating can be achieved. The plasma process will prepare the surface via etching or deposition of seed particles while the CVD process is suitable for higher throughputs which is important for industrial applications.

REFERENCE

[1]D. Wurm; M.Schmitt; R.F.Singer; "Carbon fibre reinforced magnesium alloys with improved properties through CVD fibre coating" pp.653-658, in *Magnesium Alloys and their applications*, Edited by B.L. Mordike and K.U.Kainer; Werkstoff-Informationsgesellschaft mbH, Frankfurt, Germany, 1998

[2]A.Feldhoff; E.Pippel; J. Woltersdorf; "TiN coatings in C/Mg-Al composites: microstructure, nanochemistry and function"; *Philosophical Magazine A*; **80** [3] 659-672 (2000)

[3]N.Popovska; H.Gerhard; G.Emig; "Chemical vapour deposition of thin ceramic continuous carbon fibres"; pp. 109-114; *in Werkstoffwoche'98; Band IX: Symposium 11*; Edited by H.Dimigen and W.Paatsch; Wiley-VCH Verlag GmbH; Weinheim, Germany; 1998

[4]V.Hopfe; B.Dresler; K.Schonfeld; R.Jackel; B.Leupolt; "Laser CVD of ceramic coatings on fibres"; pp. 673-677; *in Werkstoffwoche '98; Band VII: Symposium 9*; J.Heinrich; Wiley-VCD Verlag GmbH; Weinheim; Germany; 1998

[5]K.T.Kashyap; K.Govindaraju; P.V.Sharma; M.C.Mittal; "Studies on the Interface Coating of Graphite Fibers to produce Aluminium Graphite Composites"; *Journal of Materials Science Letters*; **14** [23] 1961-3 (1995)

[6]K. Kusakabe; B.-K.Sea; J.-I.Hayashi; H.Maeda; S.Morooka; „Coating of Carbon Fibers with Amorphous SiC Films as Diffusion Barriers by Chemical Vapor Deposition with Triisopropylsilane"; *Carbon*; **34** [2] 179-185 (1996)

[7]Y.Liu; D.R.Treadwell; M.R.Kannisto;B.L.Mueller; R.M.Laine; "Titanium Nitride/Carbon Coatings on Graphite Fibres"; *Journal of the American Ceramic Society*; **80** [3] 705-716 (1996)

[8]M.Karches, Ch.Bayer, Ph.Rudolf von Rohr, "A circulating fluidised bed for plasma-enhanced chemical vapour deposition on powders at low temperatures", *Surface and Coatings Technology*, **116-119** 879-885 (1999)

[9]DaimlerChrysler AG, , "Method for manufacturing of coated short fibres", German Pat. DE19828843, June. 27, 1998

[10]A. Schmidt, R. Klupsch, K. Rödiger, M. Willert-Porada, "Applicator for Combined Microwave Sintering and Thermal CVD", pp. 101-102; in *Book of Abstracts of 8th International Conference on Microwave &High Frequency Heating*, (2001)

[11]T. Gerdes, R. Tap, M. Willert-Porada, P. Bahke, CVD-Processes in MW-Heated Fluidised Bed Reactors, pp. 160-161, *in Book of Abstracts of 8th International Conference on Microwave &High Frequency Heating*, (2001)

FLASH SYNTHESIS OF TITANIUM DIOXIDE COLLOIDS

Elodie Michel, Thierry Caillot, Denis Chaumont and Didier Stuerga

Groupe d'**E**tudes et de **R**echerches sur les **M**icroondes
GERM/LRRS, UMR 5613 CNRS-Université de Bourgogne,
BP 47870, 21078 Dijon cedex, France

INTRODUCTION

Titanium dioxide has a wide range of applications : photoconductors[1] gas sensors[2], pigments, catalyst supports, dielectric materials. Numerous ways of preparation are currently available to obtain titania powders or colloids : sol-gel synthesis[3] and micro-emulsion method[4,5] hydrolysis by sulfate[6,7] or chloride[8,9] process and vapor phase hydrolysis[10,11] , some of them using organic titanates as precursors, others requiring long time of aging. Microwave processing of titania powders[12-14] has already been studied by various authors. Nevertheless, to our knowledge, no study has been driven on microwave synthesis of titanium dioxide colloidal suspensions. With our original microwave heating device, sols of titanium dioxide have been rapidly obtained by microwave-induced thermohydrolysis of $TiCl_4$ aqueous solutions.

MICROWAVE PROCESS, SAMPLE CHARACTERIZATIONS

The microwave device employed was specially designed by the authors. This microwave heating system, called **RAMO** system (French acronym of **R**eacteur **A**utoclave **M**icro**O**nde) has been described in details previously[15]. An aqueous solution containing 0.1 $mol.l^{-1}$ hydrochloric acid and 0.01 $mol.l^{-1}$ titanium (IV) chloride was poured into the autoclave reactor, a 0.4 MPa argon pressure was introduced into the system then the reactor was submitted to microwaves. Time of microwave heating varied from 20 s to 70 s ; initially, the microwave generator supplied the system with 500W power , which could be regulated so as to prevent too high pressure values within the reactor : as soon as 1.0 MPa was reached into the system, power was turned down to keep pressure to this threshold.

X-ray diffraction has been performed on powders obtained by centrifugation of the colloids. Colloidal suspensions have been characterized by photon correlation spectroscopy (PCS) ; measurements were made at 25°C in normal incidence. Particle morphology has been studied from sols by transmission electron microscopy (TEM).

RESULTS

TiO_2 colloidal suspensions were obtained from only 20 seconds of microwave heating. Hydrodynamic diameter of particles, in a colloid prepared by 30 s microwave heating of HCl 0.1 mol.l^{-1} - $TiCl_4$ 0.01 mol.l^{-1}, is shown on Fig.1. This suspension is composed of particles of narrow size distribution, centered on 27 nm. The stability of the suspension could have been checked by this technique with distribution measurements as a function of time. No shift of the peaks towards larger sizes was observed, all the size distributions were superposed so the colloid was not subjected to sedimentation.

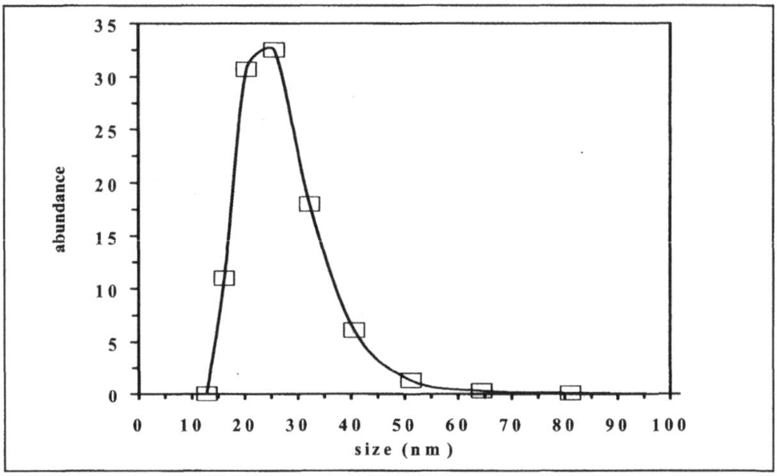

Fig.1 : PCS size distribution of TiO_2 nanoparticles in the suspension.

Moreover, our colloidal particles are crystallized, as shown by the X-ray diffraction pattern on Fig. 2 ; usual TiO2 anatase X-ray intensities are found, the (101) plane having the highest intensity, as expected. TEM experiment shows very fine particles measuring less than 10 nm in diameter. Compared to PCS values, the noticeable difference arises from the fact that hydrodynamic diameter characterizes particles in suspension, surrounded by their electrical double layer.

CONCLUSION

Nanosize TiO_2 colloidal suspensions have been synthesized in a few seconds by microwave-induced thermohydrolysis. Obtained by 500 W microwave heating, for 20 to 70 s, of $TiCl_4$ 0.01 $mol.L^{-1}$ and HCl 0.1 $mol.L^{-1}$ aqueous solution, these suspensions are constituted of very fine nanoparticles showing TiO_2 anatase structure. The interest of the RAMO system is that the association of thermal hydrolysis with microwave heating allows to cut down treatment time and that employing environmentally harmful organic reactants is avoided.

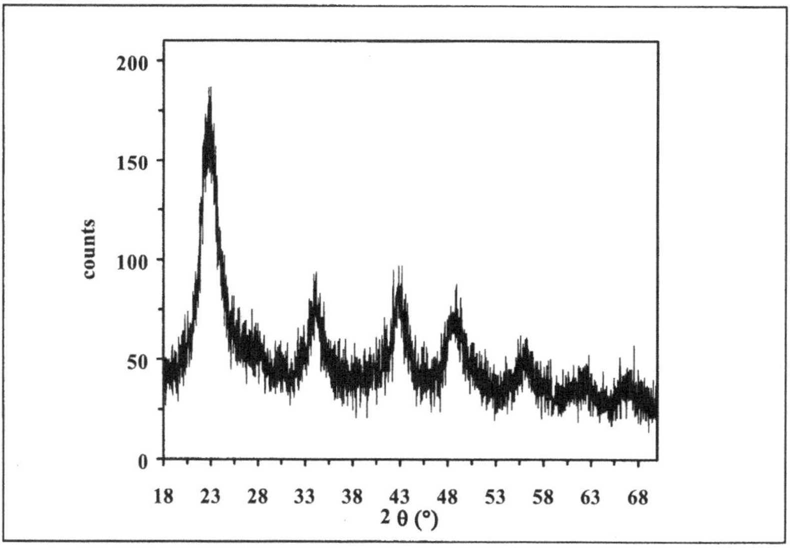

Fig. 2 : X ray diffraction pattern of the microwave-synthesized anatase.

REFERENCES

[1]Y.Li, J.Hagen, W.Schaffrath, P.Otschik and D.Haarer, "Titanium dioxide films for photovoltaic cells derived from a sol-gel process," *Solar Energy Materials and Solar Cells*, **56** 167-174 (1999).

[2]H.Tang, K.Prasad, R.Sanjinés and F.Lévy, "TiO_2 anatase thin films as gas sensors," *Sensors and Actuators B*, **26-27** 71-75 (1995).

[3]K.Terabe, K.Kato, H.Miyazaki, S.Yamaguchi, A.Imai and Y.Iguchi, "Microstructure and crystallization behavior of TiO_2 precursor prepared by the sol-gel method using metal alkoxide," *Journal of Materials Science*, **29** 1617-1622 (1994).

[4]D.Papoutsi, P.Lianos, P.Yianoulis and P.Koutsoukos, "Sol-gel derived TiO_2 microemulsion gels and coatings," *Langmuir*, **10** 1684-1689 (1994).

[5]M.Wu, J.Long, A.Huang and Y.Luo, "Microemulsion-mediated hydrothermal synthesis and characterization of nanosize rutile and anatase particles," *Langmuir*, **15** 8822-8825 (1999).

[6]E. Santacesaria, M. Tonello, G. Storti, C. Pace, and S. Carra, "Kinetics of Titanium Dioxide Precipitation by Thermal Hydrolysys," *Journal of Colloïd and Interface Science*, **111** [1] 44-53 (1986).

[7]Y.Wei, R.Wu and Y.Zhang, "Preparation of monodispersed TiO_2 powder by forced hydrolysis of $Ti(SO_4)_2$ solution," *Materials Letters*, **41** 101-103 (1999).

[8]E. Matijevic, M. Budnik and L. Meites, "Preparation and Mechanism of Formation of Titanium Dioxyde Hydrosols of Narrow Size Distribution," *Journal of Colloïd and Interface Science*, **61** [2] 302-311 (1977).

[9]E. Narita, H. Takeuchi, N. Horiguchi and T. Okabe, "Manufacture of Pure Titanium (IV) Oxide by the Chloride Process, III Precipitation of Hydrous Titanium (IV) Oxide in Hydrochloric Acid Solution at elevated temperatures," *Bulletin of Chemical Society of Japan*, **57** [5] 1388-1393 (1984).

[10]M. Visca, AND E. Matijevic, "Preparation of Uniform Colloïdal Dispersions by Chemical Reactions in Aerosols," *Journal of Colloïd and Interface Science*, **68** [2] 308-319 (1979).

[11]B.Xia, H.Huang and Y.Xie, "Heat treatment on TiO_2 nanoparticles prepared by vapor-phase hydrolysis," *Materials Science and Engineering B* **57** 150-154 (1999).

[12]J.A.Ayllon, A.M.Peiro, L.Saadoun, E.Vigil, X.Domenech and J.Peral, "Preparation of anatase powders from fluorine-complexed titanium(IV) aqueous solution using microwave irradiation," *Journal of Material Chemistry*, **10** 1911-1914 (2000).

[13]S.Komarneni, R.Roy and Q.H.Li, "Microwave-hydrothermal synthesis of ceramic powders," *Materials Research Bulletin*, **27** 1393-1405 (1992).

[14]S.Komarneni, R.K.Rajha and H.Katsuki, "Microwave-hydrothermal processing of titanium dioxide," *Materials Chemistry and Physics*, **61** 50-54 (1999).

[15]E.Michel, D.Chaumont and D.Stuerga, "Microwave Flash-Synthesis of tin dioxide sols from tin chloride aqueous solutions," *Journal of Materials Science Letters*, **20** 1593-1595 (2001)

PREPARATION OF METAL OXIDES BY POLYOL METHOD UNDER MICROWAVE IRRADIATION

Tetsushi Yamamoto, Yuji Wada, and Shozo Yanagida
Material and Life Science, Graduate School of Engineering, Osaka University
2-1 Yamadaoka, Suita, Osaka 565-0871, Japan

ABSTRACT

We applied microwave (MW, 2.45 GHz) heating to preparation of metal oxides, such as anatase TiO_2 nanocrystallites and zincite ZnO particles, by the polyol method. Anatase TiO_2 nanocrystallites were obtained in 1,4-butanediol, 1,5-pentanediol or 1,6-hexanediol under MW. The crystallite size of TiO_2 less than 10 nm could be controlled by changing the quantity of added water. Submicrometer zincite ZnO particles have been prepared in diethylene glycol under MW irradiation by the polyol method. The smaller particles were formed as the secondary aggregates under MW heating when compared to conventional heating, where the sizes of the primary crystallites were the same. The aggregation of the primary crystallites was suppressed under MW irradiation

INTRODUCTION

The polyol method was developed over the past two decades and has been applied to the preparation of submicrometer particles such as metal oxides[1,2] in high boiling point polyol media (e.g., ethylene glycol, diethylene glycol). On the other hand, the microwave (MW) technique has been applied to the polyol method for the preparation of particles because of the high boiling point of polyols and their high dielectric loss factors, ε''.[3-9]

In this paper, we show the preparation of anatase TiO_2 nanocrystallites in alkanediols ($HO(CH_2)_nOH$, n=2-6) under MW irradiation. This new method enables us to shorten the reaction time and also to control the crystallite size by changing the quantity of water as a reactant. And we also show the preparation of zincite ZnO submicrometer particles in diethylene glycol under MW irradiation. The heating rate was varied in the range of 3.0 to 14.0 $Kmin^{-1}$ in order to clarify the influence of this parameter on the particle size.

EXPERIMENTAL

Preparation of TiO_2 Nanocrystallites under MW:

MW irradiation experiments were performed using a multimode MW apparatus (Micro Denshi MMG-213VP, Figure 1). The reaction temperature was measured by a thermocouple and controlled by the repeated on-off MW irradiation. A three stub tuner was used to achieve optimal MW power absorption with a forward and reverse power monitor. The reaction was carried out using a three-neck quartz flask equipped with a water-cooled condenser outside the MW cavity and the reaction mixture was stirred by a Teflon magnetic stirrer.

Figure 1. Multimode MW apparatus, equipped with a magnetron (2.45 GHz, max. power 1.3 kW), an isolator, a forward and reverse power monitor, a three stub tuner, a wave guide and a MW cavity.

Titanium tetraisopropoxide ($Ti(OiPr)_4$, 2 mmol) was added to alkanediol (50 ml) in a 100 ml three-neck quartz flask under N_2 atmosphere. When 1,5-pentanediol or 1,6-hexanediol was used as a solvent, $Ti(OiPr)_4$ was not dissolved totally, giving a turbid solution. The solution was heated to 413 K under MW (200 W) irradiation in 3 min. Then the solution became clear, milli-Q H_2O (1-3 ml) was added, and the solution was irradiated at 900 W for 30 min, keeping the temperature at 513 K. Only in the case the quantity of added H_2O was 1 ml, after reaching at 513 K, the MW power was reduced to 700 W. The temperature was maintained at 513 K at MW 700 W (in the case of 1 ml H_2O) or 900 W (in the case of 2 or 3 ml H_2O) for 30 min. After the MW irradiation, the suspension was cooled down immediately, and diluted with ethanol (EtOH, 30 ml) for quenching further reactions, to yield a precipitate, which was separated by centrifuging at 40000 G for 10 min. The precipitate was twice washed with EtOH under ultrasonication and centrifuged at 40000 G for 10 min, and dried in air.

Preparation of Submicrometer ZnO Particles under MW:

MW irradiation experiments were performed using a multimode MW apparatus (Milestone MicroSYNTH) with a magnetron (2.45 GHz, max. power 1 kW). The reaction was carried out using a three-neck quartz flask equipped with a water-cooled condenser outside the MW cavity and the reaction mixture was stirred by a Teflon magnetic stirrer. A fiber optic sensor, placed in a Teflon coated ceramic thermowell, was inserted into the reaction solution and the reaction temperature controlled by changing a MW output power using EasyWAVE software with PID algorithms.

Zinc acetate dihydrate ($Zn(OAc)_2 \cdot 2H_2O$, 0.09 M) was added to diethylene glycol (DEG, 50 ml) in a 100 ml three-neck quartz flask under N_2 atmosphere. The solution was irradiated in the MW cavity at a heating rate of 3.0 $Kmin^{-1}$, 6.0 $Kmin^{-1}$, or 14.0 $Kmin^{-1}$. After the solution temperature reached at 453 K, this temperature was kept for 15 min under the MW irradiation. $Zn(OAc)_2 \cdot 2H_2O$ dissolved completely at 383-405 K. The solutions became white suspensions at 443-453 K. After MW irradiation, the suspension was cooled down immediately, and diluted with ethanol (30 ml), to yield a precipitate, which was separated by centrifuging at 40000 G for 10 min. The precipitate was twice washed with EtOH under ultrasonication and centrifuged at 40000 G for 10 min, and dried in air. In case of a conventional heating, a DEG (50 ml) solution of $Zn(OAc)_2 \cdot 2H_2O$ (0.09 M) was heated at a heating rate of 4.3 $Kmin^{-1}$ using a silicon oil bath, and maintained for 15 min at 453 K.

Characterization:

The XRD patterns were recorded using Rigaku MaltiFlex Cu-Kα radiation (40 kV, 40 mA) by scanning at 1° (2θ)/min. The crystallite size was calculated from the half-height width of a diffraction broading peak using the Debye-Scherrer equation: λ=0.154056 nm and the shape factor, K=0.9. Transmission electron microscopic (TEM) analyses were performed with Hitachi H-9000 operating at 300 kV. Scanning electron microscopy (SEM) images were obtained by a JEOL JSM-6700FE, working at an accelerating voltage of 15 kV. Particle size distributions of ZnO in EtOH suspension were determined by a dynamic light scattering (DLS) particle size analyzer, Horiba LB-500, equipped with a laser power 5mW at 650 nm and a silicon photodiode detector.

RESULTS AND DISCUSSION

TiO_2 Nanocrystallites:

The products, obtained in 1,4-butanediol, 1,5-pentanediol and 1,6-hexanediol at 513 K for 30 min under MW irradiation, were anatase TiO_2 (Figure 2a-d, powder diffraction file, PDF #21-1272), whereas no anatase formation in 1,2-ethanediol (ethylene glycol) and 1,3-propanediol by the powder X-ray diffraction (XRD). At the set temperature, 453 K, it took 90 min for the sample to show the XRD pattern equivalent to that obtained at 513 K for 30 min in 1,5-pentanediol (Figure 2b, 2d).

TEM image (Figure 3a) showed that the anatase TiO_2 nanocrystallites were aggregated. The 0.35 nm spacing of fringes in the HRTEM image (Figure 3b) corresponded to the d value 0.35200 nm of the (101) plane for the anatase TiO_2 (PDF #21-1272). The dark-field TEM image observed for the same areas as in Figure 1a

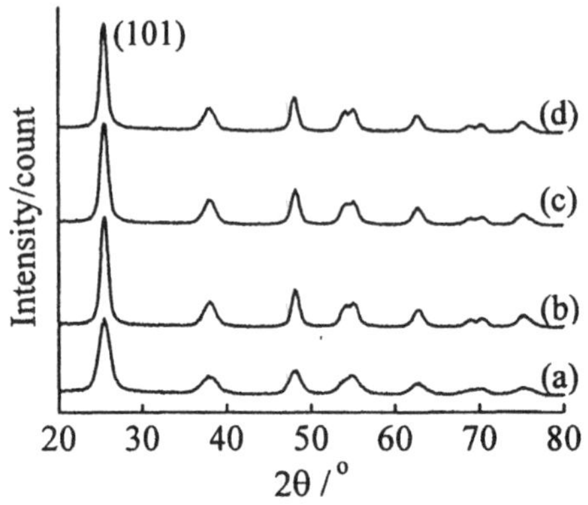

Figure 2. XRD patterns of TiO_2 obtained at 513 K for 30 min (a) in 1,4-butanediol, (b) in 1,5-pentanediol, (c) in 1,6-hexanediol, and (d) at 453 K for 90 min in 1,5-pentanediol under MW irradiation after 2 ml water was added.

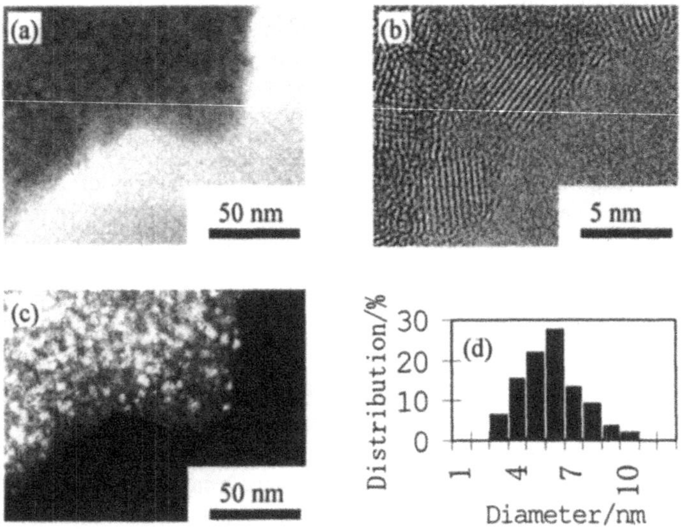

Figure 3. (a) Bright-field TEM image, (b) HRTEM image showing 0.35 nm fringes, (c) dark-field TEM image on the same area in (a), and (d) histogram of size distribution of anatase TiO_2 nanocrystallites prepared in 1,4-butanediol at 513 K.

gave a more appropriate image for determining the size distribution (Figure 3c). The distribution determined for the randomly selected 200 particles was centered at 6 nm and ranged between 3 and 10 nm (Figure 3d). The average diameter of 5.8 nm was in good agreement with the crystallite size (4.5 nm, Table 1) calculated from the (101) diffraction peak of XRD pattern (Table 1).

Using 1,5-pentanediol, the crystallites grew from 7.2 nm to 8.3 nm with the progressing time from 10 min to 30 min (Table1). The crystallite size prepared in 1,4-butanediol (5.3 nm) was smaller than that in 1,5-pentanediol (8.3 nm) and 1,6-hexanediol (8.1 nm) under the comparable conditions (Table 1). The difference in the chain length of a series of alkanediols led to the precise control of the crystallite size. When the quantity of added H_2O was increased from 1 ml to 3 ml, the crystallite size increased from 4.5 nm to 6.5 nm in 1,4-butanediol, and also increased from 5.9 nm to 10.0 nm in 1,5-pentanediol at 513 K for 30 min under intermittent MW irradiation (Table1). Therefore, the crystallite size can be controlled by changing the quântity of added H_2O.

Table 1. MW-driven hydrolysis and polycondensation reactions of Ti(O*i*Pr)$_4$ (2 mmol) in the presence of H_2O ranging from 1 ml to 3 ml

solvent /50mL	H_2O /mL	Max. Temp./K (Time/min)	Crystallite size of TiO_2/nm[a]
1,2-ethanediol	2	473 (90)	–
1,3-propanediol	2	483 (90)	–
1,4-butanediol	1	513 (30)	4.5
1,4-butanediol	2	513 (30)	5.3
1,4-butanediol	3	513 (30)	6.5
1,5-pentanediol	1	513 (30)	5.9
1,5-pentanediol	2	513 (10)	7.2
1,5-pentanediol	2	513 (20)	8.0
1,5-pentanediol	2	513 (30)	8.3
1,5-pentanediol	2	453 (90)	9.3
1,5-pentanediol	3	513 (30)	10.0
1,6-hexanediol	2	513 (30)	8.1

[a]Calculated from (101) diffraction peak of the anatase phase using the Debye-Sherrer equation.

Submicrometer ZnO Particles:

The reaction temperature was monitored by a fiber optic sensor (Figure 4). The product, obtained in diethylene glycol (DEG) at 453 K for 15 min under MW irradiation, was identified as zincite ZnO (Figure 5, PDF #36-1451) by the powder XRD patterns. The crystallite sizes of zincite ZnO calculated from XRD patterns showed slight differences between under MW irradiation (7-11 nm) and under a conventional heating (7-8 nm).

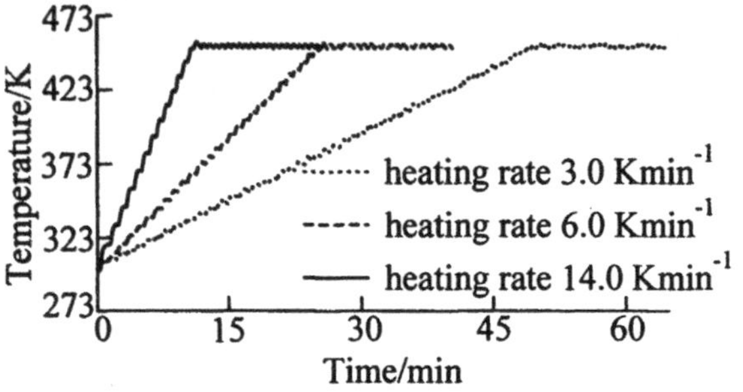

Figure 4. Temperature profiles at MW heating rates of 3.0, 6.0, and 14.0 $Kmin^{-1}$.

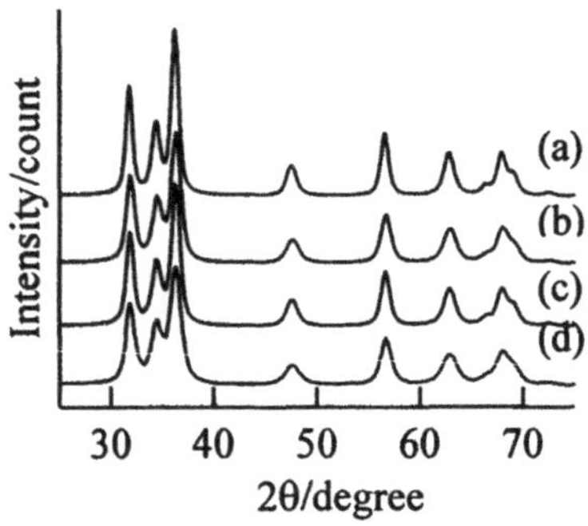

Figure 5. XRD patterns of zincite ZnO particles obtained at MW heating rates of (a) 3.0 $Kmin^{-1}$, (b) 6.0 $Kmin^{-1}$, (c) 14.0 $Kmin^{-1}$, and (d) at a conventional heating rate of 4.3 $Kmin^{-1}$.

Figure 6 shows that the spherical particles with diameter 0.1-0.25 μm obtained under MW were smaller than those with diameter 0.2-0.6 μm obtained under a conventional heating. The submicrometer spherical ZnO particles should be formed by the aggregation of primary crystallites (the average size 7-11 nm inferred from XRD line broadening). It was clarified that the smaller particles were formed under MW heating than under conventional heating, whereas the sizes of the primary crystallites constituting the particles were the same.

The sizes of submicrometer ZnO particles measured by DLS were in good agreement with the visible size of particles in the SEM images under MW heating and conventional heating (Figure 7a). Therefore, submicrometer ZnO particles should be monodispersed in ethanol. Jézéquel et al. have reported that decrease in the conventional heating rate from 14 to 6 $Kmin^{-1}$ leads to increase in the average particle size from 0.20±0.02 to 0.35±0.03 μm.[1] In present work at a conventional heating rate 4.3 $Kmin^{-1}$, an average particle size was 0.33±0.08 μm, comparable to the Jézéquel's result (0.35±0.03 μm at 6 $Kmin^{-1}$). When the heating rate was decreased from 14.0 to 3.0 $Kmin^{-1}$ under MW irradiation, the average particle sizes were almost unchanged from 0.14±0.03 to 0.17±0.04 μm, and smaller than those obtained under conventional heatings (Figure 7b). We propose that aggregation of primary crystallites was suppressed by local "hot-spots"[10] under MW irradiation in the heating rate range from 3.0 to 14.0 $Kmin^{-1}$.

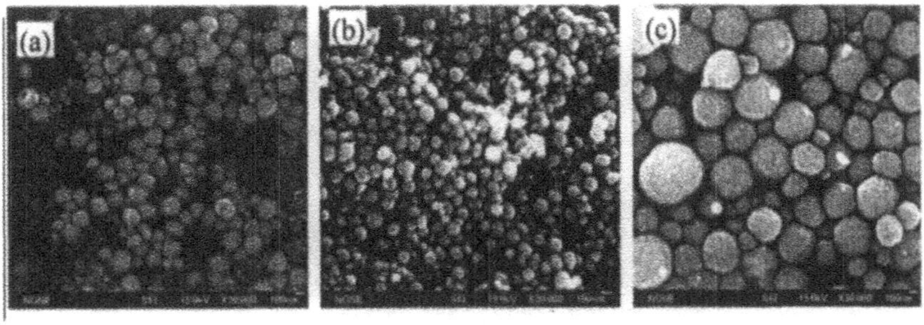

Figure 6. SEM images of ZnO particles obtained at MW heating rate of (a) 3.0 $Kmin^{-1}$, (b) 14.0 $Kmin^{-1}$, and (c) at conventional heating rate of 4.3 $Kmin^{-1}$.

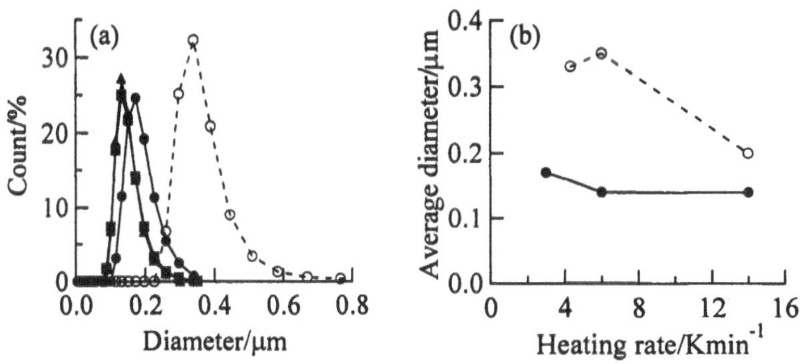

Figure 7. (a) Size distributions of ZnO particles obtained at MW heating rate of (•) 3.0 $Kmin^{-1}$, (▲) 6.0 $Kmin^{-1}$, (■) 14.0 $Kmin^{-1}$, and (○) conventional heating rate of 4.3 $Kmin^{-1}$. (b) Average diameter of ZnO particles obtained at MW heating rates (•) and conventional heating rates (○, where average diameters at heating rate of 6 $Kmin^{-1}$

and 14 $Kmin^{-1}$ were taken from reference 1).

CONCLUSIONS

We used 1,4-butanediol, 1,5-pentanediol or 1,6-hexanediol as a solvent, and succeeded in preparing size-controlled TiO_2 nanocrystallites by the hydrolysis and polycondensation reactions of Ti(O*i*Pr)$_4$ under the intermittent MW irradiation in the polyol method. The crystallite sizes range from 4.5 nm to 10.0 nm only by varying the quantity of H_2O. Selection of the polyols would give the broad range of control in the size. And we also have shown the suppression of aggregation of primary crystallites under MW irradiation heating in preparation of submicrometer zincite ZnO particles by the polyol method.

ACKNOWLEDGMENTS

We thank Dr. Takao Sakata and Prof. Dr. Hirotaro Mori of Research Center for Ultra-High Voltage Electron Microscopy, Osaka University, for TEM analysis.

REFERENCES

[1]D. Jézéquel, J. Guenot, N. Jouini, and F. Fiévet, "Submicrometer zinc oxide particles: Elaboration in polyol medium and morphological characteristics," *J. Mater. Res.*, **10** [1], 77-83 (1995).

[2]C. Feldmann and H.-O. Jungk, "Polyol-Mediated Preparation of Nanoscale Oxide Particles," *Angew. Chem., Int. Ed.*, **40** [2] 359 (2001).

[3]Y. Wada, H. Kuramoto, T. Sakata, H. Mori, T. Sumida, T. Kitamura, and S. Yanagida, "Preparation of Nano-sized Nickel Metal Particles by Microwave Irradiation," *Chem. Lett.*, **1999**, 607-8.

[4]O. Palchik, J. Zhu, and A. Gedanken, "Microwave assisted preparation of binary oxide nanoparticles," *J. Mater. Chem.*, **10** [5] 1251-4 (2000).

[5]W. Tu and H. Liu, "Rapid synthesis of nanoscale colloidal metal clusters by microwave irradiation," *J. Mater. Chem.*, **10** [9] 2207-11 (2000).

[6]O. Palchik, R. Kerner, A. Gedanken, A.M. Weiss, M.A. Slifkin, and V. Palchik, "Microwave-assisted polyol method for the preparation of CdSe "nanoballs"," *J. Mater. Chem.*, **11** [3] 874-8 (2001).

[7]R. Kerner, O. Palchik, and A. Gedanken, "Sonochemical and Microwave-Assisted Preparations of PbTe and PbSe. A Comparative Study," *Chem. Mater.*, **13** [4] 1413-9 (2001).

[8]H. Grisaru, O. Palchik, A. Gedanken, V. Palchik, M.A. Slifkin, and A.M. Weiss, "Preparation of the $Cd_{1-x}Zn_xSe$ alloys in the nanophase form using microwave irradiation," *J. Mater. Chem.*, **12** [2] 339-44 (2002).

[9]T. Yamamoto, Y. Wada, H. Yin, T. Sakata, H. Mori, and S. Yanagida, "Microwave-Driven Polyol Method for Preparation of TiO_2 Nanocrystallites," *Chem. Lett.*, **2002**, 964-5.

[10]S.A. Galema, "Microwave chemistry," *Chem. Soc. Rev.*, **1997**, 233-8.

EFFICIENT MICROWAVE SYNTHESIS OF LUMINESCENT METAL COMPLEXES

*T. Matsumura-Inoue, Y. Yamamoto and N. Yoshikawa
Nara University of Education, Takabatake, Nara-shi,
Japan, 6308528

ABSTRACT

A microwave synthetic method has been developed for novel luminescent Ir(III) complexes. The novel Ir(III) complexes give intense multi-colored luminescence and may be promising candidates for photoelectronic materials. The photochemical and electrochemical properties of the complexes are described.

INTRODUCTION

Polypyridine complexes of platinum group metals have attracted much attention for their highly luminescent character. In these complexes, polypyridine ligands, consisting of many aromatic rings, take roles for electron trapping sites and stabilize the excited electron, which generates florescence or redox reaction [1].

On the other hand, spin-orbit coupling of heavy metals contributes to the cross linking of singlet-triplet states resulting in emissions with long life. Ru(II) and Os(II) complexes are widely used as photosensitizers and energy transducers. Recently, Ir(III) complexes have been extensively studied as candidates for photoelectric materials [2,3]. Microwave irradiation is known to induce some specific effects on molecular motion, so it can be used to accelerate reaction processes [4]. We have developed an efficient synthetic method of photosensitive Ru(II) polypyridine complexes, assisted with microwave irradiation [5,6]. Conventionally, Ir(III) polypyridine complexes were prepared by melting a metal salt or metal chloride with a polypyridine ligand in a sealed tube, or a stepwise synthesis with reflux at higher temperature. The yield was extremely poor, because of complicated purifications [7]. We succeeded in obtaining purer complexes in a shorter time with a 80% or more yield by reflux synthesis under microwave irradiation [8]. Here we report the simple synthetic procedure for Ir(III) polypyridine complexes under microwave irradiation and their physicochemical properties.

EXPERIMENTAL

Reagents used were purchased from Aldrich, Eastman-Kodak, Fisher and Wako. All of them are of at least analytical reagent grade and used without further purification unless otherwise stated. Ammonium hexachloroiridate (III) $(NH_4)_3[IrCl_6]$ was of the best commercial grade. Ethylene glycol was used as a solvent for synthesis. Purity of the complexes was determined by elemental analysis, ^{1}H-NMR and mass spectrometry.

Electrochemical measurements were performed at 25℃ under an inert atmosphere. A three-electrode cell configuration controlled by a BAS-CV50W potentiostat was used. The counter electrode was a platinum wire, and the reference electrode was a BAS RE-1 Ag-AgCl. Cyclic voltammetry, CV, or differential pulse voltammetry (DPV) was obtained in a 0.8 m mol·dm^{-3} solution of the complexes in acetonitrile with 0.1 mol·dm^{-3} tetrabutylammonium perchlorate (TBAP) as a supporting electrolyte. The ferrocenium / ferrocene couple ($FeCp_2^{+/0}$) was used as a potential standard.

RESULTS AND DISCUSSION

Synthetic procedure by microwave irradiation

The microwave oven was perforated at the top to insert a reflux condenser. The microwave leakage was checked by microwave survey meter (HI-1501, HOLADAY CO, LTD), and was confirmed to be about 1 mW/cm^2 (less than 5 mW/cm^2 is the permissible leakage maximum). The synthetic procedure for tris polypyridine iridium(III) complexes is as follows: $(NH_4)_3[IrCl_6]\cdot H_2O$ (0.238 g, 0.5 m mol) and 2,2'-bipyridine (0.234 g, 1.5 m mol) were mixed in ethylene glycol (15 ml). Under nitrogen, the mixture was refluxed with irradiation of microwave at 2450 MHz. After refluxing for 15 minutes, the mixture was cooled to room temperature. Saturated aqueous solution of KPF_6 (20 ml) was added for precipitation. The precipitate was filtrated and recrystallized from acetonitrile / diethylether. Bis and mono polypiridine complexes were also prepared in a similar procedure. The yield was 80%. Thus the microwave synthetic method was highly efficient both in reaction time and yield compared with conventional method [7]. Polypyridine ligands used are listed in Figure 2.

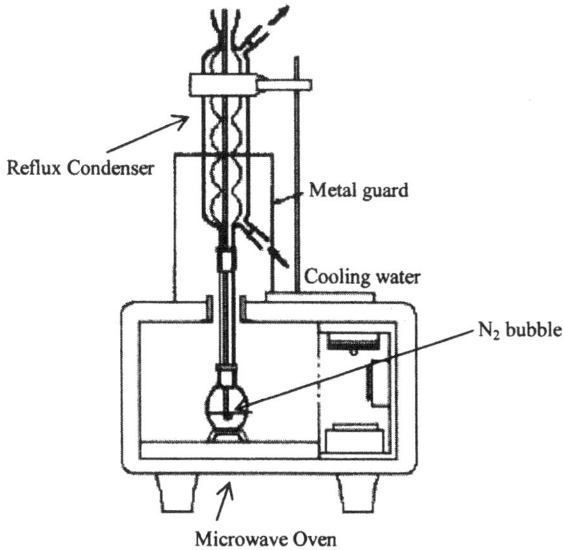

Figure 1 Experimental system for microwave synthesis

Abbreviations for these ligands are as follows; bpy = 2,2'-bipyridine, dmbpy = 4,4'-dimethyl-2,2'-bipyridine, dpbpy =4,7-diphenyl-2,2'-bipyridine, phen = 1,10-phenanthroline, dmphen = 2,9-dimethyl- 1,10-phenanthroline, phphen = 5-phenyl-1,10-phenanthroline, dpphen = 4,7-diphenyl-1,10-phenanthroline, dmdpphen =2,9-dimethyl-4,7-diphenyl-1,10-phenanthroline, tmphen = 3,7,4,8-tetramethyl-1,10-phenanthroline, bqn = 2,2'-biquinoline.

Absorption and emission spectra of Ir(III) polypyridine complexes

The absorption spectra of the complexes in acetonitrile solution give the intensive absorption bands around 270 nm and the shoulders at 378 nm and 465 nm as shown in Figure 3. The former with molar absorption coefficients, ε , of 10^4 - 10^5 $M^{-1}cm^{-1}$, correspond to a π - π * transition and the latter with ε of 10^3 - 10^4 $M^{-1}cm^{-1}$, also to a π - π * transition including slight metal orbital contribution. The energies of the latter, corresponding to HOMO-LUMO energy, range from 2.67 to 3.28 eV. Acetonitrile solution of these complexes exhibits an intense luminescence around 400 to 630 nm at room temperature, when they are excited at 318 nm as shown in Figure 3. These spectra can be assigned to a triplet π - π * emission. The emission spectra are characterized

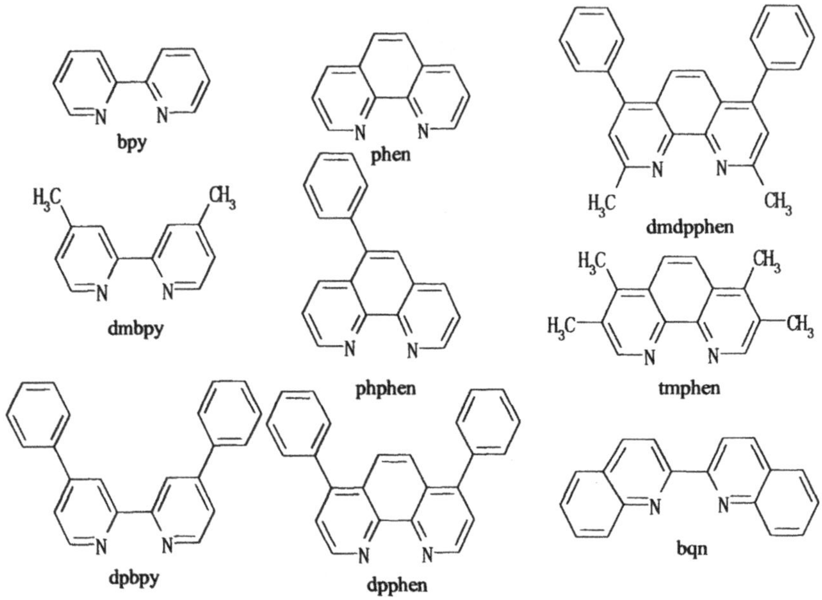

Figure 2 Molecular structures of polypyridine ligands

with high intensity and a wide range of wavelength. $Ir(bqn)_3^{3+}$ with electron withdrawing ligands shows a single emission at 630 nm. On the other hand, the complex $Ir(tmphen)_3^{3+}$ with electron donating ligands shows an emission band at 400 nm. The parameters for absorption and emission spectra of Ir(III) polypyridine complexes are listed in Table I. From these results, it is found that emission peaks for the three Ir(III) complexes with bipyridine ligands (1-3) arise around 500 to 530 nm, while those for Ir(III) complexes with phenanthroline ligands (4-8) arise at various wavelengths from 400 to 530 nm. The wavelength of emission peak is the shortest with $Ir(tmphen)_3^{3+}$, and becomes longer in the order of tmphen < dmphen < dmdpphen <phen < dpphen. Substituents of methyl group cause blue shift, while those of phenyl group cause red shift. Thus the fluorescence of Ir(III) polypyridine complexes shows a variety of colors changing from red ($Ir(bqn)_3^{3+}$), through green ($Ir(dpbpy)_3^{3+}$) to blue ($Ir(dmdpphen)_3^{3+}$. The quantum yield of typical three complexes, $Ir(bqn)_3^{3+}$, $Ir(dpbpy)_3^{3+}$ and $Ir(dmdpphen)_3^{3+}$, which give the three primary colors of luminescence, i.e., red, green and blue, are determined by using poly-

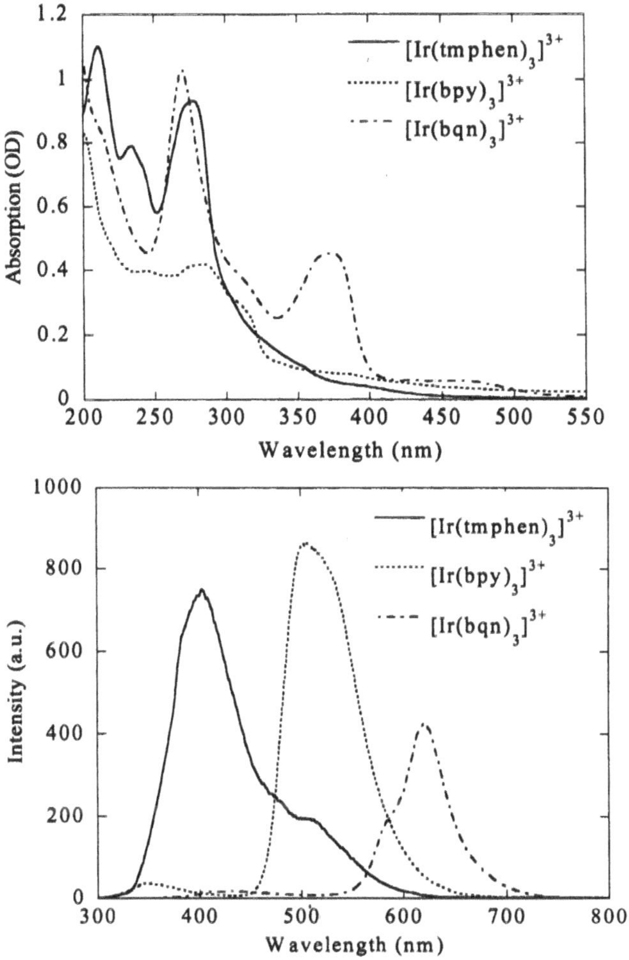

10^{-5}M-Ir (III) complex in acetonitrile

Figure 3 Absorption and emission spectra of Ir(III) polypyridine complexes

rhodamine-B and OPPA (Origo para-phenylene trimer) as standards. The quantum yield is considerably high: 0.43 for Ir(dmdpphen) $_3^{3+}$, 0.19 for Ir(dpbpy) $_3^{3+}$, and 0.12 for Ir(bqn) $_3^{3+}$. With these unique properties, intense and multi-colored luminescence, the complexes are promising for photoelectronic materials such as photosensitizers and electroluminescent device.

Table I Data for absorption emission spectra of Ir(III) Polypydine complex

Complexes	Absorption(298 K)		Luminescence[a](298 K)	
	λ max/nm	E_{ab}	λ max/nm	E_{em}
	(ε :$10^3 M^{-1}cm^{-1}$)	eV	(ε ':$10^7 M^{-1}cm^{-1}$)	eV
1 $[Ir(bpy)_3]^{3+}$	284(41.9) 379(8.1)	3.27	528(3.58)	2.35
2 $[Ir(dmbpy)_3]^{3+}$	273(50.4) 378(8.4)	3.28	525(2.00)	2.36
3 $[Ir(dpbpy)_3]^{3+}$	282(104) 398(13.8)	3.12	506(163)	2.45
4 $[Ir(phen)_3]^{3+}$	265(67.2) 397(4.1)	3.12	529(4.38)	2.34
5 $[Ir(phphen)_3]^{3+}$	270(96.8) 404(5.0)	3.07	531.5(6.48)	2.33
6 $[Ir(dpphen)_3]^{3+}$	282(112) 402(16.4)	3.08	539(18.1)	2.30
7 $[Ir(tmphen)_3]^{3+}$	277(93.3) 393(4.4)	3.16	402.5(7.50)	3.08
8 $[Ir(dmdpphen)_3]^{3+}$	291(57.4) 367(4.1)	3.38	446.5(95.6)	2.78
9 $[Ir(bqn)_3]^{3+}$	269(103) 465(5.9)	2.67	630(22.4)[b]	1.97

[a]λ_{ex} = 318 nm. [b]λ_{ex} = 383 nm., [*]arbitrary unit, Solvent : acetonitrile.

Electrochemical behavior

These Ir(III) complexes show three or four reduction waves from 0 to - 2.0 V vs Ag/AgCl and one irreversible oxidation wave. As for $Ir(dpphen)_3{}^{3+}$, a DPV gives three reduction waves at -1.04 V, -1.19 V and -1.37 V, respectively and one irreversible oxidation wave at 2.03V vs Ag/AgCl. In Table II, DPV data for Ir(III) polypyridine complexes are summarized. It is found that the reduction potentials of $Ir(dmbpy)_3{}^{3+}$with electron donating ligand are the most negative, while those of $Ir(bqn)_3{}^{3+}$ with the electron withdrawing ligand are the most positive. Thus reduction potentials of the complexes seem to be affected by the electron donor or accepter nature of ligands. The oxidation of $Ir(phen)_3{}^{3+}$, Ir $(dpphen)_3{}^{3+}$, and $Ir(bqn)_3{}^{3+}$ occurring at 2.05 V, 1.99 V and 2.03 V respectively, is similar to that of the free ligands, which suggests a ligand-based oxidation. On the other hand, the oxidation-potential of $Ir(bpy)_3{}^{3+}$ and $Ir(dmbpy)_3{}^{3+}$ shows more negative values (-0.2 V) than that of the free ligands, which suggests slight contribution of metal orbital to ligand π orbitals in the complex. By comparing the potential values with tris and bis polypyridine complexes in Table II, it is also observed that the reduction and oxidation potentials of bis complexes are nearly equal to those of tris polypyridine complexes. This trend is remarkably different from that observed with Ru(II) polypridine complexes [8], in which each oxidation potential is remarkably

Table II Electrochemical properties of Ir(III) polypyridine complexes

	Complexes	DPV V vs. Ag/AgCl				V
		Red 1st	Red 2nd	Red 3rd	Ox	ΔE (E_{OX}-$E_{RED.1}$)
1A	$[Ir(bpy)_3]^{3+}$	-1.09,	-1.28,	-1.40	2.07	3.16
1B	$[Ir(bpy)_2 Cl_2]^{+}$	-1.10,	-1.26,	-1.47	2.07	3.14
2A	$[Ir(dmbpy)_3]^{3+}$	-1.18,	-1.34,	-1.53	1.97	3.15
2B	$[Ir(dmbpy)_2 Cl_2]^{+}$	-1.16,	-1.34,	-1.54	1.95	3.11
3A	$[Ir(dpbpy)_3]^{3+}$	-1.02	-1.20,	-1.39,	1.98	3.00
3B	$[Ir(dpbpy)_2Cl_2]^{+}$	-1.02	-1.19,	-1.36	1.99	3.01
4A	$[Ir(phen)_3]^{3+}$	-1.07,	-1.24,	-1.47	2.05	3.12
4B	$[Ir(phen)_2 Cl_2]^{+}$	-1.09,	-1.26,	-1.47	2.04	3.13
5A	$[Ir(dpphen)_3]^{3+}$	-1.04,	-1.19,	-1.37	1.99	3.03
5B	$[Ir(dpphen)_2 Cl_2]^{+}$	-1.04,	-1.20,	-1.38	1.97	3.01
6A	$[Ir(bqn)_3]^{3+}$	-0.51,	-0.79,	-0.97	2.03	2.54
6B	$[Ir(bqn)_2 Cl_2]^{+}$	-0.50,	-0.78,	-0.96	2.03	2.53

DPV: differential pulse voltammetry

different between tris(A) and bis(B) polypyridine complexes. This electrochemical behavior supports that one electron is removed from π orbital in the complex (HOMO) and three or four electrons are stepwise added to the π* orbitals (LUMO). Potential difference of the oxidation wave and the first reduction wave, ΔE and the difference of LUMO - HOMO energy, $E_{\pi-\pi^*}$, are compared for these complexes. A linear relationship with a slope of 1.00 (r=0.96) has been obtained between $E_{\pi-\pi^*}$ and ΔE.

CONCLUSION

A microwave synthetic method has been developed for highly luminescent Ir(III) polypyridine complexes. The method is very efficient, providing the complexes in a short time, 15 minutes and a high yield, more than 80%. The novel Ir(III) complexes give intense luminescence whose color change from blue to red according to the ligands. Emission peaks change from 400 to 630 nm depending on the donor-acceptor nature of ligands. These properties, highly intense and multi-colored luminescence, are very novel and unique. Thus microwave develops new aspects for the synthesis of functional materials.

ACKNOWLEGEMENT

We wish to thank Dr H. Hasegawa and Professor S. Yanagida, Graduated School of Osaka University, for their help in measurement of quantum yield and helpful discussion.

REFERENCES

[1]C. Kaes, A. Katz and M. W. Hossein, "Bipyridine : The Most Widely Used Ligand. A Review of Molecules Comprising at Least Two-Bipyridine Units" *Chem. Rev*, **100**, 3553-3590 (2000).

[2]S. Lamansky, P. Djurovich, D. Murphy, F. Adbel-Razzag, H-E. Lee, C. Adachi, P. E. Burrows S. R. Forrest and M. E. Thompson "Highly Phosphorescent Bis-Cyclometalated Iridium Complexes : Synthesis , Photophysical CharacterizationandUseinOrganicLight Emitting Diodes" *J. Am. Chem. Soc.*, **123**, 4304-4312 (2001).

[3]I. M. Dixon, J. P. Collin, J. P. Sauvage, and L. Flamigni "Porphyrinic Dyads and Triads Assembled around Irdium(Ⅲ) Bis-terpyridine: Photoinduced Electron Transfer Processes" *Inorg. Chem.* **40**, 5507-5517 (2001).

[4]H. M. Kingston and S. H. J. Haswell ed., 1st ed. "Microwave-Enhanced Chemistry", Am. Chem. Soc. Washington, DC, 1997.

[5]T. Matsumura-Inoue, M. Tanabe, T. Minami, and T. Ohashi, "A Remarkable Rapid Synthesis of Ruthenium(Ⅱ) Polyoyridine Complexes by Microwave Irradiation" Chem. Lett., 1994, 2443-2446

[6]X. Xiao, J. Sakamoto, M. Tanabe, S. Yamazaki, S. Yamabe and T. Matsumura-Inoue, "Microwave synthesis and electrospectrochemical study on ruthenium(Ⅱ) polypyridine complexes" *J. Electroanal. Chem.*, **527**, 33-40 (2002).

[7]C. M. Flynn and J. N. Demas, J. Am. Chem. Soc., 96 (1974) 1959

[8]N. Yoshikawa, Y. Masuda and T. Matsumura-Inoue, "Rapid Synthesis of New Emitting Polypyridine Complexes" *Chem. Lett.*, 1206-1207 (2000).

MICROWAVE ASSISTED SYNTHESIS OF HYBRID ORGANIC-INORGANIC MEMBRANES FOR FUEL CELL APPLICATIONS

F. Bauer
University of Bayreuth
Universitätsstr. 30
95447 Bayreuth, Germany

M. Willert-Porada
University of Bayreuth
Universitätsstr. 30
95447 Bayreuth, Germany

ABSTRACT

There is still a need for the further development of membrane materials for the direct methanol fuel cell (DMFC). Existing materials properties can be improved by addition of inorganic fillers. In this paper the applicability of microwave processing was examined for its use for in-situ synthesis of inorganic filler materials in Nafion membranes. For a better understanding of the processes occuring during the exposure to microwave frequency radiation the dielectric properties of Nafion were examined under different conditions of temperature, humidity and filler material. Titanium butylate was used as precursor for titania. The inverse micelle structural model of Nafion was applied to explain the experimental results. Two different reaction routes hydrolysis and pyrolysis were carried out, respectively. The hybrid membranes were characterised by their X-ray diffraction patterns, and in case of the hydrolysis reaction route the effective methanol diffusion coefficient and the conductivity as a function of humidity was measured and compared to the parent material. At 100 °C in water saturated air the conductivity of Nafion 117 and a Nafion-TiO_2 (23 wt%) hybrid membrane were 0.1 $\Omega^{-1}cm^{-1}$ and 0.04 $\Omega^{-1}cm^{-1}$, respectively. The effective diffusion coefficients measured at 20 °C were $2.0 \cdot 10^{-10}$ m^2s^{-1} and $4.2 \cdot 10^{-11}$ m^2s^{-1}, respectively. The results seem to be promising for further tests in a single DMFC.

INTRODUCTION

Inorganic fillers in polymers are important for improvement of different properties, e.g., strength, hardness, optical properties, UV-resistance, wetting. In such composites usually besides adhesion no direct chemical interaction exists between the polymer chain and the inorganic additive. In contrast to this, in polymer electrolytes containing inorganic salts, as for example in mono- and divalent salt-PEO mixtures, (abbreviated for Poly(ethylene oxide)), a direct chemical interaction exists between the polymer molecules and the inorganic material, both contributing to charge transport and diffusion of ions [1]. Such materials, more generally known as hybrid inorganic-organic materials, are of increasing importance for advanced application in electronic and electrochemical devices, e.g., as battery or fuel cell electrolytes.

Recently, metal oxide filled ionomer membranes have been proven to show an enhanced high temperature performance when used as electrolyte for a polymer electrolyte fuel cell (PEFC) [2-5]. The organic part, usually a perfluoropolymer with sulfonic acid moieties, shows high proton conductivity based on the water content of the polymer. At higher temperatures the membrane is not capable of serving as a proton conductor due to loss of water. Therefore, in order to enable application of this kind of membranes over a wide range of temperature as well as with a variety of fuels, e.g., hydrogen, methanol and other alcohols, formation of hybrid organic-inorganic networks are attempted by different processing routes.

In case of a perfluoropolymer-sulfonate-membrane, like Nafion®, the most common processing routes are based on re-casting of a Nafion®-slurry with high specific surface area oxides like SiO_2 [4], or on soaking a cast membrane in an inorganic compound, like $ZrOCl_2$ followed by reaction with an acid, e.g., H_3PO_4 to form a Zr-phosphonate [5]. Unfortunately, no information is available at present about the contribution of the inorganic materials to the conductivity mechanism of the membrane nor to the function of the inorganic inclusions with respect to different fuels, like CH_3OH.

The goal of the work presented here is to develop different inorganic-organic hybrid materials with a composition similar to the one mentioned above, but with distinct differences regarding the bonding and interaction between the polymer chain and the sulfonic groups and the inorganic additives.

Hybrid materials composed of Nafion® 117 and TiO_2 were produced by different routes. To place the oxide within the polymeric network at sites usually penetrated by the fuel CH_3OH, metal alcoholate precursor solutions in different alcohols were employed for the soaking reaction.

Application of microwave heating is interesting because of the difference in dielectric loss of the membrane material and the metal alcoholates. Microwave radiation can therefore be used for selective heating and decomposition of the precursor, without thermal decomposition of the membrane. As shown in previous works, selective heating of an inhomogeneous dielectric material may cause temperature gradients even on a μm-scale [6,7]. The aim of the present work is to determine the effects of thermal inhomogenities on the properties of the hybrid membranes.

EXPERIMENTAL

Preparation of hybrid membranes, pre-treatment - To remove impurities Nafion 117 (DuPont, 180 μm thick membranes) was treated with boiling 10 % nitric acid for 2 h according to a method developed by Du Pont [8]. The membrane was rinsed several times in de-ionised (DI) water to remove excess acid. The membranes were stored under DI water and were dried for 1 h at 80 °C prior to use. The pre-treated Nafion membrane was swollen in a 1:3 by weight solution of titanium alcoholate $Ti(O\text{-}nBu)_4$ in ethanol or butanol (both p.a. grade) at RT for 3 h. Then the alcoholate was either hydrolysed in the moisture of ambient air (refered to as hydrolysis samples) or dried under argon at room temperature for 20 h (refered to as pyrolysis samples).

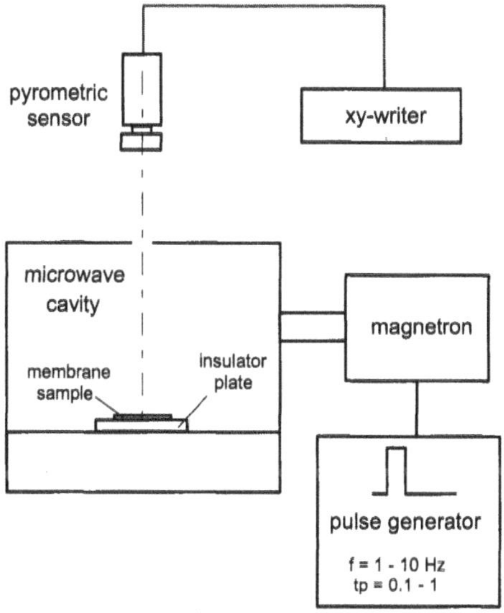

Fig. 1. Schematic representation of the experimental setup.

The experimental set-up used to carry out the pyrolysis reaction consists of a multi mode applicator (Whirlpool Pro 2880), equipped with 4x700W magnetrons with an operating frequency of 2.45 GHz. The samples were placed in Ar atmosphere on top of an insulator plate. Then they were exposed to microwave radiation for a certain time and at a maximum power level of 2800 W. The macroscopic surface temperature, measured by a pyrometric sensor, was kept low by adjusting the pulse length of microwave exposure (Figure 1).

The length as determined by the pulse to pause ratio t_p and frequency f of the microwave pulses could be set by a Fricke/Mallah microwave pulse generator from 0.1 to 1 and 1 to 10 Hz, respectively. Typically, values of 0.2 and 2 Hz were used.

The hydrolysed membranes were refluxed in 1.5-M sulfuric acid for 1 h to complete the hydrolysis and assure full protonation. Then they were rinsed several times in de-ionised (DI) water to remove excess acid. The pyrolysed membranes were examined as received after microwave treatment.

Dielectric spectroscopy - For dielectric measurements a HP 4291B dielectric analyser and a HP 4284A LCR meter were used. The samples were dried for 10 h in a vacuum oven over P_2O_5 at 80 °C, sputtered with a 25 nm gold layer on both sides and pieces of 4 mm diameter were cut out. The membranes were dried again in vacuum at 80 °C in the mounted sample cell holder (HP 16453 A), allowed to cool down to room temperature and measured in nitrogen atmosphere. In case of the humidified sample, the membrane was kept over a saturated salt solution of $Mg(NO_3)_2$ for 48 h and then quickly transfered to the sample cell. Prior to the measurements short, open and load calibration (HP4291B) and open, short compensation of the mounted cell (both instruments) were carried out.

X-ray diffraction - X-ray diffraction measurements were performed in the reflected radiation mode, using a Philips X-PERT system with a Cu Kα radiation source operating at 40 kV and 50 mA. Samples of 10 mm thickness and 25 mm length were mounted in an in-house made sample holder to keep the membranes under a constant tension. This procedure was necessary to ensure a smooth sample surface during the measurement. The measurements were carried out under ambient conditions.

RESULTS AND DISCUSSION

Prior to the microwave treatment of the precursor soaked Nafion membranes the dielectric properties of the untreated starting material Nafion 117 at the applied frequency range (2.45 GHz) were be examined. Therefore measurements were carried out on a dielectric analyser up to a maximum frequency of 2 GHz. Due to the instrumental limitations the dielectric loss at 2.45 GHz had to be estimated by extrapolating the impedance data. The membrane was measured under three different conditions, dried at 25 °C, at 85 °C and at 53 % relative humidity (rh) at 25 °C. The results are presented in Figure 2. The membrane solvent content can be described by the number λ of solvent molecules per sulfonic acid group. Under the applied conditions the water content $\lambda(H_2O)$ was 0 and 4, respectively [9].

Variations in temperature have a much lower effect on the dielectric loss of the dry membrane than variations in humidity. Uptake of about four molecules of water per sulfonic acid group results in an increase of the dielectric loss by one order of magnitude.

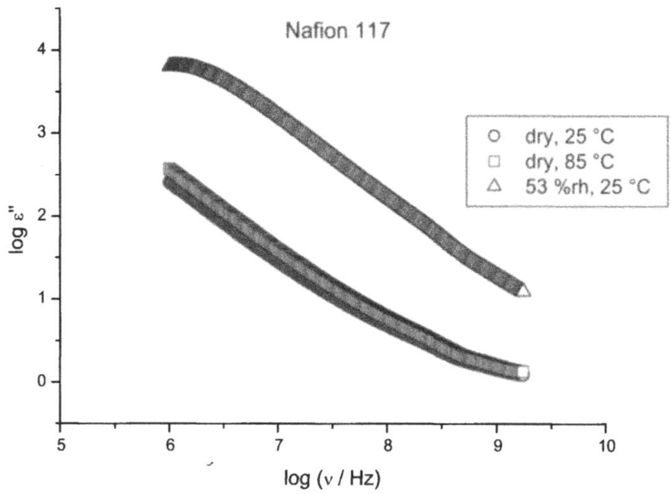

Fig. 2. ε`` *vs* frequency for untreated Nafion 117 at (O) 25 °C, dried, (□) 85 °C, dried and (Δ) 25 °C, 53 %rh.

There are two processes which contribute to the dielectric loss and therefore to the microwave heating behaviour of the membrane, conductivity and dipolar relaxation of solvent molecules and the sulfonated side chains. The dissipated power can be expressed by

$$P = \frac{\sigma_0 E^2}{1 + \omega^2 \tau^2} \quad (1)$$

$$P = \omega \varepsilon_0 \varepsilon_r^{''} E^2 \quad (2)$$

Where E and ω stand for the averaged electric field strength and angular frequency, and σ_0 and τ represent the conductivity and the averaged time between two collisions, respectively. Since the water content mainly influences the protonic conductivity the conductivity term (1) is estimated to give the prominent contribution to the heating behaviour at 2.45 GHz.

Based on these results, three different samples were treated in the microwave cavity, a hydrolysed sample with high water content, an alcoholate soaked and dried sample and an alcoholate soaked sample with high alcohol content. The temperature-time data are shown in Figure 3,4 and 5. In each case the solvent content is described by the number λ and the heating rates are listed. Due to the very high heating rate of the water saturated membrane the heating intervals were kept short. Figure 3 points out that upon repeated heating cycles the heating rate decreases indicating the membrane has lost water. The data derived from the heating experiments is listed in Table 1.

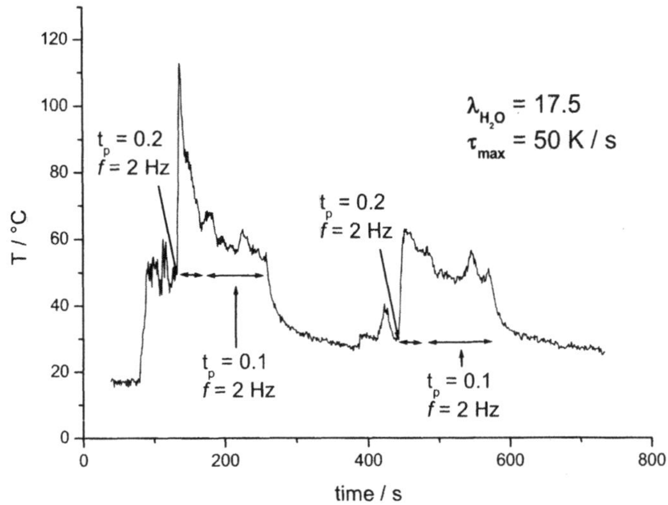

Fig. 3. Heating curve of a hydrolysed Nafion-Titanium butylate membrane containing 23 wt% titania.

Table 1. Heating rates of Nafion hybrides with different amounts of solvent

Composition	Solvent content λ $(n(solv.)/n(-SO_3H))$	Heating rate τ (K / min)
Nafion - titania/ water	17.5	$3 \cdot 10^3$
Nafion / titanium butylate	2.6	3
Nafion / titanium butylate / 1-butanol	2.6 / 16	27 (1st) 3.3 (2nd)

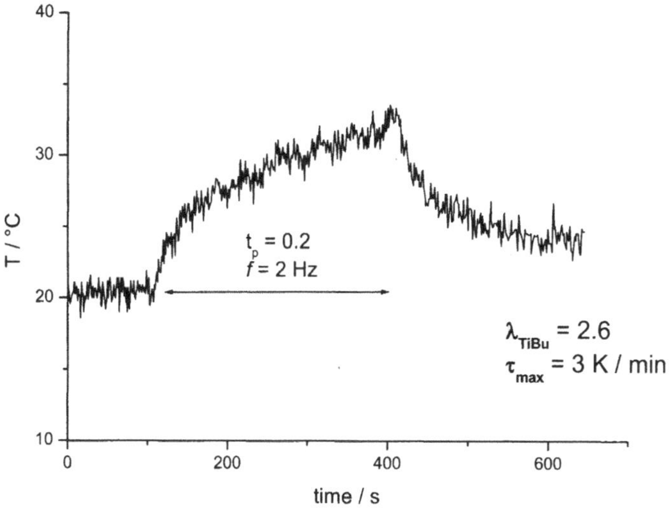

Fig. 4. Heating curve of a dried Nafion-Titanium butylate membrane containing 44 wt% titanium butylate.

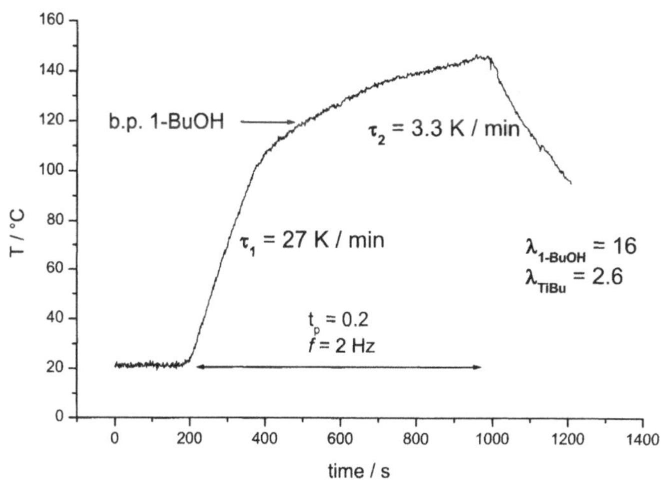

Fig. 5. Heating curve of a Nafion-titanium butylate membrane containing 28 wt% titanium butylate and 38 wt% 1-butanol.

The heating rates of the dried Nafion-titanium butylate membrane are much lower than expected. The difference between the heating rates of water and 1-butanol soaked membranes can be explained by the polarity of these solvents. In the presence of water the proton conductivity of Nafion rises significant and the dissipated power results mainly from this mechanism (according to equation 1).

The samples prepared by the two synthesis routes were dried (i.e. $\lambda(H_2O) = 0$) and analysed for their dielectric behaviour in the frequency range between 20 Hz and 2 GHz (Figure 6).

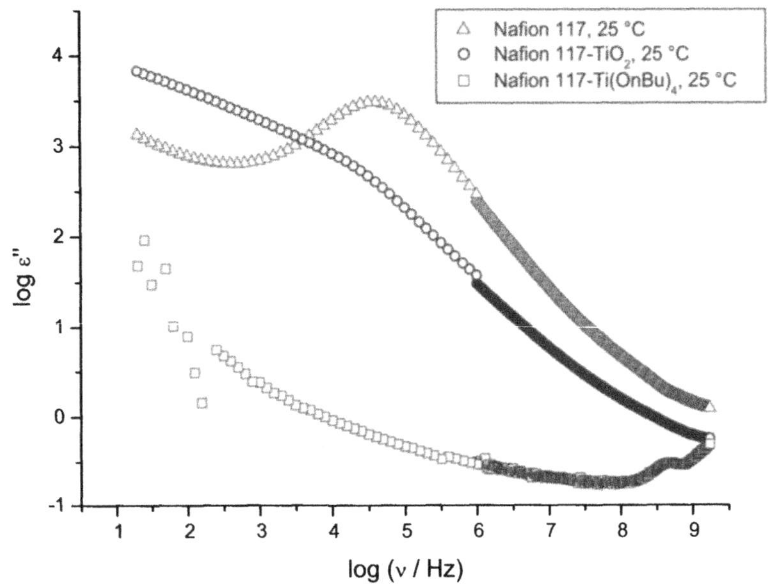

Fig. 6. ε`` *vs* frequency for dry (O) Nafion – titania (23 wt %) (□) Nafion / titanium butylate (44 wt%) and (Δ) Nafion 117 at 25 °C.

In accordance with the microwave heating experiments the extrapolated dielectric loss of the Nafion / titanium butylate hybrid is even lower than that of dry Nafion at the frequency of 2.45 GHz. Dry Nafion 117 has a relaxation with a maximum between 10^3 and 10^4 Hz, which appears attenuated in the Nafion –

titania hybrid and is absent Nafion / titanium butylate hybrid. This relaxation can be attributed to side chain movement of the sulfonic acid moieties.

The crystallinity of the samples was examined by X-ray diffractometry and compared with Nafion 117 (Figure 7).

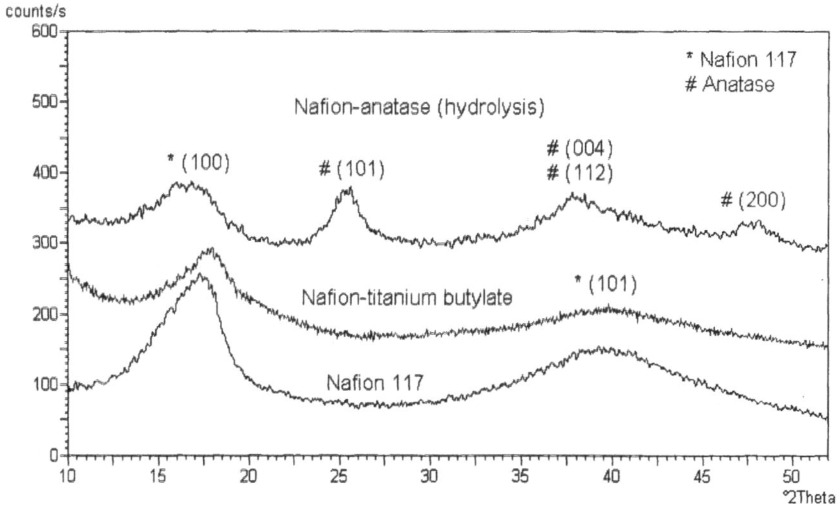

Fig. 7. X-ray patterns of Nafion 117, Nafion / titanium butylate (44 wt%) and Nafion - titania (23 wt%).

The patterns prove the formation of anatase in case of hydrolysis and the absence of an inorganic crystalline phase in case of the microwave treated Nafion / titanium butylate / 1-butanol hybrid. The pattern of Nafion 117 remains nearly unaltered after the microwave treatment (Nafion / titanium butylate). Solely a weak and broad shoulder appears besides the (100) reflection of Nafion. The grain size of the anatase phase was estimated to be between 4 and 5 nm using the Scherrer method.

Based on the results a model for the interaction of titanium alcoholate and Nafion is presented in Figure 8.

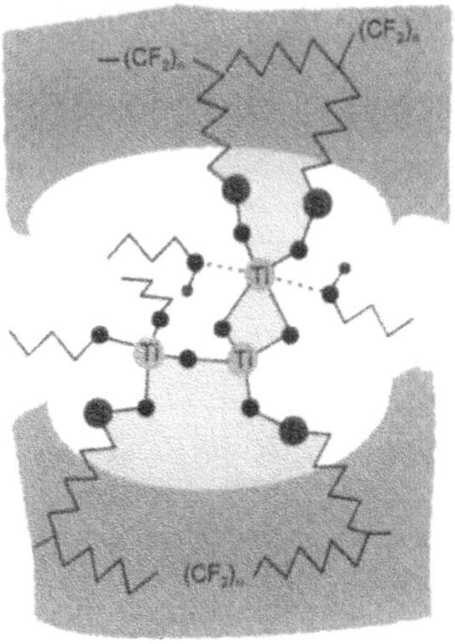

Fig. 8. Implementation of the inverse micelle structural model for the interpretation of the dielectric behaviour of the Nafion-alcoholate complex.
Ti represents the titanium atoms, S the sulfonic acid groups.

The alcoholate could act as a cross-linking reagent in the hydrophilic areas of Nafion. A condensation mechanism with participation of the sulfonic acid groups leads to the formation of an amorphous gel. Thereby the motion of side chains and also proton conduction is hindered.

CONCLUSIONS

A study of the dielectric properties of Nafion has been carried out in order to get a better understanding of the results from microwave heating experiments. It is apparent that dry Nafion is an electric insulator with low ε_r'' value. Metal alcoholates like titanium butylate even decrease ε'' due to the formation of a Nafion-alcoholate complex. Very high heating rates can be achieved using polar solvents, especially water. These solvents accumulate in the hydrophilic regions of Nafion and act as heat sources when microwave energy dissipates in the polymer. The highest heating rates of $3 \cdot 10^3$ K/min were measured for a water content of 17.5 molecules per sulfonic acid chain. With a proper choice of solvent, i.e. a thermally stable, lossy and high boiling point solvent the thermal decomposition of alcoholates in situ in the membrane could be possible.

ACKNOWLEDGEMENTS

Financial support by the Deutsche Forschungsgemeinschaft (DFG) under the contract number Wi 856/13-2 is gratefully acknowledged. The authors wish to thank C. Gainaru and Prof. E. Rössler for the high frequency dielectric measurements.

REFERENCES

[1] G.C. Farrington, H. Yang, and R. Huq, "PEO-based Polymer Electrolytes Containing Divalent cations" *Mat. Res. Soc. Symp. Proc.* **135**, 319-323 (1989).

[2] B. Baradie, C. Poinsignon, J. Y. Sanchez, Y. Piffard, G. Vitter, N. Bestaoui, D. Foscallo, A. Denoyelle, D. Delabouglise and M. Vaujany, "Thermostable ionomeric filled membrane for H2/O2 fuel cell" *J. Power Sources* **74**, 8-16 (1998).

[3] K.M. Nouel and P.S. Fedkiw, "Nafion-based composite polymer electrolyte membranes" *Electrochim. Acta* **43**, 2381-2387 (1998).

[4] V. Antonucci and A. Arico, "Polymeric membrane electrochemical cell operating at temperatures above 100 °C" *European Patent Application* **EP 0 926 754 A1**, 1-5 (1999).

[5] C. Yang, S. Srinivasan, A.S. Aricò, P. Cretì, V. Baglio and V. Antonucci, "Composite nafion/zirconium phosphate membranes for direct methanol fuel cell operation at high temperature" *Electrochemical and Solid-State Letters* **4**, A31-A34 (2001).

[6] C.-W. Schmidt, J. Petermann and M. Willert-Porada, "Prekursorpyrolyse in Emulsionen - Integrierte Herstellung von nanoskaligen Pulvern und Dispersionswerkstoffen" Werkstoffwoche **96**, 687-692 (1997).

[7] C. Gerk, C.-W. Schmidt, A. Niesenhaus and M. Willert-Porada, *in* "Microwaves: Theory and Application in Materials Processing IV", Vol. 80, p. 387-395. The American Ceramic Society, Westerville, 1997.

[8] Technical Information Nafion -"Chemical treatment of Nafion" DuPont de Nemours.

[9] T.A. Zawodzinski Jr., C. Derouin, S. Radzinski, R.J. Sherman, V.T. Smith, T.E. Springer and S. Gottesfeld,"Water Uptake by and Transport Through Nafion 117 Membranes" *Journal of the Electrochemical Society* **140** [4], 1041-1047 (1993).

MICROWAVE CATALYSIS FOR THE PRODUCTION OF FINE CHEMICALS AND PHARMACEUTICALS

G. Bond*, J. A. Gardner and J. G. McGuire
Centre for Materials Science,
University of Central Lancashire
Preston, PR1 2HE, UK

ABSTRACT

This paper describes preliminary results for comparative studies between microwave and conventionally heated heterogeneously catalysed reactions. The reactions have been chosen to be particularly pertinent to the fine chemicals and pharmaceuticals industries. These include transfer hydrogenation, esterification and acylation reactions. The experiments have been performed in both batch and flow reactors utilising a variety of microwave cavities.

INTRODUCTION

Much of the early work in the area of microwave catalysis has concentrated on reactions which are performed industrially on a large scale. These gas phase reactions include steam reforming [1] and hydrocarbon reforming [2-5]. Several authors have also studied the oxidative coupling of methane [6-12]. The overlying feature of all these studies is that there is an apparent increase in reaction rate when comparisons are made with the reactions performed using conventional heating. The majority of authors believe that the rate enhancements result from preferential heating of the reactive centre resulting in "hot spots", while there are theoretical studies that indicate that this should not be the case [13,14]. To date there is still a lack of in-situ studies to provide the experimental evidence for the observed rate enhancements in microwave gas phase catalysis.

Due to the implications of scaling-up microwave based processes the impact of microwaves is more likely to make a significant impact upon the manufacture of fine chemicals and pharmaceuticals rather than on the large scale manufacture of commodity chemicals.

This article presents some preliminary results from a number of reactions that are pertinent to the pharma and fine chemicals industries, namely, the transfer hydrogenation of nitro groups to produce amines, the esterification of long chain acids and the acylation of aromatics. These reactions have been chosen to utilise a range of solid acid catalysts covering both supported metal and oxidic catalysts. The effects of microwave heating compared to conventional heating are more subtle but none the less significant.

EXPERIMENTAL

Microwave Hardware

The Microwave experiments described in this paper have been performed in two types of focused microwave cavities. The first, a microwave single mode cavity (Industrial Microwave Systems) was fitted with a 1 kW generator. The waveguide section contained consisted of a circulator, test section, into which the sample was mounted, and a double plunge tuner. Temperature measurement was achieved using an infrared pyrometer and control was achieved using in-house software. There are problems associated with the use of surface sensitive probes, such as pyrometers, to measure temperatures in microwave heated experiments. Internal temperature gradients or "hot spot" can lead to anomalous results being obtained [15]. In the studies reported in this paper this has been overcome by performing the experiments at the boiling point of the liquid phase. This type of system has been described in greater detail elsewhere [16]. The second microwave system was a Discover (CEM) focused microwave cavity. The instrument is fitted with a 300 W microwave generator. The unique circular cavity permits the microwave energy to be focused onto the sample, with the sample being irradiated from 360 degrees. This system has added advantage over the standard single mode device in terms of ease of use, as the system is self tuning, ensuring the sample absorbs the optimum amount of energy at all times. The Discover can accommodate a range of vessels from a 125 ml round bottomed flask to a 10 ml pressure vessel capable of withstanding 500 psi. Infrared temperature measurement allows the reaction temperature to be controlled, pressure measurement is also available. Samples are stirred using a magnetic stirrer and follower.

Transfer Hydrogenation

Nitrobenzene (0.6 g) and cyclohexene (4.8 g) were reacted over a 10% Pd/C catalyst (50 mg) according to reaction scheme 1, using either isopropyl alcohol (IPA) or cyclohexane (30 ml) as the solvent. The reactions were performed under reflux in 125 ml round bottom flasks fitted with reflux condensers. Microwave experiments used the Discovery, while the conventionally heated experiments were performed using an electric heating mantle. The catalysts were only added once the system started to reflux and this was taken as being time t = 0, in this way any effect due to difference in the time taken for the two systems to attain reflux temperature was eliminated. Samples were taken periodically and the conversion of nitrobenzene to aniline was determined using gas liquid chromatography.

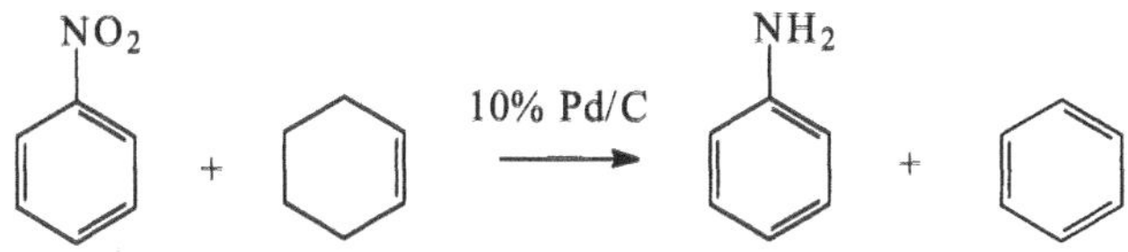

Scheme 1 Transfer Hydrogenation of Nitrobenzene

Esterification of Myristic Acid with Isopropyl Alcohol

A mixture of myristic acid and IPA in a 1:20 mole ratio was pumped using a peristaltic pump over a fixed bed of catalyst (approx 1.0 g) contained within a silica reaction vessel. The reactor was mounted in the test section of the single mode cavity. Comparative experiments were performed using a wire electric furnace and associated temperature controller to stimulate the reactions. The reaction mixture was pumped over the catalyst at a rate of 1 ml min^{-1}, while the catalyst bed was maintained at the boiling point of the IPA (83°C). The reaction mixture exiting the reactor was monitored and analysed using gas liquid chromatography. Both microwave and conventionally heated reactions were studied over a range of solid acid catalysts.

Acylation of Aromatic Hydrocarbons

The acylation of aromatic hydrocarbons has been studied using two experimental procedures. The acylation of anisole with decanoic acid has been carried out batch wise in the Discover system, while the acylation of toluene with octanoic acid was studied in a flow reactor using the single mode cavity to irradiate the catalyst bed. Comparative studies using conventional heating have been performed for both studies, an isomantle being used to provide conventional heat for the batch reactions and an electrical furnace to heat the catalyst bed in the case of the flow experiments.

The acylation of anisole was carried out under reflux at atmospheric pressure (154°C). The reaction mixture consisted of anisole (61.1 g) and decanoic acid (3.1 g), these were stirred with (1 g) of catalyst. A number of acid catalysts have been investigated, however in this review paper it is only possible to report results from zeolite Y which had previously been exchanged with lanthanum.

Acylation of toluene in the flow reactor was carried out using a reaction mixture consisting of toluene (51.9 g) and octanoic acid (0.26 g) the reaction mixture was recirculated over the catalyst bed using a pumping rate of 1 ml min^{-1}. Samples of the reaction mixture were taken periodically for analysis using gas liquid chromatography. The catalyst bed was maintained at a temperature of 110 °C throughout the experiment.

RESULTS AND DISCUSSION

Transfer Hydrogenation

Transfer hydrogenation is a particularly interesting reaction for use in conjunction with microwave heating as it permits hydrogenation reactions to be carried out without the use of molecular hydrogen. With the potential of arcing during microwave heating there is an obvious reluctance to use potentially explosive molecular hydrogen.

The conversion as a function of time for both the microwave and the conventionally heated reactions are contained in Figure 1. For both conventionally and microwave heated experiments IPA proved to be the superior solvent. In the cases of the conventionally heated experiments the rates of reaction approach zero. In contrast the microwave heated reactions continue to proceed, albeit slowly in the case of the reaction performed in cyclohexane. Considering the initial rates of reaction there is very little difference between comparative conventional and microwave reactions.

Amines are notorious for poisoning catalytic reactions; therefore, it is likely that the reactions are halted in the conventionally heated reactions due to product inhibition. The microwave reactions may continue to proceed as a result of microwaves interacting with the surface dipole of the adsorbed aniline resulting in enhanced desorption. Previous studies have shown that microwaves appear to be efficient in aiding the desorption of adsorbed molecules from surfaces.

The reasoning behind using cyclohexane as a solvent was to use a microwave transparent solvent such that more of the microwave energy was available to interact with the catalyst. Cyclohexane has the added advantage of having a similar boiling point to IPA such that the effects due to temperature can be minimalised. IPA, being a more polar solvent, is likely to be more efficient in removing the aniline from the surface of the catalyst which adds weight to the argument that the loss of catalytic activity is a result of product inhibition. It is more difficult to postulate if using a microwave transparent solvent allows more of the microwave energy to affect the product desorption. However, when cyclohexane is used as a solvent the conversion is increased by a factor of 2.1 whereas when IPA was used the increase was only by a factor of 1.3. However, several repeat reactions would have to be repeated in order to determine if this increase is significant.

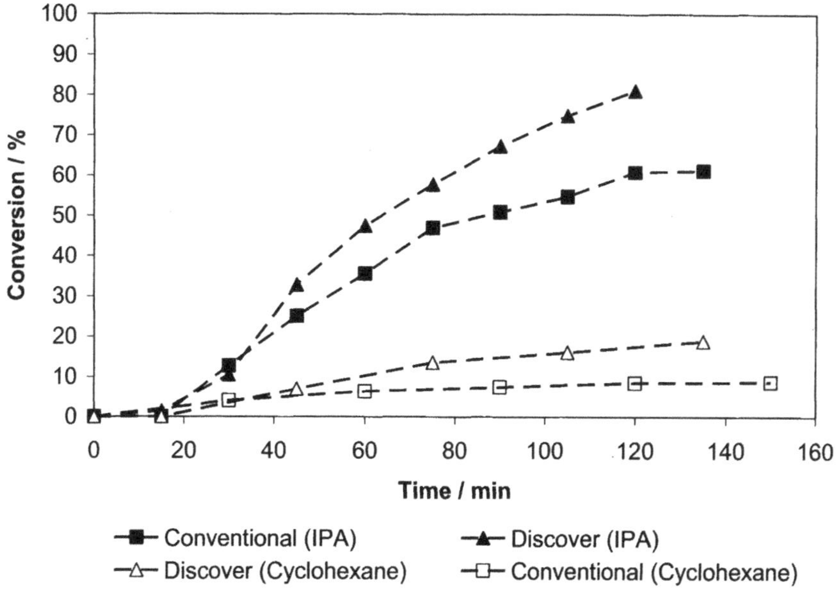

Figure 1 Conversion of nitrobenzene to aniline as a function of time for both microwave and conventionally heated experiments.

Esterification of Myristic Acid with Isopropyl Alcohol

The conversion to isopropyl myristate over both zeolite HY and Amberlyst 15 is shown in Figure 2. Both catalysts exhibit a marked increase in activity when operated under conditions of microwave irradiation.

The increased activity for the microwave experiments could originate from "hot spots". However, an alternative hypothesis may be that the microwaves dehydrate the surface of the catalyst, this is likely to occur as microwaves are known to be particularly effective for the dehydration of zeolites and other ceramic materials [17]. Esterification reactions are notorious for the water, produced as a side product, hydrolyzing the ester resulting in reduced conversions. Corma et al [18] has shown that the use of hydrophobic supports can help to minimize the back reaction. Therefore, if microwaves can maintain the surface of the catalyst in a state of dehydration then there will not be a sufficiently high concentration of water at the catalyst surface to partake in the hydrolysis of the ester. The result of this will be increased conversion with the microwave experiments.

The Amberlyst 15 catalyst when operated under conditions of microwave irradiation can be seen to increase its activity in the initial stages of the reaction. This may be due to the catalyst being insufficiently dehydrated, during the activation process, to attain optimum activity; hence, further dehydration by the microwaves results in increased catalytic activity.

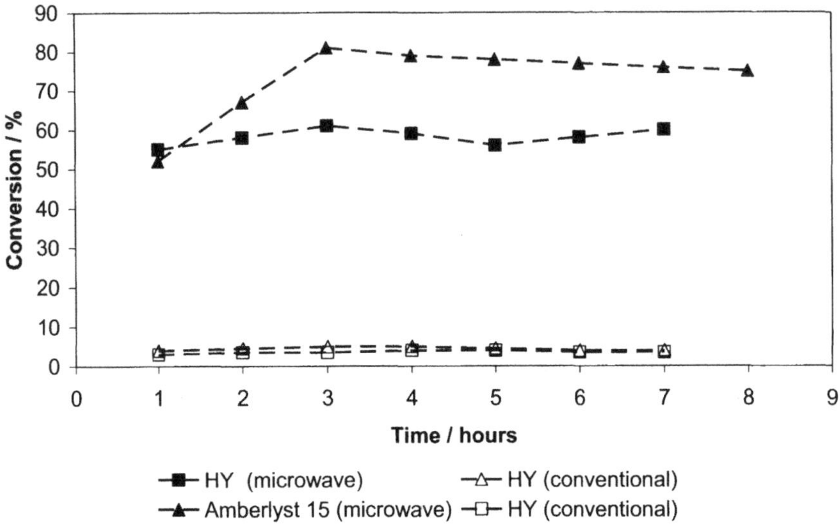

Figure 2 Conversion as a function of time for the production of isopropylmyristate over HY and Amberlyst 15 catalysts

Acylation of Aromatic Hydrocarbons

Results from the acylation of both anisole (Figure 3) and toluene (Figure 4) both confirm that microwave heating has a beneficial effect on the conversion. The origin of this enhancement could either be a result of "hot spots" which are not detected by the pyrometer, or as a result of microwaves effectively desorbing water from around the active sites on the catalyst surface. Poor conversion in acylation reactions as in esterification reaction can be attributed to water hydrolysing the product. Removing the water from the surface catalyst limits the possibility of it partaking in the hydrolysis of the product.

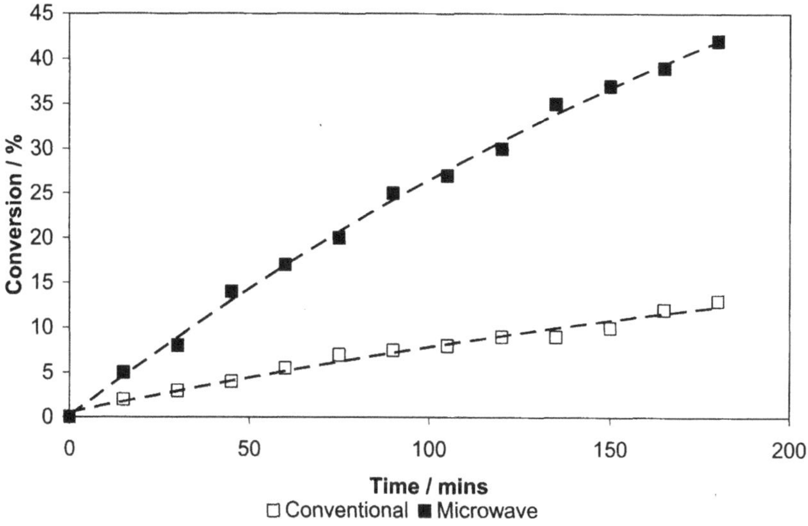

Figure 3 Acylation of anisole with decanoic acid performed under reflux.

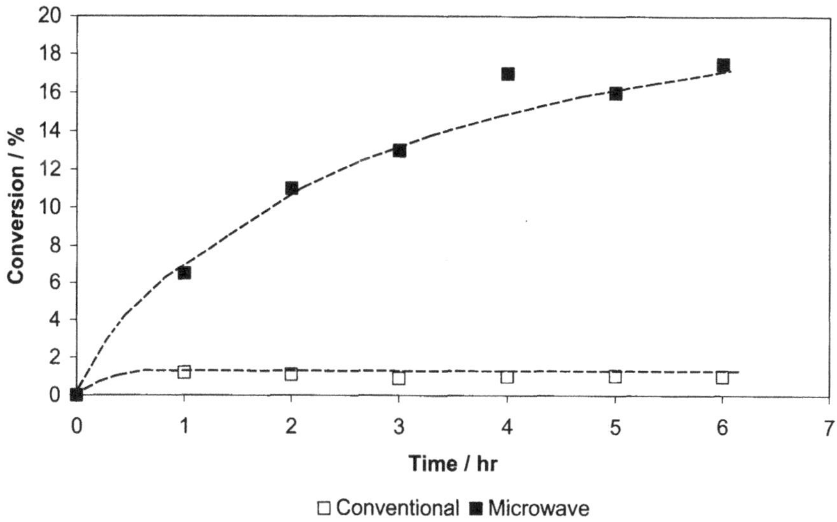

Figure 4 Acylation of toluene with octanoic acid in the flow reactor.

CONCLUSIONS

Many of the results presented in this paper are only of a preliminary nature. There appears to be evidence that the action of microwaves on catalytic reactions may be more subtle than a simple generation of "hot spots". This is particularly evident in heterogeneously catalysed reactions where the reactants are present in the liquid phase.

ACKNOWLEDGEMENTS

The authors would like to thank CEM (Microwave Technology) for their continued financial support for work in areas of microwave chemistry.

REFERENCES

1. C.J. Bool Ph.D. Thesis, University of Hull, UK 1998.
2. G. Roussy, S.Hilaire, J.M. Thiébaut, G.Maire, F. Garin, and S. Ringler, Appl. Catal. A **156** (1997) 167.
3. L. Seyfried, F. Garin, G. Maire, J.M. Thiébaut, and G. Roussy, J. Catal. **148** (1994) 281.
4. J.M. Thiébaut, G. Roussy, and M. Medjram, Catal. Lett. **21** (1993) 133.
5. J.M. Thiébaut, G. Roussy, M. Medjram, L. Seyfried, F. Garin and G. Maire, J. Chim. Phys. **89** (1992) 1427.
6. C.J. Lui, R. Mallinson, and L. Lobban, Appl. Catal. A Gen. **178** (1999) 17.
7. G. Roussy, E. Marchal, J.M. Thiébaut, A. Kiennemann G.Maire, Fuel Proc Tech. **50** (1997) 261
8. C. Chen, P. Hong, S. Dai and J. Kan, J. Chem. Soc. Faraday Trans. **91** (1995) 1179.
9. G. Roussy, J.M. Thiébaut, M. Souri, E. Marchal, A. Kiennemann G.Maire, Catal. Today **21** (1994) 349.
10. G. Bond R.B. Moyes and D.A. Whan, Catal. Today **17** (1993) 427.
11. G. Roussy, E. Marchal, J.M. Thiébaut, A. Kiennemann G.Maire, Fuel Proc Tech. **50** (1997) 261
12. S.L. Suib and R.P. Zerger, J. Catal. **139** (1993) 383.
13. W.L. Perry, D.W. Cooke and J.D. Katz, Catal Lett. **47** (1997) 1.
14. W.L. Perry, J.D. Katz, D. Rees, M.T. Paffet, and A.K. Datye, J. Catal. **171** (1997) 431.
15. L.P. Kok, M.E. Boon, and H.M. Smid, Scanning 15 (1993) 100.
16. G. Bond, R.B. Moyes, I. Theaker and D.A. Whan, Topics in Catalysis **1** (1994) 177.
17. G. Roussy, A. Zoulalian, M. Charreyre and J.M. Thiebaut, J. Phys. Chem. **88** (1984) 5702.
18. A. Corma, H. Garcia, S. Iborra and J. Primo, J. Catal. **120** (1989) 78.

APPLICATION OF MIRROR-CONFINEMENT-TYPE ECR PLASMA SPUTTERING AND MICROWAVE RADIATION TO THE LOW TEMPERATURE SYNTHESIS OF CRYSTALLIZED HIGH-K MATERIAL THIN FILMS

*S. Miyake, T. Matsumoto, A. Niinou
Joining and Welding Research Institute, Osaka University
11-1 Mihogaoka, Ibaraki, Osaka 567-0047, Japan

K. Numata, H. Saito
Kanagawa High-Technology Foundation
Kawasaki, Kanagawa 213-0012, Japan

S. Baba
Institute of Mechanical Systems Engineering
National Institute of Advanced Industrial Science and Technology
Tsukuba Science City, Ibaraki 305-8564, Japan

ABSTRACT

For low temperature preparation of crystallized high-k perovskite oxide thin films of $SrTiO_3$ (STO) and $(Ba,Sr)TiO_3$ (BST) a mirror-confinement-type electron cyclotron (MCECR) plasma sputtering system was adopted and mm-wave irradiation to these films were given as a new post-annealing process. STO films on Si substrates were amorphous-like, while those on $Pt/Ti/SiO_2/Si$ multilayer substrates could be crystallized remarkably even at as–deposited condition. At negative bias voltages to the substrate prepared films became amorphous even on the multilayer substrate and at positive ones they were crystallized. Measurement of optical emissions near the substrate surface indicated that at positive bias to the substrate mainly the neutral gas species were remarkably excited, so that excited neutral oxygen by electron collision was expected to play an important role to the film crystallization. As a post-annealing process of these films 28 GHz mm-wave radiations were applied and could lower the crystallization temperature of amorphous films compared with the conventional annealing by an electric furnace. By the mm-wave annealing crystallization temperature of STO and BST films on Si substrates indicated a low value of 573 K and 623 K, respectively. While STO films on multilayer substrates, which were sufficiently crystallized even at as-deposited condition, mm-wave irradiation drastically improved the film electrical properties. The dielectric constant of these STO films increased to a value of about 260 by the mm-wave annealing at 573 K.

INTRODUCTION

High-k materials of the ABO_3-type perovskite such as $SrTiO_3$ (STO), $BaTiO_3$ (BTO), (Ba, Sr)TiO_3 (BST) and/or (Sr, Bi)Ta_2O_9 (SBT) have been considered as candidate dielectric materials for capacitors in the next-generation dynamic random access memory (DRAM) with 1 Gbit density or higher, as well as on monolithic microwave integrated circuits (MMIC), due to their high dielectric constant.

Many studies have been performed on preparing these films but it is usually very difficult to obtain well-crystallized, high quality films at a substrate temperature lower than 900 K[1-2].Generally speaking low temperature film deposition is important from the viewpoint of preventing inter-diffusion of particles between the film and electrode as well as for utilization of thermally unstable substrates such as polymer sheets and so on.

To overcome this problem a study has been made on applying a mirror-confinement-type electron cyclotron resonance (MCECR) plasma, since it has a possibility of obtaining a high temperature and high density plasma at a vey low pressure environment, in which sufficient number of highly excited neutral and/or ionized particles can be prepared. It is expected that they will contribute to the crystallization in exchange of increasing the substrate temperature. Based on this idea we have studied STO and BST film deposition without substrate heating in a low-pressure environment of 1 - 2 x 10^{-2} Pa, and applied mm-wave radiations for low temperature post-annealing of the films[3-6].

EXPERIMENTAL PROCEDURE

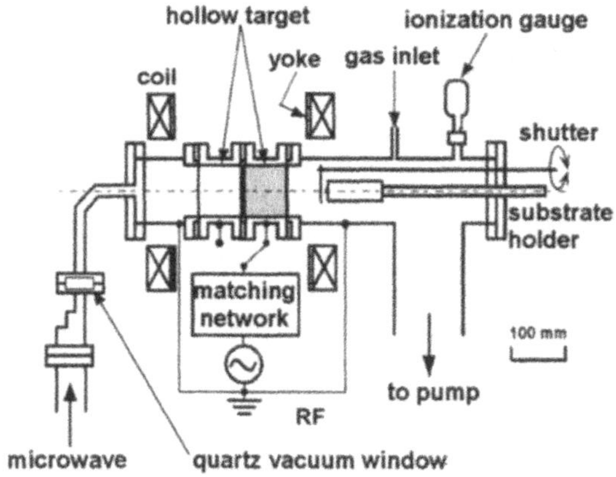

Fig.1. Schematic diagram of experimental apparatus.

Schematic diagram of the experimental apparatus is shown in Fig. 1 and the detail of the experimental procedures has been described elsewhere[3]. A microwave power of 2.45 GHz is applied in the ECR plasma production and a cylindrical sputtering target is set at the center of the mirror to which an RF power is supplied with a frequency of 13.56 MHz via a matching network. In the film deposition the microwave and RF powers were fixed at 200 W and 100 W, respectively.

For the measurement of plasma parameters an axially movable single probe was used with a Butterworth-type low-pass filter to supress RF coupling to the probe. The optical emission spectra (OES) were measured in the axial direction from the right end of the vacuum chamber.

Typically films were deposited for 60 min at 2.7 x 10^{-2} Pa in Ar and Ar/O_2 mixed gas plasmas. The substrates used were p-type Si(100) wafers of 20 mm x 20 mm x 500 μm in size with an electrical resistivity of 0.2 to 0.5 Ω.cm. The substrate was set at a position of 100 mm apart from the center of the mirror in the axial direction and its temperature measured by a thermocouple during deposition for 60 min was kept below 450 K at P_μ = 200 W

RESULTS AND DISCUSSION

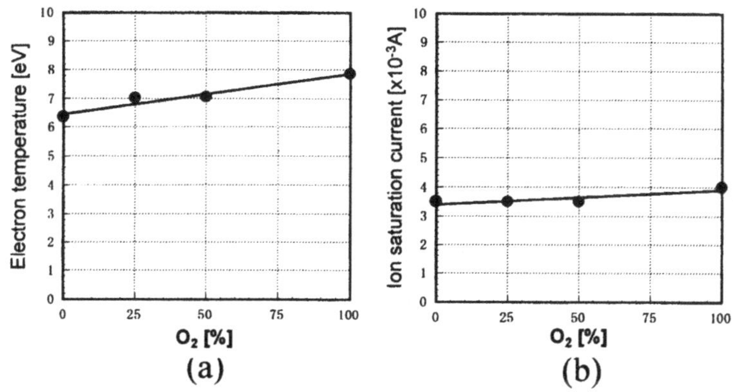

Fig.2. Variation of (a) T_e and (b) ion saturation current at substrate position versus O_2 partial pressure for P_μ = 100 W.

Figure 2 shows change of electron temperature and ion saturation current to the Langmuir probe near the substrate position. We can see that they do not change appreciably by varying the mixing rate of O_2 to Ar gas. The ion saturation current corresponds to the plasma density and a high density plasma of about 10^{11} cm^{-3} was successfully obtained due to the efficient mirror confinement of electrons between the 2 ECR zones in the axial direction [7].

Figure 3 shows XRD patterns of prepared STO films by changing the gas species and substrate materials. In the figure the substrate was set at floating potential of about +20 V. The data (a) was obtained by an Ar plasma on a Si

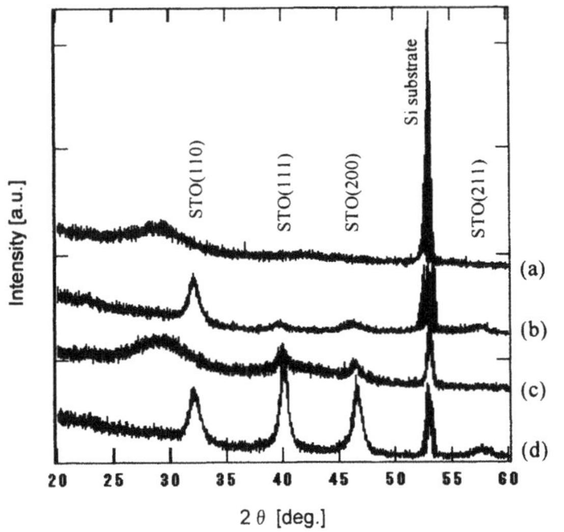

Fig. 3 XRD patterns of STO films; (a) Si substrates & Ar plasma, (b) Si & Ar/O_2, (c) Pt-multilayer & Ar, (d) Pt-multilayer & Ar/O_2

substrate, (b) by an Ar plasma on a Pt-multilayer substrate, (c) by an Ar/O_2 mixture on a Si and (d) by an Ar/O_2 mixture on a Pt-multilayer, respectively. For the same substrate the Ar/O_2 plasma is found to enhance crystallization, while for the same gas species the Pt multilayer substrate brings better crystallization. It is demonstrated that in (d) the as-deposited film is sufficiently crystallized at a substrate temperature below 450 K.

Figure 4 shows XRD patterns of STO films for various bias voltages to the Pt-multilayer substrate, where the influence of ion bombardment is clearly observed. When the substrate is set at negative potential, films become amorphous and at positive ones they are crystallized well. This feature was also observed in the case of BST film preparation[6]. We note here the data of floating potential and +20 V bias show a similar patterns as expected.

Variation of optical emissions near the substrate was measured by changing the substrate DC bias voltages. The emission intensity from excited neutral particles increased at positive bias voltages as shown in Fig. 5. While the intensity of positive ions was found to be deccreased at relatively high positive bias voltages over 40 V due to the reduction of plasma density, which was estimated from the probe measurement near the substrate surface, because the substrate had a moderate size to influence the electron flow from the plasma. This result indicates that excited state oxygen neutral species can contribute remarkably to the film crystallization.

Next a post-annealing process of these films using 28 GHz mm-wave radiation was applied and it decreased the crystal- lization temperature of amorphous films compa- red with the case by the conventional heating method. As shown in Fig. 6 amorphous STO films on Si substrates could be crystallized by the mm= wave irradiation for 120 min at a low temperature of 523 K. In the

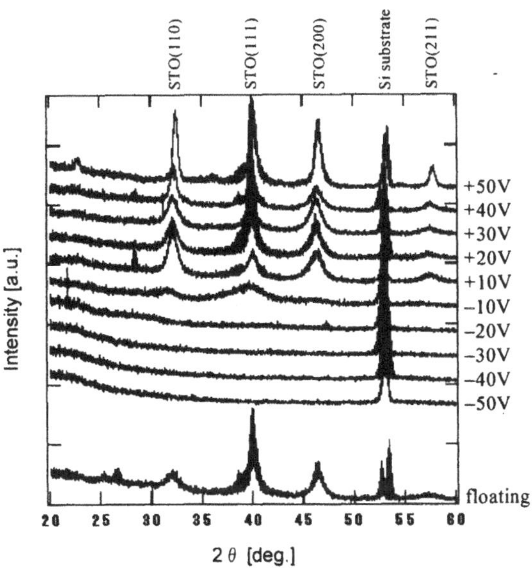

Fig. 4. XRD patterns of STO films prepared at various bias voltages to the multilayer substrate.

case of BST films on Si substrates they were also amorphous at as-deposited condition, but could be crystallized at 673 K by the mm-wave for 60min., whereas in the conventional method it was found to be 923 K[8].

On the influence of mm-wave irradiation to the film electrical properties a very attractive result was obtained as shown in Fig. 7. STO films on Pt-multilayer substrates were crystallized well at as-deposited condition (Fig. 3), and mm-wave irradiation to these films drastically improved the dielectric constant. Indeed at as-deposited condition it shows a value of about 70 and increases to a value of about 260, which is

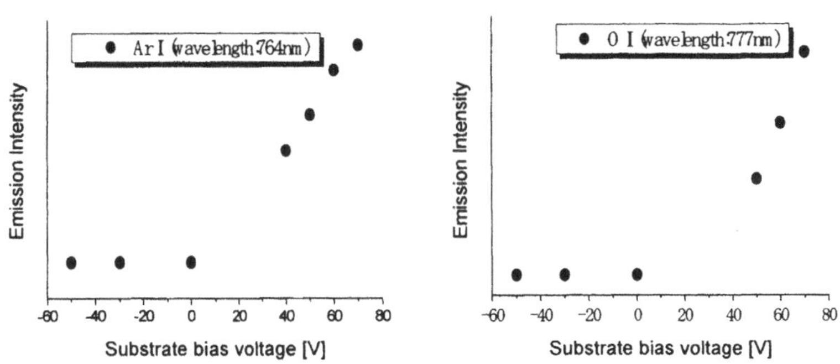

Fig. 5. Ar I and O I line emissions near the substrate surface versus bias voltages.

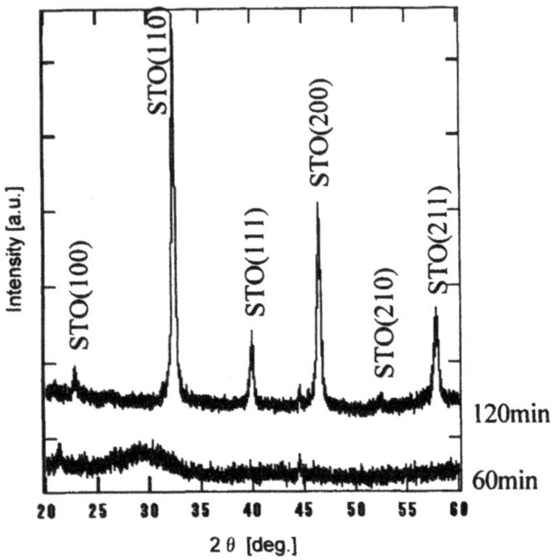

Fig. 6. XRD patterns of STO films annealed by mm-wave radiation for 60 and 120 minutes at 523 K.

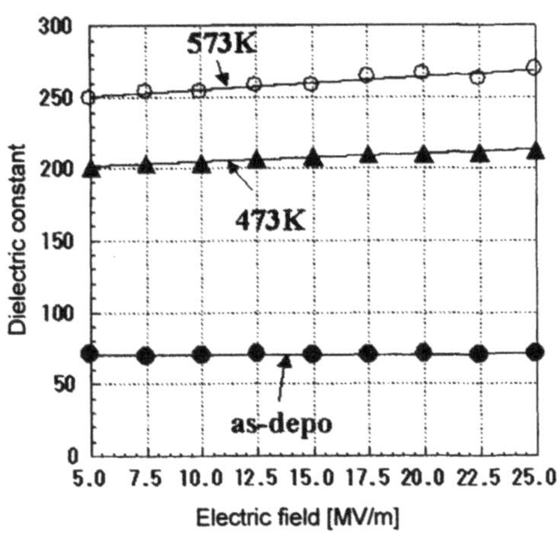

Fig. 7. Dielectric constant of STO films annealed by mm-wave radiation at 2 temperatures.

comparable with the bulk value, by the mm-wave annealing at 573 K.

As for the reason why the mm-wave irradiation could improve the dielectric constant of STO films, it has not been clarifed at all theoretically. Even experimentally our data is the only one. While on the sintering of bulk ceramics using microwave and/or mm-wave radiations, various ceramics materials can be sintered at a lower temperature than those by an electric fur- nace[9-12]. Among them it is demonstrated that the mm-wave field induces a ponderamotive force to the vacancy of ionic ceramics near the grain boundary or surface[9], by which enhancement of particle diffusion brings about reduction of the sintering temperature. This force does not appear at all in the conventional thermal heating process. We have expected that this kind of an effect would also influence the behavior of vacancy or defect of oxide thin films and improve the dielectric constant as observed in this research.

CONCLUSION

For the low temperature preparation of crystallized high-k oxide thin films of STO and BST, an MCECR plasma sputtering system was adopted and mm-wave irradiation to these films was given as a new post-annealing process.

STO films prepared on Si substrates at floating potential of +20 V were amorphous-like, while those on Pt/Ti/SiO_2/Si multilayer substrates could be crystallized sufficiently even at as–deposited condition at a substrate temperature below 450 K A negative bias to the substrates induced amorphization of prepared films even on Pt-multilayer substrates.

Measurement of optical emission near the substrate surface indicated that at positive bias to the substrate mainly the neutral gas species were remarkably excited, so that excited neutral oxygen particles were expected to play an important role to the film crysallization.

As a post-annealing process of these films 28 GHz mm-wave radiations were applied. In the case of STO films on Si substrates they were amorphous at as-deposited condition, but they could be crystallized well at a low temperature of 523 K by the mm-wave irradiation for 120 min. While for BST films they could be crystallized at 673 K whereas it was 923 K by the conventional annealing.

On the film electrical properties we could find that the dielectric constant of STO films on Pt-multilayer substrates indicated a value of about 70 at as-deposited condition and it increased to a value of about 260 by the mm-wave annealing at 573 K.

REFERENCES

[1]S. H. Park, J. Won, K. S. Lee, J. S. Choi and C. S. Park, Jpn. J. Appl. Phys., **35** (1996) 5757.

[2]S. Kim and S. Hishita, Thin Solid Films, **282** (1996) 449.

[3] S. Baba and S. Miyake, Jpn. J. Appl. Phys., **39** (2000) 4945.

[4] S. Baba, T. Ueno and S. Miyake, Trans. JWRI, **29** (2000) 15.

[5] S. Baba, K. Numata, H. Saito, M. Kumagai, T. Ueno, B. Kyoh and S. Miyake,

[6] S. Baba, K. Numata and S. Miyake, Sci. Technol. Adv. Mater., **1** (2000) 211.

[7] S. Miyake, W. Chen and T. Ariyasu, Jpn. J. Appl. Phys. **29** (1990) 2491.

[8] S. Miyake, *Proc. 3rd Int. Conf. Surf. Eng., (Chendgu, China, Oct. 11 -13, 2002).*

[9]K. I. Rybakov and V. E. Semenov, Phys. Rev., **B49** (1994) 64.

[10]S. A. Freeman, J. H. Booske and R. F. Cooper, Phys. Rev., Lett., **74** (1995) 2042.

[11]S. Miyake, Y. Setsuhara, S. Kinoshita, S. Sano, M. Kamai, T. Ohmae and N. Abe, Ceramic Trans. **80** (1997) 313.

[12]Y. V. Bykov, K. I. Rybakov and V. E. Semenov, J. Phys. D: Appl. Phys., **34** (2001) R55.

MICOWAVE PLASMA-ASSISTED PROCESSING OF NANOSCALED SUPPORTED METAL OXIDE CATALYSTS

A. Dittmar, H. Kosslick. A. Brückner, D.L. Hoang
Institute of Applied Chemistry (ACA),
D-12489 Berlin-Adlershof, Richard-Willstätter-Str. 12,
Germany

ABSTRACT

Microwave plasma enhanced chemical vapour deposition (MPECVD) has been used for the preparation of chromia supported zirconia and lanthanum doted zirconia catalysts. The influence of the plasma parameters on the dispersity of supported chromia species has been studied. The catalytic properties of obtained materials were investigated using the dehydrocyclization as test reaction. Results confirm the potential of the microwave plasma technique for the preparation of advanced catalytic materials.

INTRODUCTION

The acceleration of catalyst preparation and the improvement of catalytic properties of materials by application of microwave processes is an important challenge in the area of catalysis research[1,2]. The catalytic performance of materials is often substantially influenced by the preparation procedure. Interesting catalytic systems like supported metal and metal oxide nanocluster on different oxidic supports like ZrO_2, SiO_2 or TiO_2, mesoporous and microporous zeolitic molecular sieves supported with acidic or isolated redox active sites, or basic systems like hydrotalcites are prepared via sol-gel processes or different precipitation procedures involving hydrolysis/condensation steps in aqueous media. These systems are good absorbers for microwave radiation. Besides microwave-assisted plasma-enhanced chemical vapour deposition (MECVD) should allow the preparation of supported catalysts with highly dispersed redox active catalytic species. This technique has become an interesting field during the last few years[1,3,4]. Mainly two types of plasma will be applicable, equilibrium plasma at high temperatures and atmospheric pressure as well as non-equilibrium plasma at low temperatures and pressures. Due to the low temperature of the non-equilibrium plasma compared to other preparation methods for catalysts like incipient wetness, which require high calcination temperatures, small particles and narrow particle size distribution are expected by application of the method.

The aim of this work is to test the use of the MPECVD technique for the preparation of supported catalysts. The influence of the plasma parameters (treatment time, oxygen pressure/ flow and microwave power) on the dispersity of the supported chromium species and the catalytic properties of the obtained materials are investigated. The dehydrocyclization of octane to aromatics is used as test reaction[5].

THE MICROWAVE PLASMA

The microwave plasma is characterized by a high degree of ionisation and a very high electron density. The plasma-assisted preparation is carried out far from the thermodynamic equilibrium. There exists a high temperature difference between the “hot” electrons and the heavy particles. The formed molecule ions and electrons interact very strong with the changing field. Because of their low mass the velocity of electrons and, hence, their kinetic energy is distinctly increased. In contrast the energy of the heavy molecule ions is only slightly increased. Therefore, the temperature of the plasma remains low although it contains energetic electrons. The energy of the electrons is additionally increased by collisions with other plasma compartments. Electron temperatures of about 10 electron volts are typically obtained achieves which is sufficient for the cleavage of chemical bonds.

The conventional microwave plasma operating at a pressure of about 1 mbar is a cold non-equilibrium discharge, where the heavy parts like ions have a temperature slightly above ambient temperature. Electrons are able to take up energy from the electromagnetic field up to ten electron volts, which is high enough to break chemical bonds[6].

EXPERIMENTAL

Materials

Zirconia (ZrO_2) and lanthanum doted zirconia (La_2O_3-ZrO_2) were used as supports. The zirconia was prepared by calcination of commercially available zirconium hydroxide (MEL). The specific surface area was 33 m^2/g. Lanthanum doted zirconia was prepared by co-precipitation[7]. The obtained precipitate was calcined at 600°C. The specific surface area was 89 m^2/g. Chromium acetylacetonate, $Cr(acac)_3$, (Merck, p.a.) was used as catalytic precursor compound.

Catalyst preparation

For the MPECVD preparation of chromium(III) oxide supported zirconia and lanthanum doted zirconia. A vacuum recipient with a volume of about 50 L was used and pumped down to a base pressure of 10^{-5} mbar. A flask of quartz in its centre is mounted on a metallic rod. A 2,46 GHz magnetron together with a power supply situated on the top flange of the chamber was used for the generation of microwaves. The catalyst support powders which were slightly grinded and mixed together with $Cr(acac)_3$ are placed in the flask. The plasma gas flow was adjusted using a MKS mass flow controller and the pressure was measured using a baratron gauge. Oxygen was used as feed gas and the microwave power was set between 100 and 500 W. After MPECVD treatment and subsequent reduction in hydrogen, the samples were removed, characterized and catalytically tested in order to study the influence of the different treatments on the state of the catalysts.

Characterization

The catalysts and supports were characterized using TEM, XRD, IR, EDX, XPS, and nitrogen adsorption. The catalytic performance of catalysts was studied in the dehydrocyclization of octane at a temperature of 823 K. Prior to testing the catalyst was calcined under hydrogen atmosphere in the catalytic test apparatus. Then the catalyst was exposed to a flow of a hydrogen/octane mixture at a flow rate of 8 L/h. The catalyst amount was 100 mg. The reaction products were analysed with a Varian 3800 gas chromatograph using a CP-SIL capillary column.

RESULTS AND DISCUSSION

Chromium oxide supported catalysts

Decomposition of chromium acetylacetonate in the presence of ZrO_2 and La_2O_3-ZrO_2 leads to the precipitation of chromium oxide particles onto the inorganic supports. The chromium oxide species are highly dispersed as revealed by EDX and XPS analysis. The XPS spectra showed that the chromium remains in the valence state (III) under preparation conditions although the organic precursor has been treated in the oxygen plasma. Formation of hazardous chromate species has been avoided. However, chromium(III) oxide species could not be imaged by TEM. Therefore, the dispersity of the catalytically active chromium has been characterized by XPS using the intensity ratio of the chromium (2p3/2) and Zr (3d) peaks. The determination of the particle sizes of deposited metal oxides on oxidic supports is a generally problem which could not be solved so far. Different amounts ranging from 0.5 to 3 wt.-% could be deposited. Catalysts were also examined by XRD technique. Reflections of chromium oxide, which is typically found with less ordered nano-sized metal oxide species, were not observed in the diffraction patterns.

The decomposition of the precursor in the oxygen microwave plasma was analysed ex situ using IR spectroscopy and TG/DTA analysis. In the spectral range from 1200 to 1800 cm^{-1} appear the symmetric and antisymmetric stretching vibrations of the carbonyl groups and of the carbon-carbon double bonds in the conjugated chelate ring of the acatylacetonate. The intensities of the corresponding bands decrease distinctly after plasma treatment. The decomposition of the organic precursor starts immediately in the plasma. The organics are nearly completely removed. This could be confirmed by the combined TG/DTA measurements. The crystal structure of the supports is maintained as shown by the X-ray diffraction patterns.

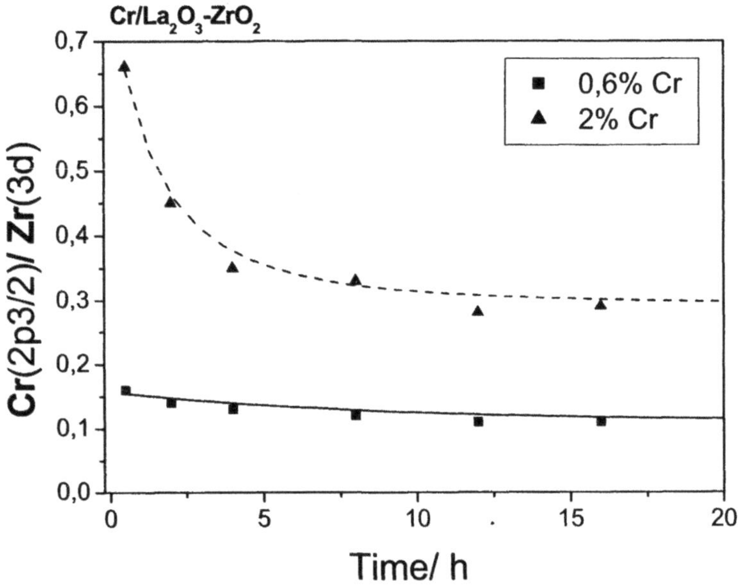

Fig. 1 Influence of the microwave plasma preparation time on the dispersity (characterized by XPS Cr/Zr intensity ratio) of chromia supported on La_2O_3-ZrO_2

The influence of the plasma parameters on the dispersity of the supported chromium oxide species has been characterized by XPS. Growing chromium (2p3/2) to Zr (3d) ratio indicates increasing dispersity with one sample.

A decrease of the Cr/Zr XPS intensity ratio is observed with increasing time of plasma treatment. The decrease is more distinct during the first few hours. With ongoing treatment the slope of the decrease is lower and saturation is obtained. Obviously, during treatment some agglomeration of particles occurs. This is faster at the onset probably due to decomposition of the acetylacetonate ligands which act as spacers (Fig. 1).

The variation of the applied microwave power shows a somewhat different influence on the dispersity. The dispersity increases with growing microwave

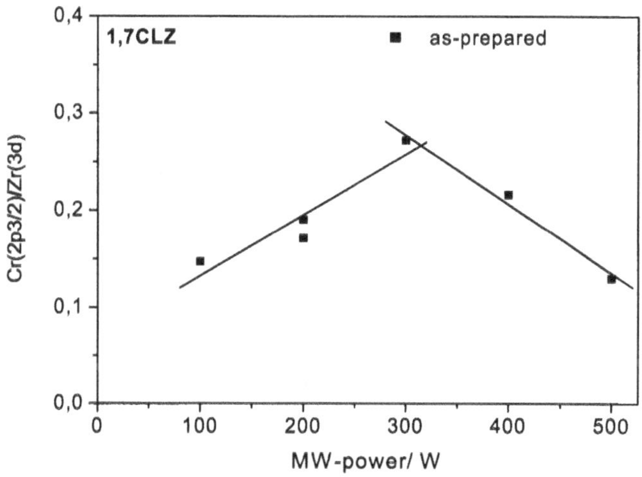

Fig. 2 Influence of the microwave power on the dispersity (characterized by XPS Cr/Zr intensity ratio) of chromia supported on La_2O_3-ZrO_2

power from 100 to 300 W. Thereafter, the dispersity decreases again when further increasing the MW power to 500 W. This course of dispersity change with variation of MW power is maintained after reduction of catalysts in the hydrogen plasma and, less distinct, after reduction in hydrogen at elevated temperature (Fig. 2).

Increasing plasma power may lead to an accelerated decomposition of the precursors giving rise to a higher dispersion on the decomposition product on the supports. Further increasing the MW radiation power beyond 300 W may lead to an enhanced agglomerization of the chromium species. As a result, an optimum MW power input is found with respect to the dispersity of supported species. Generally, the increase of the oxygen flow in the plasma causes a nearly linear decrease of the XPS dispersity.

Catalysis

The catalytic testing shows that the microwave plasma assisted prepared chromium oxide supported catalysts are active in the dehydrocyclization of octane. Deactivation is observed with growing time on stream as observed with wet chemical prepared supported chromium oxide catalysts (Fig. 3).

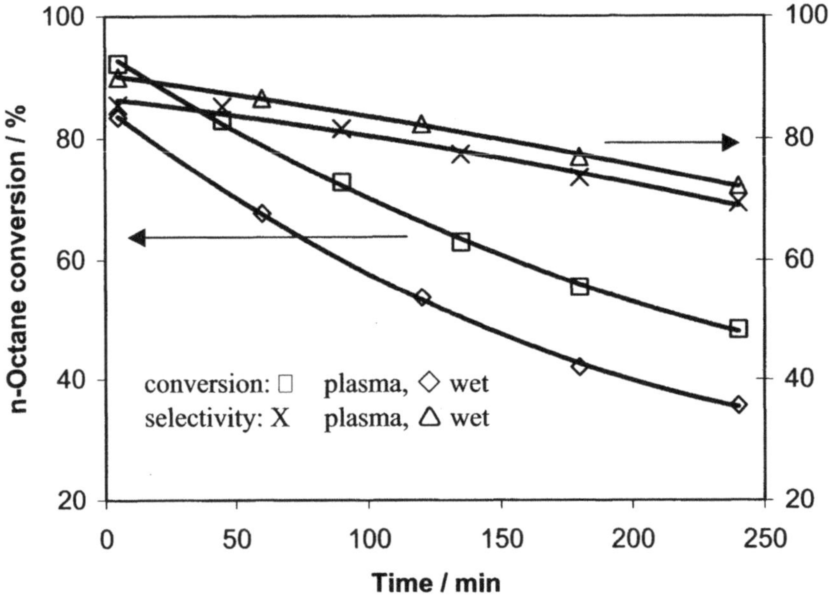

Fig. 3 n-Octane conversion and selectivity of aromatics vs. time on stream. Comparison of chromia supported catalysts prepared by the plasma and wet chemical method.

However, the plasma-chemically prepared catalysts deactivate slower than conventionally prepared catalysts. This points to a better stability of catalytically active species prepared this way. The yield of aromatics increases with the dispersity of supported chromium species. This confirms the importance of low condensed highly dispersed chromium species for the catalytic performance of these types of materials.

CONCLUSION

The studies show that the MPECVD method can be successfully applied to the preparation of supported catalysts. Obtained materials show a high dispersity of supported catalytic chromium species. Their dispersity of them can be controlled to a certain degree by selecting appropriate plasma parameters. The method can be applied to other systems because a variety of different metal organic precursors are commercially available.

REFERENCES

[1]S.L. Suib, Cattech, **6** 75 (1998).

[2]H. Kosslick, H-L. Zubowa, U. Lohse, H. Landmesser, R. Fricke, J. Caro, *Ceram. Trans.*, **89** 523 (1997).

[3]D.V. Szabo and D. Vollath, *Adv. Mater.*, **11** 1313 (1999)

[4]M.B. Kizling, S.G. Järas, *Appl. Catal. A: General*, **147** 1 (1996).

[5]Z. Paal, *J. Catal.* **105** 540 (1987).

[6]T. Kameyama, K. Sakanaka, H. Arakawa, A. Matoe, T. Tsunoda, K. Fukuda, *J. Cat. Sci.* **28** 4630 (1993).

[7]D.L. Hoang, A. Trunschke, A. Brückner, J. Radnik. H. Lieske, *Stud. Surf. Sci. Catal.* **130** 2357 (2000).

Microwave Processing of Inorganic Materials

INVESTIGATIONS OF DIELECTRIC BEHAVIOR OF MATERIALS IN THE GHZ RANGE DURING MICROWAVE PROCESSING

D.C. Dube
Department of Physics
Indian Institute of Technology
New Delhi- 11 0016(India)

M.Lanagan, H.Peng, P. Sharma, D. Agrawal
Materials Research Institute
Pennsylvania State University
University Park, PA 16802

ABSTRACT

An experimental technique is proposed which enables investigation of electrical behavior of materials in GHz region during microwave processing. The technique is based on combining a small measurement signal with the high power-processing signal and permits studies either in electric or in magnetic field in a single mode cavity. Simple arrangement for coupling and de-coupling of two signals is suggested. Experimental results obtained on manganese-zinc ferrite at ~3.6 GHz up to 1100^0 C are encouraging.

INTRODUCTION

Microwaves processing is emerging as a noble technique for the processing of advanced ceramics. A large variety of materials such as polymers [1], ceramics [2][3] and metal composition [4] have been successfully microwave processed. Microwave processing offers advantages in terms of greater phase homogeneity, higher density and better microstructure as compared to conventionally prepared materials [2][5][6]. However, there are many observations, which lack satisfactory explanation at present. For example, how does microwave absorption occur in very low dielectric loss materials like barium magnesium titanate (BMT); or, how do metallic bodies get sintered in microwave field? Some materials sinter better in the magnetic field inside the single mode microwave cavity rather than in the electric field? It is thus evident that the picture about the physical mechanisms involved in microwave processing is not clear. We have developed a method capable of studying electrical and magnetic behavior of a material in-situ i.e. when it is being actually processed in the microwave cavity. The technique combines a small (mW) measurement signal with high power (kW) processing signal. It offers the facility to put the specimen for measurement either in electric field or in magnetic field in the single mode cavity [7]. The experimental details are presented with sample data taken upto around 1100^0 C.

In recent years, a lot of interest has been shown in the measurement of dielectric properties of materials at high temperatures in the GHz region [8]-[14]. Such investigations throw light on high temperature behavior and may also prove useful in developing materials for power electronic applications and for materials to be used in high temperature conditions like in aerodynamical systems. Our proposed technique seems promising for such studies. The limitations and current problems being faced in the measurements are outlined.

EXPERIMENTAL DETAILS

For investigating electrical and magnetic properties of materials at elevated temperatures in giga-hertz region during microwave processing (in-situ), the proposed technique combines a small (mW) microwave signal of frequency (3.36 GHz) with high power (1.2 kW, 2.45 GHz) processing signal in the microwave applicator (the processing unit). The use of higher frequency for measurement enables complete de-coupling of high power microwave signal from low power measurement signal- a feature essential to avoid damage to the measurement equipment (microwave network analyzer HP 8753 E in the present case). Fig 1 shows a typical microwave-processing unit, which has been utilized to practice the idea. It operates at 1.2 kW and the dielectric sample is housed in the cavity (made out of WR340 waveguide) for processing and measurement.

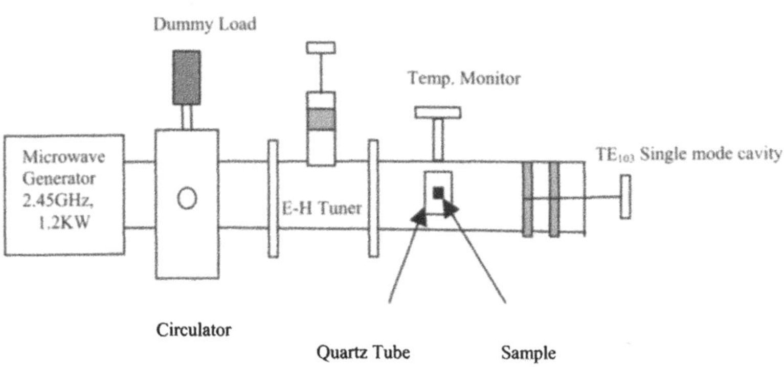

Fig 1: A typical single mode microwave processing system

Fig 2 shows the experimental setup with necessary modifications proposed and incorporated with the processing unit of Fig 1 for making these measurements.

The single mode (TE_{103}) cavity of the microwave-processing unit shown in Fig 1 has been enlarged in Fig 2 for clarity.

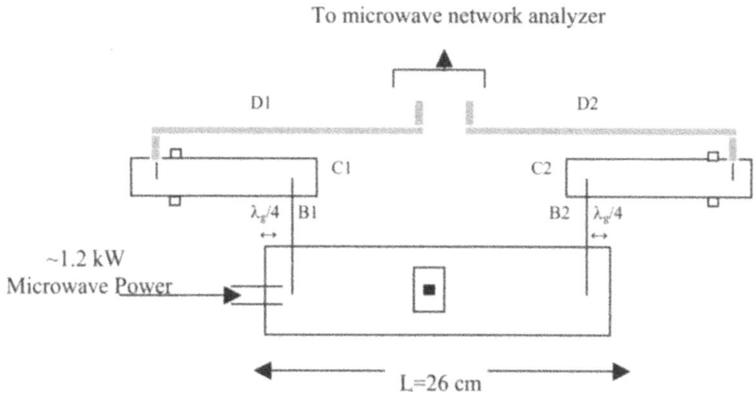

Fig 2: Experimental arrangement for evaluating dielectric parameters during microwave processing / heating of the specimen. Processor frequency 2.45GHz (TE_{103} mode) and measurement frequency 3.36GHz (TE_{105} mode).

PRINCIPLE OF OPERATION

A microwave measurement signal of 3.36 GHz from the microwave network analyzer is coupled to the processor cavity A through a C band (WR187) waveguide C1 and coaxial cable B1 (50 ohm, semi flexible). The 3.36 GHz signal excites TE_{105} mode in the processor cavity. The measurement signal is returned to the detector port 2 of the network analyzer through C band waveguide C2 and cable B2. D1, D2 are the coaxial cables of the network analyzer. The high power-processing signal is kept away from coupling to the network analyzer by the use of C band waveguides (C1 and C2) for which the cutoff frequency is 3.15 GHz and high power 2.45 GHz signal is prohibited from propagation. The set procedures for the measurements of the dielectric parameters using the cavity perturbation method may now be adopted [15].

We start by room temperature data for the resonance frequency and the half power (3 dB) bandwidth with quartz tube alone in the cavity, and then the data are recorded with the specimen in the quartz tube. The specimen is cylindrical pallet of dia ~ 6 mm and thickness ~ 2mm. The high power microwaves are now coupled to the specimen. The temperature rises and recorded in steps by a digital pyrometer, temperature below 300^0 C were estimated separately. And the corresponding changes in resonance frequency and half power width of the resonance peak are recorded in steps till the processing is complete. Since the specimen is extremely small as compared to the cavity volume and also the energy absorbed by the specimen is small, the validity conditions for a cavity perturbation method [16] are not violated.

The standard equations used in cavity perturbation technique for the measurement of dielectric parameters are [15],

$$\varepsilon' = \left[\frac{f_c - f_s}{2 f_s}\right] \frac{V_c}{V_s} + 1 \tag{1}$$

$$\varepsilon'' = \frac{V_c}{4 V_s}\left(\frac{1}{Q_s} - \frac{1}{Q_c}\right) + 1 \tag{2}$$

where ε′ is the real part of permitivity and ε″ is the dielectric loss factor. V_c and V_s are the cavity volume and sample volume respectively

The changes in the permitivity with temperature can be expressed using the above equations as,

$$\Delta \varepsilon'(T) = \left[\frac{\Delta f_r(T)}{2} * f_r(T_0)\right] * \left(\frac{V_c}{V_s}\right) \tag{3}$$

$$\Delta \varepsilon''(T) = \left(\frac{V_c}{4 V_s}\right) * \left[\Delta \left\{\frac{1}{Q(T)}\right\}\right] \tag{4}$$

where, $\Delta f_r(T)$ is the shift in the resonance frequency at measurement temperature T with respect to $f_r(T_0)$, the resonance frequency at room temperature and Q is the quality factor which is the ratio of resonance frequency to the 3 dB band width of the resonance peak.

RESULTS AND DISCUSSION

Having identified the resonance peak at the measuring frequency of 3.6 GHz, shift of resonance frequency (Δ f) and change in 3 dB bandwidth as a function of temperature of the specimen was recorded. Let us recall that the shift in resonance frequency is directly proportional to the change in dielectric constant and the changes in the loss factor of the material can be studied through the changes in the 3 dB bandwidth. In order to investigate the electrical behavior of material in E and H fields, the measurements were conducted by keeping the specimen in both fields separately [4]. To begin with, a green sample (unsintered) in the form of a small pellet (dia ~ 6 mm, height ~ 2 mm) of stochiometric manganese-zinc ferrite ($MnZnFe_3O_4$) was slipped in the applicator cavity in a quartz tube and data recorded for room temperature and also as the sample temperature rose. Fig 4(a) is the plot of Δ f data as recorded. Changes in the 3 dB bandwidth with sample temperature were used to get over all quality factor (Q) and thus loss tangent (tanδ = 1/Q) values were calculated. Since the losses of the cavity may be taken as unchanged (cavity is water-cooled), it is reasonable to assume that the variations depicted in Fig 4(b) represent the relative changes in loss tangent of the material.

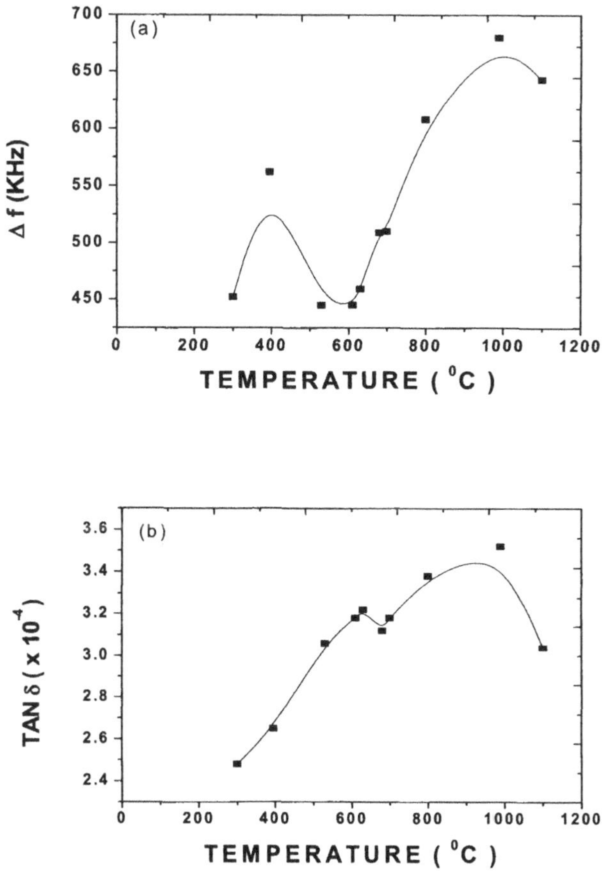

Fig 4: First Run. Shift in resonance frequency (Δf) and loss tangent (tanδ) with temperature. The specimen (green) is at E-field in processing cavity.

From Fig 4(a), it is evident that the material undergoes a phase change around 450^0, which is the Curie temperature for this ferrite. However, this phase change is not prominently reflected in loss tangent plot [Fig 4(b)]. Between 900^0 C and 1000^0 C, another transition is observed rather strongly in Fig 4(a) and 4(b). At present it is difficult to assign exact nature to this transition. This may be due to sintering or partial melting of the specimen.

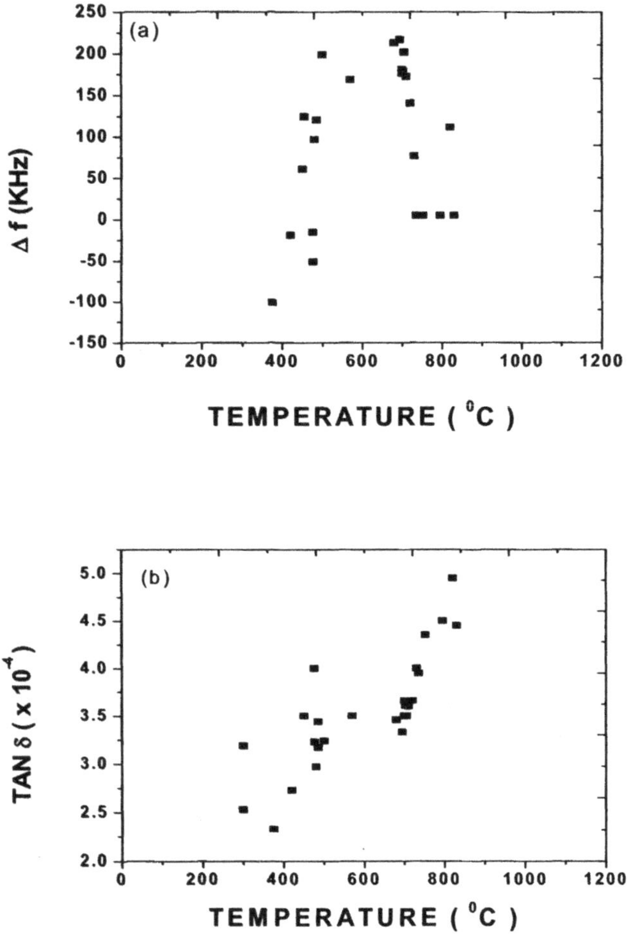

Fig 5: Third Run.Shift in resonance frequency (Δf) and loss tangent (tanδ) with temperature. The specimen (sintered) is at E-field in processing cavity.

Figs 5(a) and 5(b) depict the third run of the experiment. The character of the plots in Figs (4), and (5) are basically similar. For the second run, since the specimen was only partially sintered there was more scatter in the data. Based on our experience, the specimen may be treated fully sintered in the third subsequent run. The data points in Figs (4) (green sample) are broadly spread, while in Fig (5) (sintered sample), the variations in the Δf and $\tan\delta$ is confined to narrow temperature ranges.

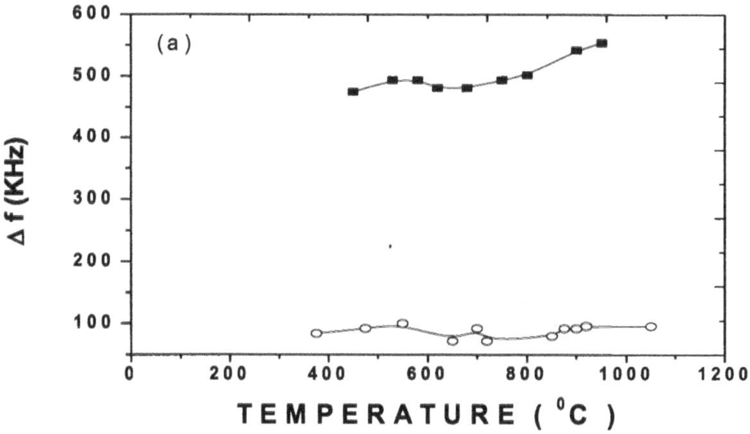

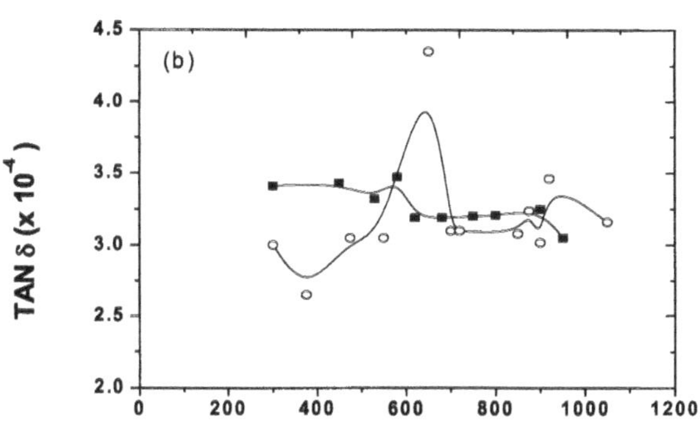

Fig 6(a) and 6(b): Shift in resonance frequency (Δf) and loss tangent ($\tan\delta$) with temperature. The specimen is at H-field in processing cavity. For the I run: closed squares and II run (sintered): open circles.

Figs 6(a) and 6(b) represent the corresponding data when the specimen lies in the high magnetic field region in the cavity. For this first run, a fresh green sample was used. These figures contain data for the second experimental run also. The third run gave data points very close to the ones obtained in second run, therefore are not plotted. These plots, which in the magnetic field, indicate that the ferrite sample get fully sintered during first run itself. This is not for the specimen in the E field. By comparing results of magnetic field, Fig (5) with that of electric field, Fig (6), one can see that the transition effects are much more pronounced when the specimen is microwave heated in the electric field. These results suggest different roles for electric field and magnetic field of the microwave field.

Few points on the difficulties experienced during the experimental deserve mention here.

The coaxial coupling probes heat-up and many times burn at the end lying in the high power cavity. Also the temperature rise of the specimen is rather fast. The possible solutions to these problems are being looked into.

The experimental results mentioned above are the first one taken on the proposed technique. The results are limited in quantity but are promising. Studies are in progress and results on other kind of materials will be available in due course of time.

Acknowledgements

The present work was carried under an Indo-US collaborated program. The financial grants received from Department of Science and Technology, New Delhi and National Science Foundation Washington are gratefully acknowledged.

References

[1]M Chen, E.J. Siochi, T.C. Ward and J.E. Mcgrath, *Polymer Engineering and Science*, **33** 1092 (1993).

[2]W. Sutton (ed), "Microwave Processing of Materials," *Materials Research Society Bulletin* **18** (1993).

[3]D.E. Clark, D.C. Folz, S.J. Oda, and R. Silberglitt (eds), " Microwaves: Theory and application in Material Processing III," *Ceramics Transactions*, Westerville, OH, American Ceramic Society (1995).

[4]D. Agrawal, *Metals World*, 672 (1999).

[5]R. Roy, J.L. Yang and S.K. Komarneni, *American Ceramic Society Bulletin*, **63** 459 (1984).

[6]R. Roy, D. Agrawal, J.P. Cheng and M. Mathis, *Ceramic Transactions*, **80** 3 (1997).

[7]J.P. Cheng, R. Roy and D. Agrawal, *Materials Research Innovations*, **5** 170 (2001).

[8]W. W. Ho, Materials Research Society Symp. Proc. **124** (1988).

[9]H. R. Garner, J.L. Kaae, R.B. Stephens, A.C. Lewis, A.J. Lieber and W.J. DeHope, *Rev. Scientific Instruments*, **61** 1832 (1990).

[10]V.V. Varden, R.D. Hollinger, V.K. Vardan and Ghodgaonkar, *IEEE Trans. Inst. & Meas.*, **40** 842 (1991).

[11]X. Weiguo and W.R. Tinga, in: Clark DE, Frank D, Sutton W H (Eds), "Microwave: Theory and application in Material Processing ," *Ceramic Transactions* **21** (1991).

[12]M. Arai, J.G.P. Binner, *Materials Research Society Symp. Proc.* **269** (1992).

[13]R. Hutchlon, M. de Jong and F. Adams, *J. Microwave Power and Electromagn. Energy*, **27** 87 (1992).

[14]V. Pohl, D. Fricke and A. Miihlbauer, *J. Microwave Power and Electromagn. Energy*, **30** 10 (1995)

[15]D.C. Dube, M.T. Lanagan, J.M. Kim and S.J. Jang, *J. Appl. Phys.*, **63** 2466 (1988).

[16]E.G. Spencer, R.C. Lecraw and L.A. Ault, *J. Appl. Phys.*, **28** 130 (1957).

MICROWAVE INTENSITY-DEPENDENT MASS TRANSPORT ENHANCEMENT IN NANOSTRUCTURED ALUMINA MEMBRANES

S.V. Egorov, A.G. Eremeev, K.I. Rybakov, V.E. Semenov, A.A. Sorokin
Institute of Applied Physics, Russian Academy of Sciences, 46 Ulyanov St., Nizhny Novgorod 603950 Russia

S.A. Gusev
Institute for Physics of Microstructures, Russian Academy of Sciences
Nizhny Novgorod 603950 Russia

ABSTRACT

New results of the comparative study of pore evolution in nanostructured alumina membranes under heating in a gyrotron microwave system and in conventional furnace are described. Microwave heating was done at different microwave power levels while maintaining constant temperature and hold time. Conventional heating was done at the same temperature but with different hold times. The pore evolution with time under conventional heating has been observed to be qualitatively similar to the change of the porous structure with an increase in the electric field strength. The dependence of the microwave-induced mass transport enhancement on the electric field strength further confirms the non-thermal nature of this phenomenon.

INTRODUCTION

The problem of microwave-enhanced mass transport in solids attracts a broad interest of researchers. The phenomenon has been investigated both experimentally[1–3] and theoretically[4–6]; however, the nature and mechanism of the effect are still far from being understood. Conceptually, two main distinctions of the microwave heating from conventional are the presence of electromagnetic field inside the material and a different temperature distribution. Respectively, in order to identify the nature of any microwave-specific effect as "thermal" or "non-thermal" one needs to differentiate between the manifestations of these two factors.

In a recent study[7,8] the authors have investigated the microwave-specific effects in nanostructured porous alumina membranes, comprising thin amorphous alumina films penetrated by nanosized tubular pore channels. Due to small thickness of the membranes, this choice of experimental object makes it possible to eliminate temperature gradients from consideration as a possible cause of microwave-specific effects. In the previous experiments it has been shown that with hold times on the order of 1 hour at 1100 °C the conventional heating of membranes did not change the porosity structure, while microwave annealing consistently resulted in a significant reduction in the surface (visible) porosity. The absence of pore closure under conventional heating was in agreement with the low value of surface diffusivity, which would not have manifested itself on the one-hour time scale of the experiments. The observed pore closure was therefore interpreted as a non-thermal microwave-induced enhancement of mass transport.

This paper presents new results of the comparative study of pore structure evolution in nanostructured membranes under microwave and conventional heating. In contrast to the previous paper, longer hold times have been implemented in the conventional heating experiments to observe the surface diffusion-induced evolution of pores. In the microwave heating experiments the experimental setup has been modified to allow processing at different microwave power levels while maintaining the same temperature. The distribution of the microwave field has been modeled in the pore geometry, and possible effects arising from the microwave field action are discussed.

EXPERIMENT

The membranes used in the experiments were amorphous alumina films penetrated by tubular pore channels. The average diameter of the pores was 60 nm, and membrane thickness was 25 μm. The average surface (visible) porosity before annealing was about 21 %. The dimensions of membrane samples were approximately 2×2 mm. Microwave annealing was done using a 10 kW / 30 GHz gyrotron system for microwave processing of materials[9]. Conventional annealing was done in a resistive furnace. The annealed membranes were characterized using scanning electron microscopy (SEM) and X-ray diffractometry (XRD). The pore evolution data were obtained by means of computer image analysis of SEM micrographs. More experimental details along with the initial results of the study are given elsewhere[7,8].

In the new series of experiments, the membrane samples were subjected to microwave irradiation at varied levels of microwave power while maintaining the same temperature, 1100 °C. This was achieved by varying the thermal insulation conditions and transiting from the heating in a multimode cavity to wavebeam irradiation. The hold time in all microwave experiments was 1 hour. The temperature was measured by a pre-calibrated Pt–Rh (type B) thermocouple.

In the experiments carried out in the multimode cavity (Fig. 1, a) the membrane sample was positioned on a sintered alumina plate of thickness 1 cm and placed in a thermal insulation arrangement made of porous alumina at a specified position inside the cavity. By varying the thickness of the top cover of the thermal insulation arrangement, different heat losses from the sample were implemented, and different levels of power were needed to maintain the same temperature.

The electric field strength at the membrane position was obtained from an additional calibration experiment. A thin titanium plate with a thermocouple was positioned exactly at the membrane position and microwave heated to 600 °C. The heating and cooling rates immediately before and after microwave shutoff were recorded and the absorbed power was calculated. From that, the amplitude of the electric field in the incident wave was obtained in the plane-wave approximation.

In the experiments carried out in the wavebeam (Fig. 1, b) the membrane sample was positioned on a sintered alumina plate of thickness 0.2 mm and placed on a partially absorbing thermal insulation plate at a specified position in the focal area of the wavebeam. By changing the material of the thermal insulation plate (and therefore its absorptivity), different heat losses from the sample were implemented, and different levels of power were needed to maintain the same temperature.

The electric field strength in the wavebeam was computed from the measured gyrotron power. The ranges of the electric field strength obtained in the multimode cavity and wavebeam experiments overlapped at about 180 V/cm. Overall, approximately a 4-fold variation in the power (2-fold variation in the electric field strength) was achieved.

The conventional furnace heating experiments were conducted at 1100 °C with varied hold times which reached 40 hours.

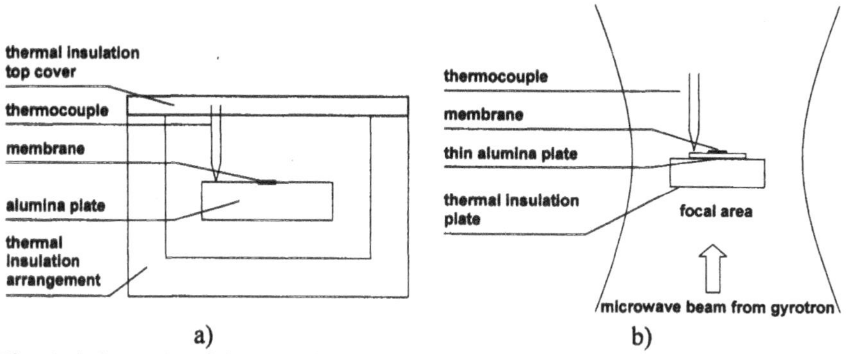

Fig. 1. Schematic of the experimental setup: a) heating in a multimode cavity; b) heating in a wavebeam.

RESULTS AND DISCUSSION

A comparison of the micrographs of the conventionally and microwave-processed membranes shows a qualitative similarity between the evolution of the porous structure with time under conventional heating and the change of the porous structure with an increase in the electric field strength. Some of the micrographs are compared in Fig. 2.

As shown previously[7,8], the pores deform due to surface diffusion and assume a bottleneck-like shape with a crater on the surface which expands over time. When an expanding crater reaches another one of a neighboring pore, the pores become interconnected on the surface, as seen at later stages of pore evolution under conventional heating (Fig. 2, b, c). Notably, similar interaction of pores is observed under microwave heating at higher electric field strengths (Fig. 2, e, f), although both the hold time and the temperature remain the same for all microwave heating experiments.

The visible porosity and pore connectivity dependencies shown in Fig. 3, 4, demonstrate similar character of the microwave (vs. the electric field strength) and conventional (vs. time) data. The visible porosity is determined by image analysis as the fraction of dark area on the micrographs. The initial decrease in the visible porosity can be explained by the pore closure due to bottleneck formation. The subsequent increase in the visible porosity apparently correlates with the onset of pore connectivity.

The observed enhancement of mass transport could be attributed to the rectified drift motion of charged vacancies in the microwave field. Partial rectification of the vacancy flows has been shown to be possible due to nonlinear ponderomotive field-vacancy interaction[4]. The distribution of the microwave electric field in the vicinity of a pore in the membrane was simulated in a quasistatic approximation by a relaxation method. The shape of a partially closed pore was obtained previously by surface diffusion simulation[7,8]. The microwave electric field vector away from the membrane was assumed parallel to the pore axis, which made it possible to use the cylindrical symmetry of the problem. Based on the electric field simulation results, the effective ponderomotive pressure of the field was calculated, which is proportional to the drop of the squared electric field normal component on the surface, $p \propto \left(E_{n\ \mathrm{ext}}^2 - E_{n\ \mathrm{int}}^2\right) 8\pi$. The pore shape and the distribution of the ponderomotive pressure along the pore surface are shown in Fig. 5. It can be seen that the ponderomotive pressure can cause a mass flow leading to additional pore closure. However, as previously noted[7,8], the efficiency of vacancy drift rectification must be close to 100 % to correlate with the value of pore closure velocity obtained from the experiments. In other words, that would mean that the alternating vacancy current induced by the microwave field along the surface

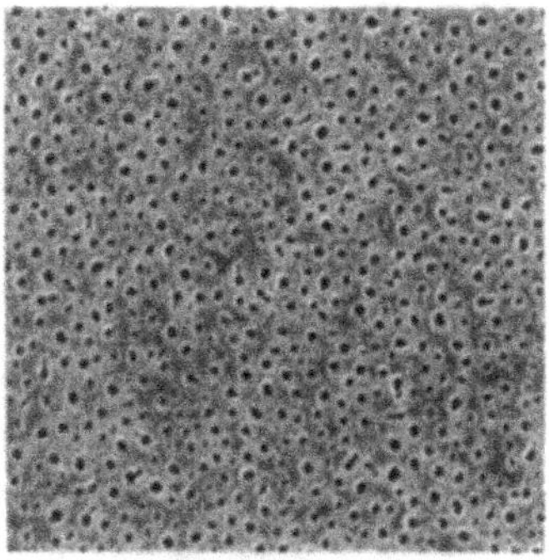

a) Conventional, 5 hours

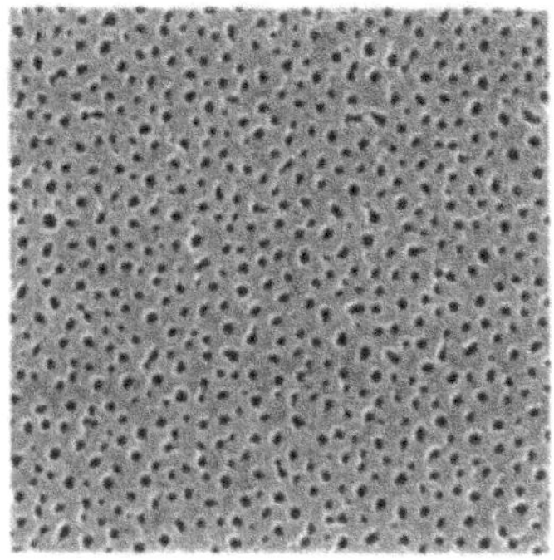

d) Microwave, 165 V/cm, 1 hour

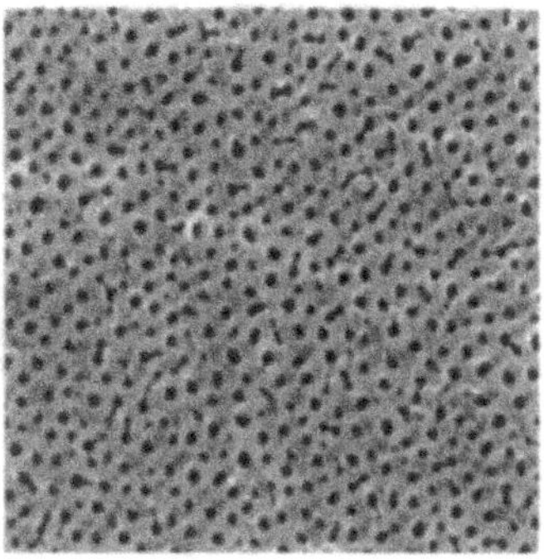

b) Conventional, 20 hours

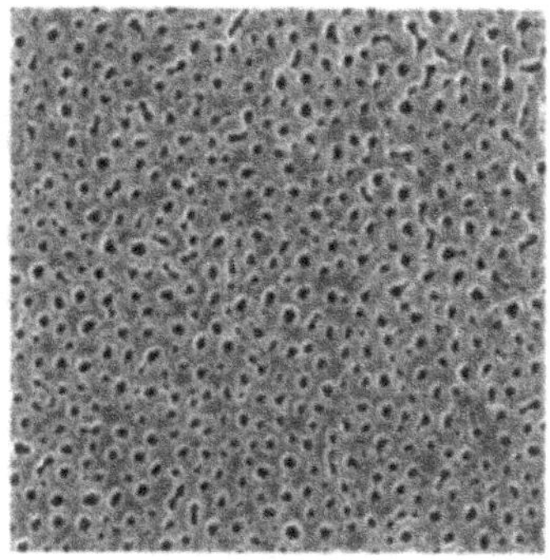

e) Microwave, 187 V/cm, 1hour

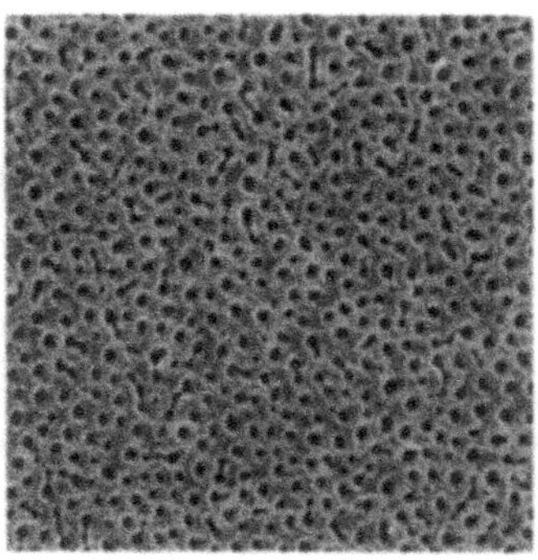

c) Conventional, 40 hours

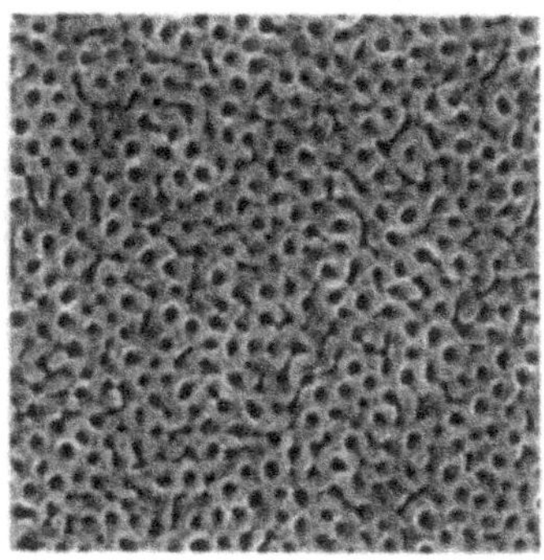

f) Microwave, 252 V/cm, 1 hour

Fig. 2. Micrographs of the membranes processed conventionally and by microwaves at 1100 °C. All image sizes are 2.28 × 2.28 μm.

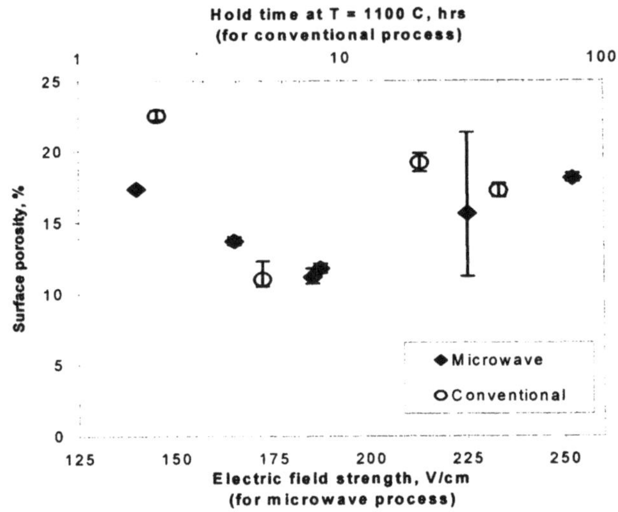

Fig. 3. Surface porosity vs. electric field strength (for microwave 1-hour process) or hold time (for conventional process) at 1100 °C.

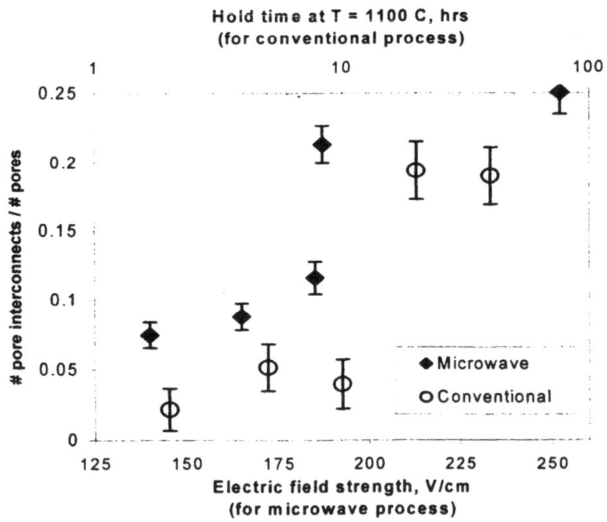

Fig. 4. Pore connectivity vs. electric field strength (for microwave 1-hour process) or hold time (for conventional process) at 1100 °C.

must be almost completely rectified, i.e., most of the vacancies on the surface must move during one half-period of the field and stop during the other half-period. The ponderomotive model[4] predicts a several orders of magnitude smaller rectification efficiency in the case of alumina, a material with low ionic conductivity. Thus, more work is needed to provide explanation for the microwave enhanced mass transport observed in the described experiments.

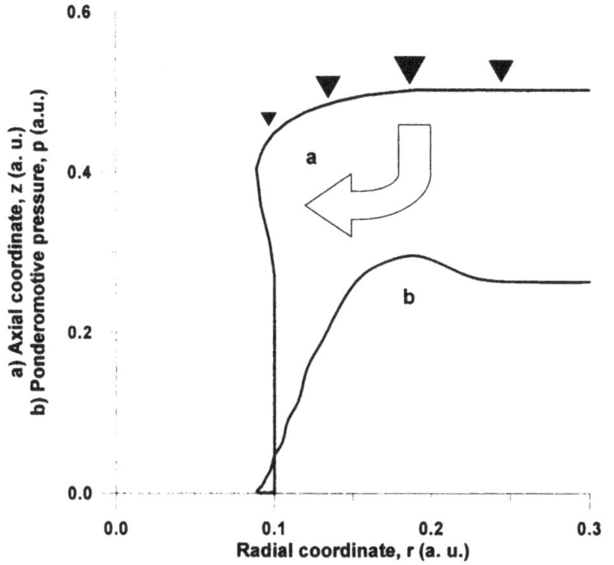

Fig. 5. Shape of a closing pore obtained by surface diffusion (a) and the calculated effective ponderomotive pressure on the pore surface for vertical polarization of the electric field (b). Black triangles illustrate schematically the distribution of the ponderomotive pressure over the pore surface, and the arrow shows the direction of possible additional microwave-induced mass transport.

SUMMARY AND CONCLUSIONS

(1) Microwave-induced nonthermal enhancement of the diffusion mass transport in nanostructured alumina membranes has been observed. The extent of the enhancement has been shown to depend drastically upon the microwave field strength.

(2) Pore evolution with time in the membranes under conventional heating at a constant temperature, including the surface diffusion-driven processes of pore closure and interconnection, is qualitatively similar to the changes in the pore

structure arising from an increase in the microwave field intensity, the temperature and duration of the microwave exposure kept constant.

(3) With the maximum electric field strength tested, the changes in the porous structure developed in 1 hour are similar to the changes resulting from 40 hours of conventional heating at the same temperature. Thus, the extent of mass transport enhancement, as compared to conventional heating, has been demonstrated to be about 1.5 orders of magnitude.

ACKNOWLEDGMENT

Partial support of this work by the Russian Academy of Sciences under the Young scientist project # 166 is acknowledged. The authors are grateful to Dr. G.L. Hornyak for providing the alumina membrane samples.

REFERENCES

[1]M.A. Janney, H.D. Kimrey, W.R. Allen, J.O. Kiggans, *J. Mater. Sci.* **32** (1997), 1347.

[2]B.A. Wilson, K.-Y. Lee, E.D. Case, *Mater. Res. Bull.* **32** (1997), 1607.

[3]A.G. Whittaker and L. Cronin, in *Proc. Int. Conf. on Microwave Chemistry, Antibes* (2000), p. 291.

[4]K. I. Rybakov and V. E. Semenov, *Phys. Rev. B* **52** (1995), 3030.

[5]J.H. Booske, R.F. Cooper, S.A. Freeman, K.I. Rybakov, V.E. Semenov, *Physics of Plasmas* **5** (1998), 1664.

[6]A. Birnboim, J. P. Calame, Y. Carmel, *J. Appl. Phys.* **85** (1999), 478.

[7]Yu.V. Bykov, S.V. Egorov, A.G. Eremeev, K.I. Rybakov, V.E. Semenov, A.A. Sorokin, S.A. Gusev, in *Microwaves: Theory and Application in Material Processing V*, edited by D. Clark et al. *(Ceramic Transactions*, vol. **111**, Amer. Ceram. Soc., Westerville, OH, 2001), p. 335.

[8]Yu.V. Bykov, S.V. Egorov, A.G. Eremeev, K.I. Rybakov, V.E. Semenov, A.A. Sorokin, S.A. Gusev, *J. Mater. Sci.* **36** (2001), 131.

[9]Y. Bykov, A. Eremeev, V. Flyagin et al., in *Microwaves: Theory and Applications in Material Processing III*, edited by D.E. Clark et al. (*Ceramic Transactions*, Vol. **59**, Amer. Ceram. Soc., Westerville, OH, 1995), p. 133.

PRECISION MICROWAVE/RESISTANCE HEATING STUDY OF ZIRCONIA: ABSENCE OF MICROWAVE EFFECT

C.C. Sorrell, O.C. Standard, N. Ehsani, K.R. Ratinac, J.R. Rider, X.Q. Zheng, and A.J. Ruys
School of Materials Science and Engineering
University of New South Wales
Sydney, NSW 2052
Australia

ABSTRACT

A precision study of the microwave hybrid (2.45 GHz) and resistance heating of 3Y-TZP was undertaken during 1994-96, although these data [1-6] generally are not acquired easily. Preliminary testing involved calibration of thermocouples within the operating cavity to ensure correct temperature measurement and demonstration of microwave absorption in the absence of a susceptor. The effects of heating rate, soak time, and soak temperature were determined in terms of measurements of thermal gradient across the sample diameter, open and closed porosities, bulk density, and grain size. Comparison of these results, combined with data from other studies of rapid resistance heating for the same material, indicates that it is not necessary to invoke a *microwave effect* to explain the data, which show no significant differences between the two modes of heating.

INTRODUCTION

Currently, there is contention concerning a posited phenomenon referred to as the *microwave effect*, which involves the suggestion of enhanced diffusion and densification due to some unexplained effects of microwave-material interactions. However, the dearth of precise densification studies accompanied by accurate temperature measurements throughout the bulk of samples heated in microwave and resistance furnaces has prevented resolution of this issue. The present work clarifies the importance of accurate temperature measurements, the critical role of thermal gradients, and the significance of kinetics in the understanding of some of the phenomena that take place during microwave heating at 2.45 GHz.

PREVIOUS STUDIES

In the present study, yttria-stabilized tetragonal zirconia polycrystal (3Y-TZP) was studied owing to its moderate microwave absorption characteristics. Several previous studies of the comparative densification by microwave (2.45 GHz) and resistance heating of similar zirconias, including 8Y-TZP [7,8] and 12Ce-TZP [9], have been reported previously. In these studies, the authors reported that equivalent bulk densities could be obtained by microwave sintering at temperatures 50°-100°C lower than by resistance heating. However, in none of these cases was the accuracy of the temperature measurements recorded. That is, the temperature gradients across the samples were not reported nor was the actual existence of microwave absorption by the samples demonstrated.

EXPERIMENTAL PROCEDURE

Furnaces

The microwave furnace* used in this work contains a multimode cavity with top, bottom, and side holes for thermocouple access. It operates at a frequency of 2.45 GHz and maximal power of 1.5 kW (all present work was done at 40-75% power). This prototype furnace was commissioned in 1992 when the problems with the available time-slicing microwave furnaces had become very apparent. Thus, this furnace is a true variable-power unit that operates in both the power-control and temperature-control modes. Forward and reflected power are detected and the outputs linearized, so allowing the forward power to be controlled. It is believed that this was the first laboratory-scale proportional-power microwave furnace to be marketed commercially. The resistance furnaces* used for comparative heating were programmable SiC and $MoSi_2$ muffle furnaces.

Temperature Calibration

The temperature calibrations were done using an experimental setup as shown in Figure 1. The accuracy of the temperature measurement by contact thermocouple was determined by detecting the latent heats of fusion of several standard metals [10,11]. The tip of the thermocouple sheath was protected by a snugly fitting, dense, thin-walled (<0.5 mm) alumina cap placed directly in a hole drilled in a cylinder of spectroscopic-grade (>99.9%) metal. The metals† used for melting point determination were Sn, Zn, Al, Ag, and Pd. Each melting point was determined in triplicate using the average of the midpoints of the thermal arrests due to melting and solidification [12]. Heating rates of 300°-1000°C·h^{-1} and cooling rates of 500°-1500°C·h^{-1} were used.

---

* Ceramic Engineering, Warragamba, NSW, Australia

† Johnson Matthey (Australia) Ltd., Thomastown, VIC, Australia

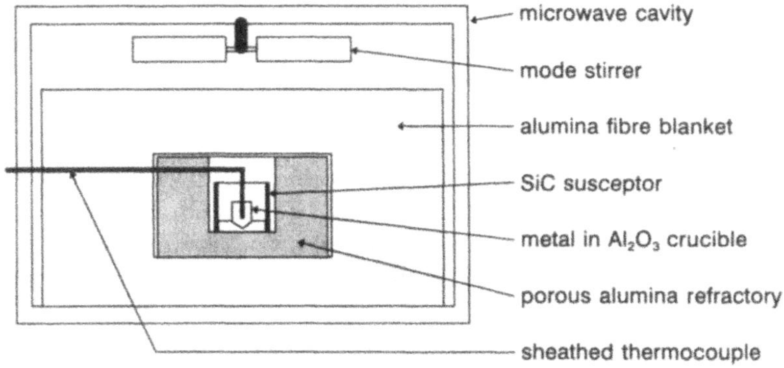

Figure 1. Experimental setup used for temperature calibration.

An example of the data for aluminum is shown in Figure 2. In the cases of all of the metals except Pd, the reported and measured melting points were within ~9°C of one another. With Pd, it is likely that high radiant heat losses and poor contact between the molten metal bead and the thermocouple resulted in a measured melting point of ~1528°C (reported melting point = 1554°C).

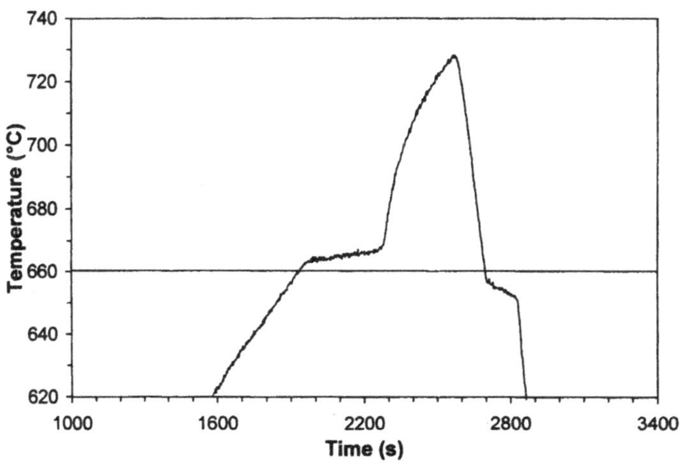

Figure 2. Typical fusion curve for Al (heating – power on; cooling – power off).

Temperature Measurement

The temperature in the microwave cavity was measured using commercial sheathed Nicrosil/Nisil (type N) thermocouples up to 1300°C and in-house platinum-sheathed Pt-Pt13Rh (type R) thermocouples for higher temperatures. The tip of the thermocouple assembly was as described previously. The end of the thermocouple assembly was grounded.

Thermal Gradient Determination

The thermal gradient work with ceramic compacts during microwave heating was done using an experimental setup as shown in Figure 3. The test samples of silicon carbide* (strong microwave absorber) and hydroxyapatite† (weak microwave absorber) were prepared by thixotropic casting [13,14], incorporating three axial wells halfway into the 60 mm length of cylinders of 40 mm diameter. The wells were located radially in the centre, scored up the surface, and at the midpoint between these two wells. Silicon carbide was heated at 1800°-2000°C·h^{-1} and hydroxyapatite was heated at 900°-1000°C·h^{-1}.

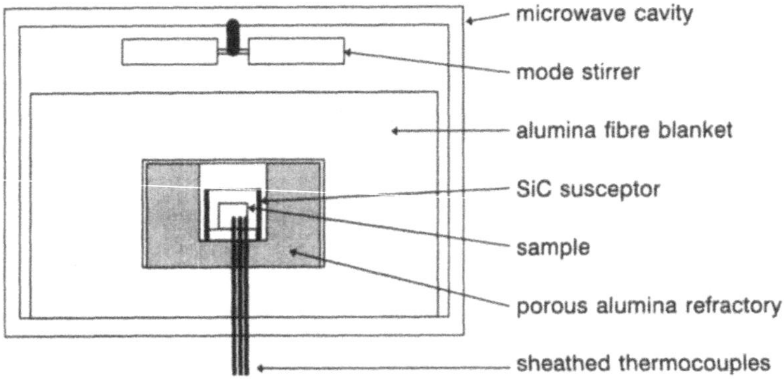

Figure 3. Experimental setup used for thermal gradient determination.

Thermal gradients using a SiC resistance furnace were determined with an identical experimental setup, except the sample was oriented with the axis being horizontal. The maximal heating rate allowed and used in this furnace was 300°C·h^{-1}.

Demonstration of Microwave Absorption

The presence of microwave absorption was determined using the setup as shown in Figure 3, with the sample being nested within a cylindrical silicon carbide susceptor. The procedure consisted of heating to 400°, 600°, or 900°C at

* 3.0 μm, Naxos Products, Auburn, NSW, Australia

† 5.1 μm (agglomerate size), Plasma Biotal Ltd., Tideswell, Derbyshire, UK

2000°C·h^{-1}, soaking for 10 min, turning off the power, removing the susceptor, turning on the power, soaking for 10 min at the resultant temperature, and heating at 2000°C·h^{-1} by increasing the power in the absence of the susceptor.

Figure 4 shows that, although susceptor removal at 400°C caused a temperature drop to <300°C, the sample retained sufficient microwave absorption to allow continued heating. The latter temperature may be compared to the calculated dielectric loss factor data [5] *vs* temperature for ZrO_2 [15] and SiC [16].

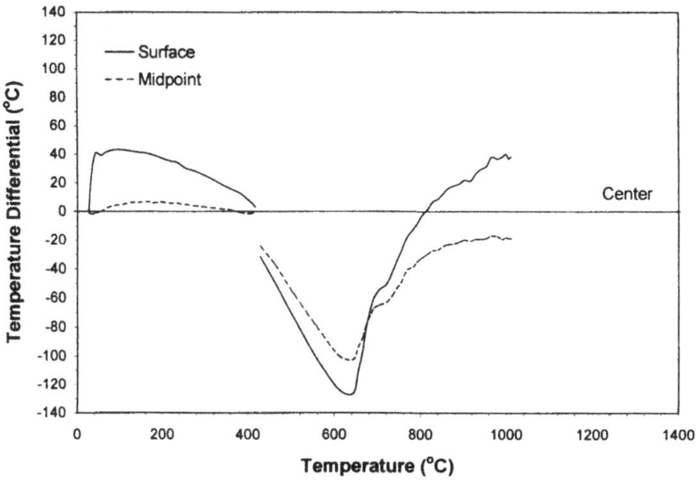

Figure 4. Effect of susceptor removal at 400°C, showing continued microwave absorption at <300°C.

The demonstration of proof of microwave absorption is critical because similar work on Al_2O_3 (Sumitomo AES-11) by the authors [17] demonstrates that this material does not absorb microwaves at 2.45 GHz until temperatures exceeding ~1400°C, which are higher than those used in most of the many reported microwave densification studies of Al_2O_3.

Densification

The densification work during microwave heating was done using an experimental setup as shown in Figure 5. The samples of 3 mol% Y_2O_3 tetragonal zirconia polycrystal* (3Y-TZP; moderate microwave absorber) were of 12 mm height and 15 mm diameter, uniaxially pressed at 10 MPa and cold

* TZ-3Y – 99.9 wt%, 25 nm, Tosoh Corp., Tokyo, Japan

isostatically pressed (CIP) at 200 MPa. A hole of 4 mm depth was drilled in the central vertical axis of the sample to allow thermocouple insertion in the bottom. A heating rate of 2000°C·h^{-1} to 1000°-1400°C in 100°C increments and soak times of 5, 15, and 30 min were used.

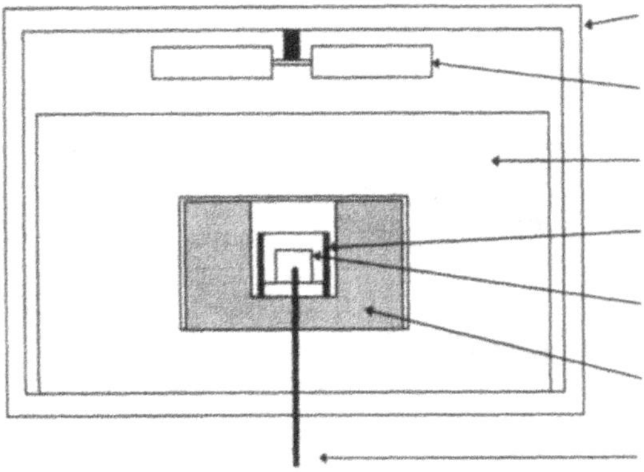

Figure 5. Experimental setup used for densification studies.

Densification using a $MoSi_2$ resistance furnace was done using a similar setup, where the thermocouple was located in the centre of four vertically oriented and tightly packed samples. The maximal permissible heating rate for this furnace was 200°C·h^{-1}. Again, 1000°-1400°C in 100°C increments were studied but a soak time of 60 min was used.

Characterization

The bulk densities, open porosities, and closed porosities were determined using the hydrostatic weighing method [18]. The grain sizes across the diameters of the samples were determined directly from fracture surfaces using a field-emission scanning electron microscope* (FESEM) at 20,000X magnification. The line-intercept method [19] (typically 10 lines crossing 200 grains) was used, without the accepted 1.56 correction factor [20].

RESULTS AND DISCUSSION

Thermal Gradients

The nature of the thermal gradients in hydroxyapatite are shown in Figure 6, which normalizes the surface and midpoint temperatures against the centre temperature. The following points can be made:

* Model S-4500II, Hitachi Ltd., Tokyo, Japan

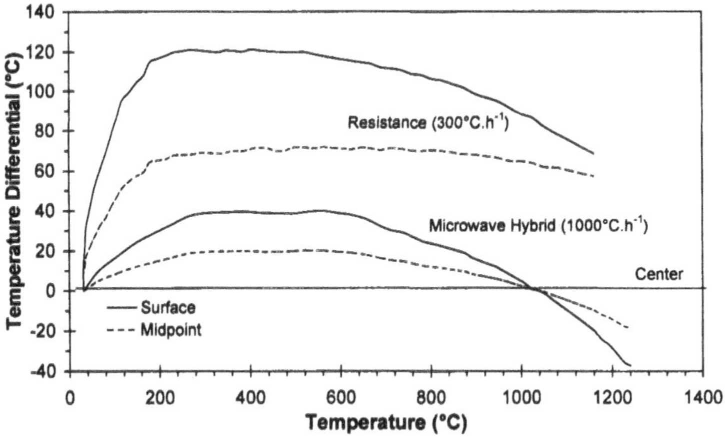

Figure 6. Normalized thermal gradients in hydroxyapatite during microwave hybrid and resistance heating.

Resistance Heating

- Heating at 300°C·h^{-1} shows a typical thermal gradient profile, where the surface is hotter than the interior.
- The size of the thermal gradient is high (as high as ~120°C across a radial inter-thermocouple distance of ~18 mm).
- The size of the thermal gradient decreases with increasing temperature as a result of increasing thermal conductivity due to densification and increasing temperature.

Microwave Hybrid Heating

- Heating at 1000°C·h^{-1} shows behavior similar to that during resistance heating as a result of the radiant heating by the susceptor at lower temperatures.
- At lower temperatures, the size of the normal thermal gradient (as high as ~40°C across ~18 mm) is less than that during resistance heating probably owing to the greater proximity of the susceptor compared to the heating elements of the resistance furnace.
- There is a conversion from a normal to a reverse thermal gradient at higher temperatures owing to the combined effects of densification and microwave absorption, the onset of which occurs at some temperature above ~300°C, as evidenced by the flattening of the curve. Owing to the similarity of the shapes of the curves for the two types of heating, the actual onset of microwave absorption cannot be determined by this method, although the reverse thermal gradient at ~1000°C demonstrates that microwave absorption is occurring at least by this temperature.

The normalized thermal gradients in silicon carbide are shown in Figure 7. The following points can be made:

Resistance Heating

- The size of the normal thermal gradient is low (~25°C across ~18 mm) owing to the high thermal conductivity.
- The size of the thermal gradient does not change with increasing temperature owing to the low amount of densification that takes place.

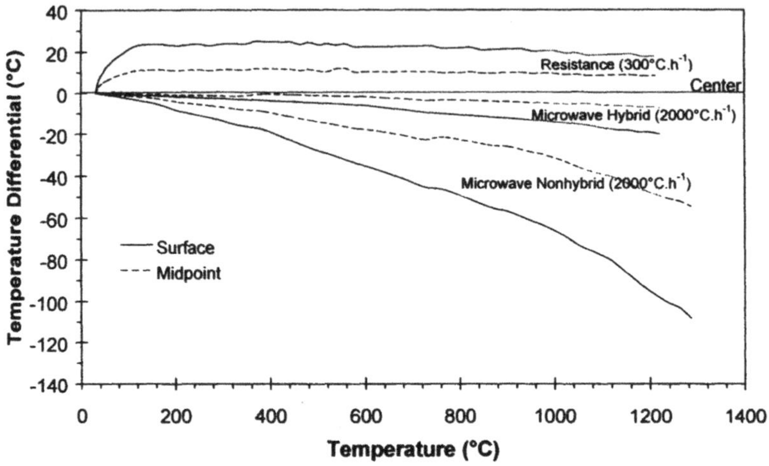

Figure 7. Normalized thermal gradients in silicon carbide during microwave hybrid, nonhyrid, and resistance heating.

Microwave Hybrid Heating

- Heating at 2000°C·h^{-1} shows a reverse thermal gradient and thus microwave absorption at all temperatures.
- The size of the reverse thermal gradient (<20°C across ~18 mm) is low at all temperatures owing to the combined effects of the microwave absorption and the radiant heating by the susceptor.

Microwave Nonhybrid Heating

- Heating at 2000°C·h^{-1} shows a significantly increased thermal gradient (as high as ~115°C across ~18 mm) owing to the radiant heat loss from the sample surface. This is enhanced by the high emissivity of silicon carbide.

The effect of this radiant heat loss from the sample surface has important implications for all materials heated in the absence of a susceptor, which serves the dual purpose of providing heating prior to microwave absorption at low temperatures and compensating for radiant heat loss at high temperatures. The preceding data allow several generalizations concerning microwave heating of microwave-absorbing materials to be made:

- Thermal gradients should be measured for the thermal history to which the sample is to be subjected.
- Thermal gradients are enhanced by high heating rates owing to the finite thermal diffusivities of materials.
- Thermal gradients are enhanced by high porosities owing to the lower thermal conductivities of porous materials compared to those of dense materials.
- Small samples, with high surface area to volume ratios, show greater surface heat losses, thus enhancing thermal gradients rather than the commonly assumed converse, which is that thermal equilibrium is achieved.

- Thermal gradient effects are enhanced for darker materials, with higher emissivities, compared to lighter materials, with lower emissivities.
- When the thermocouple is placed on the surface of the sample, wider grain size distributions will result since the larger volume of the interior will be at a temperature different from that of the surface, which frequently is the assumed processing temperature.
- The preceding effect will be enhanced even more when the thermocouple is placed in proximity of and not in contact with the sample surface since the thermal conductivity across an air gap is low.
- Noncontact temperature probes should be calibrated with thermocouples located at the identical sighting positions.
- Susceptors that do not match the surface profile of the samples are likely to cause localized variations in surface temperature and grain size.
- The measured temperature differentials in the large samples of the present work are in the range of those to which comparative densification as a function of heating mode is attributed to a *microwave effect.* The use of high heating rates; small, porous, dark samples; and uncalibrated noncontact means of temperature measurement will increase these differentials.

Densification

The present densification data were obtained using Tosoh Corp. 3 mol% Y_2O_3 tetragonal zirconia polycrystal. Chen and Mayo [21] used the same 3Y-TZP powder in a study of rapid resistance heating. Consequently, it is possible to compare these data directly with those obtained in the present work. The heating parameters used for the densification are given in Table I.

Table I. Heating parameters used during microwave hybrid and resistance heating of 3Y-TZP.

Parameter	**Present Work**		**Chen & Mayo [21]**
	Microwave Hybrid Heating	**Resistance Heating**	**Resistance Heating**
Heating Rate (°C·h^{-1})	2000	200	120 – 12,000
Soak Time (min)	5, 10, 15	60	120
Soak Temperature (°C)	1000, 1100, 1200, 1300, 1400		

It should be stressed that such comparisons are, strictly speaking, inappropriate. In order to make valid comparisons, especially when commenting upon the possible existence of a *microwave effect*, it is essential to use identical samples and experimental conditions.

The data for the bulk density as a function of sintering temperature are shown in Figures 8 and 9. These data, all of which are at temperatures well above that of the onset of microwave absorption, show no consistent variation between resistance and microwave hybrid heating of 3Y-TZP. Microwave heating gives slightly higher bulk densities at lower temperatures as well as increasing bulk densities with increasing soak times, but these effects can be explained by consideration of the kinetics of densification, as will be discussed subsequently.

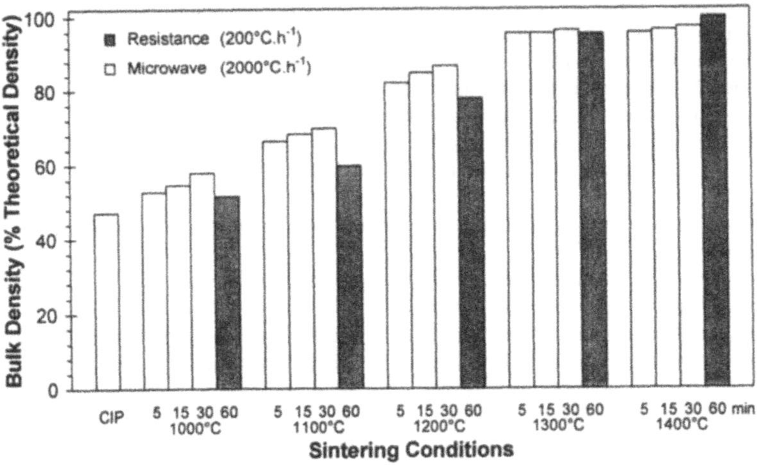

Figure 8. Bulk density as a function of sintering temperature and time for rapid hybrid microwave and slow resistance heating of 3Y-TZP.

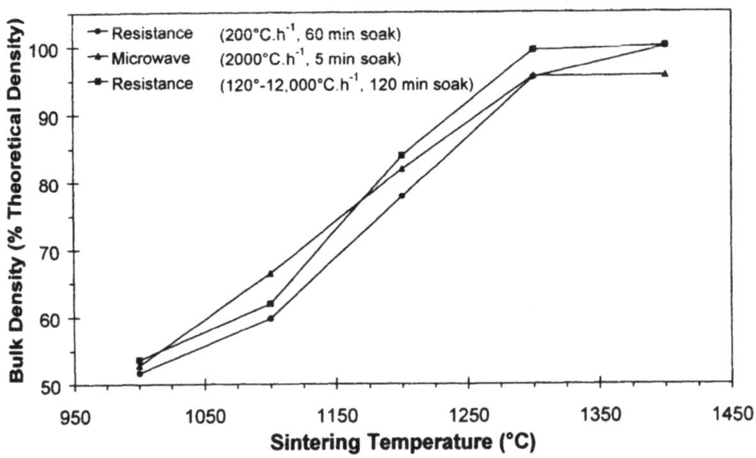

Figure 9. Bulk density as a function of sintering temperature and time for rapid hybrid microwave, slow resistance, and rapid resistance heating [21] of 3Y-TZP.

It can be seen that rapid microwave, slow resistance, and rapid resistance heating show quite similar trends, with bulk density variations no greater than ~7% at each temperature. It is significant that Chen and Mayo [21] found that the heating rate had no effect on the bulk density because most of the densification occurred during the soaking, not the heating. This is in agreement with the observations of Kim and Kim [22] for rapid sintering of the same powder.

Figure 10 shows the open and closed porosities versus the bulk density. In this case, it is very clear that there is essentially no difference between the rapid microwave heated and slow resistance heated samples. These data are important as they remove the problematic variable of temperature, which will be discussed subsequently.

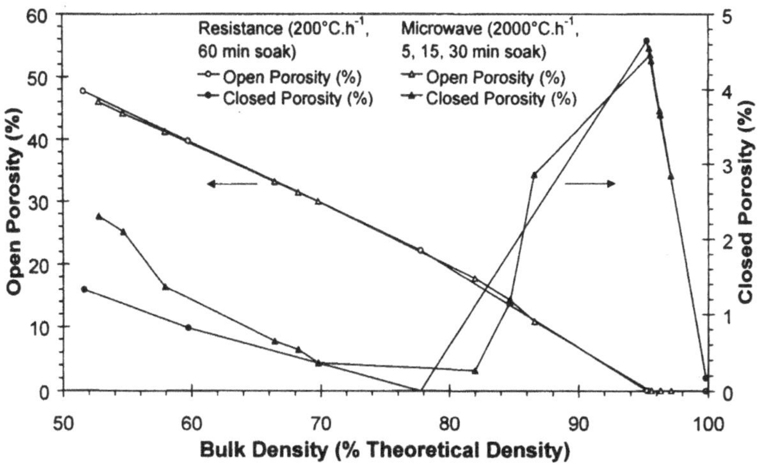

Figure 10. Open and closed porosities as a function of bulk density for rapid hybrid microwave and slow resistance heating of 3Y-TZP; the soak times during microwave heating are plotted in order of increasing time from left to right for each triplet of data points.

Kinetic Effects: It would be logical to assume that the reduced time associated with rapid microwave heating should result in grain sizes finer than those obtained with slow resistance heating. However, some researchers have observed this [23-26] and others the converse [27,28]. Dé *et al.* [29] showed that this effect was dependent on the heating temperature and soak time. Chen and Mayo [21] demonstrated convincingly that both the bulk density and grain size were unaffected by the heating rate.

Figures 11 and 12 show the grain sizes and standard deviations as a function of sintering temperature and time. These data indicate that microwave heating causes greater grain growth at higher temperatures as well as increasing grain sizes with increasing soak temperatures, but these effects also can be explained by consideration of the kinetics of densification, as discussed subsequently.

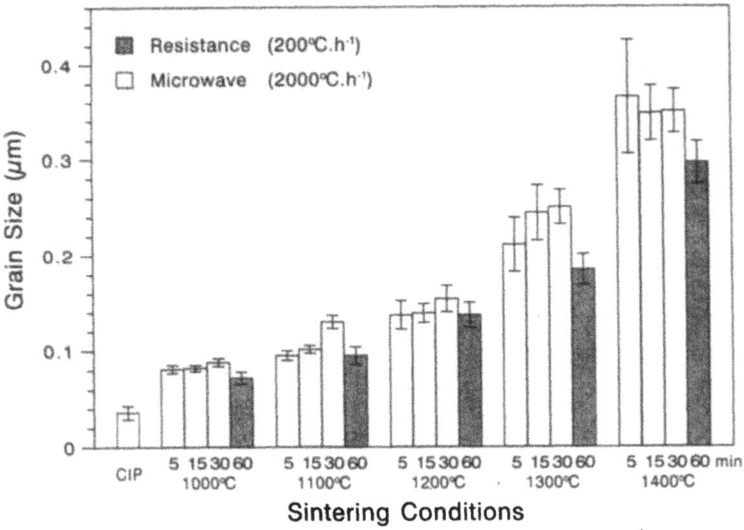

Figure 11. Grain size as a function of sintering temperature and time for rapid hybrid microwave and slow resistance heating of 3Y-TZP.

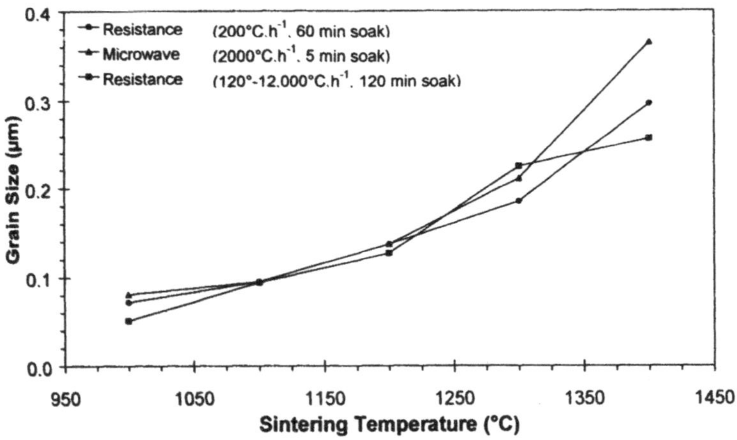

Figure 12. Grain size as a function of sintering temperature and time for rapid hybrid microwave, slow resistance, and rapid resistance heating [21] of 3Y-TZP.

Figure 8 shows that 3Y-TZP achieves >90% bulk density by 1300°C for both heating methods. Therefore, rapid grain growth associated with separation of the grain boundaries from the pores [30] can occur by 1300°C. At this stage, there is grain growth but little or no increase in bulk density. Figure 11 confirms that the grain sizes for the samples densified by both methods at the two highest temperatures show increased grain sizes. However, while Figure 8 indicates that the bulk densities achieved by both heating methods are similar, Figure 11 shows that microwave heating results in considerably larger grain sizes. The Chen and Mayo [21] results now allow the present data to be understood. Since ~90% bulk density is achieved very quickly during microwave hybrid heating (5 min at 1300°C), then the grain growth can be attributed to the soak times at the maximal temperature, even though they are of duration ≤30 min. Since ~90% bulk density is achieved slowly during resistance heating (at 1200°C, 81% of the total densification takes place during the 120 min soak [21]), then there is less time for grain growth, despite the longer soak time of 60 min in the present work.

Consequently, it is important to note that rapid microwave heating may be disadvantageous microstructurally if careful attention is not paid to kinetics. This observation suggests that considerable flexibility in microstructural evolution can be achieved with suitable variation in microwave heating schedules. In order to achieve the maximal densification while avoiding exaggerated grain growth, careful attention must be paid to the heating rate used to reach ~90% bulk density and the subsequent soak time. Thus, a slower heating rate may be advantageous in terms of both achieving the maximal bulk density and minimizing the thermal gradients, which increase at higher heating rates [17].

The importance of these kinetic considerations is emphasized by the divergence of the curves at the higher temperatures shown in Figure 12. These data also demonstrate that rapid hybrid microwave heating, slow resistance, and rapid resistance heating show similar trends.

However, it is important to point out a discrepancy in the Chen and Mayo [21] data at 1400°C. The larger grain size in the sample microwave sintered at 1400°C compared to the grain sizes of the resistance heated samples potentially could be attributed to a *microwave effect* (*i.e.*, enhanced grain growth at lower temperatures during microwave heating [31]). While this support is in the form of only a single data point at 1400°C, its validity is contradicted by the following:

- In the data of Chen and Mayo [21], the grain size increase from 1200° to 1300°C is greater than that from 1300° to 1400°C, which is contrary to the universally observed trends in grain growth of ceramics (as is shown in the resistance heating data in the present work).
- Since the critical parameter is the soak time after achievement of closed porosity (bulk density ≥90%), then the longer soak time of 120 min used by Chen and Mayo [21] should result in a grain size at 1400°C larger than that obtained during resistance heating, with a soak time of 60 min at 1400°C, in the present study. The result is the reverse of that expected.

These considerations lead to the conclusion that the grain size at 1400°C measured by Chen and Mayo [21] is erroneously low. If this is the case, then this data point should lie at the largest grain size for the three curves, which supports the previous comment that the critical consideration of grain growth is simply the soak time at bulk density ≥90%, not the mode of heating.

Uncoupling of Temperature Effect: Since it is clear that accurate temperature measurements are not undertaken by many researchers, this aspect remains critical to the substantiation or discrediting of the existence of a *microwave effect*. A means of resolving this obstacle is shown in Figure 10, where removing the temperature variable demonstrated no difference in the open and closed porosities as a function of bulk density for resistance and microwave heating.

This strategy also is shown in Figure 13, which plots grain size as a function of bulk density for three different heating histories. The rapid grain growth of 3Y-TZP at bulk density >95% is clear. However, it is plainly apparent that there is no significant difference between the samples heated by the three different methods. This is striking in light of the fact that two different heating modes and three different thermal histories were used by two independent research groups and the results are nearly identical. Since both the trends and experimental values for bulk density *vs* open and closed porosities (Figure 10) and grain size (Figure 13) are strikingly similar, this indicates that the heating mode is not significant and that apparent anomalies in the physical properties data of microwave absorbing materials do not require a *microwave effect* for explanation.

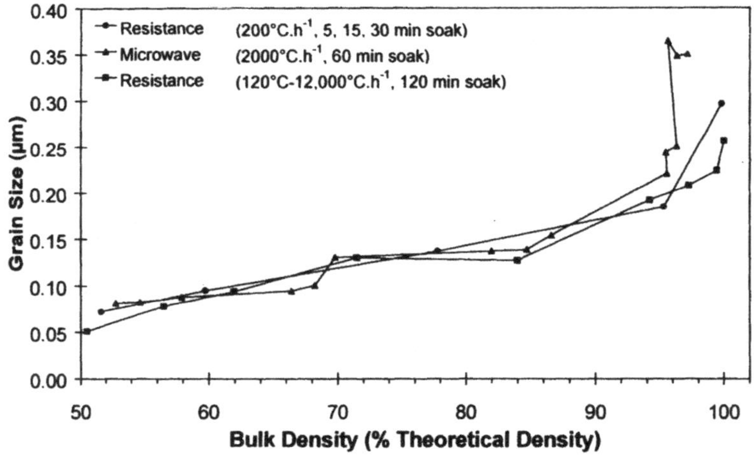

Figure 13. Grain size as a function of bulk density for rapid hybrid microwave, slow resistance, and rapid resistance heating [21] of 3Y-TZP; the soaking times during microwave heating are plotted in order of increasing time from left to right for each triplet of data points.

SUMMARY

The present work on 3Y-TZP indicates that it is not necessary to invoke a *microwave effect* to explain apparent anomalies in the densification data of ceramics. This work supports the conclusion that the discrepancies between microwave heating and resistance heating are due to incorrect temperature measurement and the consequent inability to compare equivalent data. In some cases, such as Al_2O_3, microwave absorption may not be taking place at all. Rapid heating in a microwave furnace results in the rapid attainment of high bulk densities (≥90%), which result in rapid grain growth according to conventional models. These data and those of others [21] indicate that the mode of heating has little or no effect on the ultimate bulk density, grain size, open porosity, and closed porosity, as shown in Figures 8-13. It is the kinetics that are affected. Thus, achievement of microstructural control requires sufficient knowledge and careful monitoring of the kinetics.

REFERENCES

[1]N. Ehsani, A.J. Ruys, O.C. Standard, and C.C. Sorrell, "Advanced Microwave Sintering of Ceramics I. Temperature Measurement," *Mater. Eng.*, **8** [2] 89-95 (1997).

[2]O.C. Standard, A.J. Ruys, and C.C. Sorrell, "Advanced Microwave Sintering of Ceramics II. Temperature Calibration," *Mater. Eng.*, **8** [2] 97-106 (1997).

[3]C.C. Sorrell, N. Ehsani, A.J. Ruys, and O.C. Standard, "Absence of Microwave Effect in Ceramics: Precise Temperature, Thermal Gradient, and Densification Determination in a Proportional Power Microwave Furnace," pp. 471-86 in *Ceramic Microstructures: Control at the Atomic Level.* Edited by A.P. Tomsia and A.M. Glaeser. Plenum Press, New York, 1998.

[4]O.C. Standard and C.C. Sorrell, "Densification of Zirconia – Conventional Methods," pp. 251-300 in *Zirconia Engineering Ceramics: Old Challenges – New Ideas* [*Key Eng. Mater.*, **153-54**, 251-300 (1998)]. Edited by E. Kisi. Trans Tech Publications, Zurich, Switzerland, 1998.

[5]C.C. Sorrell, O.C. Standard, N. Ehsani, K.R. Ratinac, C.M.G. Harding, and A.J. Ruys, "Advanced Microwave Sintering of Ceramics: Densification and Thermal Effects of Oxide Ceramics in a Proportional-Power Microwave Furnace," pp. 333-66 in *Ceramic Interfaces: Properties and Applications.* Edited by R.St.C. Smart and J. Nowotny. IOM Communications Ltd., London, 1998.

[6]C.C. Sorrell, O.C. Standard, N. Ehsani, C.M.G. Harding, and A.J. Ruys "Assessment of Thermal and Kinetic Effects during Comparative Microwave and Resistance Heating," pp. 665-80 in *Advances in Science and Technology, 14. Ceramics: Getting into the 2000's – Part B. Proceedings of the World Ceramics Congress, Part of the 9th CIMTEC – World Ceramics Congress and Forum on New Materials.* Edited by P. Vincenzini. Techna Srl, Faenza, Italy, 1999.

[7]M.A. Janney, M.L. Jackson, and H.D. Kimrey, "Microwave Sintering of ZrO_2-12 mol% CeO_2," pp. 101-108 in *Ceramic Transactions, Volume 36. Microwaves: Theory and Application in Materials Processing II.* Edited by D.E. Clark, W.R. Tinga, and J.R. Laia, Jr. American Ceramic Society, Westerville, OH, 1993.

[8]M.A. Janney and H.D. Kimrey, "Diffusion-Controlled Processes in Microwave-Fired Oxide Ceramics," pp. 215-27 in *Materials Research Society Symposium Proceedings, Volume 189. Microwave Processing of Materials II.* Edited by W.B. Snyder, Jr., W.H. Sutton, M.F. Iskander, and D.L. Johnson. Materials Research Society, Pittsburgh, PA, 1991.

[9]M.A. Janney, C.L. Calhoun, and H.D. Kimrey, "Microwave Sintering of Zirconia-8 Mol% Yttria," pp. 311-18 in *Ceramic Transactions, Volume 21. Microwaves: Theory and Application in Materials Processing.* Edited by D.E. Clark, F.D. Gac, and W.H. Sutton. American Ceramic Society, Westerville, OH, 1991.

[10]Anonymous, "Calibration of Thermocouples," pp. 112-22 in *Manual on the Use of Thermocouples in Temperature Measurement (ASTM Special Technical Publication 470B).* ASTM, Philadelphia, 1981.

[11]A.E. Heywood, "Thermocouples and Resistance Thermometers," pp. 22-32 in *Temperature Measurement and Control*. Chameleon Press, London, 1978.

[12]W. Hume-Rothery, J.W. Christian, and W.B. Pearson, *Metallurgical Equilibrium Diagrams*. Institute of Physics, London, 1952.

[13]A.J. Ruys, S.A. Simpson, and C.C. Sorrell, "Thixotropic Casting of Fibre-Reinforced Ceramic Matrix Composites," *J. Mater. Sci. Lett.*, **13** [18] 1323-25 (1994).

[14]A.J. Ruys, J. Kerdic, and C.C. Sorrell, "Thixotropic Casting of Ceramic-Metal Functionally Gradient Materials," *J. Mater. Sci.*, **31** [16] 4247-55 (1996).

[15]M. Arai, J.G.P. Binner, G.E. Carr, and T.E. Cross, "High Temperature Dielectric Property Measurements of Engineering Ceramics," pp. 483-92 in Ref. 7.

[16]W. Xi and W.R. Tinga, "A High Temperature Microwave Dielectrometer," pp. 215-24 in Ref. 9.

[17]C.C. Sorrell, O.C. Standard, N. Ehsani, K.R. Ratinac, J.R. Rider, X.Q. Zhen, and A.J. Ruys, unpublished data.

[18]Anonymous, *Australian Standard AS 1774.5–1989. Refractories and Refractory Materials – The Determination of Density, Porosity and Water Absorption*. Standards Australia, Homebush, NSW, 1989.

[19]Anonymous, *American Society for Testing and Materials ASTM E 112–88. Standard Test Methods for Determining Average Grain Size*. ASTM Committee on Standards, Philadelphia, PA, 1988.

[20]M.I. Mendelson, "Average Grain Size in Polycrystalline Ceramics," *J. Amer. Ceram. Soc.*, **52** [8] 443-46 (1960).

[21]D.-J. Chen and M.J. Mayo, "Rapid Rate Sintering of Nanocrystalline ZrO_2-3 mol% Y_2O_3," *J. Amer. Ceram. Soc.*, **79** [4] 906-12 (1996).

[22]D.-H. Kim and C.H. Kim, "Entrapped Gas Effect in the Fast Firing of Yttria-Doped Zirconia," *J. Amer. Ceram. Soc.*, **75** [3] 716-18 (1992).

[23]A. Dé, I. Ahmad, E.D. Whitney, and D.E. Clark, "Microwave (Hybrid) Heating of Alumina at 2.45 GHz: I. Microstructural Uniformity and Homogeneity," pp. 319-28 in Ref. 9.

[24]D.K. Agrawal, Y. Fang, D.M. Roy, and R. Roy, "Fabrication of Hydroxyapatite Ceramics by Microwave Processing," pp. 231-36 in *Materials Research Society Symposium Proceedings, Volume 269. Microwave Processing of Materials III*. Edited by R.L. Beatty, W.H. Sutton, and M.F. Iskander. Materials Research Society, Pittsburgh, 1992.

[25]M.C.L. Patterson, P.S. Apté, R.M. Kimber, and R. Roy, "Mechanical and Physical Properties of Microwave Sintered Si_3N_4," pp. 301-10 in Ref. 24.

[26]J. Cai, C.Y. Song, B.S. Li, X.X. Huang, J.K. Guo, and Y.L. Tian, "Microwave Sintering of Zirconia Toughened Mullite (ZTM)," pp. 545-50 in *Materials Research Society Symposium Proceedings, Volume 347. Microwave Processing of Materials IV*. Edited by M.F. Iskander, R.J. Lauf, and W.H. Sutton. Materials Research Society, Pittsburgh, 1994.

[27]C.E. Holcombe and N.L. Dykes, " 'Ultra' High-Temperature Microwave Sintering," pp. 375-86 in Ref. 9.

[28]A. Dé, I. Ahmad, E.D. Whitney, and D.E. Clark, "Effect of Green Microstructure and Processing Variables on the Microwave Sintering of Alumina," pp. 283-88 in Ref. 8.

[29]A. Dé, I. Ahmad, E.D. Whitney, and D.E. Clark, "Microwave (Hybrid) Heating of Alumina at 2.45 GHz: II. Effect of Processing Variables, Heating Rates and Particle Size," pp. 329-39 in Ref. 9.

[30]R. Morrell, *Handbook of Properties of Technical & Engineering Ceramics. Part 1: An Introduction for the Engineer and Designer*. HMSO, London, 1989.

[31]M.A. Janney, H.D. Kimrey, M.A. Schmidt, and J.O. Kiggans, "Grain Growth in Microwave-Annealed Alumina," *J. Amer. Ceram. Soc.*, **74** [7] 1675-81 (1991).

MICROWAVE SINTERING OF LARGE-SIZED NTC THERMISTOR CERAMICS

Aimin Chang, Jianwen Zhuang and Wen Yang
Xingjiang Institute of Physics
Urumuqi,830011
P.R.China

Yanrong Li
University of Electronics Science &Technology
Chendu , 610054
P.RChina

ABSTRACT

In this report, microwave sintering of large-sized (50 mm in diameter and 50 mm in thickness) crack-free Co-Mn-Ni spinel oxide NTC (Negative Temperature Coefficient) ceramics has been achieved successfully by the adoption of $MgAl_2O_4$-$LaCrO_3$ (MALC) susceptor in the temperature range of 1100-1200 °C. From SEM observations, microwave sintered samples showed higher homogeneity and smaller grain size in microstructure than conventionally sintered identical samples. From resistance vs. temperature measurement, the results showed that NTC thermistor elements from microwave sintered NTC ceramics had higher production rate than that from conventionally sintered NTC ceramics, this is due to the higher homogeneity in microwave sintered samples than conventionally sintered samples. It needs shorter time in microwave sintering run for only 4 hrs compared to 48 hrs in conventional sintering.

INTRODUCTION

NTC thermistors are semiconducting ceramics fabricated in a variety of shape,and sizes from a solid solution of the Mn-Ni-Co spinel oxide system[1]. Conventional sintering is conducted in the range of 1150-1200°C with a long sintering time, and there exists inhomogeneity in the large–sized NTC ceramics after sintering. Microwave sintering shows advantage for the fabrication of NTC ceramics for its volumetric heating[2]. It is expected that NTC ceramics would form more homogeneous microstructure during microwave sintering process, and then result in the increase production rate of NTC elements from NTC ceramics. Microwave sintering of many electronic ceramics has been reported these years[3-5].However there are fewer reports for the microwave sintering of NTC ceramics. This is mainly obstructed by the thermal cracks easily occurred in the microwave sintering of NTC ceramics, and thermal runaway induced by inaccurate temperature measurement[6]. In this study, the appropriate insulator and susceptor and also accurate temperature measurement are adopted, in order to obtain the suitable microwave sintering technology for NTC ceramics.

EXPERIMENTAL PROCEDURE

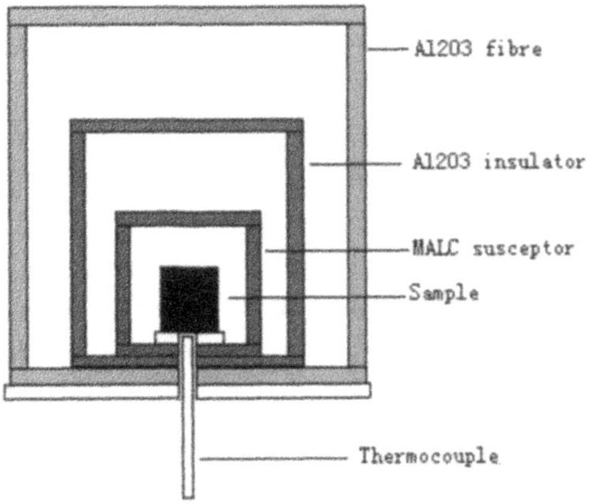

Figure 1. Set-up of insulating system.

The powders of CoMnNiO from coprecipitation method are pressed into cylinder shape (50mm in diameter and 50 mm in thickness). After cold isostatic pressing at the pressure of 380Mpa, the two sets of samples are conventionally sintered and microwave sintered for comparison. The susceptor adopted in microwave sintering are made of $MgAl_2O_4$-$LaCrO_3$ materials[7], and the insulating system are shown in Fig.1. The typical sintering schedule of conventional sintering and microwave sintering are shown in Fig.2. The NiCrAl alloy tube sheathed thermocouple was used to measure temperature in microwave sintering. The tip of the thermocouple was placed near the surface of the sample. The samples are soaked at the peak temperature for 20 mins and 2 hours for microwave sintering and conventional sintering, respectively. The total sintering time for conventional sintering was 48 hours. The heating rate is 10°C/min for microwave sintering, and 2°C/min for conventional sintering. After chipping and pasting, the resistance vs. temperature (R-T) characteristics of thermistor elements were measured.

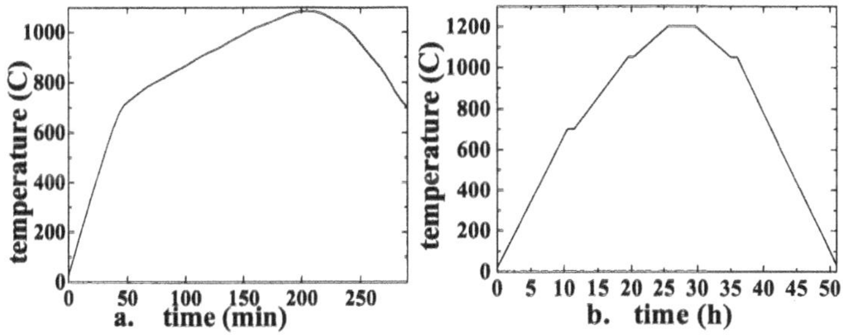

Figure 2. Typical Sintering Schedule of (a) microwave sintering and (b) conventional sintering

RESULTS AND DISCUSSION

Successful Microwave Sintering of Large-sized NTC Ceramics

Thermal cracks of NTC oxide ceramics are due to their low thermal conductivity and high thermal expansion coefficient in the rapid heating and cooling during the microwave sintering process even by the adoption of SiC susceptor. In this study, the samples for microwave sintering are placed in $MgAl_2O_4$-$LaCrO_3$ susceptor which is enclosed by Al_2O_3 ceramic and Al_2O_3 fiber materials. A Al_2O_3 chip is placed between the sample and the insulator to guarantee the homogeneous heating to the sample. This made the surface and the bottom of the sample in the same thermal environment. A small hole in the bottom lid allowed the sheathed thermocouple to be placed near the bottom of the sample in order to measure the surface temperature of the sintered sample. With this insulating system, crack free and large-sized NTC ceramics were successfully microwave sintered ,as shown in Figure 3.

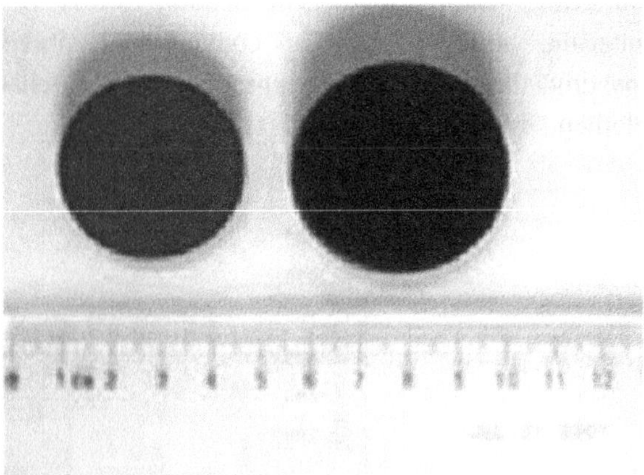

Figure 3. NTC ceramic sample with 50 mm in diameter by microwave sintering (left: after microwave sintering; right:before microwave sintering).

Microstructure and Density Comparison

Figure 4 are the SEM micrographs of microwave sintered and conventionally sintered samples at various temperatures. Microwave sintered samples showed better homogeneity than the conventionally sintered samples. This is due to the volumetric heating of microwave sintering. The grain size of microwave sintered samples is smaller than the conventionally sintered samples due to shorter sintering time which restrain the growth of grains during the microwave sintering.

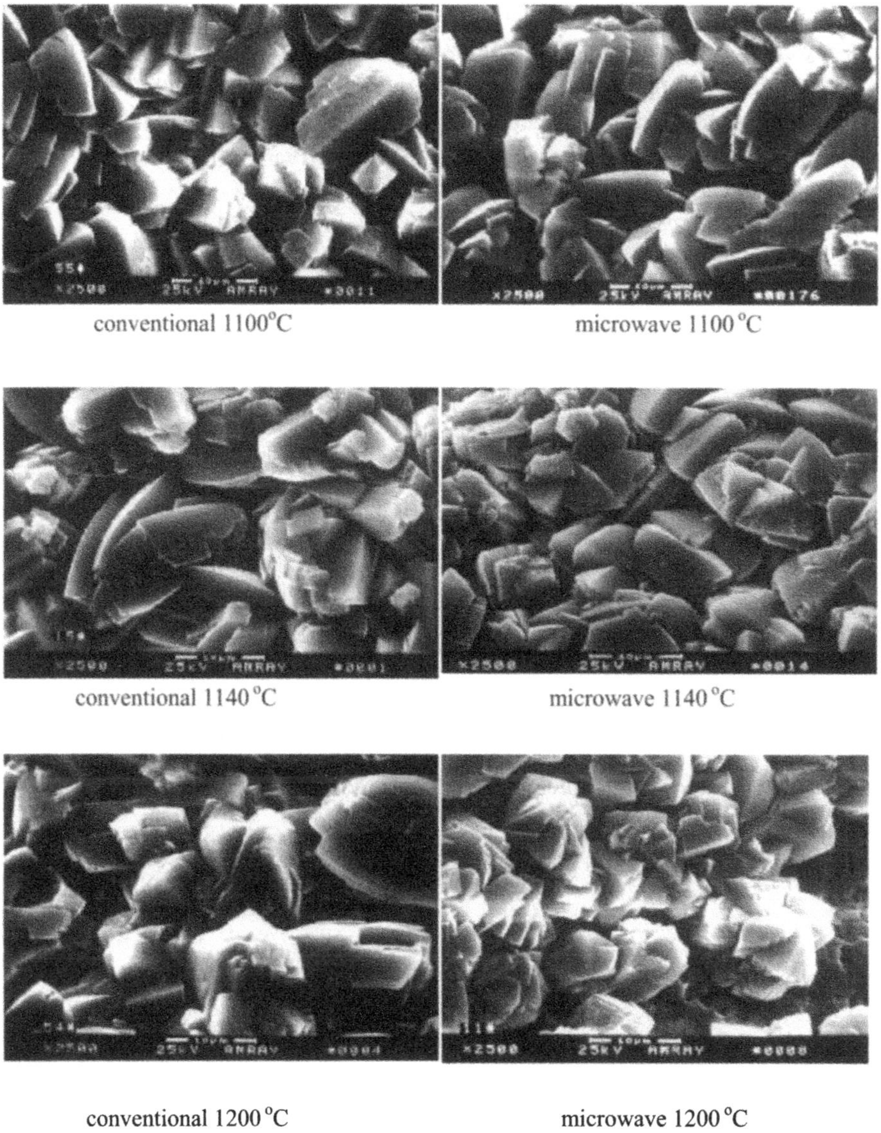

Figure 4. SEM of samples by microwave and conventional sintering at various temperatures.

From Figure 5, it showed that the density of microwave sintered sample is lower than that of the conventionally sintered identical samples at the same sintering temperature. This is due to the microwave sintered sample experienced shorter time than conventional sintered sample. Thus the air in microwave sintered samples is not easily driven off compared to conventional sintered samples.

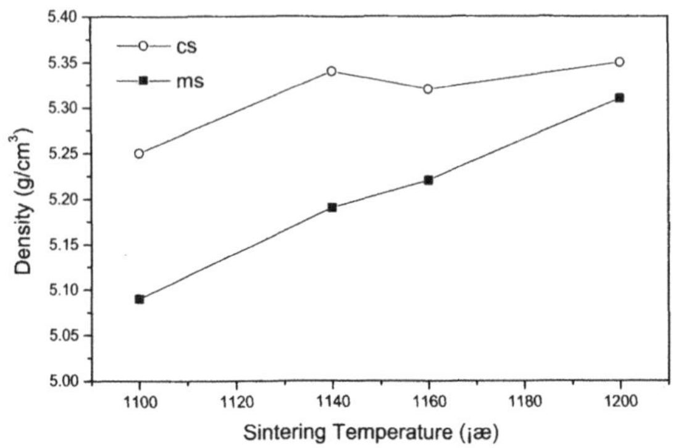

Figure 5. Density of microwave and conventionally sintered sample.

Electrical Properties of Sintered NTC Ceramics

The resistivity (ρ) for NTC thermistors is defined by[8]:

$$\rho = \rho_0 \exp\left(\frac{Q}{kT}\right) = \rho_0 \exp(B/T) \tag{1}$$

where Q is the site activation energy (eV), T the temperature (K), k the Boltzmann constant, and ρ_0 the preexponential term independent of temperature, and B=Q/k is the energetic constant of the material in kelvin.

Figure 6 showed the distribution of B value and resistance of NTC elements of microwave sintered samples. The homogeneity of microwave sintered NTC ceramics are also reflected in the electrical properties. R_{25oC} is the resistance of NTC elements measured at 25°C, the NTC elements with $R_{25oC} \pm 1\%$ homogeneity are improved to 65% from 28% in conventionally sintered samples. The homogeneity of $R_{25oC} \pm 3\%$ are improved to 90% compared to 63% in conventionally sintered samples, as shown in Figure 7.

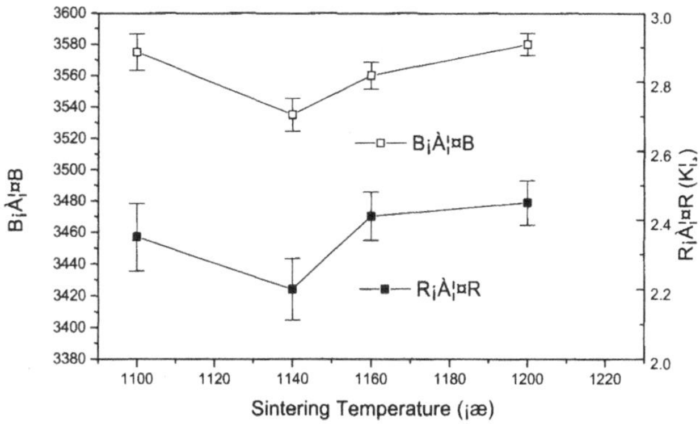

Figure 6. Distribution of B values and Resistance of NTC elements from microwave sintered samples.

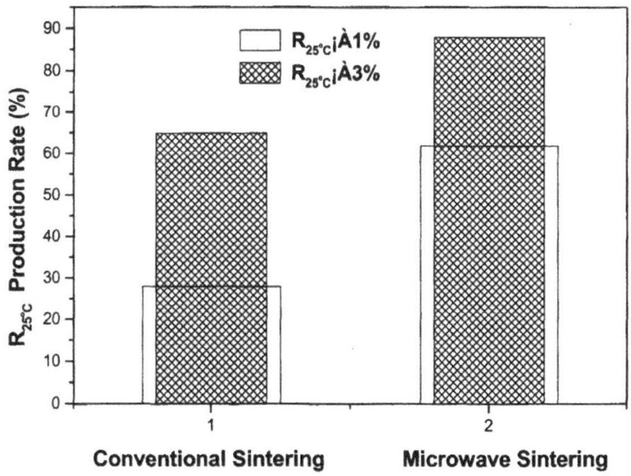

Figure 7. NTC elements production rate comparison.

CONCLUSIONS

This study demonstrated that it is feasible to use microwave sintering to fabricate crack free and large-sized NTC ceramics by the adoption of $MgAl_2O_4$-$LaCrO_3$ susceptor. Compared with conventional sintering, microwave sintered samples had homogeneous microstructure and fine grain size, and especially an improvement of homogeneity in B constant and resistance of NTC elements. Thus the production rate of NTC elements has been improved significantly by the application of microwave sintering technique.

ACKNOWLEDGEMENT

The author gratefully acknowledges the support of K.C.Wong Education Foundation, Hong Kong.

REFERENCE

[1]J.G.Fagan and V.R.W.Amarakoon, “Reliability and Reproducibility of Ceramic Sensors: Part I, NTC thermistors”, 72[1]70-79(1993).

[2]W. H.Sutton, “Microwave processing of cearmic materials”, *American Ceramic Society Bulletin*, 68[2]376-85(1989).

[3]L.Sheppard, “Manufacturing Ceramics with Microwaves: The potential for Economic Production”, Am.Ceram.Soc.Bull.,67[10]:1556-61(1988)

[4]D.Stein etal."Microwave Processing-An Emerging Industry Technology?",Mat Res.Soc.Symp.Proc. Vol.347(1994)

[5]Laurel M.Sheppard, “Innovative processing of advanced ceramics”, American Ceramic Society Bulletin, 72[4]48-53(1993)

[6]A.M.Chang, “Microwave proceesing of electronic ceramic materials”, doctoral dissertation, University of Electronics Science &Technology (2002)

[7]A.M.Chang,Y.Han,J.Wu,M.D.Tao, “Insulator Applied for Microwave Sintering of Electronic Ceramics”, China Pat.ZL97117256.0, Dec,2000.

[8]P.Fau,J.P.Bonino,J.J.Demai and A.Rousset, “Thin Film of Nickel manganese oxide for NTC thermistor applications”, *Applied surface Science*, 65/66,319-24(1993)

ENHANCEMENT OF CUTTING PERFORMANCE OF CEMENTED CARBIDE CUTTING TOOLS BY MICROWAVE TREATMENT

S.Aravindan
Research Fellow
Department of Mechanical & Control Engineering
Tokyo Institute of Technology
TOKYO, 152-8552 JAPAN
E-MAIL:aravindhans@hotmail.com

J.Ramkumar, S.K.Malhotra and R.Krishnamurthy
Manufacturing Engineering Section,
Department of Mechanical Engineering
I.I.T Madras, Chennai – 600 036, India

ABSTRACT

Glass fiber reinforced plastics (GFRP) find wide applications, particularly in the aerospace and automotive industries. In order to fabricate structural components of GFRP to near-net shape, machining of GFRP composites, such as cutting and drilling are necessary. The heterogeneity/non-homogeneous nature of GFRP makes machining a difficult task. The non-homogeneous nature of these materials certainly poses problems, such as rapid tool wear and associated deterioration of the quality of the manufactured component. The present study attempts to enhance the cutting performance of tungsten carbide (WC) inserts by way of microwave post-sintering treatment.

INTRODUCTION

Fiber-reinforced composites are finding structural applications due to their excellent mechanical and physical properties. Machining of GFRP is needed for the production of near-net shape geometry. Production of damage free surfaces in GFRP when machining is difficult due to contact of tool with relatively softer matrix and hard abrasive fibers of the composite material. Such variation results in rapid tool wear and reduced tool life. Improvement of cutting performance and quality of the machined surface are major requirements in cutting of GFRP [1&2].Microwaves are effectively used in the processing activities of metals, ceramics, semiconductors and composites by researchers [3-5]. Application of microwave radiation in the production of cemented carbide drills offers a great potential to the cutting industry by providing methods for shortening the process sequence [6-8].The present study emphasizes microwave-assisted post-sintering treatment of cemented carbide (WC) drill inserts for possible enhanced metallurgical and mechanical properties of inserts, facilitating improved machining performance. Details are presented in the following sections.

EXPERIMENTAL PROCEDURE

Cemented carbide drill inserts (K-20) were exposed to microwave-assisted post sintering treatment, using a multimode microwave furnace (2.45 GHz), for microwave hybrid heating technique. Three sets of samples (each sample consists of 5 cemented carbide inserts) were subjected to microwave exposure for 1800 seconds. The shank portion of drill inserts was effectively sealed from microwave exposure in order to avoid microwave reflection using alumina and silicon carbide fencing arrangement. Treated and untreated cutting tools were studied through optical microscopy and micro-hardness test to assess the mechanism behind the change in characteristics.

To evaluate the cutting performance of microwave irradiated cutting inserts, drilling was carried out on GFRP composites. The trials were conducted and the best cutting speed and feed were fixed as 2000 rpm and 0.06mm/rev. During drilling thrust and torque were measured through electrical resistive temperature compensated strain gauge type tubular dynamometer. Acoustic Emission (A.E.) signals emanated during drilling was acquired as a measure of the response of the work material. A.E. signals were acquired by AET 5500 system using a broad band sensor (100 kHz-2 MHz). The signals from the sensor were magnified using a pre-amplifier and stored in a data logger. The acquired signals were analysed by a suitable pc based data acquisition system. Tool wear and hole geometric deviation by tool makers microscope were made as off-line measurement.

RESULTS AND DISCUSSION

Micro structural observations

Plain cemented carbide cutting tools consist of cemented carbide grains and cobalt (Co) as a binder. Microwave exposure of cemented carbide tools facilitated changes in metallurgical characteristics, due to the absorption of microwaves by WC. Changes in microstructure of the WC inserts are observed owing to absorption of microwaves by WC grains. Liquification of Co, depletion of Co at surface, upliftment of cemented carbide and formation of complex carbide phase at the surface are the major effects associated with microwave treatment of cemented carbide tools [9]. Typical optical micrographs of treated and untreated specimens are presented in Fig.1(a & b).Compared to untreated, more cemented carbide grains are observed over the surface due to the depletion of surface Co resulting in formation of WC skeleton matrix. Pooling of WC is observed in the micrographs of treated tool.

(a) (b)

Fig.1. Typical micrographs of (a) microwave treated tool (b) untreated tool

Micro hardness survey

The influence of microwave treatment on mechanical characteristics of tools can be evaluated through hardness survey. Typical variation of hardness of both treated and untreated tools are presented in Fig 2. The possible reason for the enhancement in hardness of treated tools may be due to the WC grain upliftment and complex carbide formation.

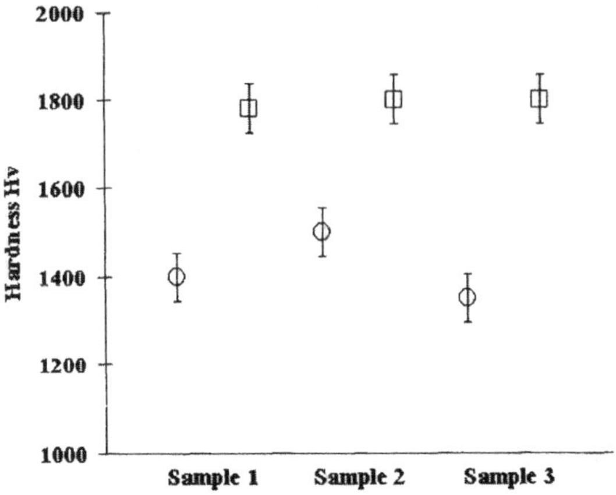

Fig.2. Typical variation of hardness of treated and untreated cutting tools (Square form -treated, round form-untreated)

Performance of treated tools

Microwave treated tools are subjected to metallurgical and mechanical/physical changes. Such changes enhance the cutting performance. The improved hardness over the surface and the softer bulk of cutting tools resulted due to microwave treatment can be effectively used to machine heterogeneous composites. Treated tools are used to drill holes in GFRP. During drilling, measurement of thrust, torque, and acquisition of acoustic emission signals were carried out. Measurement of tool wear and hole shrinkage were carried out as off-line quality assessment of produced holes. For comparison, normal untreated tools were also used to perform the drilling process.

Cutting Forces

Typical variation of cutting forces (thrust & torque) during drilling of GFRP using microwave irradiated and untreated tools with number of holes drilled is presented in Fig.3 (a & b). Due to the tool wear, the cutting forces increase with the number of holes irrespective of treatment. There is considerable amount of reduction in cutting force components (thrust & torque) for a microwave treated tool, due to the improved properties of treated tools. Rate of increase in cutting force with number of holes for treated tools is comparatively at a lower order as illustrated in figure.

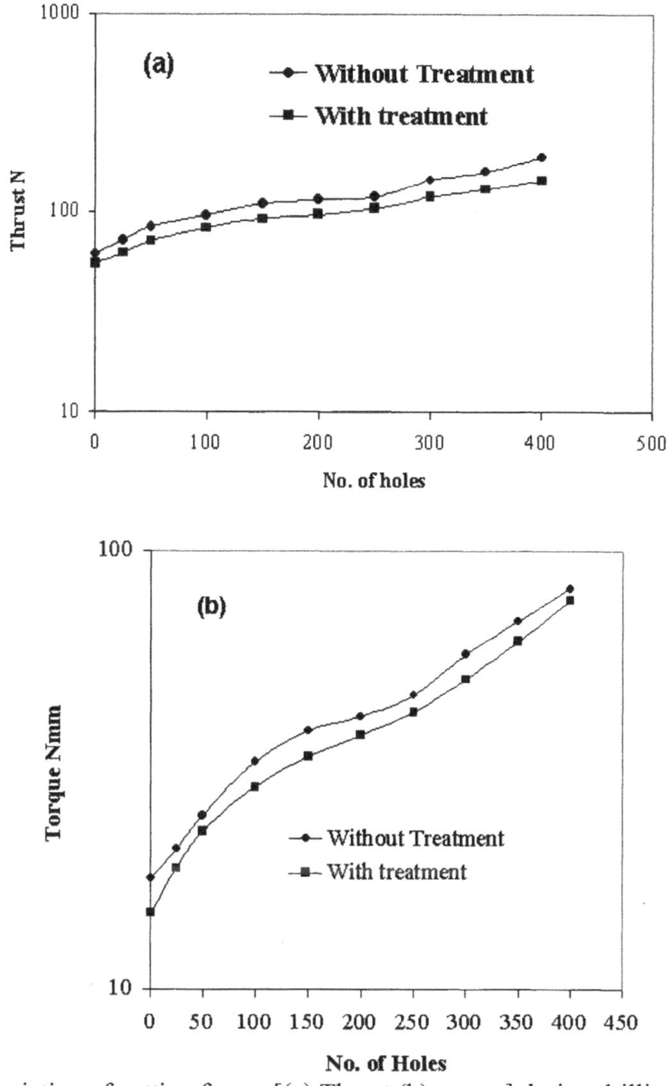

Fig.3. Variation of cutting forces [(a) Thrust (b) torque] during drilling of GFRP

Tool Wear

Rapid tool wear is the major undesirable effect encountered during drilling of composites. Typical variation of tool wear of treated and untreated tools with the number of holes produced is illustrated in Fig.4. Increase in tool wear is observed in both cases of treated and untreated tools, with the number of holes. The rate of increase in tool wear of untreated tools after around 250 holes is comparatively larger, as illustrated in figure, due to the accumulation of thermal-assisted material degradation. Tool wear of treated tools is on the order of lower magnitude and even after 250 holes the tool wear progresses steadily. During drilling of heterogeneous materials, cutting tools are often subjected to force variation because they encounter soft matrix and hard fibers. Such variation in cutting forces results in reduced tool life due to increased tool wear. Combination of relatively softer bulk and harder surface of the treated inserts has given adequate strength to the cutting inserts to accommodate the fluctuation of cutting forces, during machining of non-homogeneous material.

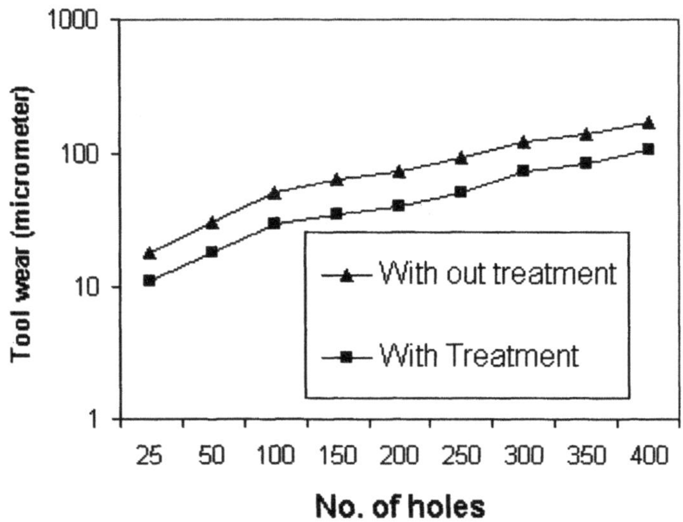

Fig.4. Variation of tool wear

Hole shrinkage

Quality of the surface produced is a good indicator of the machining process. The rapid tool wear resulted in deformation/deviation in hole size. Due to reduced tool wear of treated tools, the hole deviation/shrinkage of holes produced by treated tools is comparatively minimum. The maximum value for the hole tolerance allowed is 100 microns as per data reported. The untreated tool has reached the maximum value with 80 holes and the microwave treated is found to have lower amount of hole shrinkage.

Monitoring of A.E. signals

Acoustic emission is defined as the transient elastic wave generated by the rapid release of energy from a localized source or sources within a material when subjected to a state of stress. A.E. signals emanating from GFRP during drilling are acquired using AET 5500 system. Typical r.m.s values of the acquired signals are presented in Fig.5. The increase in r.m.s. values with the number of holes confirms the continuous mode of fracture of work material during drilling. Treated tools exhibit comparatively lower order r.m.s. value, which confirms the quality of cutting tool and the manufactured component.

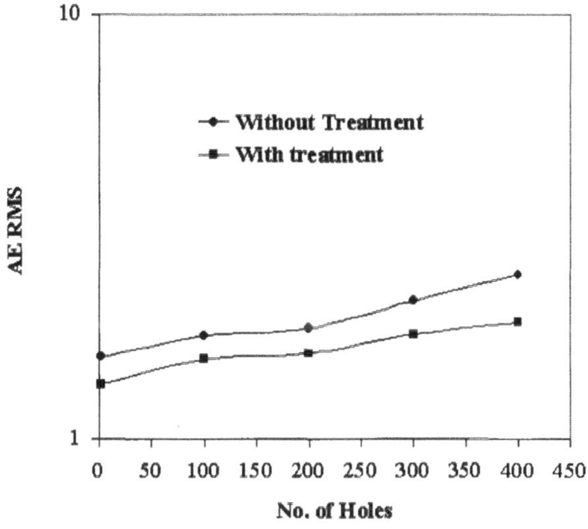

Fig.5. Variation of R.M.S. value of A.E.signal

CONCLUSION

Microwave treatment of cemented carbide inserts has resulted in favorable material response. These effects resulted in softer bulk and harder surface of cutting inserts which is highly suited for machining of heterogeneous material like polymeric composites. Microwave treated tools exhibit enhanced cutting performance through reduced cutting forces, reduced tool wear and reduced hole shrinkage.

REFERENCES

[1]Santhanakrishnan., R.Krishnamurthy and S.K.Malhotra, "Machinability characteristics of Fiber Reinforced Plastic composites," *Journal of Mechanical Working technology,* **17**, 195-204 (1988).

[2]W.Konig, Ch.Wulf, P.Grass and H.Willerschield., "Machining of Fiber reinforced plastics," *Manufacturing technology CIRP Annals*, **34** [2] 537-548 (1985)

[3]W.H.Sutton, "Microwave processing of ceramic materials," *American ceramic Bulletin* **68**[2] 376-386 (1989).

[4]D.E.Clark *et al*, "Recent developments in microwave processing of ceramics,"*MRS Bulletin*, Nov, 40-46 (1993).

[5]K. Thompson, J.H. Booske, D.F. Downey, Y. Gianchandani, R. Cooper, "RF and Microwave Annealing for Ultra Shallow Junction Formation," *Rapid Thermal and Other Short Time Processing Technologies II Symposium, Proceedings of the Electrochemical Society*,Washington DC, March 24-28 (2001).

[6]Willert-Porada *et al*, "Microwave sintering of hard metals," *International Journal of Refractory metals & Hard Materials*, **16** 409-416 (1998).

[7]R.Roy, D.Agarwal and J.P.Cheng, "Extra ordinary successes with microwave fields for synthesis and sintering of ceramics, metals and composites,"-*Proceedings of symposium on Microwave effects and applications,* Kokushikan University, Tokyo 5-9 (2001).

[8]J.P.Cheng, D.K.Agarwal, S.Komarneni, M.Mathis and R.Roy, "Microwave processing of WC-Co composites and ferroic titanates," *Material Research Innovation,* **1**, 44-52 (1997).

[9]J.Ramkumar,S.Aravindan,S.K.Malhotra and R.Krishnamurthy, "Enhancing the metallurgical properties of WC insert (k-20) cutting tool through microwave treatment," *Material Letters*,**53** 200-204 (2002).

THE EFFECT OF REACTION PARAMETERS ON MICROWAVE – INDUCED COMBUSTION SYNTHESIS OF Al_2O_3 – TiC COMPOSITE POWDERS

Duangduen Atong
National Metal and Materials Technology Center
114 Science Park, Paholyothin Road
Klong 1, Klong Luang
Pathumthani, 12120 THAILAND

David E. Clark
Virginia Polytechnic Institute and State University
Department of Materials Science and Engineering
Blacksburg, VA 24061 USA

ABSTRACT

Al_2O_3-TiC powders were produced by a combustion synthesis technique using microwave heating (MH) and microwave hybrid heating (MHH). Previous work has been done to identify the most important reaction parameters that influence microwave-ignition of this powder using Plackett-Burman design. They were amount of excess Al and Al_2O_3, particle size of C and Al_2O_3, and heating method. In this study, the effect of those important reaction parameters on the ignition behavior and characteristics of the resulting powders was extensively evaluated by the central composite design (CCD). The empirical models relating these important parameters and their interactions to the responses were then developed.

INTRODUCTION

Ceramic composites have been used in the cutting tool industry for years. A major improvement in ceramic cutting tools was an Al_2O_3-based composite containing 1 to 5 microns diameter titanium carbide particles uniformly dispersed throughout the matrix. Most commercial materials are nominally Al_2O_3-30 wt% TiC (25vol%TiC). Currently, commercial Al_2O_3–TiC composites are manufactured primarily by pressureless sintering or hot pressing of TiC and Al_2O_3 powders. However, the additional finishing of the cutting edges of hot pressed samples and the need for Hot Isostatic Pressing (HIP) for final densification of the pressureless sintered samples are the major production cost of these tools [1]. Recently, the synthesis of Al_2O_3–TiC composites by the aluminothermic reduction of TiO_2 in the presence of carbon has drawn much attention as an alternative to conventional processes [2-4]. The process utilizes the heat released by an exothermic reaction between reactants to prepare materials. This process is commonly referred to as self-propagating high temperature synthesis (SHS) [5] or more simply, combustion synthesis [6]. Generally, the SHS process is ignited at the surface of the material by thermal radiation from a laser beam, a heating coil or an isothermal furnace. The surface temperature is raised to the ignition temperature (T_{ig}) where the materials react exothermically. The heat generated at the surface allows the adjacent materials to react and a combustion wavefront propagates across the sample converting reactants to products.

A completely different approach for igniting the SHS process occurs when microwaves are used as the heating source. Microwaves tend to heat the entire sample nearly uniformly due to its internal heating mechanisms. The surface of the sample radiates thermal energy leading to an inverted temperature gradient where the interior of the sample is hotter than the surface. In this approach, the sample ignites in the center and a combustion wavefront propagates outward in a radial manner, which ensures complete combustion of the material.

The union of two technologies: microwave processing and combustion synthesis has been found to reinforce each other in the fabrication of composites by allowing for enhanced process control and superior product quality, as compared to conventional combustion synthesis [7-8]. However, the use of microwave energy to synthesize ceramic powders has received limited attention [9-11].

Previous work has been done on producing Al_2O_3-TiC powders by a combustion synthesis technique using microwave heating (MH) and microwave hybrid heating (MHH) [1]. A sixteenth-run Plackett-Burman design was performed to analyze the significance of reaction parameters on ignition behavior and characteristics of the resulting powders. The reaction parameters considered included compositions of the reactants, particle sizes of reactants, heating method, and microwave power. The results showed that the most significant factors affecting the investigated responses were amount of excess Al and Al_2O_3, particle size of C and Al_2O_3, and also heating method.

In this study, the thirtieth-run central composite design (CCD) was used to perform a comprehensive evaluation of the effects of those 5 important parameters identified by Plackett-Burman design. The empirical models based on statistical analysis relating the important reaction parameters to ignition behavior and characteristics of the resulting powders were developed.

EXPERIMENTAL

Al_2O_3-TiC powders were synthesized by the SHS method from a mixture of TiO_2, C, Al and Al_2O_3 reactants using microwave energy, according to equation (1).

$$3TiO_2 + 3C + (4 + x)Al + yAl_2O_3 \rightarrow 3TiC + (2 + y)Al_2O_3 + xAl \qquad (1)$$

The central composite design (CCD) was chosen for conducting sequential experiments to study the effect of reaction parameters on the ignition time and temperature, combustion time and temperature, and characteristic of resulting powders. Based on the results of the sixteen-run Plackett-Burman design, the most significant parameters affecting the investigated responses were amount of excess Al and Al_2O_3, particle size of C and Al_2O_3, and also heating method (MH and MHH). In this study, the qualitative parameter, heating method, is of interest. Therefore, two experimental sets for MH and MHH were then conducted in parallel to each other. Thus there were only four parameters to be investigated in each set of experiment. Each reaction parameter was evaluated at five different levels (-2, -1, 0, 1, 2). The five levels for each of the four parameters are shown in Table I. Other (less significant) factor levels were set at fixed levels based on the literature and economics. The particle size of TiO_2 and Al were chosen at 2-3 μm (99% rutile, Alfa Aesar) and 10 μm (99.7%, Alfa Aesar), respectively. The samples were ignited under microwave power of 800 watts operating at 100% duty cycle. The number of experimental runs required for four parameters at five levels was 30. The actual factor levels for a thirty-run CCD design are shown in Table II. This design was used for both experimental sets, MH and MHH.

The reagent powders were mixed and pulverized in a porcelain mortar and then sieved through a 35-mesh screen. Then the powders were dried in an oven at 120°C for 2 hours prior to ignition. For MH, the 5g-powder mixture was filled in a fused quartz crucible and placed inside the insulation, which was then positioned at the center of the microwave oven. For the MHH case, the crucible filled with powders was encased in an Alfrax-66-40% SiC susceptor, which was then enclosed in the insulation. An inconel-shielded K-type thermocouple (0-1250°C) was used to determine ignition temperature (T_{ig}). The combustion temperature (T_c) was monitored by a two- color infrared pyrometer (800-2300°C). The microwave heating setup for this work can be found elsewhere [9]. The temperature versus time data was recorded. The powders produced were then characterized by x-ray Diffraction (XRD) and Scanning Electron Microscope (SEM) techniques. The Coulter method, ultrapycnometer and Brunauer, Emmett, and Teller (BET) method were used to analyze the particle size, density, and surface area of powders, respectively. Analysis of variance (ANOVA) was used to provide a quantitative measure of confidence of collected data. Empirical models relating the important factors to responses were developed.

Table I. Factors and Investigated Levels of Thirty-Runs CCD

Factor	Level				
	-2	-1	0	1	2
C Particle Size	0.017 μm (99.1%)[1]	0.027 μm (>99%)[1]	0.037 μm (>99%)[1]	0.042 μm (99.9%)[2]	0.075 μm (>99%)[1]
Al_2O_3 Particle Size	0.49 μm (99.8%)[3]	<10 μm (99.7%)[4]	20-38 μm (99.9%)[2]	44-53 μm (99.9%)[2]	63-74 μm (>99%)[5]
Excess Al	0%	10%	20%	30%	40%
Excess Al_2O_3	0%	10%	20%	30%	40%

1- Cabot Corp., 2 - Alfa Aesar, 3 – Alcoa, 4 – Aldrich, 5 – Fisher

Table II. Actual Factor Levels for a Thirty-Run Central Composite Design for the Microwave-Induced Combustion Synthesis of Al_2O_3-TiC Powders

Run	Factor Levels			
	C Particle Size	Al_2O_3 Particle Siz	Wt% Excess	Wt% Excess Al_2O_3
1	0.027	<10	10	10
2	0.027	<10	10	30
3	0.027	<10	30	10
4	0.027	<10	30	30
5	0.027	44-53	10	10
6	0.027	44-53	10	30
7	0.027	44-53	30	10
8	0.027	44-53	30	30
9	0.042	<10	10	10
10	0.042	<10	10	30
11	0.042	<10	30	10
12	0.042	<10	30	30
13	0.042	44-53	10	10
14	0.042	44-53	10	30
15	0.042	44-53	30	10
16	0.042	44-53	30	30
17	0.037	20-38	20	20
18	0.037	20-38	20	20
19	0.037	20-38	20	20
20	0.037	20-38	20	20
21	0.037	20-38	20	20
22	0.037	20-38	20	20
23	0.017	20-38	20	20
24	0.075	20-38	20	20
25	0.037	0.49	20	20
26	0.037	63-74	20	20
27	0.037	20-38	0	20
28	0.037	20-38	40	20
29	0.037	20-38	20	0
30	0.037	20-38	20	40

RESULTS AND DISCUSSION

Al_2O_3-TiC powders were produced by a combustion synthesis technique using microwave heating (MH) or microwave hybrid heating (MHH) as a heating source. A clear difference in ignition and combustion time was observed for these two ignition methods. It was found that approximately 7sec to 1.50min were required for MH ignited samples to reach ignition temperatures while MHH ignited samples required longer times, 20sec to 6 min. The MHH samples ignited at higher temperature than MH samples and resulted in lower combustion temperature. The confirmation of the product phases made by x-ray diffraction analysis is illustrated in Figure 1.

The characteristics of synthesized Al_2O_3–TiC powders of the thirtieth-run CCD design showed no significant difference in the properties of powders ignited by these two methods. Pycnometry analysis showed that the densities for MH and MHH ignited powders were in the range of 3.37-4.10 g/cm^3 and 3.4-4.20 g/cm^3, respectively. The specific surface areas and the particle sizes for MH ignited powders were in the range of 0.77-3.05 m^2/g and 0.55-2.11 microns, and were 0.90-3.76 m^2/g and 0.45-1.77 microns for MHH samples. The SEM micrograph of these MH and MHH ignited powders were given in Figure 2.

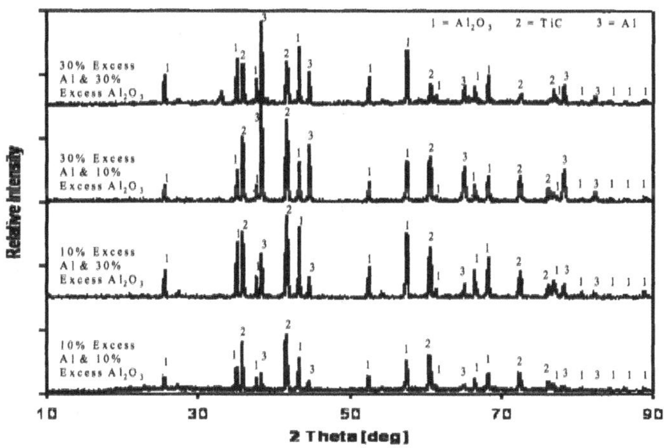

Figure 1. X-ray diffraction patterns for the MH ignited Al_2O_3-TiC powders.

The JMP®Start Statistics software was used to perform a multiple regression analysis for each of the responses measured. The regression variables were the four reaction parameters aforementioned. Analysis of variance (ANOVA) linear regression was performed, and significance levels for each linear (e.g. x_1, x_2,...), quadratic (e.g. x_1^2, x_2^2,...), and crossproduct (e.g. x_1x_2, x_1x_3,...) factors were thus determined. A model that described the response over the applicable ranges of the factors was then constructed. A full quadratic response-surface model in four variables consists of the following 15 terms:

$$Y = B_0 + B_1X_1 + B_2X_2 + B_3X_3 + B_4X_4 + B_{12}X_1X_2 + B_{13}X_1X_3 + B_{14}X_1X_4 \\ + B_{23}X_2X_3 + B_{24}X_2X_4 + B_{34}X_3X_4 + B_{11}X_1^2 + B_{22}X_2^2 + B_{33}X_3^2 + B_{44}X_4^2 \quad (2)$$

where, Y is the investigated response, B_o is the average of all the observations, B_i, B_{ij}, and B_{ii} are the effect of parameters in linear, crossproduct, and quadratic terms and X_i, X_{ij}, X_{ii} are the values of the parameters in linear, crossproduct, and quadratic term, respectively.

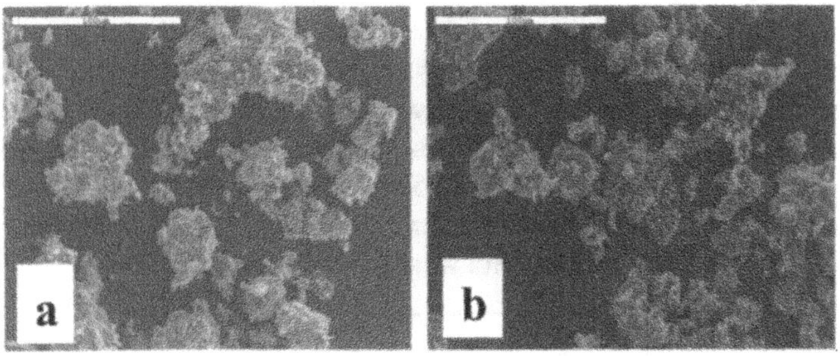

Figure 2. SEM of the a) MH and b) MHH ignited powders (bar = 100 μm).

The significance at the 0.5 levels (above 95%) of each parameter on the ignition behavior and characteristic of MH and MHH ignited powders are presented in Table III. The higher the significance level, the greater the certainty that the parameter is an important factor contributing toward the investigated response value. A dash represents a significance value below 90%, which is not considered significant, and those parameters are deleted from the model. The final regression models for predicting all the investigated response values for MH and MHH experiments are summarized in Table IV. For example, suppose the objective was to use only 10% Al (-1 level) and 10%Al_2O_3 (-1 level), and the only available carbon and alumina were the intermediate size (0 level). Then the ignition time of the sample ignited by MH could be estimated by using the first relationship shown in Table IV. to be 206 seconds (t_{ig} = 159.83 – 25.83 (-1) –20 (-1) + 45.83 (0) + 48.33 (0) = 205.66 seconds).

The relationship between the effect of the parameters and the investigated responses can also be illustrated graphically by the prediction profiler plots. The prediction profiler plots for ignition time, combustion temperature, powder density, and agglomeration size of Al_2O_3-TiC powder ignited under (MH) are displayed in Figure 3. It can be seen from the plots that, as the particle size of carbon increased, the ignition time decreased under MH, but not significantly.

Table III. Relative Significance of Reaction Parameters on all Investigated Responses for both MH and MHH experiments

MH	Significance (%)							
Parameter	Ignition Time (sec)	Ignition Temperature (°C)	Combustion Time (sec)	Combustion Temperature (°C)	Powder Density (g/cc)	Particle Size (μm)	Specific Surface Area (m^2/g)	Agglomeration Size (μm)
Average	>99.99	>99.99	>99.99	>99.99	>99.99	>99.99	>99.99	>99.99
X_1	(93.00)	-	(90.36)	(90.83)	95.27	-	95.28	99.55
X_2	-	-	-	-	-	-	-	99.97
X_3	99.95	96.43	99.91	98.97	>99.99	99.28	99.98	-
X_4	>99.90	99.80	>99.90	99.97	95.27	95.12	97.33	99.88
X_1^2	-	-	-	99.86	98.08	97.52	99.73	-
X_1X_2	-	-	-	-	-	-	-	-
X_2^2	-	-	-	99.98	98.56	-	-	-
X_1X_3	-	-	-	-	-	-	-	-
X_2X_3	-	-	-	-	-	-	95.69	-
X_3^2	-	-	-	-	-	96.11	95.93	-
X_1X_4	-	-	-	99.32	-	-	-	-
X_2X_4	(94.43)	-	(91.45)	99.21	-	-	-	-
X_3X_4	(92.33)	-	-	>99.90	-	-	-	-
X_4^2	-	98.87	-	-	-	(90.07)	-	-
MHH								
Average	>99.99	>99.99	>99.99	>99.99	>99.99	>99.99	>99.99	>99.99
X_1	97.06	-	97.03	-	95.84	96.62	99.69	99.78
X_2	(91.80)	-	(91.02)	-	-	-	-	99.93
X_3	99.93	99.18	99.92	99.94	>99.99	98.25	>99.99	-
X_4	99.96	99.85	99.96	99.96	97.89	99.04	99.96	99.97
X_1^2	-	-	-	-	-	99.18	99.63	-
X_1X_2	-	-	-	-	-	-	-	-
X_2^2	-	(93.29)	-	>99.9	95.57	-	-	-
X_1X_3	-	-	-	-	-	-	-	-
X_2X_3	-	-	-	-	-	-	98.53	-
X_3^2	-	-	-	-	-	99.95	99.97	(91.98)
X_1X_4	-	(92.29)	-	99.79	-	-	-	-
X_2X_4	-	-	-	99.86	-	-	-	-
X_3X_4	-	-	-	>99.90	-	-	97.23	(94.54)
X_4^2	-	-	-	-	-	98.46	98.73	-

X_1 - C particle size, X_2 - Al_2O_3 particle size, X_3 - Wt% excess Al, X_4 - Wt% excess Al_2O_3

( )- grey zone where $0.05 \leq$ prob(2-tail) ≤ 0.10 (90% <significance < 95%)

Table IV. Summary of the Regression Models for Predicting the Ignition Behavior and Characteristic of a) MH and b) MHH Ignited Al_2O_3-TiC Powders

(a) MH	Prediction Equations
Ignition Time (sec)	$t_{ig} = 48.83 - 5.83X_1 + 13.17X_3 + 16.08X_4 - 7.5X_2X_4 + 6.88X_3X_4$
Ignition Temperature (°C)	$T_{ig} = 145.67 + 30.08X_3 + 48.75X_4 + 35.77X_4^2$
Combustion Time (sec)	$T_c = 57.83 - 6X_1 + 14X_3 + 18.03X_4 - 7.63X_2X_4$
Combustion Temperature (°C)	$T_c = 1929.17 + 23.20X_1 - 37.79X_3 - 61.38X_4 + 47.05X_1^2 + 59.05X_2^2 - 49.44X_1X_4 - 48.31X_2X_4 - 117.06X_3X_4$
Powder Density (g/cc)	$\rho = 3.76 + 0.02X_1 - 0.17X_3 - 0.02X_4 - 0.02X_1^2 - 0.02X_2^2$
Particle Size (μm)	$P.S. = 1.39 - 0.18X_3 - 0.12X_4 - 0.14X_1^2 - 0.12X_3^2 - 0.10X_4^2$
Specific Surface Area (m²/g)	$Sp.S.A. = 1.16 - 0.17X_1 + 0.38X_3 + 0.19X_4 + 0.27X_1^2 + 0.17X_3^2 + 0.21X_2X_3$
Agglomeration Size (μm)	$A.S. = 52.76 + 4.81X_1 + 6.70X_2 - 5.75X_4$
(b) MHH	**Prediction Equations**
Ignition Time (sec)	$t_{ig} = 159.83 - 25.83X_1 - 20X_2 + 45.83X_3 + 48.33X_4$
Ignition Temperature (°C)	$T_{ig} = 383.17 + 61.46X_3 + 78.13X_4 - 37.28X_4^2 - 46.94X_1X_4$
Combustion Time (sec)	$T_c = 163 - 26.61X_1 - 19.80X_2 + 45.63X_3 + 48.79X_4$
Combustion Temperature (°C)	$T_c = 1726.17 - 55.83X_3 - 58.67X_4 + 66.96X_2^2 - 58.38X_1X_4 - 61.38X_2X_4 - 125.25X_3X_4$
Powder Density (g/cc)	$\rho = 3.80 - 0.03X_1 - 0.17X_3 - 0.03X_4 - 0.02X_2^2$
Particle Size (μm)	$P.S. = 1.51 - 0.11X_1 - 0.13X_3 - 0.14X_4 - 0.14X_1^2 - 0.20X_3^2 - 0.12X_4^2$
Specific Surface Area (m²/g)	$Sp.S.A. = 1.05 + 0.22X_1 + 0.37X_3 + 0.29X_4 + 0.21X_1^2 + 0.28X_3^2 + 0.17X_4^2 + 0.22X_2X_3 + 0.19X_3X_4$
Agglomeration Size (μm)	$A.S. = 54.32 + 5.24X_1 + 6.02X_2 - 6.74X_4 - 2.5X_3^3 - 6.64X_3X_4$

X_1 - C particle size, X_2 - Al_2O_3 particle size, X_3 - Wt% excess Al, X_4 - Wt% excess Al_2O_3

The temperature profiles of $2Al_2O_3$ -3TiC -20%excess Al – 20%excess Al_2O_3 with a variation of carbon particle size are shown in Figure 4. It was observed that the slowest ignition time occurred when the smallest carbon particle size was used for both MH and MHH techniques. When carbon is small, it may be possible for a continuous carbon phase to act as a barrier to titania-aluminum particle contact, which is believed to be necessary for the initial reaction to TiO_2 and Al. When the carbon size is increased, the titania-aluminum particle contact is greater which may lead to decreased ignition time. However, the sample with intermediate carbon particle size was ignited faster than one with largest carbon particle size. This may be because it provided the best reactant particle arrangements and particle contact between titania-aluminum, which resulted in fastest ignition time. In the MH case, the sample with the largest carbon particle size was observed to have highest combustion temperature. The reason for this was thought to be due to a decrease surface area of the larger carbon particle size, decreasing heat loss. When using the MHH technique, a variation of carbon particle size did not change combustion temperature significantly. The susceptor in this case provided another heat source and helped reduced heat loss from the surface of the sample.

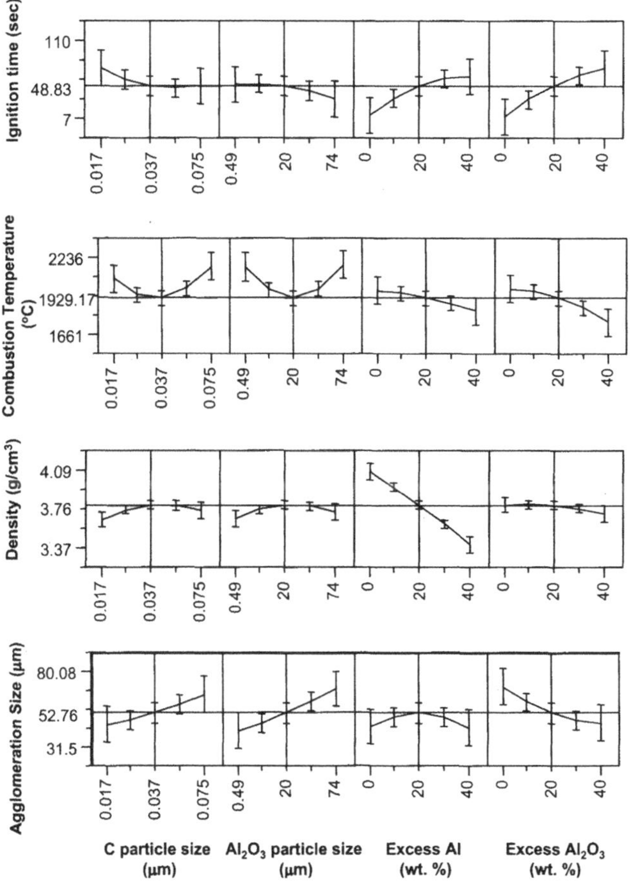

Figure 3. Prediction profiler plot of ignition time, combustion temperature, powder density, and agglomeration size of Al_2O_3-TiC powders ignited under MH (Bar represents the standard deviation).

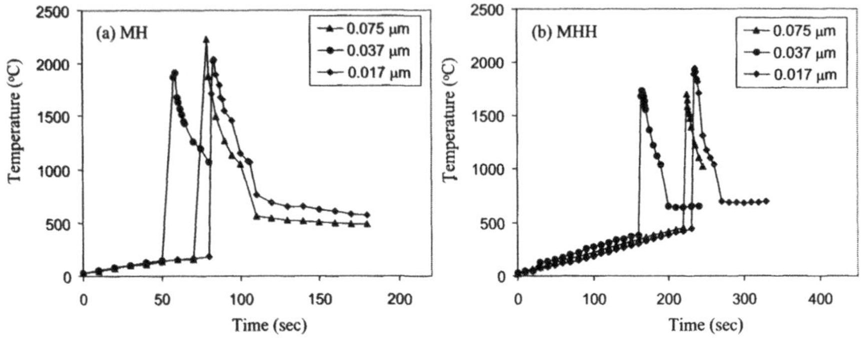

Figure 4. Temperature profile of $2Al_2O_3$ -3TiC -20% excess Al -20% excess Al_2O_3 with varying carbon particle size ignited by (a) MH and (b) MHH.

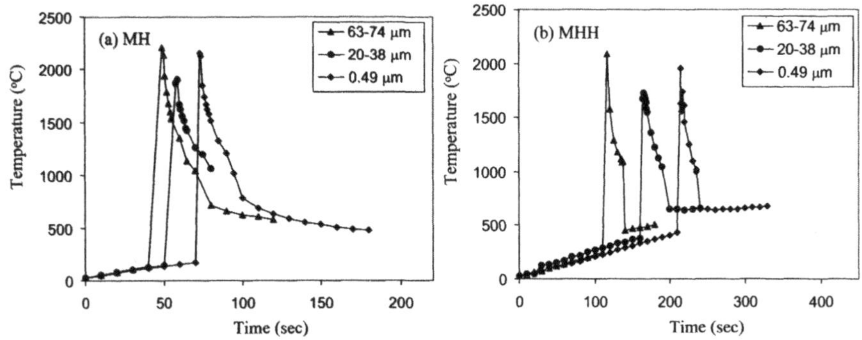

Figure 5. Temperature profile of $2Al_2O_3$ -3TiC -20% excess Al -20% excess Al_2O_3 with varying alumina particle size ignited by (a) MH and (b) MHH.

Figure 5 shows the temperature profile of $2Al_2O_3$ -3TiC -20%excess Al – 20%excess Al_2O_3 with variation of alumina particle size. As the alumina particle size increased the ignition time and temperature decreased. This was because the rest of the reactants could act as a continuous phase and thus increased the particle contact between aluminum and titania, resulting in faster ignition. Samples with smaller alumina particle size had lower combustion temperatures, presumably due to the reaction taking place over a longer time where heat losses become more significant. However, the sample with an intermediate alumina particle size (20-38 microns) was observed to have a lower combustion temperature than the one with the smallest alumina particle size (0.49 microns). A possible explanation for this was that the titania-aluminum particle contact network was disrupted when intermediate size alumina was used. With Al_2O_3 particle size (20-38microns) being close to Al particle size (10microns), smaller TiO_2 particles (2-3microns) would be distributed in the intersticies of packed larger particles of both Al_2O_3

and Al. Melting occurred locally and only the Al in direct contacted with TiO_2 reacted. On the other hand, using the smallest alumina, both TiO_2 (2-3microns) and Al_2O_3 (0.49 microns) would fill the voids between larger Al particles (10microns). This structure resulted in longer ignition time. However, Al melting occurred uniformly throughout the sample and thus the combustion temperature was higher than for the former case where an intermediate Al_2O_3 was used. These trends were also in agreement with the prediction profiles in Figure 3.

The effects of diluent addition on ignition behavior of $2Al_2O_3$-3TiC-Al were also shown in Figure 3. The addition of diluents (Al and Al_2O_3) to the reactants increased ignition temperature (not shown), and decreased the combustion temperature. In addition, the ignition time increased as the amount of the diluents increased. This was due to the decreased heat generation rate. In addition, it is thought that the high microwave reflectivity of the Al partially shielded the sample. In other words, sample containing large amounts of Al had a chance to form an interconnected network of metal particles, which would reflect microwaves and resulted in a shallow microwave penetration depth. Samples having small amounts of Al particles allowed microwaves to penetrate into the whole sample far greater than those having large amounts of Al. Furthermore, as the amounts of Al increased, the amounts of the rest of the reactants (TiO_2, C, Al_2O_3) decreased. The decreased amount of carbon, which is effectively heated with microwave, could result in less efficient absorption of microwave radiation. The Al_2O_3 particles also may act as a barrier to reactant particle contact. Alumina has low electrical conductivity and very low loss tangents thus allowing microwaves to pass through with very little absorption. These trends were similar in both MH and MHH techniques.

The effect of the addition of Al on the final product density was much more significant as compared to the other effects (Figure 3). The density of product decreases with increasing amount of Al. As the amount of Al increases, the combustion temperature decreases. The product would take a shorter time to cool and therefore a liquid phase would be present in the product for shorter periods. This resulted in the less densified product. It was noted that the effect of amount of Al_2O_3 on the product density also showed a similar trend but to a lesser extent.

The effects of various parameters on the agglomeration size of ignited powders are also shown in Figure 3. It can be seen that the larger reactant particles produced coarser reacted powders. The addition of excess Al_2O_3 to the reactants decreased the agglomeration size. The reason for this decrease may be that as the amount of Al_2O_3 increased, the combustion temperature is lowered below the melting point of the alumina phase and thus there is less liquid phase present during the reaction. On the other hand, the agglomeration size did not change significantly with the amount of excess Al.

Some interactions between the parameters also have significant effects on the investigated responses. The interaction between the amount of excess Al_2O_3 and other parameters was found to have more of an effect on the responses than any other combinations of the interaction. Figure 6a shows the interaction profile plots of amount of excess Al_2O_3 and its particle sizes. The strong interaction between these two parameters is illustrated. Fastest ignition time is observed when using small Al_2O_3 particle and small amounts of Al_2O_3. With larger additives of Al_2O_3, larger Al_2O_3 particles resulted in faster ignitions. Figure 6b shows that there was no interaction between the amount of excess Al_2O_3 and its particle size on the agglomeration size. The agglomeration size decreases with increasing amount of excess Al_2O_3 regardless of particle sizes used. As Al_2O_3 particle size was increased, the resulting agglomeration size increased.

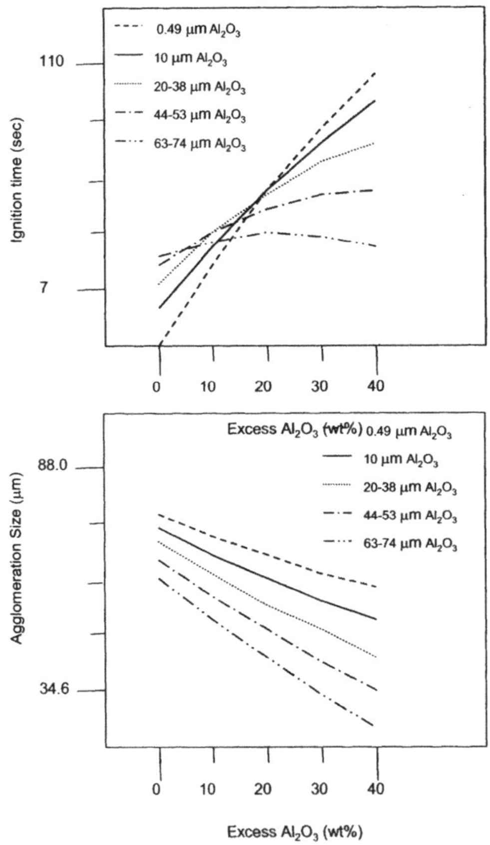

Figure 6. The effect of the amount of excess Al_2O_3 and its particle size on a) the ignition time and b) agglomeration size of Al_2O_3-TiC powders.

(a)

(b)

SUMMARY

Al_2O_3-TiC powders were produced by a combustion synthesis technique using MH or MHH as a heating source. A clear difference in ignition time and temperature was observed for the two heating methods. The MHH samples required longer time to reach ignition temperature. Thus they ignited at higher temperature than MH samples and resulted in lower combustion temperatures. The results showed no significant difference in the characteristics of powders ignited by these two heating methods. The thirtieth-run CCD was performed to investigate the effect of amount of excess Al and Al_2O_3, particle size of C and Al_2O_3 on ignition behavior and characteristics of the resulting powders. The addition of diluents to the reactants increased ignition time and temperature, while decreasing the combustion temperature. An increased particle size of alumina lowered the ignition time and temperature. While for the particle size of carbon, the ignition time and temperature of the ignited sample decreased with an increased particle size of carbon under MHH only. Under MH, an increased particle size of carbon did not change the ignition time and temperature significantly. It was also observed that the sample with the largest alumina or carbon particle size had the highest combustion temperature. The density of product decreased with increasing amounts of Al. The addition of excess Al_2O_3 to the reactants resulted in decreasing the agglomeration size. The investigated responses also were influenced by the interactions between these parameters. In some cases, the interaction effects were significant, but in no case was the interactive effect as large as for individual parameters. Various empirical models relating these important parameters to the investigated responses were developed. This provided the information required for adjusting the setting of parameters in order to obtain the desired products.

ACKNOWLEDGEMENT

The authors would like to thank EPRI for their financial support for this project.

REFERENCES

1. J.G. Baldoni and S.T. Buljan, "Ceramics for Machining," Amer.Ceram.Soc.Bull., 76[2] 381-87 (1988)
2. S. Adachi, et.al, "High-Pressure Self-Combustion Sintering of Alumina-Titanium Carbide Ceramic Composite," J.Am.Ceram.Soc., 73[5], 1451-52 (1990)
3. C.R. Bowen and B. Derby, "The formation of TiC/Al2O3 microstructures by microstructures by SHS reaction," J.Mater.Sci., 31, 3791-3803 (1996)

4. A.O. Kunrath, T.R. Strohaecker, J.J. Moore, "Combustion Synthesis of Metal-Matrix Composites: Part III, The Al-TiC-Al2O3 System," Scripta Materialia, 34[2], 189-194 (1996)
5. A.G. Merzhanov and I.P. Borovinskaya, Self- Propagating High-Temperature Synthesis of Refractory Inorganic Compounds," Dokl. Acad. Sci USSR (Chem.), 204[2], 429-432 (1972)
6. J.B. Holt, "Combustion Synthesis of Refractory Materials," Lawrence Livermore National Laboratory Report No. UCRL-53258 (1982)
7. D.E. Clark, I. Ahmad and R.C. Dalton, "Microwave Ignition and Combustion Synthesis of Composites," Materials Science and Engineering, A144, 91-97 (1991)
8. Duangduen Atong and David E. Clark, "Microwave-Induced Combustion Synthesis of TiC-Al_2O_3 Composites," Ceramic Engineering and Science Proceedings, 20[4], 111-118 (1999)
9. Duangduen Atong and David E. Clark, "Microwave-Induced Combustion Synthesis of Al_2O_3-TiC Powders," Microwaves: Theory and Application in Materials Processing V (Ceramic Transactions V.111), 197-206 (2001)
10. S. Anthonysamy, et.al, "Combustion Synthesis of Urania-Thoria Solid Solutions," J. Nuclear Materials, 278, 346-357 (2000)
11. V. Chandramouli, et.al, "Microwave Synthesis of Solid Solutions of Urania and Thoria – A Comparative Study," J. Nuclear Materials, 254, 55-64 (1998)

MICROWAVE SINTERING OF A COMBUSTION SYNTHESIZED AlN POWDER

Cheng-Yu Hsieh, Chun-Nan Lin, Hung-Jia Chen, and *Shyan-Lung Chung
Department of Chemical Engineering
National Cheng Kung University
Tainan, 70101, Taiwan.
R.O.C.

Jiping Cheng and Dinesh K. Agrawal
Microwave Processing and Engineering Center
Material Research Institute
Pennsylvania State University
State College, PA, 16801
U.S.A.

ABSTRACT

We report our preliminary study on microwave sintering of a combustion synthesized AlN powder. The combustion synthesis method was previously developed by Chung et al. at National Cheng Kung University, and the microwave sintering experiments were carried out at the Microwave Processing and Engineering Center in the Pennsylvania State University. The samples were sintered by microwave at 2.45 GHz in a multi mode cavity. The power used ranged from 200W to 1 KW, and the sintering temperature ranged from 1600°C to 2000°C. The percent theoretical densities of some of the sintered specimens reached 99% and higher. These specimens were also characterized by XRD and SEM. The thermal conductivity of the specimens was also measured. These results will be presented and the effects of several important experimental parameters on the sintering will be discussed.

INTRODUCTION

Aluminum nitride (AlN) has been acknowledged as an important industrial material because of the unique combination of its properties such as high thermal conductivity, high electrical resistivity; low thermal expansion coefficient, low dielectric constant, good thermal shock resistance and good corrosion resistance(1). It has been considered for many applications such as electronic substrates, microelectronic packaging materials, fillers in high thermal conductivity materials, heat sinks and crucibles for containing and processing molten metals and salts(2).

We have recently developed a combustion synthesis method for the synthesis of aluminum nitride(3). This method has advantages including simple processing, fast production, low production cost, high product purity, and being capable of mass production. It has thus a high potential to be used as a method for commercial production of the AlN powder. Since sintering is a necessary step in the fabrication of many AlN related ceramic products, understanding of the sintering behavior and the properties of the sintered specimens is essential to the applications of the AlN powders produced by a newly developed method. Besides, it is also desirable to reduce the cost of sintering, and this is especially important for the case of AlN. Due to its covalent bond nature, AlN is usually sintered with relatively severe conditions such as a sintering temperature of 1800°C and a soaking time of 1 hr or higher (4). Sintering thus results in a major fraction of the production cost of many AlN ceramic products.

Application of microwave for sintering ceramic materials has recently attracted much attention(5). Advantages of microwave heating and sintering include fast heating rate, energy efficiency, volumetric and uniform heating, enhancement of mass transport during sintering, and superior properties of the sintered products(5,6). For the sintering of AlN ceramics, it has been reported (5,7) that lower sintering temperatures, shorter soaking times, and better sintered specimens properties can be realized by using microwave as compared to the conventional sintering techniques. One of our research programs has thus been aimed at studying the microwave sintering of the AlN powders synthesized by our newly developed combustion synthesis method. In this work, we report our preliminary results of this study.

EXPERIMENTAL

Combustion Synthesis of AlN

All of the AlN powders used in the present study were prepared by our newly developed combustion synthesis method(3). Aluminum powder (250g) and a small amount of ammonium chloride powder (0.1-3wt%) were thoroughly mixed and then poured into a perforated aluminum container, which was cylindrical in shape with 80 mm in diameter and 100 mm in height. The container holding the reactants was then placed in a reactor, which has been described and shown schematically in our previous study(8). The reactor was evacuated (to 600 Pa) and then was refilled with nitrogen gas to the desired pressure (0.05-0.5 MPa). The combustion reaction was ignited by heating the top surface of the reactant powders with an electric current (~1200W) passing through the tungsten heating coil. It took ~60 sec to ignite the combustion reaction and the heating power was turned off after ignition. A regulated flow of nitrogen was supplied to the reactor to maintain a constant pressure inside the reactor. It took ~30 min for the combustion reaction to complete. The products were taken out of the reactor after cooling to near the room temperature. The conversion (defined as the mass fraction of the Al converted to AlN) was higher than 99.9% as determined by the method described previously(9).

Listed in table I are the properties of the AlN prepared in the present study. The product as synthesized was crushed to under 50 mesh and was then attrition milled using ZrO_2 balls(diameter : 5 mm)with acetone as the milling fluid. The purity level of the as-synthesized AlN is basically determined by that of the Al powder which is used as the reactant. Due to the lack of proper equipment for crushing, grinding and subsequent drying, significant increases in the amounts of impurities resulted with a heavy grinding, which is the case of powder AlN-1 listed in Table I. To reduce contaminations, a light grinding was also taken and the resultant powder is designated as AlN-2, which is seen to have lower impurity contents but a larger mean particle size than that of AlN-1. AlN-1 and AlN-2 were used in two sets of experiment, i.e., experiment I and experiment II, respectively. In experiment I, two different sintering temperatures, i.e., 1700 and 1800°C, with three different soaking times, i.e., 15, 30, and 60 min were used. In experiment II, four different sintering temperatures, i.e., 1600, 1800, 1950 and 2000°C, with only one soaking time, i.e., 30 min were used.

Microwave Sintering of AlN

In the present experiment, samples of different levels of Y_2O_3 additions from 0-10 wt% were fabricated by uniaxial pressing. The compact was ~ 15.5

mm in diameter and ~ 3 mm in thickness and the green density was ~ 64% of the theoretical density. The compact was placed in a BN crucible, which served as a microwave absorber to preheat the compact. The BN crucible was packaged in a insulation packet.

Table I. Properties of the AlN Prepared in the Present Study

Type of AlN	Average particle size D_{50} (μm)	Si (wt%)	O (wt%)	Fe (wt%)	Zr (wt%)
as-synthesized	-	<0.035	0.2	<0.03	<0.01
AlN-1	0.9	0.04	4.0	0.256	0.528
AlN-2	3.0	0.03	3.0	0.087	0.215

The microwave furnace used in the present study consists of a 2.45 GHz microwave generator with a power output continuously adjustable from 0 - 6 KW, and a water-cooled multi mode cavity. As shown in figure 1, a quartz tube which was insulated by fiberfrax insulation packages, was placed vertically at the center of the cavity. The BN crucible was placed on the perforated BN support, which was at the center of the quartz tube and the top end of the quartz tube was covered with a copper lid. The quartz tube was evacuated to 1 torr by flushing with nitrogen between evacuations. After the evacuations, the reactor was back filled with nitrogen at the desired pressure (0.1MPa) and the heating process was performed in a flowing nitrogen gas atmosphere with a flow rate of 750 ml/min. Variation of the temperature at the top surface of the sample was measured by using an optical pyrometer through the quartz window at the center of the copper lid.

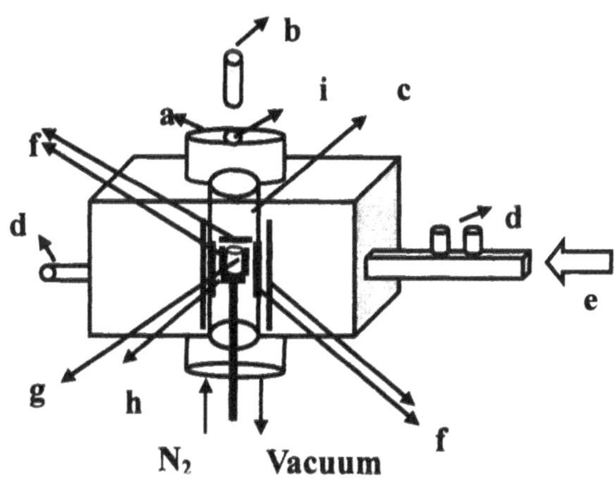

Figure 1 Schematic illustration of the microwave cavity.
a: view window ; b: optical pyrometer ; c: quartz tube (D_{in}: 62mm) ; d: tuner ; e: microwave source ; f: insulated package ; g: BN crucible containing AlN green compacts ; h: BN support.

RESULTS

Experiment I

In this set of experiment, AlN-1 powder was used for microwave sintering. Figure 2 shows the shrinkages as a function of Y_2O_3 content at various sintering conditions. The measured densities and the calculated percent theoretical densities of the sintered specimens are also provided and marked on the respective data points. As shown in this figure, a nearly fully densification can be achieved with a sintering temperature of 1800°C and a soaking time of 30 min when the Y_2O_3 content is 5wt%. A sintering temperature of 1700°C is obviously not high enough to achieve a high densification when the soaking time is 30 min or shorter. A sintering temperature of 1800°C and a soaking time of 15 min is also seen to be not long enough to achieve a high densification. With a sintering temperature of 1800°C and a soaking time of 30 or 60 min, the shrinkage is seen to increase rapidly as the Y_2O_3 content increases from 0 to 3 wt%, but is relatively unaffected with further increase in the Y_2O_3 content. An increase of the soaking time from 30 to 60 min has little effect on densification when the sintering temperature is 1800°C. A sintering temperature of 1800°C and a soaking time of 30min with a Y_2O_3 content of 3-5 wt%thus can be considered as a set of optimum sintering conditions for the AlN-1 powder.

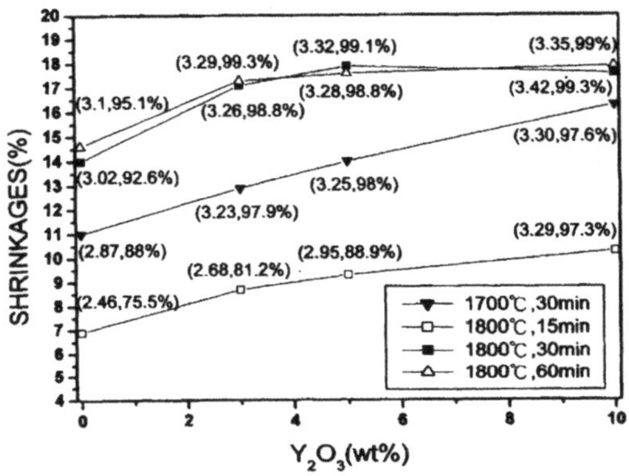

Figure 2 The shrinkage as a function of Y_2O_3 content under various sintering conditions for AlN-1 powder.

Experiment II

In this experiment, AlN-2 powder was used for microwave sintering under conditions of 4 different sintering temperatures and a soaking time of 30 min. As shown in figure 3, the shrinkages (and the densities) obtained at the sintering temperature of 1600°C are much smaller than those obtained at 1800°C and higher, indicating that 1600°C is not high enough as a sintering temperature for the AlN-2 powder. The shrinkage is seen to increase with increasing sintering temperature at all the Y_2O_3 contents used. For the three higher sintering temperatures (i.e., 1800, 1950 and 2000°C), the shrinkage increases rapidly with increasing Y_2O_3 content when the Y_2O_3 content is less than 3wt%.The increases become relatively small as the Y_2O_3 content is increased beyond 3wt%.

Phase Analysis with XRD

Typical XRD patterns (Fig 4.) shows that $Al_5Y_3O_{12}$ is the only second phase detected at 3wt% Y_2O_3 content specimen. $YAlO_3$ and $Al_2Y_4O_9$ are being to be detected as the Y_2O_3 content increased to 5wt%. $Al_2Y_4O_9$ becomes the major second phase and, in some cases, Y_2O_3 is also detected as the Y_2O_3 content is increased to 10wt%.

Microstructure Analysis

Figure 5 shows typical SEM photographs of the fractured surfaces of the specimens composed of AlN-1 powder sintered under various conditions. The specimens with Y_2O_3 contents of 0, 3, 5, or 10wt%contain numerous pores and have little grain growth when they are sintered at 1800°C with a soaking time of 15 min. This can be seen in Fig. 5(a), where the grain size ~ 1μm, similar to the particle size of the starting powder. As shown in figure 5(b), 5(c) and 5(d), the specimens are nearly fully dense when they are sintered at 1800°C with a soaking time of 30 min. The grains are seen to be well-defined hexagonal in shape and uniform in size distribution (the grain size ~ 2μm).

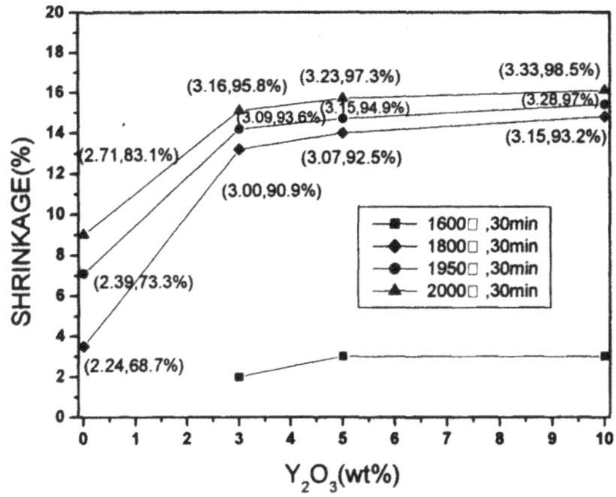

Figure 3 The shrinkage as a function of Y_2O_3 content at various sintering temperatures with a soaking time of 30 min for AlN-2 powder.

Figure 6 contains two SEM photographs of the fractured surfaces of the specimens composed of AlN-2 powder and sintered at 2000°C with a soaking time of 30 min. The specimens are also seen to be nearly fully dense with grain sizes ranging from 5 to 10μm. The grain shapes are, however, more like sphere rather than hexagonal. Back-scattered electron SEM photographs of the fractured and polished surfaces of the specimens shows that the second phases are formed along the grain boundaries. This also can be seen in figure 6.

Thermal Conductivity Measurement

Listed in Table II are the thermal conductivities of the specimens sintered under various conditions. For the AlN-1 specimens sintered at 1800°C with a soaking time of 30 min, the thermal conductivity increases significantly (from 90 to 122w/m·k) as the Y_2O_3 content is increased from 3 to 5wt%. As the Y_2O_3 content is further increased (from 5 to 10wt%), the increase in the thermal conductivity is relatively small (from 122 to 126 w/m·k). For the AlN-2 specimens sintered at 2000°C with a soaking time of 30 min, the thermal conductivity is almost unaffected by the increase of the Y_2O_3 content from 3 to 5wt%. The thermal conductivity, however, decreases (from 132 to 122w/m·k) as the Y_2O_3 content is increased from 5 to 10wt%.

DISCUSSION

Combustion synthesized powders have been claimed to have superior sintering properties as compared to those produced by conventional methods(10). Because the materials experience large temperature gradients and high heating and cooling rates during the SHS processes, the powders produced are expected to have high defect concentrations, thus being more reactive for sintering. Xu et al.(11)studied the microwave sintering of a SHS-produced AlN powder and found that a high densification (98.7%of theoretical density) can be achieved by sintering the powder (with 3wt%of Y_2O_3 added) at 1600°C for 4 min. In the present study, the AlN-1 powder, which has a similar particle size distribution with that used by Xu et al, is seen to be unable to be sintered to high densities (98.0%) at 1700°C for 30 min with 3wt% Y_2O_3 added (see Fig.2). When using a sintering temperature of 1800°C and a soaking time of 15 min, the maximum theoretical density obtained was 97.3%in which 10 wt% of Y_2O_3 was added. A sintering temperature of 1800°C, a soaking time of 30 min and a 3-5wt% Y_2O_3 content are thus considered to be optimum conditions to achieve high densifications for the AlN-1 powder. The AlN-2 powder is seen to be less sinterable than the AlN-1 because the former has much larger particle sizes than the latter. When sintered at 1800°C with a soaking time of 30 min and a 5wt% Y_2O_3 content, the specimens from the AlN-2 powder reach 92.5% of theoretical density, compared to 99.1%for the specimens from the AlN-1 powder. A sintering temperature of 1950°C or higher is thus required to achieve high densifications (e.g., 99%or higher) for the AlN-2 powder when a soaking time of 30 min is used.

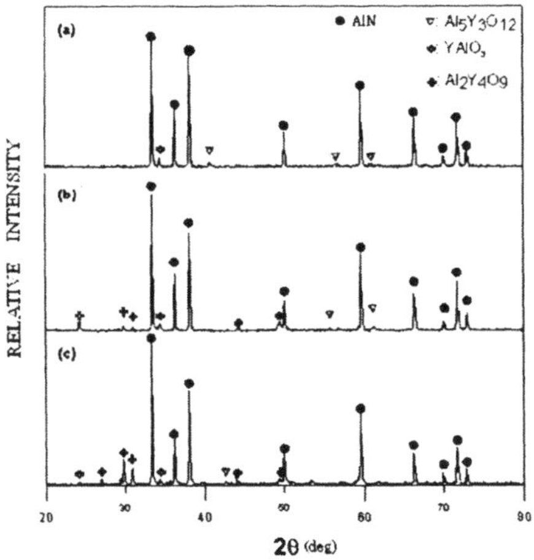

Figure 4 XRD patterns of the specimens composed of AlN-2 powder and sintered at 2000°Cwith a soaking time of 30 min, contents of Y_2O_3: (a), 3wt% ; (b), 5wt% ; and (c), 10wt%.

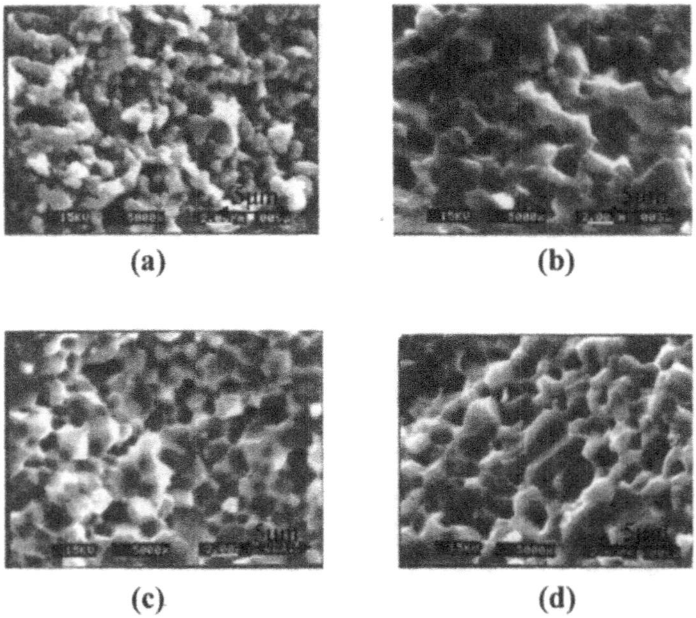

Figure 5 SEM photographs of the fractured surfaces of the specimens composed of AlN-1 powder and sintered at 1800°Cwith different soaking times and Y_2O_3 contents: (a), 15 min and 5wt% ; (b), 30 min and 3wt% ; (c), 30 min and 5wt% ; and (d), 30 min and 10wt%.

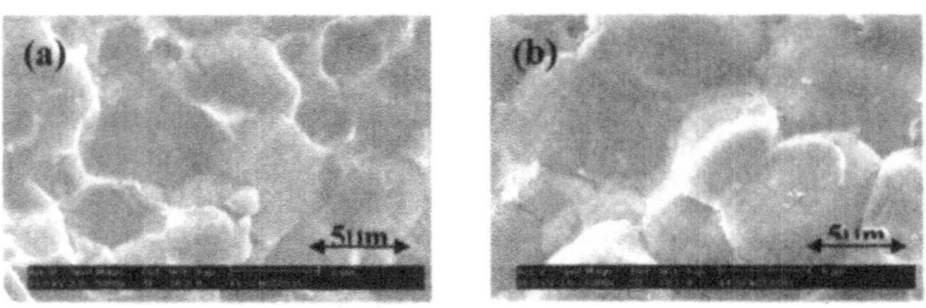

Figure 6 SEM photographs of the fractured surfaces of the specimens composed of AlN-2 powder and sintered at 2000°C with a soaking time of 30 min, Y_2O_3 contents : (a), 5wt%and (b), 10wt%.

Table II. Thermal Conductivities of the Specimens Sintered at 1800 or 2000 °C with a soaking time of 30 min

AlN powder	sintering temperature (°C)	Y_2O_3 content (wt%)	density (g/cm^3)	linear shrinkage (%)	thermal conductivity (W/m·k)
AlN-1	1800	3	3.26	17.1	90
	1800	5	3.32	17.9	122
	1800	10	3.42	17.6	126
AlN-2	2000	3	3.16	15.1	131
	2000	5	3.23	15.7	132
	2000	10	3.33	16.1	122

When using a 3 or 5 wt% Y_2O_3 content, the specimens made from the AlN-2 powder possess higher thermal conductivities than those from the AlN-1 powder although the densities of the former are lower than the latter (see Table II) . As pointed out by many researchers(12), the oxygen content within the AlN grains poses the most detrimental effect on the thermal conductivity. The difference in the oxygen content between the AlN-1 and AlN-2 powders (see Table I) may thus be a primary factor causing the difference in thermal conductivity between their sintered specimens.

It has also been pointed out that cation impurities have significant effects on degradation of the thermal conductivity of AlN ceramics(13). Although the AlN-1 powder contains much higher cation concentrations than the AlN-2 powder does, the concentration of each cation of the two powders exceeds its respective dissolving limit in the AlN lattice. The cation impurities may thus pose an equal effect on degradation of the thermal conductivities of the two types of specimens made from AlN-1 and AlN-2 powders.

Second phase distribution has also been pointed out (14)to affect the thermal conductivity of AlN ceramics. As mentioned previously, the second phases are formed along the grain boundaries. This may thus be another factor causing low thermal conductivities of the specimens in the present study. Baranta et al.(15)studied the effect of second phase on the thermal conductivity of AlN ceramics and found that high contents of sintering aid caused a decrease in thermal conductivity. Although the thermal conductivity of a specimen is affected by many other factors (e.g., oxygen content and degree of densification), this effect would be more pronounced for the specimen with larger grains. Since the specimen with larger grains has less grain boundary area, the second phase present along the grain boundary would be thicker, thus causing a greater extent of decrease in thermal conductivity for such a specimen. This may explain the opposite effects on the thermal conductivities of the AlN-2 specimens, when the Y_2O_3 content is increased from 5 to 10wt%. (see Table II and note that as mentioned previously, the specimens made from AlN-2 have grain sizes ranging from 5 to 10μm, and those made from AlN-1 has a grain size of ~2μm.)

The thermal conductivities measured in the present study are relatively low as compared to those reported in many other studies where the AlN powder from Tokuyama Soda, Grade F has been mostly used. Nevertheless, by employing proper equipment for processing the powder produced by our combustion synthesis method, significant improvements of the thermal conductivity can be expected.

CONCLUSION

A combustion synthesized AlN powder was preliminarily tested for its sinterability and the thermal conductivities of the sintered specimens were measured. The AlN powder(AlN-1)with an average particle size(D_{50})of 0.9 μm could be sintered to ~ 99 %of theoretical density at a sintering temperature of 1800°C with a soaking time of 30 min and a Y_2O_3 content of 3-5 wt%. The AlN powder with a D_{50} of 3.0 μm (AlN-2) was seen to be less sinterable than AlN-1. In this case, 97.3 %of theoretical density was obtained at a sintering temperature of 2000°C with a soaking time of 30 min and a Y_2O_3 content of 5 wt%. The thermal conductivities reached 122 and 132 W/m . k for the specimens made from AlN-1 and AlN-2, respectively, when 5 wt%of Y_2O_3 was added. The higher thermal conductivity of the AlN-2 specimen is attributed to its lower oxygen content in spite of its lower densification. The thermal conductivities are relatively low as compared to those reported in many other studies and this is considered to be caused by high oxygen contents, high contents of other impurities, and that the second phases exist along the grain boundaries. A significant improvement in thermal conductivity is expected by employing proper grinding equipments with which AlN powders with higher purity can be produced.

ACKNOWLEDGMENT

Support of this research by the National Science Council of the Republic of China under Grant No. NSC 90-2214-E-006-006 is gratefully acknowledged.

REFERENCES

[1]G.A. Slack, R. A. Tanzilli, R. O. Pohl and J. W. Vandersande, "The Intrinsic Thermal Conductivity of AlN," J. Phys. Chem. Solids, 48 [7] 641 (1987).

[2]L.M. Sheppard, "Aluminum Nitride: A Versatile but challenging Material," Ceram. Bull., 69 [11] 1801 (1990).

[3]C.N. Lin and S. L. Chung, "A Combustion Synthesis Method for Synthesis of Aluminum Nitride Powder Using Aluminum Container," J. Mater. Res., 16 [12] 3518 (2001).

[4]C.F. Chen, M. E. Perisse and A. F. Ramirez, "Effect of grain boundary phase on the thermal conductivity of aluminum nitride ceramics," J. Mater. Sci., 29, 1595 (1994).

[5]J. Cheng, D. Agrawal, Y. Zhang, B. Drawl and R. Roy, "Fabricating Transparent Ceramics by Microwave Sintering," Am. Ceram. Soc. Bull., 9, 71 (2000).

[6]G.F. Xu, I. K. Lloyd, Y. Carmel, T. Olorunyolemi and O. C. Wilson, "Microwave sintering of ZnO at ultra high heating rates," J. Mater. Res., 16 [10] 2850 (2001).

[7]Z. Xie, J. Yang and Y. Huang, "Densification and grain growth of alumina by microwave processing," J. Mater. Lett., 37,215 (1998).

[8]W.C. Lee, C. L. Tu, C. Y. Weng and S. L. Chung, "A novel process for combustion synthesis of AlN powder," J. Mater. Res., 10 [3] 774 (1995).

[9]C.N. Lin and S. L. Chung, "Combustion Synthesis of Aluminum Nitride Powder Using Additives," J. Mater. Res., 16 [8] 1928 (2001).

[10]O.R. Bermann and J. Barrington, "Effect of Explosive Shock Waves on Ceramic Powders," J. Am. Ceram. Soc., 49 [9] 502 (1966).

[11]G. Xu, H. Zhuang, J. Cai, and S. Xu, "Microwave sintering of SHS-produced AlN powders," J. Mater. Syn. Proc., 4(5), 307 (1997).

[12]G.A. Slack, "Nonmetallic Crystals with High Thermal Conductivity," J. Phys. Chem. Solids, 34, 321 (1973).

[13]P.S. Baranda, A. K. Knudsen and E. Ruh, "Effect of Silica on the Thermal Conductivity of Aluminum Nitride," J. Am. Ceram. Soc., 76 [7] 1761 (1993).

[14]Y.D. Yu, A. M. Hundere, R. Hoier, R. E. Dunin-Borkowski and M. A. Einarsrud, "Microstructure characterization and microstructural effects on the thermal conductivity of $AlN(Y_2O_3)$ ceramics," J. Eur. Ceram. Soc., 22, 247 (2002)

[15]P.S. Baranda, A. K. Knudsen and E. Ruh, "Effect of Yttria on the Thermal Conductivity of Aluminum Nitride," J. Am. Ceram. Soc., 77 [7] 1846 (1994).

DENSIFICATION MECHANISMS IN SiC_f/SiC COMPOSITES BY MICROWAVE ENHANCED CHEMICAL VAPOUR INFILTRATION

D.Jaglin[†], B.Vaidhyanathan and J.Binner[†]
University of Loughborough
Institute of Polymer Technology
and Materials Engineering
Loughborough
Leicestershire LE11 3TU, UK

C.Prentice, B.Shatwell
QinetiQ Ltd
Cody Technology Park
Farnborough GU14 0LX, UK

ABSTRACT

The densification mechanisms of reinforced silicon carbide fibre-silicon carbide matrix composites processed using microwave enhanced chemical vapour infiltration have been investigated using a non-destructive evaluation technique, x-ray absorptiometry, and scanning electron microscopy. Due to the inverse temperature profile generated within the preform, preferential densification of the composite from the inside out was observed. An average relative density of ~55% was achieved in 24 h across the composite, the centre densifying up to 70%. The intratow porosity was efficiently filled within 8 h whilst the intertow pores mostly remained free of matrix material even after 24 h.

INTRODUCTION

The demand for structural ceramics such as fibre-reinforced ceramic matrix composites has grown substantially in the past few years and has led to a rapidly expanding interest in chemical vapour infiltration (CVI) [1-4]

Existing conventional CVI techniques suffer from the disadvantage that the preform is hottest at the surface. This can result in non-uniformly densified composites, in particular in isothermal CVI [5-7], since the deposition of the matrix material occurs preferentially near the outer surface, leading to the blockage of the

[†] formerly at Nottingham University

surface channels which, in turn, prevents the gaseous reagents from penetrating the preform. It is minimisable by lowering the reaction temperatures, however, long processing times, up to several months, are therefore common [2,8-13], in addition to interruptions by removal for machining of the dense crust formed on the exterior of the preform [11]. Techniques designed to alleviate these problems result in the use of temperature and pressure gradients [14-17]. However, when a thermal gradient is combined with a forced-gas flow, non-uniformly densified composite can also result, even though higher densities have been achieved. Disadvantages also include more complex temperature and gas flow controls as well as greater complexity whilst processing multiple parts [18-19].

The search for more efficient methods that yield rapid and complete densification continues and the use of microwave heating in the processing of ceramic materials has been recognised as a promising alternative. Microwave radiation can be used to generate a controllable inverse temperature profile during the heating of ceramic fibre preforms [20-22]. This allows the chemical vapour infiltration of the fibre preforms to be performed from the inside out, resulting in deeper infiltration of the matrix [5,22-26]. Microwave enhanced CVI (MECVI) also avoids surface pore closure and reduces deposition times to an order of magnitude shorter than that of conventional CVI with the production of uniformly and highly densified fibre-reinforced ceramic matrix composites. Initial experiments on the SiC_f/SiC system were achieved and preforms were infiltrated up to ~70% of the theoretical density [21] with the centre density reaching ~100%. The infiltration rate of SiC was mainly affected by the temperature and its profile across the sample, pressure and reactant concentration.

EXPERIMENTAL

An MECVI system featuring a purpose-built multimode microwave cavity operating at 2.45 GHz (Figure 1) has been used to infiltrate a range of silicon carbide fibre preforms with the aim of producing dense SiC_f/SiC composites. The output power of the magnetron was adjustable in the range 0.5 to 5 kW.

Samples consisted of 30-layer Nicalon NL-202 fibre fabric stitched together. The diameter was 48 mm with a thickness of ~10 mm. Prior to stitching, the fibre discs were heated in air to remove the sizing agent from the fibres. Methyltrichlorosilane (MTS) was used as the precursor material for SiC deposition; when heated to about 1000°C it decomposes into SiC and HCl according to the reaction:

$$CH_3SiCl_{3(g)} + nH_{2(g)} \rightarrow SiC_{(s)} + 3HCl_{(g)} + nH_{2(g)} \qquad (1)$$

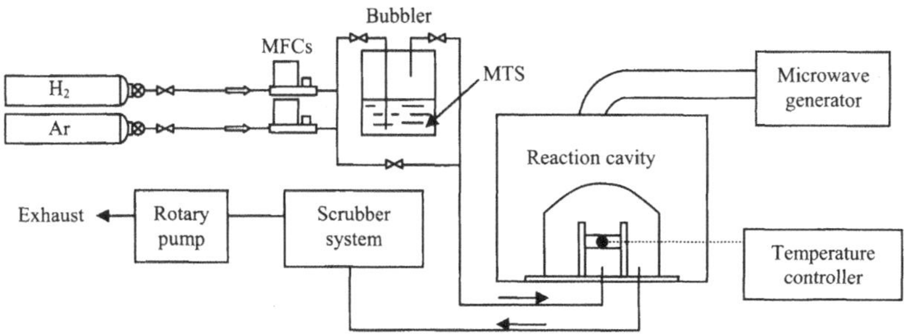

Figure 1: Schematic diagram of the MECVI reactor system.

The MTS was carried into the reactor by passing the hydrogen carrier gas through the precursor vapour delivery unit maintained at room temperature. Control of the MTS concentration was achievable by setting the hydrogen flow rate and the total pressure of the system. Infiltration experiments were conducted for periods spanning 1 to 24 h at 70 kPa using a constant hydrogen flow rate of 300 ml min^{-1}. Temperature was measured with a 3-mm inconel shielded type-K thermocouple inserted halfway through the thickness of the sample by a 3-mm hole drilled at the centre of the sample. The possibility of electromagnetic interaction with thermocouples was investigated by comparing measurements with values obtained via optical thermometry (Accufiber optical fiber thermometer, Model 100, Luxtron, USA). Experiments showed that the temperature difference between the two was less than 20 K.

Characterisation techniques essentially comprised scanning electron microscopy (SEM), dielectric measurement via a cavity perturbation technique and x-ray absorptiometry (XRA). The latter was used as a non-destructive, density evaluation technique; it is based on scanning a specimen with a beam of x-rays and detecting the transmitted beam that has been attenuated by the sample. The resultant image is a map of the photon attenuation. All other factors being equal, this can be directly correlated to spatial variations in density within the sample.

RESULTS

The plot of the relative temperature difference between the centre and points across the SiC fibre preforms for 2 central temperatures ranging from 1173 to 1273 K are shown in Figure 2. The bell-shaped curves are typical of the inverse temperature profiles generated by volumetric microwave heating.

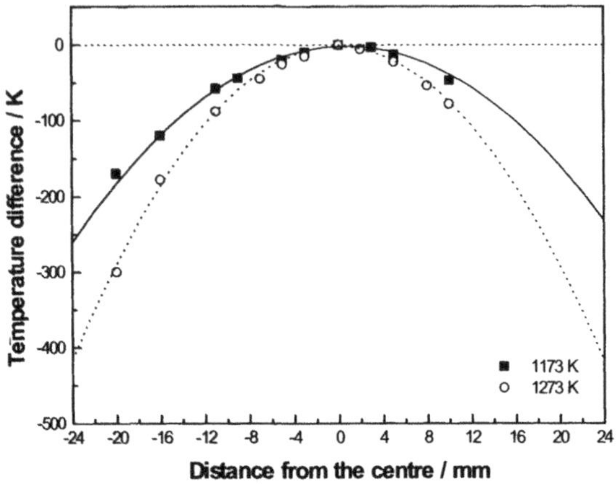

Figure 2: Temperature profiles recorded across the diameter of 30-layer fibre preforms (10 mm thick) in air for central setpoint temperatures of 1173 and 1273 K.

An average relative density of ~55% was achieved after 24 h of MECVI, representative of a gain of 26 vol.% over the initial fibre preform. The densification rate ($\% h^{-1}$) is also represented as a function of the processing time (Figure 3). It is fairly constant at $1.4\% h^{-1}$ for the first 4 h, but then steadily decreased with infiltration time. The centre was densified up to 70% of the theoretical density for a SiC_f/SiC composite (Figure 4). After 16 h, the density profile grew up as a bell-shaped curve showing a maximum at $1.8 \, g \, cm^{-3}$. A peak value of $2.2 \, g \, cm^{-3}$ was obtained at the centre after 24 h.

The XRA scans are presented in their raw versions (Figure 5) and show the initial green state and the state of infiltration after 24 h. Because the XRA can only produce scans on a grey scale, enhancement of the scans into a colour version was developed in order to evaluate infiltration patterns between different scans, however, it was not possible to show them in this paper due to printing limitations. It could be clearly seen that the density pattern developed with the infiltration time, the densifying region first becoming broader and denser near the centre. The pattern was then modified as the relative density increased rather uniformly across the sample.

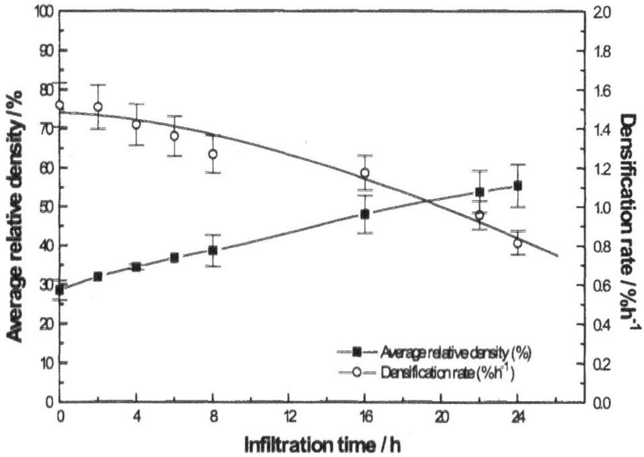

Figure 3: Average density variation of 30-layer fibre preforms as a function of the MECVI processing time.

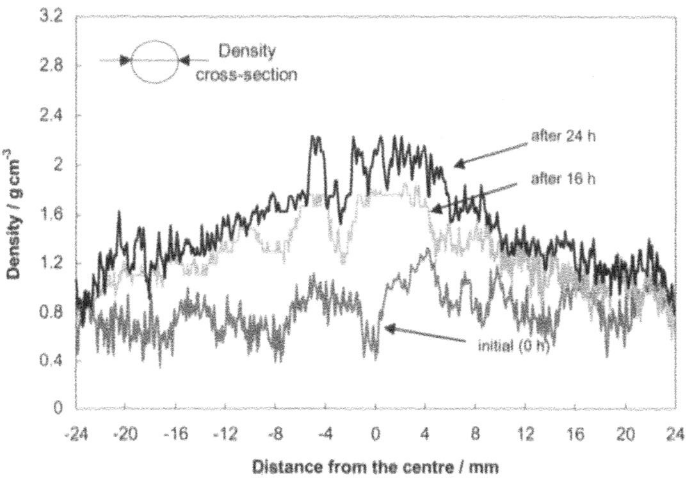

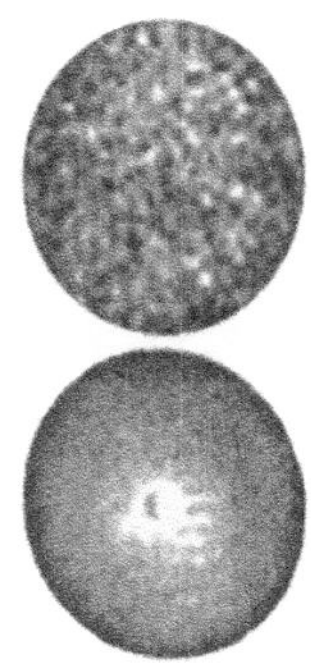

5(a)

5(b)

Figure 4: Density profile obtained from XRA data across after 24 h of MECVI.

Figure 5: XRA scans for 30-layer Nicalon preforms (a) green (b) infiltrated for 24 h by MECVI. Enhanced scans (colour) are not presented.

Figure 6 to 9 present SEM micrographs of polished cross-sections at low and high magnification of the region near the preform centre after 0, 4, 8 and 24 h of MECVI process. The fabric layers appeared as longitudinal fibre tows and sectioned transverse fibre tows (Figure 6). Intertow gaps could be as large as 250 μm with intratow gaps varying in size up to 10-15 μm. In the early stages, deposition took place around cylindrical filaments with relatively little overlap occurring (Figure 7). After 8 h, the intratow porosity appeared to become isolated, though this is difficult to ascertain due to the two-dimensional nature of the section (Figure 8). The preform centre after 24 h contained little intertow porosity (Figure 9), but the extent of deposition decreased rapidly towards the edge of the preform. This is demonstrated by the average thickness of deposition on the fibres as a function of time and location within the preform (Figure 10). At the preform centre, the layer of deposited matrix could be up to 20 μm thick (Figure 11), whilst only 2-3 μm was deposited at the edges.

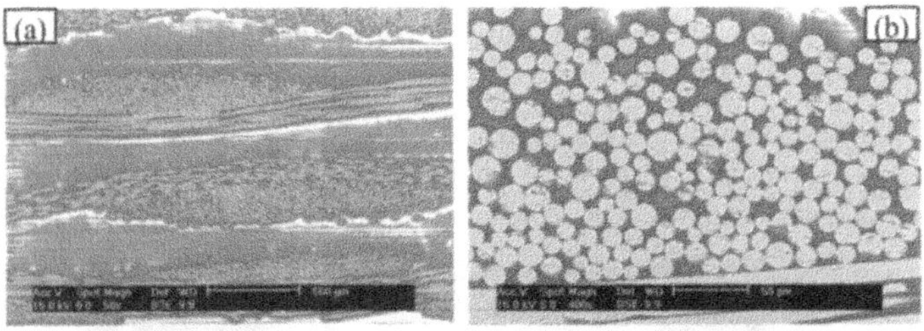

Figure 6: (a) and (b) BSE micrographs showing the initial 30-layer Nicalon fibre preform near the centre axis.

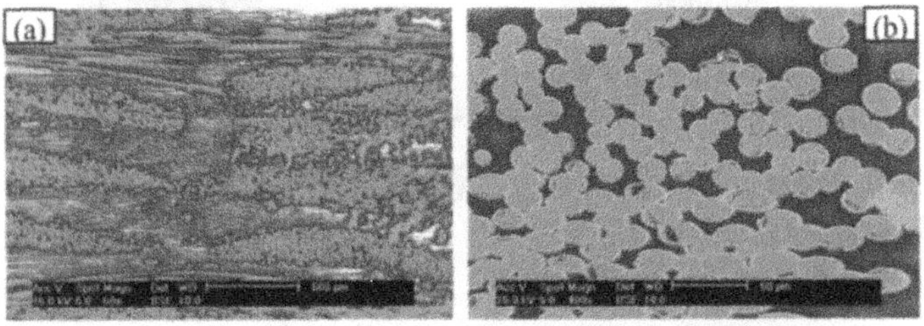

Figure 7: (a) and (b) BSE micrographs showing the state of densification for a 4-hour infiltrated preform near its centre axis.

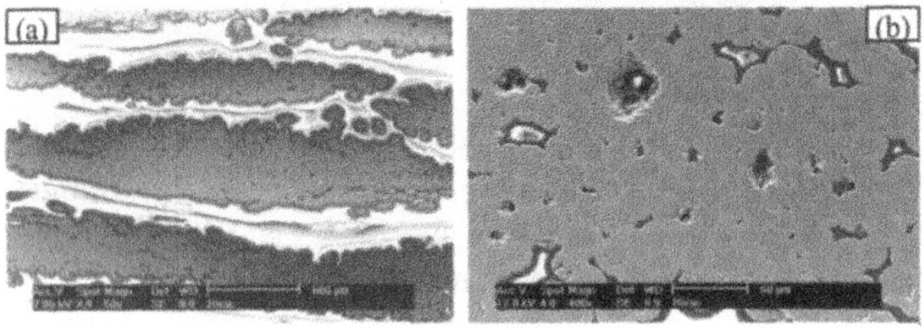

Figure 8: (a) and (b) SE and BSE micrographs showing the state of densification for a 8-hour infiltrated preform near its centre axis.

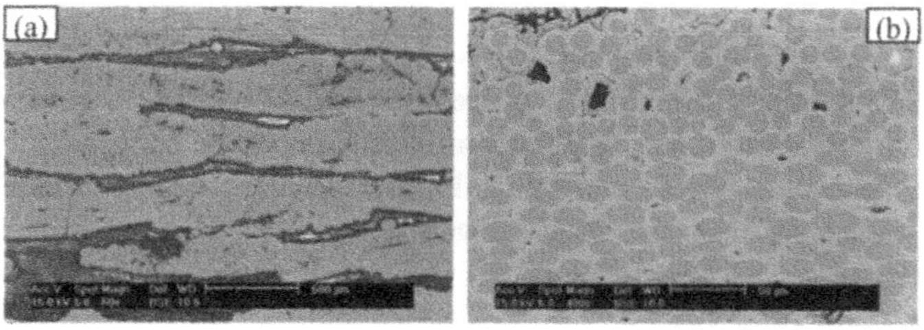

Figure 9: (a) and (b) BSE micrographs showing the state of densification for a 24-hour infiltrated preform near its centre axis.

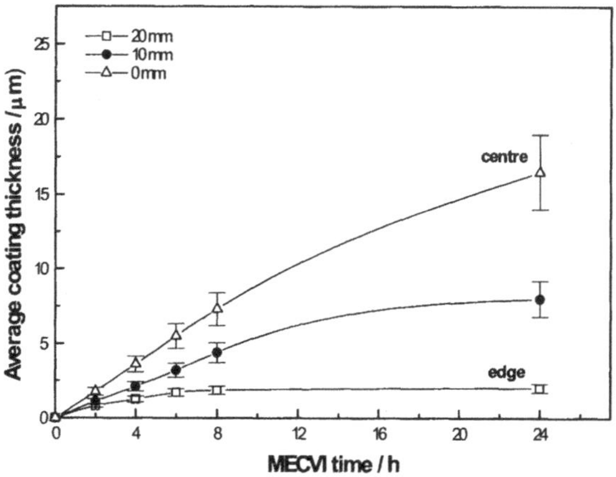

Figure 10: Average coating thickness measured on the outer fibres of the tows as a function of the MECVI processing time for different locations across samples of series S10 (0, 10 and 20 mm from the central axis).

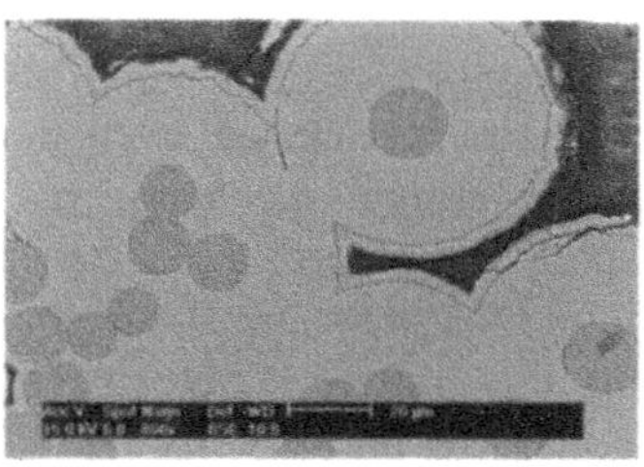

Figure 11: BSE micrograph showing the infiltration of the outside of a tow where the coating thickness on the fibres can reach 20 μm.

DISCUSSION

The overall deposition of SiC from the MTS-H_2 system was a linear function of time only during the early stages of the MECVI process (Figure 3). A transition to a slower rate of deposition is typically seen in CVI processes [27] since some of the fine porosity within the fibre tows becomes either fully infiltrated or the tow surface becomes sufficiently sealed so that the high-surface-area fibres are no longer available. This occurred in MECVI after ~8 h of the process. It is believed that the intratow porosity around the central axis was no longer accessible for the reactant gas species (Figure 8) since the connectivity of the residual porosity had been significantly reduced as described by Figure 12 (stage 3).

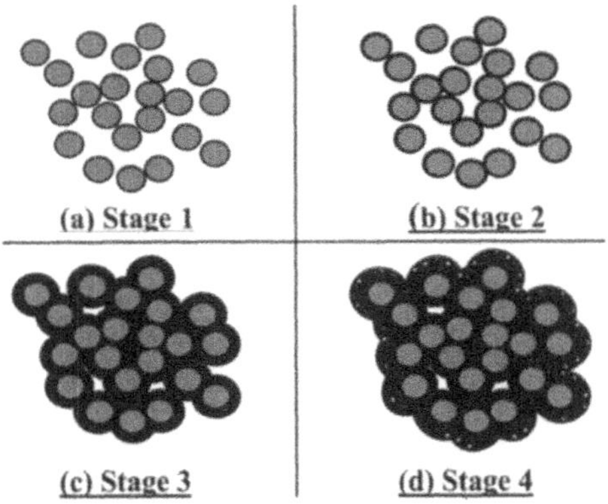

Figure 12: Schematic diagram of the progression of the matrix infiltration in a fibre tow: (a) Initial transversal cross-section of the fibre tow ($t_i = 0$ h), (b) SiC deposition near the central axis ($t_i = 0 \rightarrow 4$ h), (c) connectivity of the residual porosity is significantly reduced (t_i ~8 h), (d) the exterior of the tows is deposited with SiC matrix, leaving the intratow porosity unchanged.

Where intratow porosity was not accessible near the centre, even though tow interiors did not completely densify, infiltration around the tows was initiated (Figures 11 and 12, stage 4), the hot region spreading progressively outwards. The hot spot size, where the majority of the deposition took place, remained small for the first 4 h, but then increased, also indicating that intratow porosity was being infiltrated further out. When nearly half of the intratow porosity had been filled and the overall relative density of the preform had reached ~55%, only intratow porosity near the edge of the densifying region and intertow infiltration across the preform remained available for infiltration. Thus one should expect lower infiltration rates and longer processing times if further infiltration is carried out.

The progression of the densification originated initially at the centre of the preform with the densifying area increasing in size with infiltration time. The available intratow porosity kept on decreasing around the centre until after 24 h of processing when it had spread up to a distance 10 mm away from the central axis (Figure 10). This suggested that a 'reaction front' exists, which corresponds to the infiltration progression of the intratow porosity through the preform, as is found with forced-flow, thermal gradient CVI (FCVI) process [17].

Temperatures were recorded across the fibre preforms and showed that the initial inverse temperature profile flattened with time, indicative of temperature levelling. The thermal conductivity is not believed to have played an influential role in MECVI since the intertow pores remained as large as 100 μm. The stoichiometry of the SiC deposited varied from a Si-rich matrix (more absorptive) at the edges to a C-rich matrix (less absorptive) at the centre, resulting in more microwave energy being dissipated where a Si-rich SiC matrix was deposited. Permittivity changes were also suggested to be responsible for the temperature profile variations as the densification and deposition of matrix material modified the dielectric properties, but these results will be presented in a later publication.

CONCLUSION

An average relative density of ~55% was achieved in 24 h across the composite, the centre densifying up to 70%. Densification progressed inside out, following a 'reaction front' for the intratow porosity whilst the intertow pores mostly remained free of matrix material even after 24 h. Stoichiometry of the deposited matrix and permittivity variations are thought to be responsible for the temperature profile flattening.

REFERENCES

1 Sheppard L.M., Ceram.Bull. **69** 4 666-673 1990

2 Starr T.L., Smith A.W., Vinyard G.Y., Ceram.Eng.Sci.Proc. **12** 9-10 2017-2028 1991

3 Park H.D., Jeong H.J., Choi Y.M., Lee J.D., Ceram.Trans. **46** 155-163 1994

4 Sheldon B.W., Besmann T.M., J.Am.Ceram.Soc. **74** 12 3046-3053 1991

5 Devlin D.J., Currier R.P., Barbero R.S., Espinoza B.F., Ceram.Eng.Sci.Proc., **14** 9-10 761-767 1993

6 Evans J.N., Gupta D., MRS Symp.Proc., **189** 101-107 1991
7 Morell J., Economou D.J., Amundson N.R., J.Mater.Res., **8** 5 1057-1067 1993
8 Muhlratzer A., cfi / Ber. DKG **4** 30-35 1999
9 Tsamopoulos J.A., Gupte S.M., J.Electrochem.Soc. **137** 11 3675-3682 1990
10 Besmann T.M., Ceram.Trans. **58** 1-12 1995
11 Sotirchos S.V., AIChE J. **37** 9 1365-1378 1991
12 Sotirchos S.V., Ofori J.Y., AIChE J. **42** 10 2828-2840 1996
13 Sasaki M., Suzuli T., Hirai T., J.Mat.Proc.Tech. **2** 2 109-115 1994
14 Roman Y.G., Stinton D.P., Besmann T.M., J.Phys IV **c2** suppl.J.Phys.II 1 689-695 1991
15 Besmann T.M., Sheldon B.W., Kaster M.D., Surf.Coat.Tech. **43-44** 167-175 1990
16 Besmann T.M., Lowden R.A., Stinton D.P., High Temperature Ceramic Matrix Composites Woodhead Publishing Ltd. 215-229 1993
17 Lackey W.J., Starr T.L., Fibre-reinforced ceramic composites: Materials, processing and technology, ed. Mazdiyasni K.S. Noyes Publications 397-450 1990
18 Starr T.L., Smith A., Besmann T.M., McLaughlin J., Sheldon B., High Temperature Ceramic Matrix Composites Woodhead Publishing Ltd. 231-239 1993
19 Chiang Y-M., Haggerty J.S., Messner R.P., Demetry C., Ceram.Bull. **68** 2 420-428 1989
20 Morell J.I., Economou D.J., Amundson N.R., J. Electrochem. Soc., **139** 1 328-336 1992
21 Yin Y., Binner J.G.P., Cross T.E., Ceram Trans., **80** 349-356 1997
22 Ofori J.Y., Sotirchos S.V., J.Mater. Res., **11** 10 2541-2555 1996
23 Gupta D., Evans J.W., J.Mater.Res. **6** 4 810-818 1991
24 Golecki I., Mat.Sci.Eng. **R20** 37-124 1997
25 Ahmad I., Paquette L, Silberglitt R., U.S. Patent 5 238 710 24 August 1993
26 Day P.S., Skamser D.J., Jennings H.M., Johnson D.L., Proceedings of the Conference on Composites and Advanced Ceramic Materials-B, Ceram.Eng.Sci.Proc., 908-915 1994
27 Probst K.J., Besmann T.M., Stinton D.P., Lowden R.A., Anderson T.J., Starr T.L., Surf.Coat.Tech, **120-121** 250-258 1999

ACKNOWLEDGEMENT
The authors wish to acknowledge the funding provided by QinetiQ with the MoD Corporate Research programme, the EPSRC ROPA scheme, the University of Nottingham Scholarship Scheme and the assistance of Rolls Royce.

DIRECT MICROWAVE COUPLING TO METALS AT ELEVATED TEMPERATURES

*E. B. Ripley
Y-12 National Security Complex
Bldg. 9201 MS 8096
5000 Bear Creek Rd.
Oak Ridge, TN 37831

P. A. Eggleston
Y-12 National Security Complex
5000 Bear Creek Rd.
Oak Ridge, TN 37831

T. L. White
Oak Ridge National Laboratory
Building 9201-2, Dock 63
Bear Creek Road
Oak Ridge, TN 37831

ABSTRACT

The Technology Development division at the Y-12 National Security Complex (NSC) in Oak Ridge, Tennessee is currently developing microwave metal melting technology to support several Y-12 NSC missions. We have observed, on many occasions, that the reflected microwave power from multimode microwave cavities drops just before the metal charge in the susceptor crucible melts (forward microwave power is kept constant). This effect is seen regardless of the type of metal being melted. In some experiments, detailed infrared temperature monitoring has shown that the temperature of the metal charge rapidly increases as the microwave absorbing ceramic crucible temperature drops near the metal's melting point. One possible interpretation of these observations is that the metal has enhanced microwave absorption near its melting temperature. However, at elevated temperatures, metal resistivities only increase linearly with temperature due to electrons scattering off the crystal lattice[1]. In contrast, we have observed that this enhanced microwave absorption by near-molten metals is a highly nonlinear phenomena not predicted by linear scaling. In order to understand the cause of the enhanced absorption, we are planning to conduct some low-power microwave resistivity measurements on several conventionally heated metals through their melting point to try and document the enhanced absorption we observe in the high power microwave furnaces.

INTRODUCTION

For the past decade, the Technology Development Group at the Y-12 National Security Complex (NCS) in Oak Ridge, TN, has been processing metals using microwaves. This work is counter-intuitive given common misconceptions about the behavior of metals in microwave fields. The work presented here is based on the simple principals outlined below.

In order to successfully heat and melt metals using microwaves without the normally associated problems, three basic ingredients are required. These are a multimode microwave cavity, a microwave-absorbing ceramic crucible, and a thermally insulating casket that is microwave-transparent. The metal charge is placed in an open ceramic crucible (no lid) and the insulating casket is positioned to completely cover the open crucible. The casket and crucible assembly is then placed into a high-power multimode microwave cavity that is capable of uniformly heating the crucible to the desired temperature as shown in Fig. 1. When microwave energy is applied to the cavity, the energy is strongly absorbed by the crucible. The metal charge in the crucible is quickly heated through radiation, conduction, and convection with the heated crucible walls. The thermally insulating casket increases the energy efficiency of the microwave system by trapping the heat generated in the crucible. In this way, metal objects that could not be directly heated by microwave energy can be melted easily and efficiently.

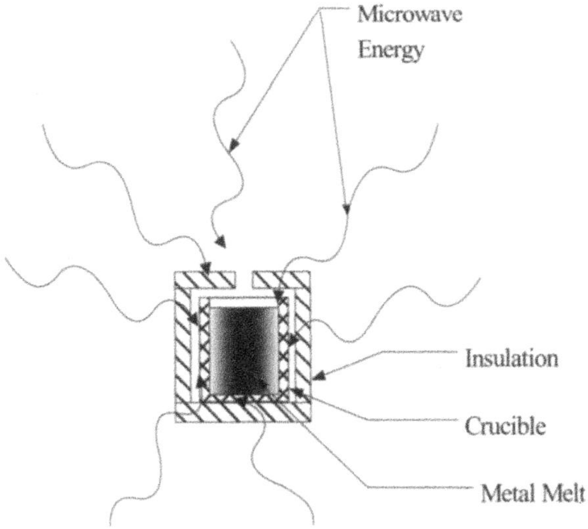

***Figure 1**. Microwave energy can easily be transferred to metal shapes using a susceptor crucible.*

EXPERIMENT

It has been known for a long time that metals do not couple with microwaves at room temperatures. Since metals are electrically conductive they reflect the incident energy. However, the electromagnetic field generated by the microwave field will allow for the loosely bound electrons to move, and concentrate at surfaces, edges and points. This results in discharge of the energy in the form of arcing[2]. Unfortunately unprotected metals in microwave fields at room temperature will exhibit this behavior. Many ceramic systems have dielectric loss factors that allow them to absorb microwave fields at room temperatures, and continue to heat to very high temperature[3,4]. The electromagnetic waves in a microwave cavity will either be absorbed in a material, reflected from a material or the cavity wall, or be converted to a different form of energy.

The microwave units that were used for this paper are relatively standard 2.45 GHz multi-mode cavities. The first is one of our own designs, which has vacuum capability as well as complete atmosphere control. A pair of 6 kW COBER S6F Industrial Microwave Generators provides the microwaves to the cavity. The waveguides are equipped with dual couplers, and power meters, and a set of quarter wave tuning stubs. One waveguide is directed into the cavity in transverse magnetic (TM) mode, the second enters the cavity in transverse electric (TE) mode. The second unit has a much larger cavity and has a power capability of 30 kW. The general design features of the larger cavity are similar to the smaller unit. Due to experimental convenience as well as issues of size and scale, most of the information used in this report is from experiments conducted in the smaller unit.

In Fig. 2 below you will see a plot of data from a 500-minute uranium casting run. Note that there are periods of thermal arrest at around 772°C, and 1132°C. These correspond with the solid phase changes that take place in the metal. At 460 minutes into the run the temperature of the melt begins to increase rapidly even though the microwave net power is held constant for about 20 minutes. One possible interpretation of these observations is that the metal has enhanced microwave absorption near its melting temperature. Prior runs done with empty crucibles show no anomalous heating at these temperatures, so there is no possible temperature dependent absorption due to the crucible itself. The crucible is not contributing to the observed heating at 1132°C. Also, it is well known that the microwave heating patterns in multimode cavities vary with frequency. Most high power microwave sources exhibit some frequency variation with power. However, since the power was held constant for about 20 minutes after the onset of the overheating, the mode pattern on the crucible and metal charge remained constant. Ultimately, the power was reduced to avoid overheating the metal charge. Therefore, since the microwave heating uniformity did not change before or after the rapid temperature rise, a change in the microwave heating uniformity cannot be responsible for the overheating. When looking at both the crucible and the metal charge using a two-color pyrometer, we observe that, from room temperature to about 900°C the crucible has the highest temperature. From around 900°C until the end of the run, the metal charge has the highest temperature. From this point the temperature of the metal charge begins to rapidly increase until the power is decreased at the end of the run. At elevated temperatures, metal resistivities only increase linearly with temperature due to electrons scattering off the crystal lattice[1]. In contrast, we have observed that this enhanced microwave absorption by near-molten metals is a highly nonlinear phenomena not predicted by linear scaling.

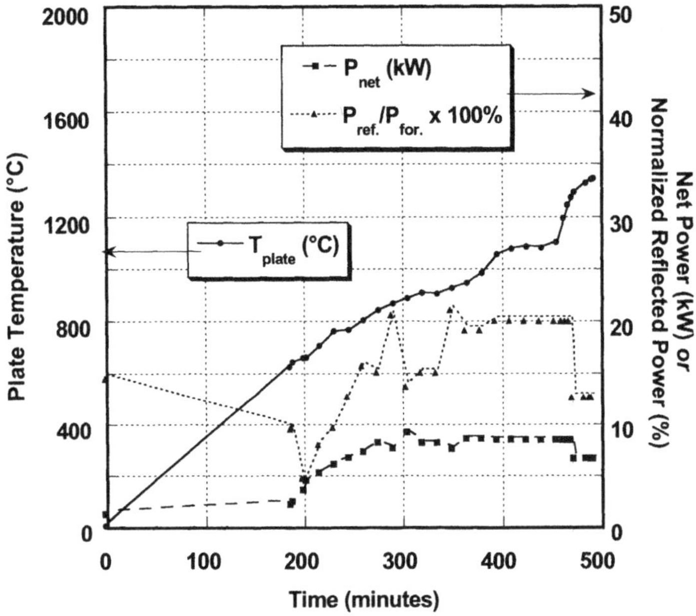

Figure 2. *The Uranium run showing thermal arrest corresponding to a phase change at 772 °C, and anomalous absorption at the melting point of 1132 °C.*

To date we have successfully melted aluminum, tin, copper, brass, steel, stainless steel, bronze, uranium, titanium, platinum, as well as several other alloys. Temperatures of 2800°C have been reached using these techniques. Castings range in size from a few grams to several hundred kilograms. One exciting thing about this type of metal processing system is its unique flexibility. Metals can be processed in air, argon, nitrogen, and other atmospheres with a simple press of a button. Unlike induction heating, there are no induction coils to change out when the casting stack is changed.

A variety of casting sizes, materials and processing atmospheres can be used, without time-consuming set up changes. Powder metallurgy parts sintered using microwaves seem to compare very favorably with traditionally processed PM parts. Techniques for more complex metallurgical process such as oven brazing, alloying, and directly converting oxides to metal have been successfully accomplished using this same basic equipment setup. It appears that the types and number of processes possible using this technique is limited only by the creativeness of the engineers and operators.

SUMMARY AND FUTURE WORK

There appears to be a strong temperature dependence on microwave energy absorption for uranium metal near its melting point. In order to understand the phenomenon of enhanced absorption of microwave energy by near-molten metals, we are attempting to quantify the phenomena by directly measuring the microwave absorption of several metals of interest as a function of temperature through the metals' melting point. Using a small, low power, single-mode, microwave cavity and a conventional lab metal melting furnace, we plan to develop a modest database of selected metals that exhibit an enhanced absorption of 2.45-GHz microwave energy at near-molten temperatures. The cavity is pictured in Fig. 3.

This experimental cavity is designed to measure the amount of microwave absorption in the solidified (near molten) end as a function of temperature at 2.45 GHz using a low-power microwave network analyzer. The process is conducted under a controlled atmosphere to avoid surface contamination. From this simple test setup, we can determine the metal resistivity, as a function of temperature. The open end of the cavity is inserted into a conventionally heated liquid metal bath that is allowed to solidify. Upon conventional re-heating, we measure the cavity "Q" or quality factor, which indirectly measures the resistivity of the metal melt.

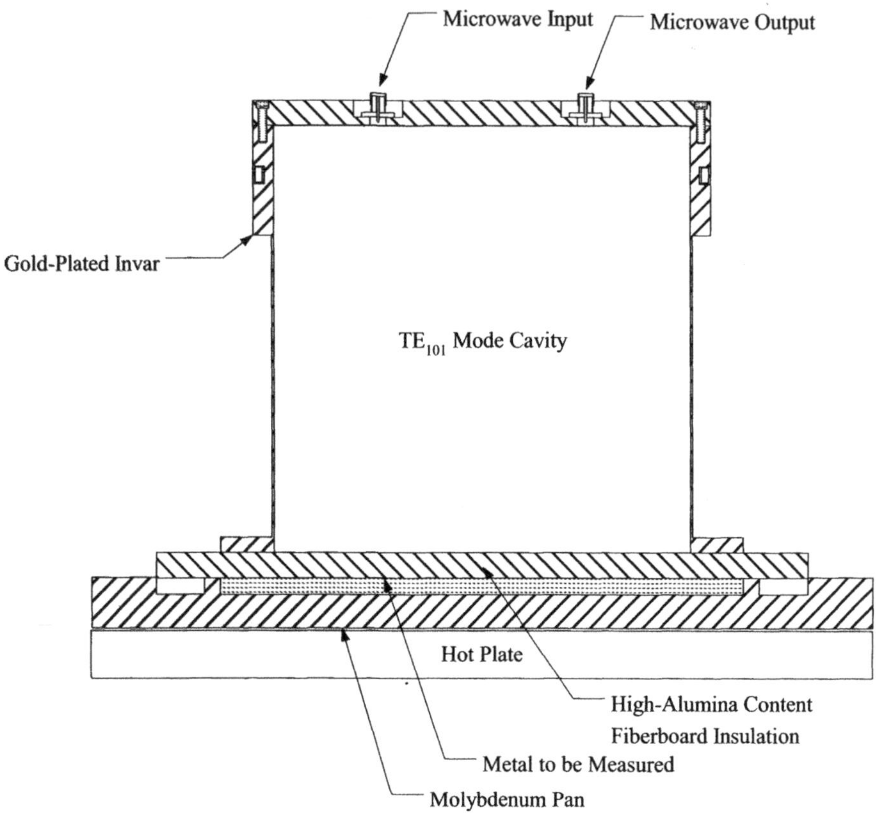

***Figure 3**. The TE_{011} mode microwave cavity used to make resistivity measurements.*

REFERENCES

[1]L.H. Van Vlack, "*Elements of Materials Science and Engineering*," Third Edition, Addison-Wesley, pp 140-145, Reading Mass., (1975).

[2]A. C. Metaxas and R. J. Meredith, "*Industrial Microwave Heating,*" Peter Peregrinus Ltd. 1988

[3]D.E. Clark and W.H. Sutton, D.A. Lewis, *"Microwaves: Theory and application in Materials Processing IV,*" American Ceramic Society, Westerville, OH.

[4]R.F. Schiffmann, "*Principals of Industrial Microwave and RF Heating*," First World Congress on Microwave Processing, Jan 5-7, 1997.

MICROWAVE ASSISTED COMMINUTION OF ORES AT ECONOMICALLY VIABLE ENERGY INPUTS

S. Bradshaw and T.V. Chow Ting Chan
University of Stellenbosch
Stellenbosch
South Africa

S. Kingman and K. Jackson,
Nottingham Mining &Minerals Centre
University of Nottingham
Nottingham, NG7 2RD, UK

N. Rowson[3] and R.Greenwood
School of Chemical Engineering
University of Birmingham, Edgbaston
Birmingham, B152TT, UK

ABSTRACT

Experimental work has shown that mineral ores exposed to microwave radiation at 2.45 GHz for less than 1 s and with volumetric power dissipation of the order of 1 X10^9 W m^{-3} undergo significant reductions (> 50%) in strength as determined by point load tests. Comparative drop weight tests showed that reductions in required comminution energy of over 30% could be achieved for microwave energy inputs of less than 1kWht^{-1}. In addition to reductions in required comminution energy, other benefits of microwave treatment are discussed. A finite difference time domain code was used to simulate microwave heating of ores in TE_{10n}, TM_{020} and TE_{111} single mode cavities. The largest maximum power density (7.6 x 10^9 W m^{-3} for 10 kW input power) was achieved in the TE_{10n} applicator. The TM_{020} and TE_{111} applicators gave better radial uniformity but with lower maximum power density.

INTRODUCTION

Comminution forms a large proportion of the capital and operating costs of any mineral processing plant and for those treating hard ores, up to 70% of the total power draw of the plant is attributable to comminution. In addition to trying to reduce energy consumption there are many other incentives for optimising the process, including increased mill capacity, reduced mill wear, improved liberation and reduced production of slimes. Of the many approaches that have been taken to improve the efficiency of comminution, one that has seen considerable research is thermally assisted comminution, in which heating and quenching are used to induce fracture in the ore. Various studies have demonstrated that while thermally assisted comminution can significantly reduce the energy required for grinding, the input of thermal energy required is economically unfavourable.

Microwave heating can be used successfully to heat many valuable minerals, while in general gangue materials (the minerals of little value) are not heated. This suggests that microwaves can achieve differential heating of mineral phases and that thermal stresses generated in this way might be sufficient to cause intergranular cracking. This would lead to a reduction in the grinding energy requirement and to improved liberation.

Most of the testwork carried out to date on microwave-assisted comminution has been at high specific energy consumption and low power density. This has largely been predicated by the use of multi-mode microwave cavities. Whilst the influence of microwave radiation from this type of cavity has been shown to have a significant influence in reducing the Bond Work Index of ores, the benefits have not been sufficient to justify use of the technology[1].

In contrast with the approach above, application of microwave energy at high power density and with short exposure time has been the focus of recent work by the current authors[2]. Finite-difference simulations were performed in 2 dimensions to examine the thermo-mechanical response of ores to microwave treatment. The ore was simulated as 1 mm^2 grains of pyrite within a microwave-transparent calcite matrix. Thermal stresses were calculated for various power densities over a range of exposure times and uniaxial compressive strength tests were simulated. Simulation at a power density of $1x10^9 W/m^3$ showed a reduction in strength from 125MPa to 80MPa in 30 seconds microwave exposure. Samples treated at $1x10^{11} W/m^3$ showed a theoretical reduction in strength from 125MPa to 60MPa in 0.05 seconds. This showed conclusively and for the first time that significant strength reduction could be obtained at high power input yet with relatively low energy consumption, which has important implications for the economic viability of the process.

The work reported in this paper had two components. The experimental work aimed at achieving significant strength reduction of an ore at a specific microwave energy input that is economically viable. At the same time simulations were done to model the power dissipation in the ore and to evaluate applicator design for optimum throughput.

SIMULATION OF MICROWAVE APPLICATORS

The Concerto 2.0® FDTD (finite difference time domain) code marketed by Vector Fields was used as the primary simulation code. Simulations were done on an Athlon AMD 1800+ machine with 1024 MB DDR RAM. Convergence of solutions was checked by comparing S_{11} results for a pulse input with those from the finite element frequency domain code ANSYS®. Simulations were then done at a single frequency (2.45 GHz) and convergence was assumed when the desired S_{11} value was reached.

The ore was simulated as a matrix of quartz ($\varepsilon = 23 - 0.1j$) containing cubic pyrite grains ($\varepsilon = 60 - 14j$). The range of pyrite grain sizes used was 1 – 8 mm. The total volume of grains was 5% and the grains were packed on both regular and irregular grids. Because the aim of the work was to find the power density within the pyrite grains, use of a mixing rule to obtain a mixture property was not appropriate. Pyrite and quartz were chosen, as their dielectric properties are representative of absorbing and non-absorbing minerals found in many ores.

Simulations were made for the following cavities: TE_{10n} in WR 430 waveguide (approximating the experimental cavity), TM_{020} and TE_{111}. The TE_{10n} applicator with the load is shown in Figure 1. Note that the ends of the applicator are closed (unlike the experimental cavity, and that the ends of the applicator have been shortened. This was done to reduce the computational requirements.

HEATING AND CHARACTERISATION OF SAMPLES

A South African copper carbonatite ore was tested. It consisted primarily of calcite ($CaCO_3$) and dolomite ($CaMg(CO_3)_2$), with abundant magnetite occurring as discrete, liberated grains (often >300μm). The principle copper-bearing sulphide minerals were found to include cubanite ($CuFe_2S_3$), chalcopyrite ($CuFeS_2$), bornite (Cu_5FeS_4) and chalcocite(Cu_2S).

The microwave heating system used a 3-15kW, 2.45 GHz generator (Sairem Ltd). The applicator consisted of an 82 mm ID tube mounted in the broad face of WR430 waveguide (terminated with a sliding short), and extended 300 mm on either side of the waveguide. An automatic E-H tuner gave impedance matching. Samples (1 kg in the heating zone) were placed in 75 mm ID cylindrical glass tubes and were exposed in the cavity for varying time-periods and at different power levels. The work of Whittles et al.[2] suggested that a specific minimum microwave input power of approximately 10 kW kg^{-1} would be required to achieve significant strength reduction. A preliminary economic analysis indicated that the microwave energy input should be less than 1 kWh t^{-1}. These two bounds defined the operating window, limiting exposure time to less than 1 s. The residence time was controlled by using a pneumatic piston in which the speed of the return stroke could be controlled.

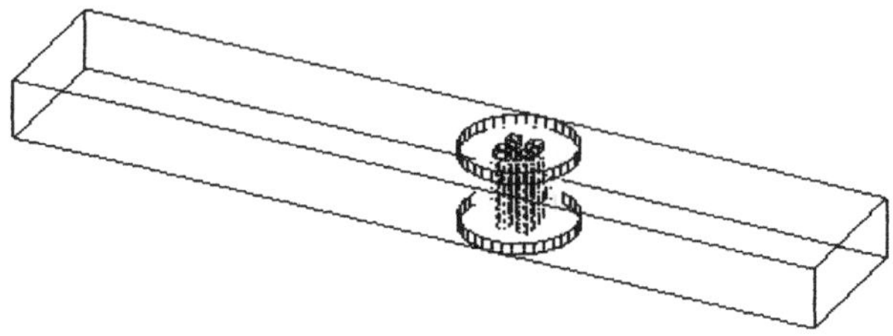

Figure 1. TE_{10n} applicator showing detail of load.

Samples were tested by two methods. Point load tests were used to determine uniaxial compressive strength. The test has recently been adapted for prediction of comminution behaviour in rocks[3]. In the test an increasing force applied to the sample that is held between a pair of specially shaped, hardened-steel tips. The force required to break the sample is reported in terms of the Is(50), an approximation to the compressive strength. Drop weight tests were also used[4]. These provide effective characterisation of an ore for crushing but not for further grinding. Note that chipping and abrasion occur in tumbling mills and that additional tests would be required to obtain a comprehensive assessment of the benefits of microwave treatment. The point load test gives the specific comminution energy (ECS) required to reduce a proportion of the feed to $1/10^{th}$ of its original mean size.

RESULTS AND DISCUSSION

Figure 2 shows reduction in strength (% of original Is(50)) as a function of microwave exposure time for material in the size range -53 +45mm. Microwave treatment reduces the strength of the ore significantly (40%) even for exposures as short as 0.1 seconds. It is especially important that significant strength reductions were obtained when using high power after such short exposure time and that increasing the exposure time (> 0.5 s) in such cases gave little further improvement. Qualitatively similar results were obtained for smaller size fractions as well.

Calculations were made to determine the reduction in ECS achieved as a function of the fraction of feed material reduced to 10% of its original size. It was found that, over the range of size reduction for which this test is valid, the ECS could be reduced by close to 30%. As mentioned earlier, this test provides effective characterisation of an ore for crushing, but not for grinding. After crushing, this type of ore would always be ground to –200 μm in tumbling mills. To assess how this grinding operation would benefit from microwave treatment would require additional pilot plant testing. It would be expected that the reductions in ECS would be much greater for grinding material that had been exposed to microwave treatment.

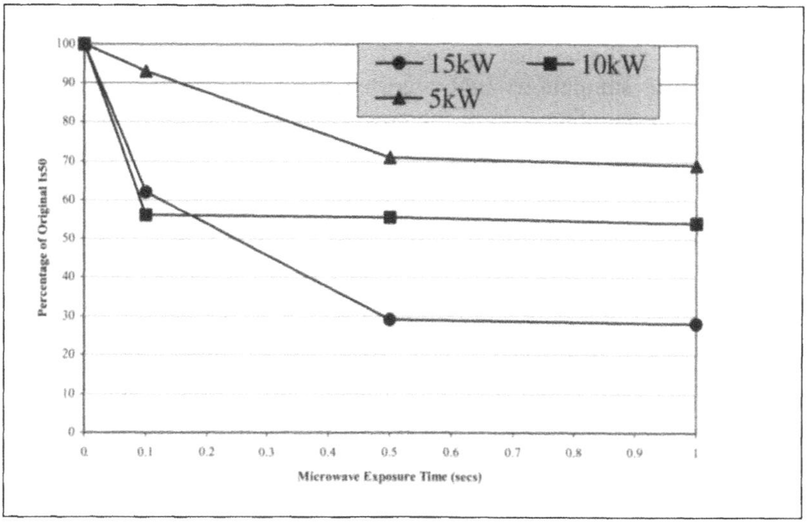

Figure 2. Plot of Percentage of Original Is50 as a function of Microwave Exposure Time for Material -53+45mm.

It is possible to correlate the impact parameter (which is determined from the drop weight test) with the Bond Work Index (BWI), which is a measure of the grindability of an ore, but the correlation is not good[4]. Using this correlation, it was found that the BWI of the microwave treated material is effectively zero, while that of the untreated was about 8.5 kWh/t. The overall implication is that microwave treatment has reduced the strength of the ore very significantly. The reduction in ECS or BWI that can be achieved needs to be balanced against the microwave energy input in order to assess the viability of the technology. For 1 kg of ore treated at 15kW for 0.2 seconds, the microwave energy input is 0.83 kWh t^{-1}. While this is not optimal, it is clear that it is an economically viable energy input (with an acceptable capital cost). More importantly, it is expected that the real benefit of microwave-assisted comminution will be enhanced liberation. Even 1-2% increases in the recovery of a valuable mineral in a large tonnage operation could lead to a significant increase in profit. This is in addition to any potential energy savings.

Figure 3 shows the variation of power density (in the vertical plane) obtained in the TE_{10n} applicator for 3 mm grains. After scaling for S_{11} to simulate the effect of perfect impedance matching, the maximum power density was found to be 6.3 x 10^5 W m^{-3} for a 1 W input. Note that in the experimental work it was found that reflected power could be reduced to less than 5% of forward power, justifying the scaling used. The effect of greater microwave input power could be obtained by simple linear scaling. Note that the result Figure 3 was obtained for an applicator with closed ends and one that is considerably shorter than the one used in experiments. Other simulations (not reported here) showed that extending the applicator 300 mm on either side of the waveguide resulted in the power density being concentrated within the waveguide, with a maximum value similar to that shown above.

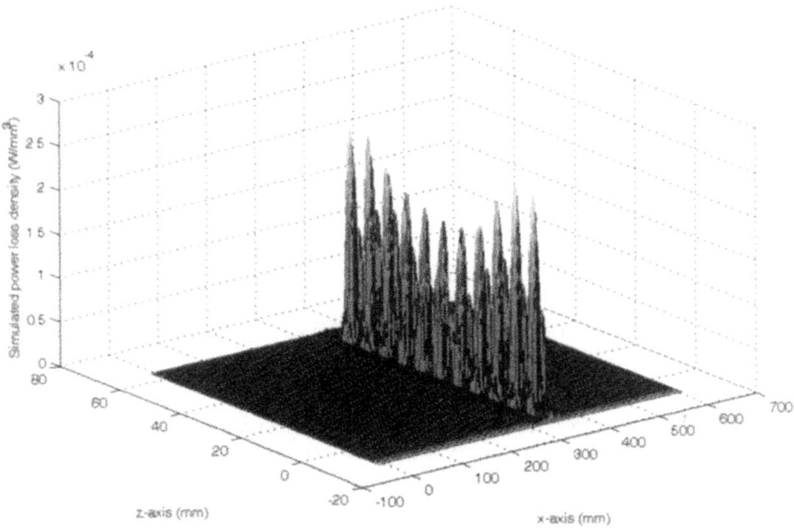

Figure 3. Axial variation of power density in TE_{10n} applicator.

The radial variation of power density at the half height of the waveguide is shown in Figure 4. This shows that uniformity is poor across the waveguide, as expected. Although lack of *axial* uniformity is unimportant for a continuous flow system, lack of *radial* uniformity is. This suggests that the TE_{10n} applicator is capable of generating high volumetric power dissipation. However, processing uniformity might be compromised with the design unless excessively high power levels were used that ensured that the minimum power density were met everywhere across the radius. The simulated maximum power density agreed well with an estimate obtained for the experimental work. In that case it was calculated that the maximum power density for a 1 kg sample containing 5% by volume of absorbing material should be on the order of 1×10^{10} W m^{-3} for a 10 kW power input.

It was found that there is a variation of maximum power density with grain size showing that there is an optimum size. This means that it will be more difficult to dissipate enough power to cause intergranular fracture in finely disseminated minerals particularly as it is known that lower thermal stresses would be in that case. This has implications for the amenability of a given ore to microwave treatment. Also note that simulations of grain sizes less than 1 mm require more computer memory than was available (1024 MB).

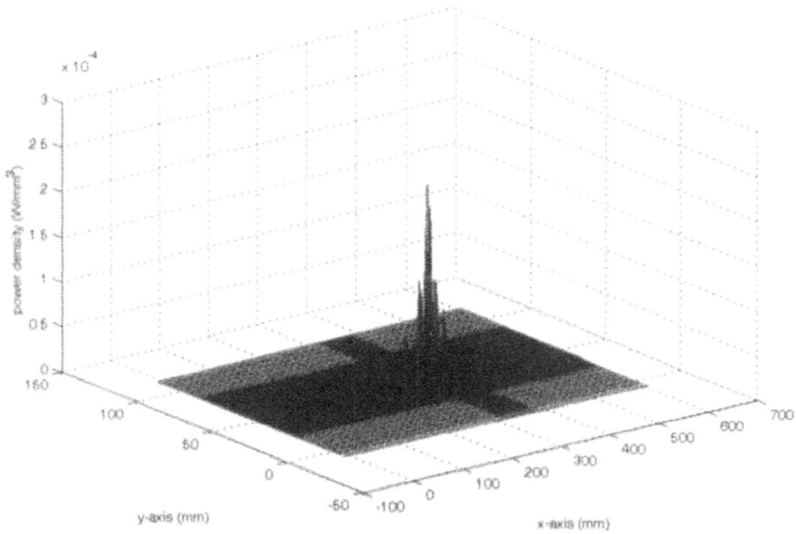

Figure 4. Radial variation of power density at mid height of the TE_{10n} applicator.

In an effort to find a cavity with better radial uniformity, a TM_{020} applicator was designed. This was done by sizing the cavity analytically for a solid load of pyrite and then adjusting the cavity diameter to ensure resonance (with the desired modes) with the correct load under numerical simulation. The load radius was

27.5 mm and the cavity radius 86.8 mm. The power density can be seen in Figure 5. After scaling for S_{11} the maximum power density was found to be an order of magnitude lower (260×10^3 W m^{-3}) than for the TE_{10n} cavity, which implies greater uniformity. Clearly there is a trade-off between achieving a high enough power density yet also achieving a heating zone of sufficient size to treat a meaningful throughput. Note that the ideal cavity would be one in which the zone of high power density is thin in the direction of material flow yet large in cross sectional area perpendicular to that direction. This also means that extending the TM_{020} applicator axially would be disadvantageous, as the available power would simply be dissipated over a greater length of material.

A TE_{111} applicator was designed and simulated and the results are shown in Figure 6. Although the loaded mode is now an HE mode, the heated zone remains lozenge-shaped, as desired. The intention was that this applicator would concentrate the electric field in volume that was relatively broad and axially limited. The maximum power density after scaling was 310×10^3 W m^{-3}. Although this is lower than that obtained for the TE_{10n} cavity, this has been achieved with better uniformity although the uniformity has not yet been quantified. At this stage it remains unclear how the TE_{111} applicator could be adapted to continuous feed while retaining the desired modal pattern.

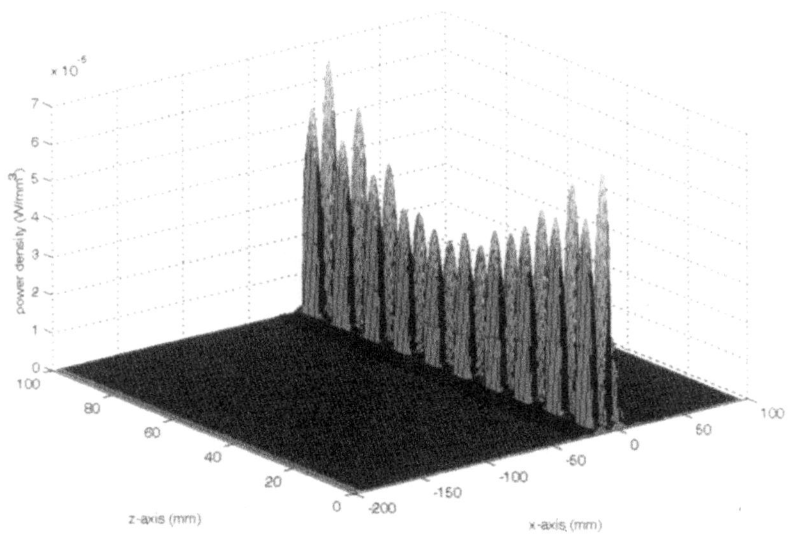

Figure 5. Axial power density in TM020 applicator.

CONCLUSIONS

It has been shown that significant reductions in strength (>50%) can be achieved in very short microwave exposure times (<1s) at high power (>10 kW kg^{-1}). Drop weight tests showed that reductions in required breakage energy of over 30% could be achieved for modest microwave energy inputs of less than $1kWht^{-1}$. Further work is suggested at the tonne-per-hour level to provide better assessment of the process benefits. FDTD simulations of different applicators showed that the TE_{10n} gave the highest power density but with poor radial uniformity. There was good agreement with the estimated experimental power density. TM_{020} and TE_{111} applicators gave better radial uniformity but with lower maximum power density. Simulations showed that there is an "optimal" particle size at which power density is a maximum. The work reported in this study is the first in which microwave assisted comminution has been achieved at an energy consumption that is economically meaningful, and at conditions which have potential for scale up to plant production rates. Future work includes commissioning a continuous system (1t/hr), detailed modelling and pilot studies of downstream benefits and work on the engineering problems of treating large tonnages.

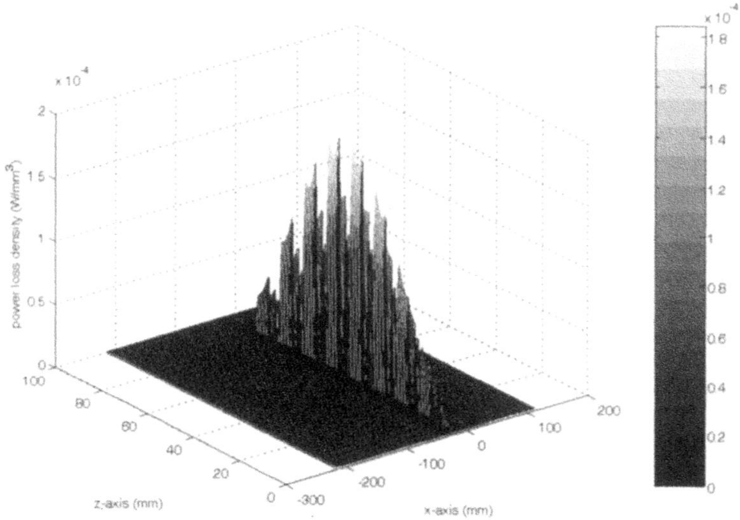

Figure 6. Power density in a TE_{111} cavity.

ACKNOWLEDGEMENTS

- Prof Howard Reader for electromagnetic knowledge
- EPSRC for financial support
- Postgraduate students at Nottingham Antonio Cumbane and Aled Jones
- David Whittles at the University of Nottingham for help and guidance with FLAC

REFERENCES

[1]Kingman, S.W., Vorster, W. and Rowson, N.A., 2000. The Influence of Mineralogy on Microwave Assisted Grinding, Minerals Engineering Vol. 13, No.3, pp313-327.

[2]Whittles, D., Kingman, S.W. and Reddish, D., 2003. The influence of power density on microwave assisted breakage. International Journal of Mineral Processing vol.64 pp 71-91

[3]Bearman, R.A., Briggs, C.A. and Kojovic, T., 1997. The Application of Rock Mechanics Parameters to the Prediction of Comminution Behaviour, Min. Eng., 10(3) 255-264.

[4]Napier-Munn, T.J., Morell, S., Morrison, R.D. and Kojovic, T., 1996. Mineral Comminution Circuits. Their Operation and Optimisation. JKMRC Monograph Series in Mining and Mineral Processing 2, University of Queensland, Australia.

Systems and Diagnostics

A TECHNICAL REVIEW OF EQUIPMENT AND METHODS FOR WAVEGUIDE POWER MEASUREMENT IN MICROWAVE HEATING APPLICATIONS

John F Gerling
Gerling Applied Engineering, Inc.
PO Box 580816
Modesto, CA 95358
USA

ABSTRACT

Various methods for waveguide power measurement were developed in the 1930's that have changed little over the years and are still in common use today. However, more recent improvements and advances in measurement signal processing technologies offer significant enhancements to the practical utilization of traditional power sensing methods. Before choosing a power sensing method for a particular application, the importance of coupling factor and directivity as performance parameters should be well understood. Measurement accuracy depends on a) the accuracy of calibration of the stated coupling factor, and b) the error introduced by the effects of reverse power as a result of finite directivity.

Three waveguide power sensing methods stand out as the most commonly used in laboratory and industrial heating systems, each having characteristics preferred for different applications.

Aperture Array Coupler – The two basic types of aperture array couplers are a) *Broadwall* couplers in which two parallel waveguide sections are coupled to each other along the broad wall by an array of apertures spaced ¼-guide wavelength apart, and b) *Cross-guide* couplers in which two perpendicular waveguide sections are coupled to each other by a single aperture.

Resistive Loop Coupler – Waveguide power is coupled magnetically and electrically to a sensor element consisting of a resistor-terminated loop and a probe.

Multi-probe Reflectometer – Three or four non-directional antenna probes detect voltages in the waveguide and enable the determination of standing wave reflection coefficient and relative phase angle as well as forward and reverse power.

The detected power signal must be converted to any of several forms as may be preferred for utilization in the application. The most popular signal conversion methods are diode detectors and thermocouple devices. Recent advances in detector circuitry offer significant improvements by combining a variety of useful features in cost-effective packages. Methods of calibration, an important factor in the accuracy of any measurement, are briefly reviewed.

INTRODUCTION

Measurements of microwave power became necessary simultaneously with the advent and use of high power microwave energy as early as the 1930's.[1] Several methods for power measurements were devised then and in the decades to follow, most of which are quite obsolete by today's standards. Of these methods, only a few are suitable for high power measurements (i.e. a few hundred Watts and above). For applications in microwave heating, the selection is further narrowed to methods suitable for operation within the ISM (Industrial, Scientific and Medical) frequency bands centered at 915 MHz (896 MHz in the UK) and 2450 MHz.

The methods of choice for most microwave heating applications are mainly those utilizing waveguide components for power coupling, along with various types of sensors for power signal processing. More recently, the integration of microprocessors and sophisticated electronic circuitry have greatly simplified the power measurement task and reduced costs while maintaining or improving accuracy and reliability. Fundamental differences in performance, reliability, size and cost necessitate careful consideration when selecting the measurement method best suited for a particular heating application.

High Power Coupling Methods

Water loads were devised very early and quickly became the most widely used method for high power calibration.[2] The primary advantage is that power is absorbed directly into the water itself, thus acting as both a load and a means for reasonably accurate power measurement by calorimetry. However, water load calorimeters are mostly unsuitable for industrial heating applications due to their bulkiness and the complexity of ancillary equipment.

An interesting yet mostly impractical method is that of force-operated power indicators. Such instruments operate by interaction of the electromagnetic field with a moving object, such as a rotating vane. Several designs and methods have been devised, although most are highly sensitive to external effects of mechanical vibration making them unsuitable for general use.[3]

Various types of waveguide power couplers were devised as a means to divide the power between the main waveguide line and a branch line, ideally with no losses or reflections. A sensing device is then used to convert the microwave field in the branch line to a dc signal having a known mathematical relationship to microwave power. Waveguide couplers generally are either directional, meaning preferential to power flowing in one direction, or non-directional.

Another common waveguide power measurement device is the multi-port reflectometer. Detailed discussions of multi-port reflectometers can be found in early literature, but they did not gain popularity as waveguide power measurement devices until the 1970's with the advent of inexpensive desk-top computers.[3] Designs having three or four non-directional probes to sense the voltage at different points in the waveguide, along with appropriate signal processing, deliver information relating to the standing wave reflection coefficient and relative phase angle in addition to forward and reverse power.

WAVEGUIDE COUPLERS

Non-directional couplers typically employ a simple probe that protrudes into the waveguide to sense the e-field voltage at a single point in the waveguide. The voltage detected is the vector sum of voltages due to both forward and reverse power, thus the non-directionality of such devices. One can then see the inherent sensitivity to the relative phase of the standing wave resulting from the presence of both forward and reverse power. Consequently, non-directional couplers are best used in the absence of reverse power or where the relative phase of an anticipated standing wave is constant.

Early designs of directional couplers included relatively simple arrangements between adjacent waveguide sections connected by one or more apertures as well as resistive-loop type couplers. Three directional coupling types as shown in Figure 1 have proven to be quite popular for microwave heating applications.

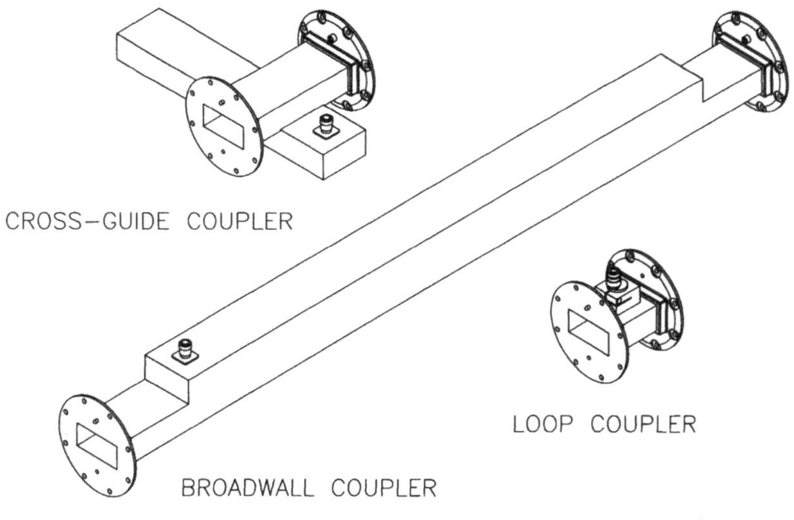

Figure 1 – Common waveguide directional coupler types.

Cross-guide Couplers

Cross-guide couplers consist of two perpendicular waveguide sections which are coupled to each other by a single aperture (two apertures for bi-directional couplers). The aperture is located such that a unidirectional circularly polarized H-field wave in the main waveguide couples a similar circularly polarized H-field in the branch waveguide which then excites unidirectional wave propagation.[4] Directivity is obtained by the fact that H-field polarization is opposite in direction between forward and reverse propagating waves, preventing a reverse wave from coupling through an aperture located for the forward wave and vise versa. Cross-guide couplers offer superior flatness of coupling factor over a broad band, although directivity is generally less than other coupler types.

Broadwall Coupler

In broadwall couplers, two parallel waveguide sections are coupled to each other along the broad wall by an array of apertures spaced ¼-guide wavelength apart. Energy propagating in one direction in the main waveguide section couples through each aperture and propagates initially in both directions in the branch waveguide. However, as a result of the aperture spacing, reverse propagating energy in the branch waveguide cancels while forward energy is additive.[2,3,4] Broadwall couplers offer superior directivity of 40 dB or more, but their large size is a significant disadvantage in most industrial applications.

Resistive Loop Coupler

A resistive loop type coupler is where the branch line is a sensor element consisting of a small strip of metal terminated at one end by a resistor and the other end by the output connector. Again taking advantage of the change in polarity between forward and reverse waves, the loop is oriented in such a manner as to couple only to the transverse magnetic field polarized in the preferred direction.[2,4] Loop couplers typically have the smallest size of all waveguide coupler types and offer reasonably good coupling factor and directivity performance.

DIRECTIONAL COUPLER PERFORMANCE

The operation and performance of a waveguide directional coupler are generally expressed in terms of a) accuracy and frequency sensitivity of its coupling factor, and b) directivity. It is important to understand the potential error in coupling factor that can be introduced due to finite directivity.

Performance Parameters

Coupling factor is defined as the ratio between the measured (forward) wave in the main waveguide and the coupled output wave.[2,4] When connected as shown in Figure 2(a) to a source and perfect loads, coupling factor is expressed in decibels (dB) as

$$CF = 10\log(\frac{P_F}{P_{3F}}) = 20\log(\frac{V_F}{V_{3F}}) = 20\log(\frac{1}{S_{31}}) \tag{1}$$

where P, V, and S represent power, voltage and s-parameter measurements, respectively. This equation represents a positive value for coupling factor, although it is often given as a negative value by inverting the logarithm function.

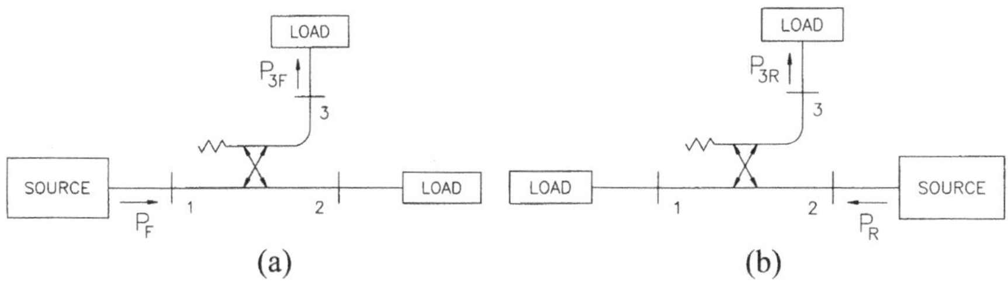

Figure 2 – Directional coupler connections for measuring (a) coupling factor and (b) directivity.

Directivity is then defined as the ratio between the forward coupled and reverse coupled waves.[2,4] When connected as shown in Figure 2(b) and assuming P_R is equal to P_F, directivity is expressed in decibels as

$$D = 10\log(\frac{P_{3F}}{P_{3R}}) = 20\log(\frac{V_{3F}}{V_{3R}}) = 20\log(\frac{S_{31}}{S_{32}}), \quad \text{or} \quad D = CF_R - CF_F \tag{2}$$

Error Due to Directivity

As mentioned earlier, the finite directivity of a directional coupler gives rise to the potential for power measurement error due to interference from a reverse propagating wave. The *indicated* reading from the output of the branch line is derived from the *vector sum* of the forward coupled and reverse coupled wave. The vector diagram in Figure 3 shows that the reverse wave can be at any phase relative to the forward wave. Thus, the reading is sensitive to both relative phase of the reflection and coupler position along the waveguide.

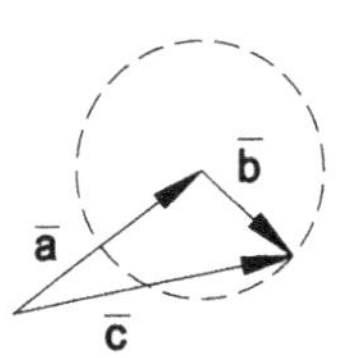

$\bar{a}$ = voltage at coupler probe output due to true forward power

$\bar{b}$ = interfering voltage from reverse power due to finite directivity

$\bar{c}$ = total actual voltage at coupler probe output

$\bar{c} = \bar{a} + \bar{b}$

Figure 3 – Vector diagram of directional coupler output voltage.

From Figure 3 it is easy to see that the limits of the *true* reading can be found by adding and subtracting the absolute value of the interfering voltage from the actual output voltage. Thus,

$|\bar{c}|$ is calculated from the indicated power reading

$|\bar{b}|$ is calculated from known reverse power and coupler directivity

Limits of $|\bar{a}| = |\bar{c}| \pm |\bar{b}|$

Assuming waveguide impedance is constant, the actual voltage at the coupler probe output is calculated from the *indicated* forward power as

$$V_{Ind} = V_F \cdot 10^{\left(\frac{-CF}{20}\right)} = \sqrt{P_F} \cdot 10^{\left(\frac{-CF}{20}\right)} \tag{3}$$

If reverse power is present and known (and assumed correct), the *interfering* voltage is calculated as

$$V_{Int} = V_R \cdot 10^{\left(\frac{-(CF+D)}{20}\right)} = \sqrt{P_R} \cdot 10^{\left(\frac{-(CF+D)}{20}\right)} \tag{4}$$

The limits of actual forward power can then be found by adding and subtracting the interfering voltage from the indicated voltage,

$$P_{F\max} = V_{F\max}^2 = (V_{Ind} + V_{Int})^2 \cdot 10^{\left(\frac{CF}{10}\right)} \tag{5}$$

$$P_{F\min} = V_{F\min}^2 = (V_{Ind} - V_{Int})^2 \cdot 10^{\left(\frac{CF}{10}\right)} \tag{6}$$

A quick analysis of the exponents reveals that the term for coupling factor (CF) can be eliminated, implying that the error in power measurement due to finite directivity is entirely independent of coupling factor.

Figures 4 thru 6 illustrate the effects of the interference and magnitude of the resulting measurement error for typical cases encountered in microwave heating processes. Figure 4 represents a typical case where the load is reasonably well matched. The range of error is minimal when using a coupler with 40 dB directivity and within acceptable limits with a 25 dB directivity coupler.

Figure 5 represents a case where the load is poorly matched, as may be the case during initial startup and tuner adjustment. Again, the error limits are minimal for a 40 dB directivity coupler, but the limits for a 25 dB directivity coupler are significantly greater. The system operator should not be surprised to see large changes in the forward power reading during tuner adjustment.

Power reflected from the process load is often measured on a continuous basis. If the load is well-matched and reflected power is minimal, the error caused by interference form the forward wave can be many times the indicated reading as illustrated in Figure 6. A common mistake especially in research applications is to infer a process effect based solely on a change in the reflected power reading without considering the potential magnitude of the interference.

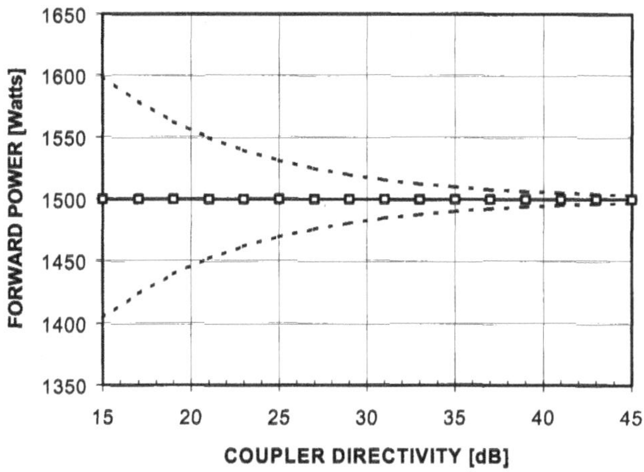

Figure 4 – Minimum and maximum *true forward* power limits for a typical case with a matched load (1500 Watts *indicated forward* power, 50 Watts *true reverse* power)

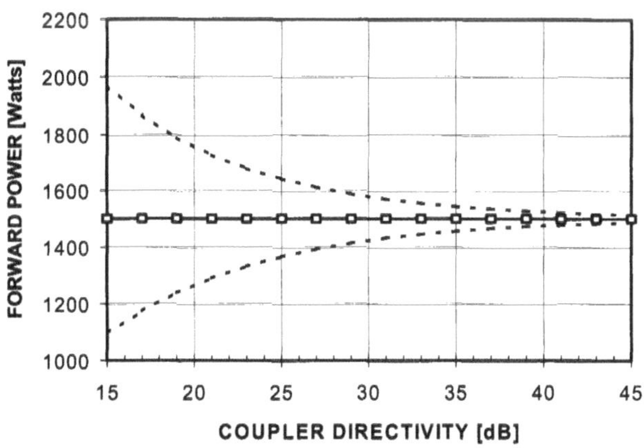

Figure 5 – Minimum and maximum *true forward* power limits for a typical case with a poorly matched load (1500 Watts *indicated forward* power, 1000 Watts *true reverse* power)

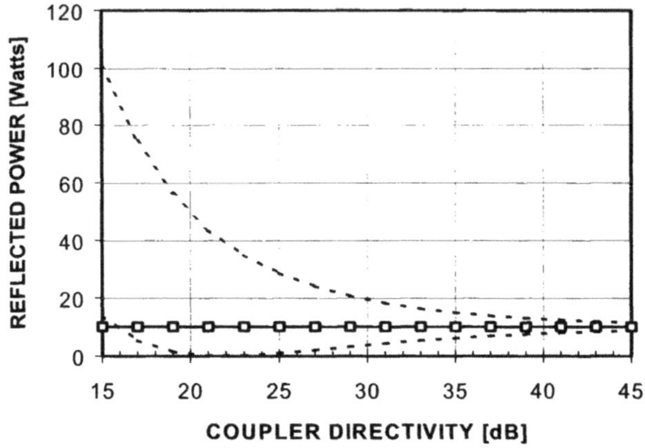

Figure 6 – Minimum and maximum *true reflected* power limits for a typical case with a well matched load (10 Watts *indicated reflected* power, 1500 Watts *true forward* power)

Table 1 provides a comparison of performance characteristics between the three common coupler types. Note that the frequency response of the cross-guide coupler is much better than that of the loop coupler, but directivity is not as good. As such, the cross-guide coupler is preferred for well matched, broad band applications such as telecommunications whereas the loop coupler is preferred for heating applications. The broadwall coupler offers superior performance in both coupling factor and directivity compared to the other coupler types, but size and cost limit its usefulness and desirability.

Table 1 – Performance Comparison for WR284 Waveguide Directional Couplers

Coupler Type	Mfr	Operating Band (GHz)	Coupling Factor Frequency Response				CF Mean Accuracy (+/- dB)	Minimum Directivity (dB)
			Mfr Specification		Within ISM Band			
			+/- dB	% of Band	+/- dB	% Error		
Loop	A	2.60-3.95	0.75	25%	0.22	5.3%	0.5	25
	B	2.60-3.95	1.0	20%	0.37	8.9%	N/A	25
	C	2.4-2.5	0.2	100%	0.20	4.7%	0.2	28
Cross-guide	A	2.60-3.95	1.0	100%	0.07	1.7%	1.0	20
	B	2.60-3.95	1.5	100%	0.11	2.6%	0.5	20
Broadwall	A	2.60-3.95	0.5	100%	0.04	0.9%	1.0	40
	B	2.60-3.95	0.5	100%	0.04	0.9%	0.5	40

SENSORS

The microwave output signal from the branch line of a directional coupler is of little use for power measurement purposes until it has been converted to a form which can be easily displayed and/or recorded. Several types have been devised, each having characteristics preferred for specific applications.

Bolometric types, including thermistors, operate on the basis of the dc resistance temperature coefficient of a power absorbing element. Bolometers are favored for use as laboratory measurement standards due to their inherent accuracy and stability.[3] Thermistors have been commonly used commercially in conjunction with electronic power meters, but more recently they have been replaced by diode and thermocouple devices which offer greater sensitivity, dynamic range and power handling capability.[1]

Detector diodes operate by rectifying the microwave signal directly to provide a pulsed dc voltage signal. At the low end of the operating range, the output signal follows a "square-law" transfer function which is conveniently proportional to waveguide power. Diodes are quite popular in commercial and industrial applications for their simplicity and low cost. However, as semiconductor devices they have an inherently high temperature coefficient which limits their accuracy under varying ambient conditions.[1,3]

Thermocouple sensors are popular for high performance measurement requirements due to their higher sensitivity, ruggedness and accuracy. The principle of operation is based on the voltage generated due to the differential temperature between hot and cold junctions of dissimilar metals. The effect is inherently "square-law" in operation and truly averaging, making thermocouple sensors ideal for all types of signal waveforms. The only drawback to these sensors is high cost which limits their practicality in typical industrial heating applications.

REFLECTOMETERS

Multi-port reflectometers have gained popularity in industrial heating applications for their ability to provide information about the complex impedance of a load in addition to forward and reverse power. Commercially available types utilize three or four non-directional probes to detect the voltage at different positions along the waveguide. Various models with integrated electronics for signal detection and conditioning deliver the individual probe output signals, while others utilize microprocessors for signal analysis. Several mathematical methods are available for complex impedance analysis using three and four probe models with equal or random probe spacing.[5,6] The most sophisticated products are used in conjunction with automatic impedance tuners and predictive s-parameter algorithms for very high speed performance.

CALIBRATION

A critical factor in the usefulness of any directional coupler is its calibration accuracy. Reasonably accurate calibration at low power is possible using a standard network analyzer having suitably high dynamic range. However, high power calibration requires the use of a high power measurement standard. The

National Institute of Standards and Technology (NIST) maintains standards up to 1 kW at up to 1 GHz, but in the 1 GHz to 98 GHz frequency range the available standards are limited to 10 mW.

Thus, alternate methods must be employed for high power calibration at 2.45 GHz using NIST-traceable standards. A simple method is by substituting a previously calibrated coupler with the coupler under test. A more cumbersome yet highly accurate method is the "cascaded-coupler" technique.[7] Commercially available calorimeters utilize the resistance substitution method which compares the electrical heat loss to microwave power absorption.

A calibration routine should be established depending on the type of coupler used and requirements for measurement accuracy in a particular application. Aperture array couplers have fixed geometries and thus require no periodic calibration. Resistive loop couplers are adjustable and therefore should be routinely calibrated at least once every 1-2 years. Devices employing electronic circuitry to derive the output signal are more subject to drift and should be calibrated a minimum of once every year.

SUMMARY

Several methods for high power measurements in waveguide are available, but only a few are recommended for industrial heating applications. Directional and non-directional couplers offer a range of performance characteristics for simple waveguide power measurements, while multi-port reflectometers add the capability to deliver information on the complex impedance of a load. Directivity and interference from reverse power should be well understood when interpreting the output reading from a directional coupler. Calibration must be routinely maintained to ensure measurement accuracy.

REFERENCES

[1] "Fundamentals of RF and Microwave Power Measurements," Hewlett-Packard Application Note 64-1A, 1997

[2] C.G. Montgomery, "Technique of Microwave Measurement," MIT Radiation Laboratory Series, Vol. 11, McGraw-Hill, New York, 1947

[3] A. Fanton, "Radio Frequency & Microwave Power Measurement," Peter Peregrinus Ltd., London, 1990

[4] R. Meredith, "Engineers' Handbook of Industrial Microwave Heating," Institution of Electrical Engineers, London, 1998

[5] P.I. Somlo and J.D. Hunter, "Microwave Impedance Measurement," Peter Peregrinus Ltd., London, 1985

[6] R. Caldecott, "The Generalized Multiprobe Reflectometer and Its Application to Automated Transmission Line Measurements," IEEE Transactions on Antennas and Propagation, Vol. AP-21, No. 4, July 1973

[7] J.W. Allen, "Switched-Coupler Measurements for High-Power RF Calibrations," NIST

DESIGNING INDUSTRIAL MICROWAVE HEATING SYSTEMS FOR SAFE OPERATION: BATCH OVENS

Robert F. Schiffmann
R.F. Schiffmann Associates, Inc.
149 West 88 Street
New York, NY 10024

ABSTRACT

While a great deal of effort in the design of industrial microwave heating equipment is concentrated upon heating of the load and the achievement of the desired material properties, often not enough is focused upon its safe operation. This paper will focus upon batch type ovens and examine various aspects of equipment safety and suggest ways in which they may be achieved.

At the last AMPERE, the author presented a paper "Fires in Microwave and RF Heating Systems: Causes and Prevention". There are many other safety concerns that are of serious potential harm to the personnel, equipment and facility. Among the items to be discussed:

- The construction of access doors and windows.
- Designing openings for the introduction of hot air and other gases.
- Interlocking for access and cleaning.
- Personnel safety requirements in large batch ovens.
- Arc detection requirements.
- Dummy load requirements.
- Reducing exposure to microwave leakage.
- Compatibility of materials of construction and processing.

INTRODUCTION

Industrial microwave processing systems have been in use since the mid 1950's and their applications have spanned a wide range of industries including food, rubber, pharmaceutical, ceramics, chemical, and many more. When designing these systems, microwave engineers focus upon:

- Efficient heating of the load
- Maximizing energy coupling to the load
- Achieving the desired product quality

- Maximizing throughput
- Costs of the system
- Application of auxiliary heat sources
- more

While significant attention is paid to safety issues regarding worker exposure to microwave radiation, often there is not an equally serious concern for other worker safety issues.

Considerable attention has been paid to issues regarding the safe use of microwave heating for many years. Two early publications (1,2) concerned safety aspects for domestic microwave ovens. As for industrial microwave systems, the safety literature is very limited. A paper delivered at the Second World Congress discussed "Fires in Microwave and RF Heating Systems: Causes and Prevention" (3) that was the result of two serious fires occurring in one microwave and one RF industrial heating system. Subsequently, serious design flaws were uncovered by this author, regarding the safe operation of a large microwave batch oven. This has led to the consideration of the many non-radiation safety issues that should be considered when designing industrial microwave systems, and this paper focuses upon industrial microwave batch ovens. These ovens are characterized as multimode resonant cavities with one or more access doors. They are designed to have the load placed in the cavity; the door closed (either manually or automatically); the microwave power energized for a specific amount of time, and then shut off; the door opened and the load removed from the oven. The next load may be placed in the cavity immediately thereafter, or after some time. Such ovens often employ auxiliary heating including hot air, infrared or steam, or combinations thereof. They may also use atmospheres other than air and even vacuum. Most batch oven cavities are only a few times larger than the load and so, may be approximately one or two meters long on a side. However, there are also very large batch ovens that may be several meters on each side, large enough for a workman to enter the cavity and even for a fork lift truck to be driven into the cavity to deposit or remove the load. Such very large ovens present unusual safety problems, as described below.

OBJECTIVE

The overall objective of this publication is to provide information for the avoidance and elimination of injury to workers; damage to the equipment and facility; and, liability for the manufacturer. It will not deal with issues such as improving microwave distribution or coupling efficiency.

GENERAL EQUIPMENT REQUIREMENTS

Cavities and Waveguides (4)

The multimode resonant cavities generally used for batch ovens are constructed of steel – either stainless or carbon, or aluminum. The choice is largely based upon processing conditions and construction costs. There is an exception in situations such as the U.S. Food and Drug administration's (FDA) requirement for only stainless steel.

Aluminum has excellent electrical and microwave characteristics, but does not tolerate temperature well. Even at temperatures as low as 120°C, aluminum will distort over time, but this may be years rather than hours. Therefore, durability is a consideration where elevated temperatures are involved for constructing cavities of aluminum.

Stainless steel has excellent temperature and mechanical characteristics but is somewhat microwave lossy. It distorts readily with temperature, so welding can be a challenge. However, it remains the material of choice for a wide range of high temperature applications.

Carbon steel suffers from lower electrical conductivity, and easily forms unattractive oxide coatings. It can, however, be coated with a variety of materials that enhances its use as a microwave construction material.

All seams and joints must be welded, preferably continuously rather than spot welding. This is to ensure that joints do not separate which can cause serious microwave leakage problems. It is suggested that these be internal welds for better cavity efficiency.

Interface points for such things as waveguides and chokes must be flat, flush and electrically conductive to prevent microwave leakage. They may be bolted in place, with bolts being placed on less than half-wavelength centers.

Waveguide sections should be mechanically strong and supported, if necessary, to avoid undue stress on joints and flanges, as these may break causing microwave leakage.

Circulators

Also called iso-circulators, or junction circulators, these microwave ferrite devices are placed between the power source and the cavity and provide total protection to the magnetron from even 100% reflected power, which otherwise would damage or destroy the tube. It is important, however, that the circulator and power source be matched in terms of capacity. For example, a 75-kilowatt generator requires a circulator of a 75-kilowatt capacity. If, for example, a 30-kilowatt circulator were used, very high reflected power could overload and damage the circulator and ultimately the power source. (4)

(Note that the nomenclature for circulators is somewhat misleading. Circulators often need a load for magnetron protection; together they become an isolator.

Quite often they are specified as to total power handling capacity, so a "75-kilowatt circulator", under this nomenclature, could only tolerate 25 kilowatts of reflected power with a 50 kilowatts of forward power.)

Access Doors

These are constructed of the same material as the rest of the cavity. Their function is to allow an operator to place into and remove from the cavity, the load to be processed. There are two major classifications of doors: manual and automatic.

Manual Doors

The doors must be structured to prevent leakage of microwave energy. Such structures as ferrite-filled quarter wave chokes are useful. (5) It is important that these chokes operate successfully after long time use. They must be kept clean, and a periodic cleaning schedule, commensurate with the processing schedules, should be employed.

When auxiliary heat is used in the cavity, the door must be of sturdy enough construction to prevent warping, which can increase microwave leakage and worker exposure. A strong, positive latch system, such as used in industrial refrigerators, is useful to maintain good door to cavity/choke contact. Many of the doors employed in batch ovens are quite heavy, so a sturdy hinging system must be employed to ensure tight door closure even after thousands of hours of use.

An absolutely essential feature is to have at least two independently wired interlocks that will intercept the microwave power when the door is opened. They also prevent the microwave power from being turned on unless the door is firmly closed and latched. This is usually shown on the generator as a warning or "Ready" light. By being independently wired, should one interlock switch fail, the other will continue to operate. A preferable system may be to follow the design required for domestic microwave ovens in the United States (6): which requires that each oven be provided with two interlocks and an interlock monitor, commonly a third "monitor" interlock. The first two, primary and secondary interlocks, are usually normally open switches, while the monitor interlock is a normally closed switch. In this configuration should either of the first two fail, the other will short and blow the fuse, to prevent microwave power generation.

Automated Doors

Beside most of the same requirements as manually operated doors, these doors have another important requirement: that they stop closing when they contact any part of the operator's body or any other object placed in their path. They must react, in a manner similar to elevator doors, to a minimal amount of force, opening fully until they are re-activated. The stop mechanism should have at least two independent interlocks.

This problem is further exaggerated in very large walk-in type ovens, where the doors are very large and extremely heavy – perhaps weighing several tons, and could kill or severely injure a worker in its path when closing, unless a positive stop mechanism is employed. It is also useful to employ an audible warning and/or flashing lights as the door closes.

Another potential problem encountered with very large batch ovens in that the choke seal, contacted by the door when it closes is likely to be located in the floor. There is a danger of damage to this choke structure if forklifts, rolling carts, or similar are employed in order to bring loads in and out of the oven. Ramps over the choke should be employed in these situations. Also there is a danger of damage to the choke due to external forces that may damage the floor. These floor chokes must be cleaned frequently to remove any dirt or foreign material which may interfere with its functioning. Finally, it is essential that microwave leakage be monitored frequently, over the entire length of the choke with the door fully closed and the oven energized with a minimal load.

Windows

Windows allow an operator to view the product in the oven as it is being microwaved. In order to prevent microwave leakage through the window, they usually employ a thin perforated metal sheet with small holes, although there have been other structures employing metal tubes of one inch or slightly larger cross section, about 6 inches in length. All of these structures must keep the microwave leakage level at acceptable levels, e.g., 1 milliwatt per cm^2 at a distance of 5 cm from the windows surface (this is only a suggestion, not a standard).

In almost every case, the metal screen is covered, usually inside and outside with glass, quartz or high temperature clear plastic. This serves multiple functions:

- Prevents insertion of foreign objections through the screen
- Prevents loss of hot air through the screen. When the ambient temperature in the cavity is high, this could be a significant personnel hazard.
- Prevents the loss of moisture or other vapors, some of which may be dangerous, through the screen.

When there is high temperature inside the oven during processing, it is necessary to use tempered glass or quartz to deal with the mechanical stresses due to heat-induced warping, expansion and shrinking, as well as the heat itself.

Introducing Gases, Instruments and Materials

It is often necessary to introduce materials, gases, or objects into the cavity, for example, temperature probes. When doing this it is important to prevent microwave leakage, which can cause worker exposure to microwave radiation. The simplest and most common way to do this is to use a circular or rectangular waveguide beyond cutoff, typically, approximately one inch or narrower in diameter and four to six inches long. (It may be wider for ovens operating at 915 MHz). While this is effective there are certain safety precautions:

- Metallic devices such as thermocouples or screwdrivers should not be inserted in the tube as these create a coaxial structure that may cause serious microwave leakage problems. To prevent this the tube can be capped when not being properly used. Or, instead of being a straight tube, it may have one or more right angle bends making it impossible to insert these foreign objects.
- Gases, steam, vapors, etc. may escape through the tube. In order to prevent this, the end should be capped when not being properly used for temperature measurement, for example. This cap, which may also include a sealing bladder, may be permanently installed and have an opening of a size to fit the device being properly inserted into the tube.

Safe Access into the Cavity

It is often necessary for workers to have access to the interior of the cavity: to place or remove product to be processed, or for cleaning or repair. This will be done through the open access door(s). These doors must be properly interlocked, or described above, in order to prevent microwave exposure. All appropriate OSHA (in the U.S.A.) lockout/tag out procedures must be followed (4).

Any cleaning that is done should not use abrasive cleaning agents to avoid scratching the interior surface of the cavity. Should flammable solvents be employed, the cavity must be fully vented before microwave power can be restored. Any standing water in the oven must be removed and the cavity left completely dry; a water drain in the floor of the oven can be used for this purpose.

SAFETY ISSUES WITH LARGE (WALK-IN) OVENS

Large, walk-in ovens are employed for the processing of large or numerous items. Workers frequently enter these ovens in order to place or remove the load, which may be on a rolling cart. Sometimes the worker may bring in or remove the load by means of a forklift. There are other occasions such as cleaning, maintenance or repair in which one or more workers may need to be inside the oven cavity. All of these situations present unique problems to avoid workers being accidentally exposed to the microwave power.

As a first line of defense, there should be warning lights to indicate to other plant personnel that someone is in the oven. There should also be a positive shut-off to prevent the doors closing and the generators being energized. The safety interlocks on the doors will do this, but an independent backup is best. Emergency "Stop" buttons should be placed both inside and outside the oven .

In the unlikely, but by no means impossible event, that the oven door closes and the power is turned on while the worker is inside, it is imperative that the worker be able to shut off the power and get out of the oven before suffering injury. An easily accessible shut off mechanism is required. It is likely to be very dark in the oven so safety switches are not useful. A very useful device is a rope or several ropes along the entire length of the interior cavity walls, at easily accessible heights. By pulling on any one rope the worker is able to shut off the power, open the door, sound an audible alarm and energize flashing danger lights. I would also be sensible to have an automatic time delay of 10 second, for example, between the time the door is fully closed and the time at which the microwave power can come on.

Two other serious problems encountered with this type of oven is a result of the very large heavy doors and the use of the bottom choke seal in the floor. Both these issues have been dealt with earlier in this text.

Many of these ovens employ auxiliary heating systems which require special safety precautions:

- All auxiliary heating must be shut down and the internal temperature reduced to a safe ambient temperature before the door may be opened to allow a worker to enter. In the United States, OSHA limits the maximum surface temperature of parts to 60°C so they may be manually handled. Only in serious emergencies, such as a fire, should it be possible to open a hot oven, and that should require a separate set of controls.
- In some cases, atmospheres other than air, or vacuum, may be used inside the oven during processing. In such cases, the oven cavity must be fully purged to restore the ordinary air atmosphere before the door is opened and a worker may enter.

DESIGNING FOR HIGH VOLTAGE SAFETY

The voltages required to power modern day microwave power sources are high voltages. These voltages can be lethal. Sufficient provision should be designed into each piece of high voltage gear to assure that accidental exposure to high voltage is prevented. Typically a double redundant safety interlock system is employed in a microwave power source that will inhibit personnel exposure to the high voltage conditions. All servicing of high voltage microwave power sources should be left to properly trained high voltage technicians or the manufacturer's authorized service group. (4)

All high voltage components should be sufficiently isolated so that it is impossible for accidental contact or exposure. Enclosures housing the high voltage devices must be interlocked so that if anyone opens the enclosure, the power will immediately interrupted. In many cases, the generator housing is locked with a key in order to prevent any unauthorized personnel from opening the enclosure.

Another important consideration is to prevent water from entering or condensing inside the high voltage enclosure. Many plants, especially for food processing, use large quantities of hot water for cleaning. Unless the high voltage cabinet is made especially watertight, this can create a serious hazard. The most common and effective technique is to locate the generators remotely from the oven, in a separate room. This is typically done with microwave bacon cookers, where the generators are located on a floor above the cavity and interconnecting waveguide brings the microwave power into the oven.

As for condensation, industrial magnetrons are water-cooled and this can cause water to condense within the cabinet. To prevent this, it is important that the cooling water and its quality be maintained within the manufacturer's specifications. Cooling water that is too cold is likely to condense inside the generator cabinet; water that is too warm will not provide sufficient cooling.

Still another potential problem is faulty placement of components. In one case I investigated, an under-capacity circulator was placed above the high voltage power supply. The circulator was water-cooled and had it failed due to large reflected power, it was possible that the cooling components would rupture spilling large volumes of water on to the generator. It is imperative that such construction be avoided. Note that some manufacturers include the circulator and water load inside the generator cabinet. The head of engineering for one firm stated, "I would prefer to engineer my way around a leak scenario than have the circulator outside the generator."

COMPATIBILITY OF MATERIALS OF CONSTRUCTION AND PROCESSING

There are a number of issues that must be considered in the design of a microwave processing oven and one of them relates to the compatibility of the materials being processed and the materials of construction. For example, in the United States, it is mandatory that all food processing cavities and metal structures within the cavity be manufactured of stainless steel. Another requirement is that any glass that is present, in the door windows, for example, be tempered and shatter proof to prevent glass pieces falling in the food.

In ovens where a corrosive atmosphere is present, the cavity must be constructed of a non-corroding metal, such as stainless steel (7), and it may even be necessary to coat the interior walls with a corrosion resistant coating. Volatiles may condense on the inner walls of the cavity and other internal structures. These may represent a fire, corrosion or other hazard. It is imperative that the interior of the oven be examined daily and any condensate be removed. This includes the waveguide windows that may otherwise be blocked and reduced power into the oven or create intense heating at that surface. (7)

Polymeric materials are frequently used as mechanical supports for the load. (8) Teflon ® or similar materials are clearly the choice. They are microwave transparent, tolerate temperatures over 200°C and are reasonably priced. All polymers exhibit runaway heating when they soften due to temperature, therefore, it is advisable to avoid lower temperature plastics. Even polypropylene, which may be useful up to 120°C can be a problem after long time since its glass transition temperature decreases overtime. This is described in greater detail in reference 3.

Most importantly, never use Nylon® in a microwave cavity. Both its permittivity and loss tangent increase dramatically with temperature, leading to runaway heating and, since it is flammable, fire! (9)

DEALING WITH HIGH TEMPERATURE SYSTEMS

Some ovens are used to process materials at high temperatures. This can cause the exterior wall of the oven to be above the safe temperature limit (60°C in the United States) for contact by the workers. In these cases it is usually necessary to double-wall the oven, with insulation between the walls, to prevent worker injury.

Note that insulation may be used at lower temperatures, 50°C for example, to prevent heat loss and warming the processing area.

ARC DETECTION

The presence of arcs within an oven can cause damage to the cavity or internal structure, and may even lead to a fire. In order to prevent serious problems it is advisable to place arc and fire detectors within the oven, capable of scanning the entire work area. Detectors of the ultra-violet type are recommended, as they are able to see through smoke.

When an arc is detected, the microwave power should immediately be shut off and the source of the problem investigated and removed. Both visible and audible alarms should be installed to alert workers of the problem.

CONTROLLING WORKER EXPOSURE TO MICROWAVE LEAKAGE

Several actions should be taken on a daily basis:

- All access doors, surfaces and components must be designed to keep leakage at or below acceptable standards
- Choke seals and contact point around doors should be kept clean
- Any damage to the door or choke should be repaired at once or the defective part replaced

- Leakage testing should be done on a frequent mandatory basis

MISCELLANEOUS ISSUES

- Flammable and explosive solvents should be avoided, even when cleaning. If used, they must be thoroughly removed before powering up the system.
- The cavity work area must be kept clean and free of tramp metals and other sources of arcing.

CONCLUSION

It is not possible to define all the issues that may create hazards when operating an industrial microwave processing system. However, the issues described above are all to be taken seriously and the safety suggestions followed. In my forty plus years of working with industrial microwave processing I have encountered nearly all described. Several led to catastrophic results including serious worker injury, the total destruction of the processing system and the processing plant. Hopefully, following these suggestions will avoid such calamities in the future.

REFERENCES

(1) R.F. Schiffmann, Food Product Design, Weeks Publishing, Northbrooke, Illinois (1994).

(2) R.F. Schiffmann, Leaders Product Liability, Law and Strategy, New York, NY (1994).

(3) R.F. Schiffmann, Proceedings of the Second World Congress on RF and Microwave Processing, The American Ceramic Society, Westerville, Ohio (2000).

(4) M. Yonnone, private communication (2002)

(5) R. Meredith, Microwave heating applicators 1: multimode ovens 5.8.2 Door chokes, pp 184, Engineering Handbook of Industrial Microwave Heating, The Institution of Electrical Engineers, U.K. (1998).

(6) The Code of Federal Regulations (CFR) Part 1030.30, "Performance standards for microwave and radio frequency emitting products", Center for Devices and Radiological Health.

(7) R.V. Decareau and R.A. Peterson. Applicators and processing systems 6.6.7 Materials of construction, pp 186, Microwave Processing and Engineering, Ellis Horwood Ltd., Chichester England (1986).

(8) R. Meredith, and Microwave heating applicators 1: multimode ovens 5.10.1 Conveyor types and materials, pp 190, Engineering Handbook of Industrial Microwave Heating, The Institution of Electrical Engineers, U.K. (1998).

(9) A.R. von Hippel, Dielectric constant and loss vs. temperature, pp 412, Dielectric Materials and Applications, The M.I.T. Press, Cambridge, Massachusetts, (1954)

RECENT DEVELOPMENT OF MICROWAVE KILNS FOR INDUSTRIES IN JAPAN

M. Sato, T. Mutoh, T. Shimotuma, K.Ida, O.Motojima, M.Fujiwara
National Institute for Fusion Science
322-6 Oroshi-cho, Toki-shi, Gifu-ken, 509-5292, Japan

S.Takayama, M. Mizuno, S. Obata, K.Ito, T.Hirai, T. Shimada,
Gifu Prefectural Institute for Ceramics Research and Technology
3-11 Hoshigadai, Tajimi-shi, Gifu-ken, 507-0811, Japan

ABSTRACT

The microwave sintering for ceramics is going to take off from the laboratory to the real industries in Japan. The National Institute for Fusion Science and Gifu Prefectural Institute of Ceramics has been successful in bridging the science and technology by developing an idea of a self-consistent isothermal susceptor[1] [2], which was presented in the 2nd World Congress, to sinter almost all kinds of ceramics at 2.45GHz uniformly and rapidly. We designed and manufactured more than ten microwave kilns for industries in 2001, including a pure microwave roller hearth. In this paper, we will present the concept of the isothermal susceptor, experimental results and the kilns.

INTRODUCTION

The condition for "stress free sintering" is to maintain a uniform temperature distribution in the body of the work piece (sintered material) during microwave heating. Thermal energy escapes from the surface, when it faces the lower temperature material. The energy flow produces a temperature gradient in it. The hybrid heating kilns were developed. The gas or electric power rises the temperature of susceptor in proportion to the work piece[3]. The hybrid heating requires a feed back controlled according to the information of temperature just on surface of work piece.

We have an idea of a self-consistent susceptor that has certain dielectric constant and loss-tangent to be heated by microwave in keeping the equal temperatures and equal rising rates with the work pieces. The energy flow by radiation /convection balances between the work pieces and the inner most wall of the susceptor. The thermal energy flux balances on the surface of work pieces, i.e. zero in total [(1),(2)]. Under such an ideal thermal insulation, work pieces get energy only by electric field of microwave and heat themselves uniformly. The stress free system realizes the rapid sintering.

Instead of the Self-Consistent Susceptor

Microwave field distribution in the work piece is another key word for "stress free sintering". The 2.45GHz low frequency microwave does not show any non-uniform heating in our experiment when more than two magnetrons are operated simultaneously with mode stirrers.

Through the development of the microwave kilns, energy consumptions were measured and compared to the conventional (traditional) systems. The bodies of the kilns consume almost 95% of entire power, both in traditional and in microwave kilns. However, microwave kilns need only 1/2 ~1/5 energy for the unit weight of products. Roughly speaking, it is the rapidity of microwave heating that saves the energy. It inhibits the grain growth and numbers and size of pores. Now, many industries pay attention and are planning to introduce the microwave kilns into their factories in Japan.

BATCH KILNS

Batch kilns were built. The smallest was 0.3 x 0.3 x 0.3m with 3~5kW microwave for laboratories and the largest was 1 x 2 x 1m with 90kW for factory as shown in Figure 1.

Basic Structure of Microwave Batch Kiln

Figure 2 illustrates microwave kiln with isothermal barrier. The blankets have cascaded structures that consists of thermal insulator made of alumina-silica fiber blanket / fiberboard and inner most isothermal Barrier (Susceptor) that is made of material with having identical microwave absorption to work piece. They are built in the metallic chamber. The waveguide cavities, on which 2.45 GHz magnetrons are mounted, are installed on the metallic wall directly Metal fans rotate in the chamber to work as mode stirrers. The diameters of the fans are 0.45~0.5m. Rotating speed is 10 rpm.

Energy Balance in Microwave and Conventional Kilns

Energy consumption measured in a kiln operating at 20kW/2.45GHz microwave was compared to the conventional kiln with propane gas burners. Both kilns have identical internal volume of $0.15m^3$. Typical heat curves are shown in Figure 3 for the sintering of 8kg porcelains. The temperature rises linearly up to 1300EC in 2 hours in microwave kiln. Conventional kiln needs several steps over 7.5 hours. The energy exhausts are 63kWh (54000kcal) and 335kWh (1206000kcal) in microwave and in conventional kilns respectively. Microwave kiln spends only 1/5 of energy to sinter the same weigh of work pieces.

Figure 1. An example of batch microwave kiln for factory use, 90kW.

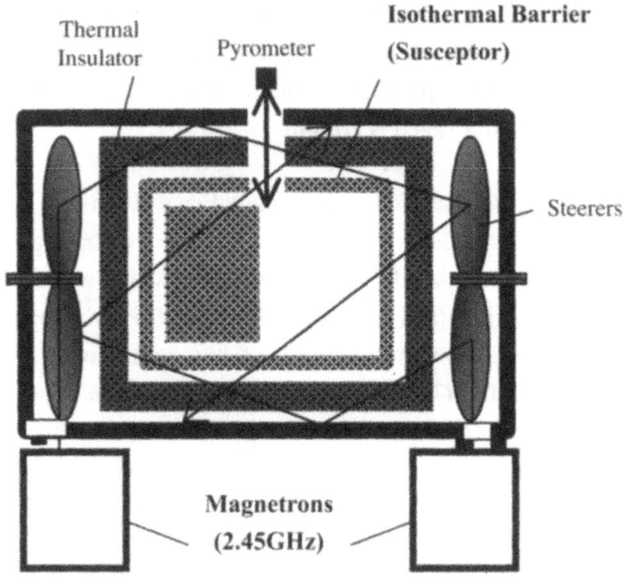

Figure 2. Schematic view of a microwave kiln with isothermal barrier

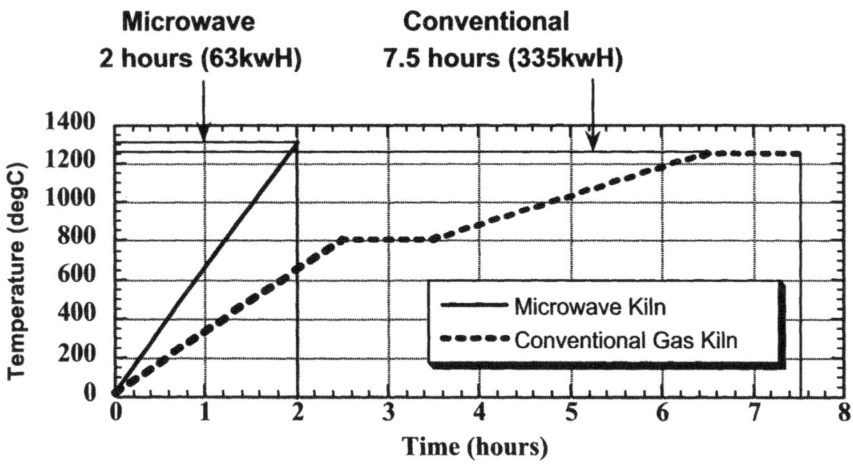

Figure 3. Comparisons of Heat Curves in Microwave and Conventional Kilns

The power balances estimated on both kilns are as shown in figure 4. Only 1% and 4.3 % of the overall energy is spent for sintering in microwave and gas kilns respectively. 45% of energy is lost in the collector (anode) of the magnetrons. 48% is lost into the exhaust gas in conventional kiln. They are the largest energy losses. The rest of energy is spent in the wall and equipment of kilns.

A few % of energy do not play any important role for energy saving. A reduction of processing time, for example to 1/4, contributes a lot to the energy savings. The key to the rapid sintering is "stress free heating" by maintaining a uniform temperature distribution. The temperature uniformity is, for an example, +/-3EC at 1200EC in microwave kiln. It was +/-19EC in gas kiln. It was measured by shrinkages of so called "Measure Ring".

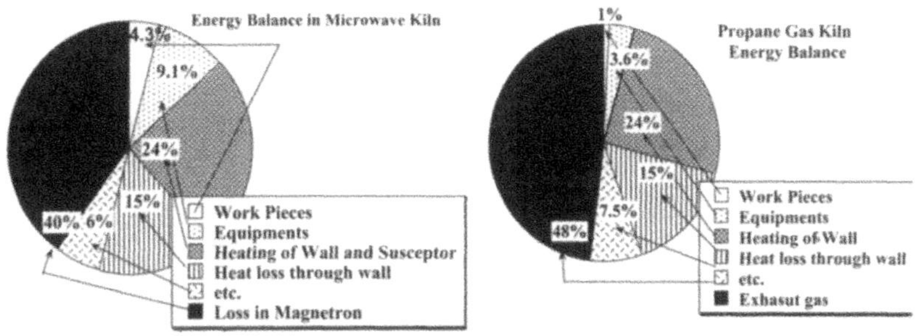

Figure 4. Energy Balance Estimations in Microwave (left) and Conventional gas (right) kilns.

Sintering of Ceramics in Microwave Batch Kiln

The ceramics, from old to modern such as alumina, zirconia, silicon nitride and silicon carbide, were sintered rapidly in the kilns without cracks or deformations. Beta-needle crystal structure was separated in silicon nitride at the low temperature of 1500EC.

In case of alumina, especially large size work pieces, it takes 3 days for degassing and 2~3 days for sintering in the factory by using a conventional kiln. It was reduced to 1 day for degassing and 3~8 hours for sintering in the prototype industrial kiln shown in Figure 1. The energy spent was 1200kWh(4320000kcal) in a batch cycle. It is 1/10 of conventional kiln of the same size. In addition to rapid processing, finer grain structure was obtained in the rapid sintering of 3.5 hours for 99.5% pure alumina. The bending stress increased to 500MPa in microwave sintering from 380MPa in conventional sintering. But, for the 99.99% 100micron-alumina powders, it was almost equal to the conventional. The deformation by microwave was 1/3 in comparison to the conventional sintering as shown in Figure 5. The near net sintering is very attractive for the industry.

		Microwave	Conventional
Dimensions D x H(mm)		355 x 81	355 x 81
Bulk Density (kg/m^3)		3.95 x 10^3	3.95 x 10^3
Deformation (mm)	DL	0.102	0.305
	DH	0.203	0.711

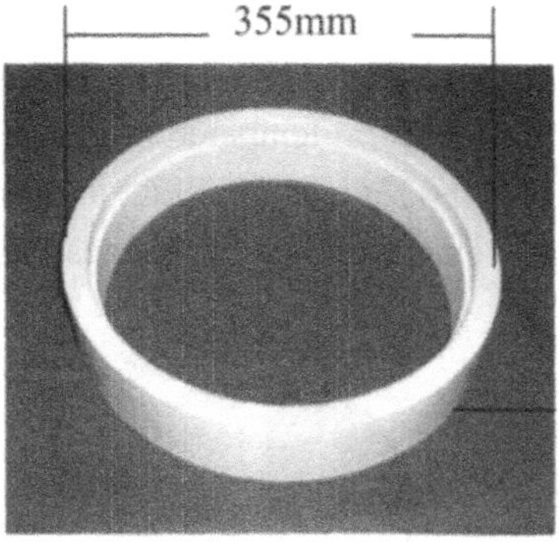

Figure 5. Example of industrial product by microwave (Hard Wear for IC Production)

MICROWAVE ROLLER HEARTH

Concept and Designs of Pure Microwave Continuous Kiln

A microwave roller hearth has been built with the concept of continuous microwave kiln with isothermal barrier. The cross section, heating zone and overall length are 0.27 x 0.3m, 6m and 14m, respectively. The maximum processing weight is 50 kg/hour. The 60 magnetrons generate 53 KW at 2.45 GHz. to the best of our knowledge, this is the first pure microwave continuous kiln to succeed in the full sintering of ceramics such as alumina, zirconia, ferrite and old ceramics in industrial scales.

The concept of self-consistent susceptor was applied. In the batch kiln the temperatures of work pieces and isothermal barrier synchronize temporally. As the temperature profile of the isothermal barrier becomes steady and fixed in the tunnel, work pieces must move to synchronize in space and time in it. The time axis in the heat curve in batch kiln corresponds to the position along the tunnel.

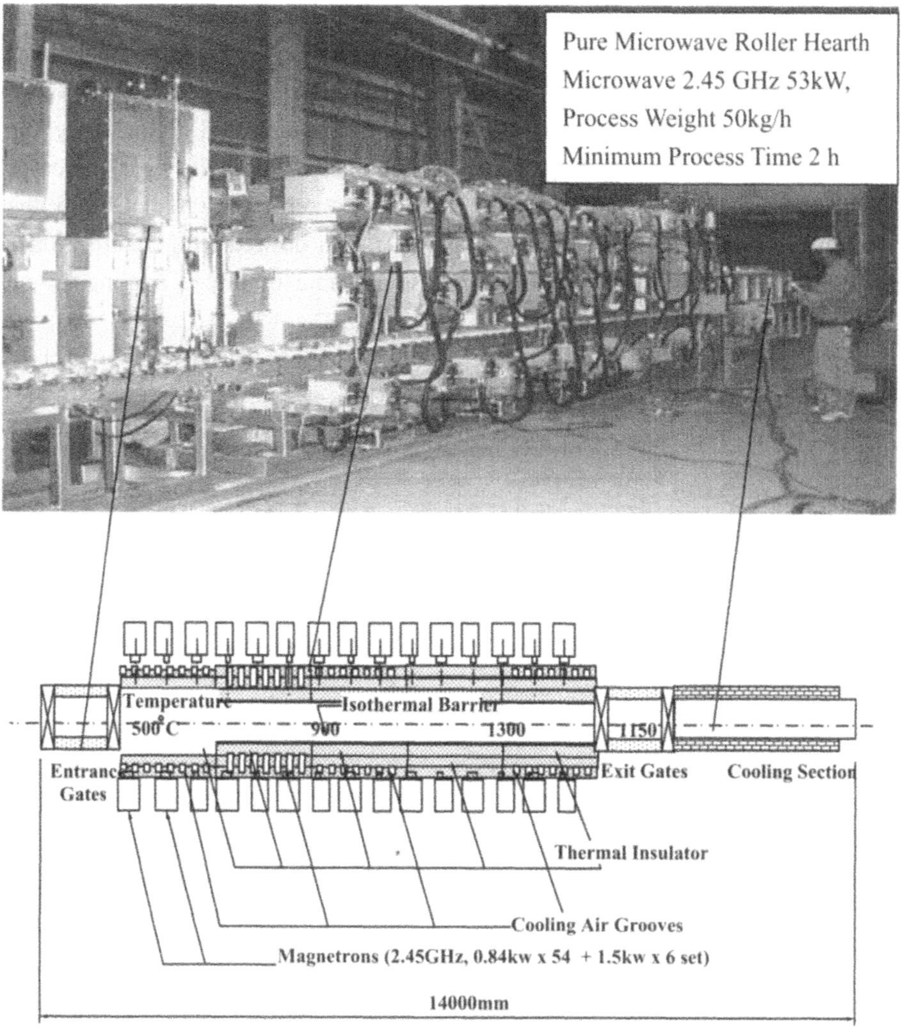

Figure 6. Outside and schematic views of pure microwave roller hearth.

It means that the temperature profile in the tunnel kiln must be similar to the heat curve of work piece in batch kiln. The work pieces must be carried with optimized speed to synchronize with that temperature profile. If the speed is too slow or too fast, the temperature is not uniform in work pieces. It breaks the stress free sintering condition.

The wall consists of three layers, i.e., innermost isothermal barrier, thermal insulator and the outer thermal insulator with grooves for forced cooling air as illustrated in Figure 6. Innermost layer is made of mullite coating (with 2~5 mm?). (Thermal insulator thickness varies from 60mm to 120mm along the tunnel.) Figure 7 shows microwave dissipation in the wall [(2)]. It increases exponentially with temperature. If the wall can be kept at low temperature, the microwave dissipation stays low. It suggests that a desired temperature profile will be obtained by variations of thickness and rate of forced cooling air on thermal insulator.

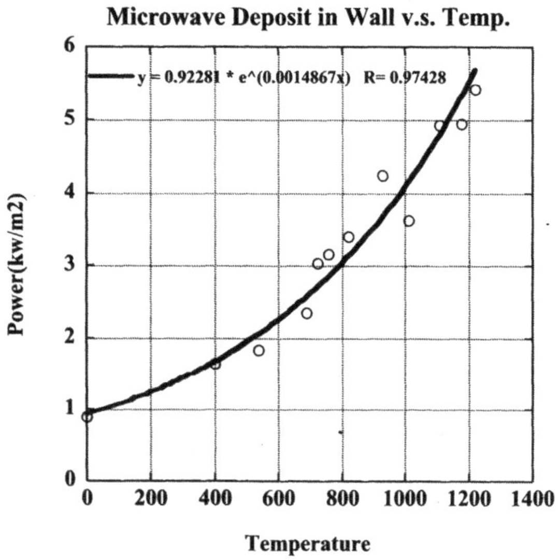

Figure 7. Microwave depositions on isothermal barrier and insulator performances of the roller hearth.

Total power of 53 kW microwave injected from the 60 ports distributed all along the tunnel. The temperature varies almost linearly from 500EC at the entrance to 1300EC in the hottest zone. Profile can be modified from linear to S-shape by adjusting the forced cooling air blowing in the grooves in the outer thermal insulators. The microwave loss is estimated in the sintering zone by the data in figure 7. The efficiency for sintering and typical process time is 7% and

2 hours that are almost same in microwave batch kiln. In microwave roller hearth, process time takes 2-4 hours for fine ceramics, such as Alumina, Zirconia, Ferrite, Aluminum Nitride and Silicon Nitride, etc. It takes more than days in conventional kiln for such ceramics.

The microwave roller hearth can be applied to different kinds of ceramics just by changing the speed and power. For example, Ferrite was processed in 4 hours from room to room temperature and then, by changing the speed and power, old ceramics were processed in 2 hours. Material performances are better with microwave heating than by conventional heating.

CONCLUSIONS

- The delivery time reduced to 1/5 in comparison to conventional heating.
- The process time determines energy consumptions in sintering a unit weight of products.
- The energy consumption reduced to 1/5~1/2.
- Now, many industries pay attention and are planning to introduce the microwave kilns into their factories in Japan.
- It gives big profit to not only to industries, but also reduces the green house effect on the earth.

REFERENCES

[1] M. Sato et al., "Insulation Blankets of 2.45GHz Microwave Sintering of Traditional Ceramics", Proc. 2nd World Congress on Microwave & Radio Frequency Processing, MWP-CE-13, Florida, 2000.

[2] S. Takayama et al., "Microwave Sintering of Ceramics at 2.45Ghz in the "Adiabatic Cascade Insulator", Proc. 8th International Conference on Microwave and High Frequency Heating, Bayreuth Germany, 2001

[3] S.J. Oda, I.S.Balbaa, B.T.Barber, "The Development of Microwave Heating Applications at Ontario Hydro's Research Division" Material Research Society Symposium Proceedings Vol.189, P391 (1990)

HEPHAISTOS – A NOVEL AUTOMATED MICROWAVE PROCESSING SYSTEM FOR CARBON FIBRE REINFORCED PLASTICS (CFRP)

L. Feher
Forschungszentrum Karlsruhe
IHM
D-76021 Karlsruhe, Germany

M. Thumm
Forschungszentrum Karlsruhe
IHM
D-76021 Karlsruhe, Germany
and Universität Karlsruhe,IHE
D-76128 Karlsruhe, Germany

ABSTRACT

The performed experimental work shows that the development of a specific sophisticated high frequency processing technology is promising for the automated fabrication of CFRP [1][2][3]. The development has been assisted by extensive numerical simulations. Composite structures, even if low-cost produced on an industrial scale are not just an extension of their metal counterparts. The use of microwaves shows new approaches on the process interaction for the fibre materials, precursors, resin systems, lay-up preparation etc. with direct consequences on the adjustable material properties to set new standards for the requirements of e.g. vibration properties, durability, damage tolerance, maintainability, reparability and crashworthiness. A novel microwave system and concept is presented.

INTRODUCTION

In a general sense, microwave heating at 2.45 GHz and 915 MHz ISM (Industrial, Medical, Scientific) frequencies has been established as an important industrial technology since more than 50 years ago. Successful application of microwaves in industries has been reported e.g. by food processing systems, domestic ovens, rubber industry, vacuum drying etc. As a matter of increasing oil and energy prices in the upcoming future, sustainable concepts and technologies will gather more and more importance to conserve mobility at an affordable price if resource and energy consumption can significantly be reduced .

Due to a strong decreasing price of carbon fibres in the next future, another application field of laminated materials within the aerospace and transportation industry can be recognized. The most commonly used advanced composite fibres are carbon and graphite, Kevlar and boron. Carbon fibres are manufactured by pyrolysis of an organic precursor. In the last decades considerable progress in advanced composite technology has been made. However, the full potential in the design, the manufacturing process and the application of these materials has not been realized [4]. A specific bottleneck for wide spread application is the price/kg in comparison to aluminium, which is not yet competitive due to high fabrication costs (material costs just about 30% of the actual end price). The highest potential on cost reduction is carried by the manufacturing process which implies substantial long time and high energy consumption, as well as a low degree of automation. The most commonly used processing method of composite materials is the simultaneous application of heat and pressure in autoclave systems.

HEPHAISTOS – The name is the philosophy

The cure cycle control is automated by the novel micro-/millimetre-wave technology developed at FZK and has to permit an automatic monitoring and control of the cure cycles of composite materials. The main hardware contribution of the system development is realized for a specific modular applicator containment providing an excellent homogeneous electromagnetic field distribution, which is an easy to use upgrade tool for existing conventional autoclave systems. The fabrication process can be performed pressurized up to 5 bar or at standard conditions. The process is measured remotely by infrared sensors or shielded low-cost thermocouples.

Fig.1: Hephaistos (Greek God), painting ca. 525 BC. He is also known as Hephaestus and Vulcan (Roman). His attributes in iconography include the axe and tongs.

This automated microwave processing system is named after Hephaistos, who is the builder and craftsman for the Greek gods. Fig.1 shows an about 2500 years old picture of Hephaistos, where the relation of Hephaistos to transportation and airframe design (wings) is obvious. The name HEPHAISTOS (**H**igh **E**lectromagnetic **P**ower **H**eating **A**utoclave **I**n**S**e**T** **O**ven **S**ystem) integrates the full philosophy of the microwave approach. The current system integrates advantageously the basic processing steps as tooling, tempering of the resin, the impregnation of the fibres, pre-forming techniques as well as finally the process curing of the high performance laminate.

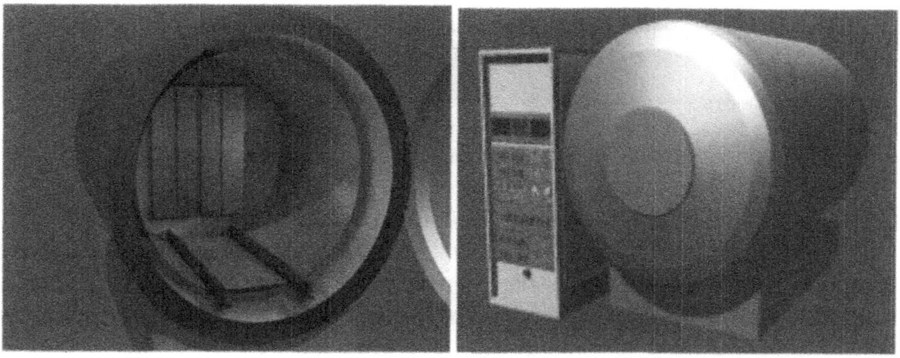

Fig.2: A conventional standard autoclave is changed by the HEPHAISTOS-CA upgrade to a full electromagnetic "cold oven" System

Fig.2 visualizes the concept to equip a standard medium-sized autoclave with microwave technology. The most notable effect processing CFRP materials with micro-/millimeter-waves is their volumetric heating, offering the opportunity of very high heating rates. In comparison to conventional heating where the heat transfer is diffusive and depends on the thermal conductivity of the material, the microwave field penetrates the material and acts as an instantaneous heat source at each point of the sample.

Homogeneous field distributions for batch processing are essential for commercialized industrial systems. Monomode applicators involve only controllable field properties in small specific applicator regions, whereas multimode ovens promise, especially for frequencies higher than 2.45 GHz, the possibility for low field fluctuations in larger regions. Simulations and experiments have shown, that achieving high quality homogeneous field distributions over most of an applicator volume is not trivial, even at frequencies in the millimetre-wave regime (30 GHz) [5].

Fig.3: Upgraded small standard autoclave at DLR Braunschweig with an HEPHAISTOS-SA System

An industrial microwave oven for production needs is successfully designed, if the available processing volume containing uniform field properties is about the size of the whole cavity and reflections are minimized. Conceptions of novel millimetre-wave technology [6] have been transferred by intense DELFI simulations to 2.45 GHz technology to realize these demands and to provide a large part capability for CFRP processing with the HEPHAISTOS System.

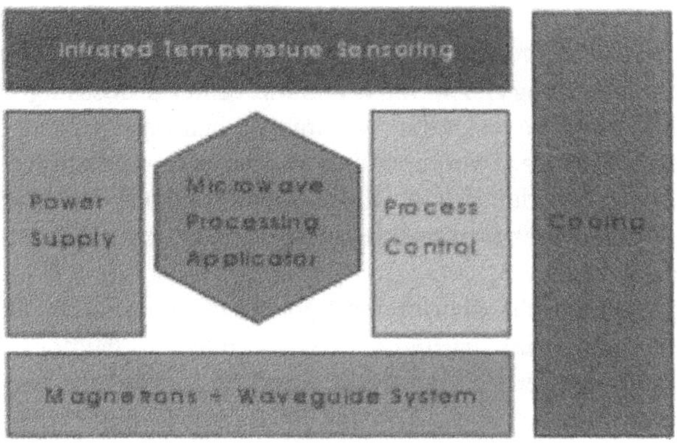

Fig.4 : In-Principle Set-up of the HEPHAISTOS System

Fig.3 shows an upgraded HEPHAISTOS test system, whereas Fig.4 views the schematic set-up of the system components.

CFRP Production

CFRP can be distinguished generally in duroplastics and thermoplastics depending on the choice of matrix material, which is used in advanced composites to interconnect the fibrous reinforcements. They are as varied as the reinforcements. Resins or plastic materials, metals, and even ceramic materials are used as matrices. Today, epoxy resin is the primary thermoset composite matrix for airframe and aerospace applications, especially for wet injection techniques. In all thermoset materials, the matrix is cured by means of time, temperature, and pressure into a dense, low-void-content structure in which the reinforcement is aligned in the direction of anticipated loads.

An important element in determining the material behaviour is the composition of the matrix that binds the fibres together. The selected matrix formulation determines the cure cycle and affects such properties as creep, compressive and shear strengths, thermal resistance, moisture sensitivity and ultraviolet sensitivity. For a proper composite forming, a vacuum bag technology is applied achieving a low atmospheric pressure of 1 bar on the samples. The vacuum bag additionally reduces volatiles of the resin matrix system and emerging voids during the injection of the viscous resin in the preform weave. Injection processes have been enhanced and accelerated with millimetre- and microwaves by pretempering the weave to 80°C and decreasing the viscosity of the injected resin [7][8].

Fig. 5: Composite laminate part fabrication

As a result, Fig.5 (left) shows a large sized CFRP plate fabricated in the HEPHAISTOS-BA microwave system.

DELFI and THESIS – The Simulation Base

Fibre reinforced composite materials are in general anisotropic materials if one considers their mechanical, thermal and electromagnetic properties[2]. The fibres are embedded in an essentially homogeneous matrix. Their material properties have to be described by tensors with respect to the orientation of the fibres. It turned out, that a precise modelling capability of the related electromagnetic and thermal parameters is essential. Demands for design and development of CFRP processing technology are to model and simulate

- Applicator Design
- Waveguide Systems
- Fibre/matrix composition of the laminate
- Material dynamics
- Hot spot formation
- Process/temperature control
- Exothermal heat wave propagation
- Tooling arrangements

Two basic codes have been worked out. To model system technology and optimize electromagnetic materials-field interaction, the DELFI-Code was developed

DELFI-Code: (**D**istribution of **EL**ectromagnetic **FI**elds)

- Explicit 3D FDTD-Code solving the full time dependent Maxwell´s equation.
- Arbitrary geometries (Applicator, load, waveguides).
- Arbitrary materials.
- Reduced memory needs, fast and stable algorithm.
- Code for complex microwave system design and engineering.

The Thesis-Code, originally developed for microwave sintering, calculates the material and process dynamics for the curing of CFRP composites [9].

THESIS-Code: (**THE**rmodynamic **SI**ntering **S**imulation)

- 3D FDTD-Code solving the nonlinear time dependent heat equation for processed materials.
- Material modelling: Easy implementation of temperature dependent parameters (experimental fingerprint).
- Boundary conditions: Conduction, Convection, Thermal Radiation, Microwave Heating, Hybrid Heating.
- Code for conventional oven design, Microwave/Hybrid Heating, Investigations on materials dynamics (e.g. densification, hot spots, temperature gradients).

Anisotropic material and field properties are currently implemented.

Fig. 6: Left: Energy density calculated for the plate of Fig.5 with the DELFI-Code

Right: Initial temperature distribution of this plate calculated with the THESIS-Code

In Fig.6, the microwave energy density around the plate of Fig.5 processed in an HEPHAISTOS-BA system is visualized. The calculated fluctuations are less than 3% to the average value showing the high homogeneity of the applicator.

To have a complete and user-friendly tool, a Windows-based interactive interface is added to the THESIS-Code. The input section includes the full set of dielectric and thermal parameters, constant and temperature dependent. Two different libraries have been created: the Materials Library includes the set of parameters for a specific material, the Processes Library includes the full process definition used for materials processing, i.e. heating rates, holding temperature, frequency, geometry specifications as well as chemical reaction contributions. All the physical parameters are accessible during the simulation by 2-dimensional graphs and 3-dimensional slides within the volume data of a processed sample.

As a conclusion, modelling and simulation have been playing an important role to achieve homogeneous field and temperature conditions for high quality processing of CFRP. The electromagnetic DELFI modelling optimized applicator geometry and multi-waveguide coupling for a mostly homogeneous field distribution within the processing area. The use of mode-stirring devices is not necessary. In the experiments, as a result no arcing can be detected any more at the carbon fibres. The THESIS simulations resulted in precise predictions on the heat propagation, optimized process control and showed the dynamics of the curing process within the laminate. Material investigations for test specimens are currently carried out for certification.

REFERENCES

[1] L. Feher, M. Thumm, Millimeter-Wave Processing of Composite Materials , Proc. 2nd IEEE International Vacuum Electronics Conference, Noordwijk, Netherlands, 2001, pp.83-84

[2] L. Feher, M. Thumm, Application of Higher Frequency Microwaves for Composite Materials Heating/Processing, Proc. Displays and Vacuum Electronics, ITG, Garmisch-Partenkirchen, 2001, ITG 165, pp.363-368

[3] J.B. Wei, T. Shidaker, M. Hawley, Recent Progress in Microwave Processing of Polymers and Composites, Trends in Polymer Science, V.4, No.1, Jan. 1996, pp.18-24.

[4] Michael C.Y.Niu, Composite Airframe Structures, Hong Kong Conmilit Press Ltd., 1992, pp. v

[5] L. Feher, M. Thumm System design development for microwave and millimeter-wave materials processing, SPIE 16th Annual International Symposium on Aerospace, Simulation and Controls, Orlando, U.S.A., 2-4 th April 2002

[6] L. Feher, G. Link, M. Thumm: Optimized design of an industrial millimeter wave applicator for homogeneous processing of ceramic charges, 6th Int. Conf. on Microwave and High Frequency Heating, Fermo, Italy, Sept. 9-13, 1997, pp. 443-446.

[7] C. Hunyar, L. Feher, M. Thumm, Processing of Carbon-Fiber Reinforced Composite (CFRP) Materials with Millimeter-Wave Technology, 8th Conference on Microwave and High Frequency Heating, Bayreuth, Germany, 2001, Book of Abstracts, pp.170-171

[8] L. Feher, M. Thumm, Industrial Higher Frequency Microwave Processing of Composite Materials, 8th Conference on Microwave and High Frequency Heating, Bayreuth, Germany, 2001,Book of Abstracts, pp.156-157

[9] L. Feher, G. Link, M. Thumm: The MiRa/THESIS-Code package for resonator design and modelling of millimeter-wave material processing, 1996 Spring Meeting of the Materials Research Society, Microwave Processing of Materials V, San Francisco, 1996, Symposium Proc., Vol. 430, pp. 363-368

DILATOMETER SYSTEM WITH HYBRID HEATING TECHNIQUE FOR DETAILED INVESTIGATIONS ON SINTERING OF CERAMICS

Guido Link and Manfred Thumm[#]
Forschungszentrum Karlsruhe GmbH
Institut fuer Hochleistungsimpuls- und Mikrowellentechnik
P.O.Box 3640
76021 Karlsruhe, Germany
[#] also University of Karlsruhe, IHE

ABSTRACT

A versatile dilatometer set-up has been developed which allows detailed investigations on sintering of ceramics under various heating conditions. Linear shrinkage of zirconia and alumina ceramic powder compacts has been measured during millimeter wave heating, conventional heating as well as during a combination of both heating methods, so called hybrid heating. Simultaneous investigations on temperature gradients in the vicinity of the sample help to understand the different results.

INTRODUCTION

Dilatometry is a widely used technique in materials science and development, to measure thermal expansion coefficients, phase transformations or the linear shrinkage during sintering as a function of process temperature. The basic information such a dilatometer gives is the variation in sample length during heating or cooling through a programmed cycle. This information can be very effectively used to optimize the process parameters for a sintering process.

Although the accuracy of inductive displacement transducers used in such instruments is extremely high, one of the major drawbacks for exact measurements is the fact that one has to make sure that the sample is in thermal equilibrium with its environment. This gives strong restrictions to the process parameters with respect to the heating rates. With conventional heating, that means infrared heating, we can observe a sample surface with temperatures higher than in the volume. This results in heat flow from the sample surface into the volume until sample is in thermal equilibrium. Since thermal conductivity of powder compacts can be rather low thermal equilibrium can only be achieved when heating rates are sufficiently small. With pure mm-wave heating the problem appears the other way round. If the sample shows low dielectric loss, resulting in volumetric heating, there is permanent heat flow from the inside out where heat is lost from the sample surface by radiation and convection. But in contrast to conventional heating, temperature gradients cannot be avoided by slow heating. Here a real thermal equilibrium can never be achieved [1].

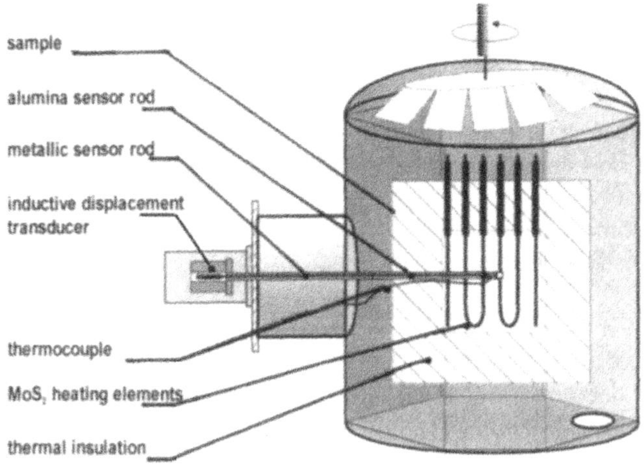

Figure 1. Scheme of the dilatometer system in the mm-wave applicator

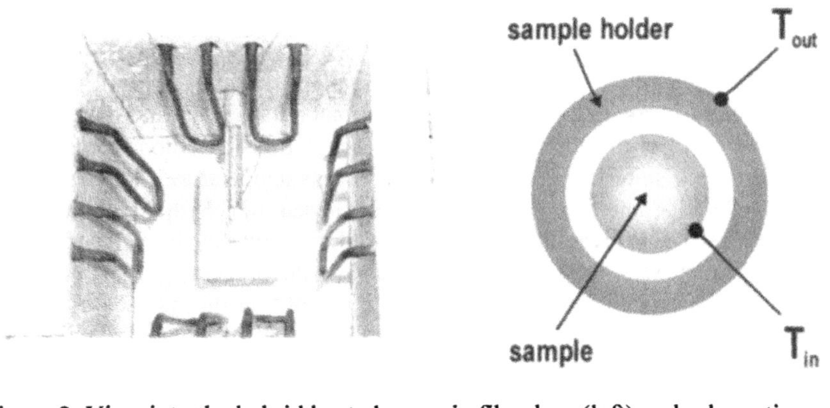

Figure 2. View into the hybrid heated ceramic fiber box (left) and schematic cross-section of the sample holder indicating the location of temperature measurement (right)

There are different ways to overcome this problem. One is using a microwave transparent box of ceramic fiber boards for thermal insulation of the dilatometer sample holder, as it has been performed up to now [2]. Another solution is the combination of conventional and microwave heating technique, so called hybrid heating. Microwaves are used to heat the volume of a sample, whereas the conventional power compensates radiation losses from the sample surface, if it is used to control the temperature of the applicator walls to identical values [3]. So temperature gradients during a sintering process can be actively controlled to zero. In addition it is a versatile tool to investigate the influence of different heating methods on the sintering process of ceramic compacts in a single system. The compact 30 GHz, 15 kW gyrotron installation of the Forschungszentrum Karlsruhe, Germany has been equipped with such a dilatometer system (see Fig. 1 and 2) for detailed investigations in sintering of various functional and structural ceramics by means of millimeter wave (mm-wave) radiation.

EXPERIMENTAL RESULTS AND DISCUSSION

Large efforts have been undertaken to study the intrinsic behavior of the dilatometer under various heating conditions for calibration purposes. This behavior is determined by the temperature evolution in the sensor head. If residual temperature gradients are present, the sensor rod may show a thermal expansion different to the sample holder. But this will result in a rather small intrinsic dilatometer signal superimposed to the large signal coming from the ceramic sample under test due to shrinkage during sintering.

It was found that with pure mm-wave heating temperature profiles strongly depend on the heating rates and the dielectric loss behavior of samples itself, so that a calibration of the dilatometer system is unsatisfactory [2]. However with hybrid heating, where temperature profiles can be actively controlled, a proper calibration is feasible.

The sintering behavior of different types of functional and structural ceramics has been investigated in the prescribed dilatometer setup. The left graphs in Figure 3 and 4 show the relative linear shrinkage as a function of temperature with different heating rates for ceramic powder compacts made from nanocrystalline yttria stabilized zirconia (YSZ) delivered from HITK, Germany and an ultra-fine alumina TM-DAR from Taimei Chemicals, Japan. In case of nanocrystalline YSZ samples the dilatometer curves indicate that sintering with mm-waves starts at lower temperatures if heating rates are increased (Fig. 3). This might be interpreted as an increasing non-thermal microwave effect, which has been reported from many authors for different ceramic materials [4]. If such an effect exists, which is caused by the interaction of the electromagnetic fields with the dielectric material, then one could imagine, that with increased heating rates, which result from an increased field strength, sintering is even more enhanced and therefore starting at lower temperatures as is has been observed. Another indication for such an effect might be the fact that with conventional heating sintering happens at a temperature more than 100 °C higher compared to mm-wave heating at identical heating rates.

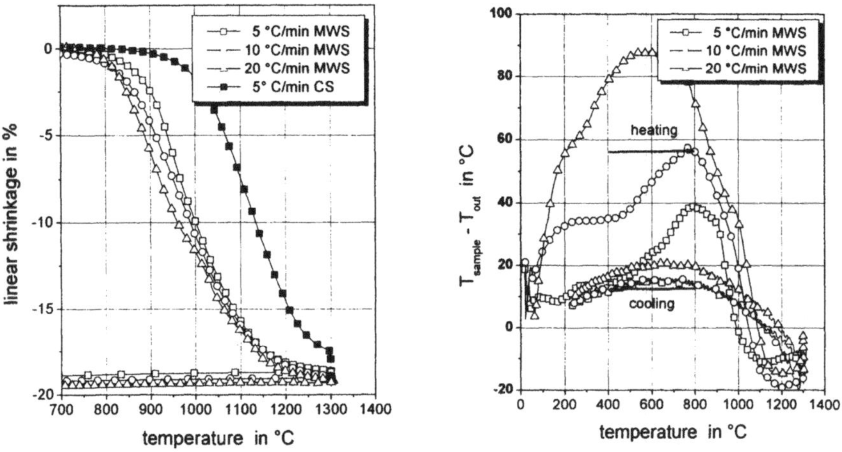

Figure 3. Linear shrinkage (left) and temperature gradients (right) during mm-wave (MWS) and conventional sintering (CS) of nanocrystalline YSZ.

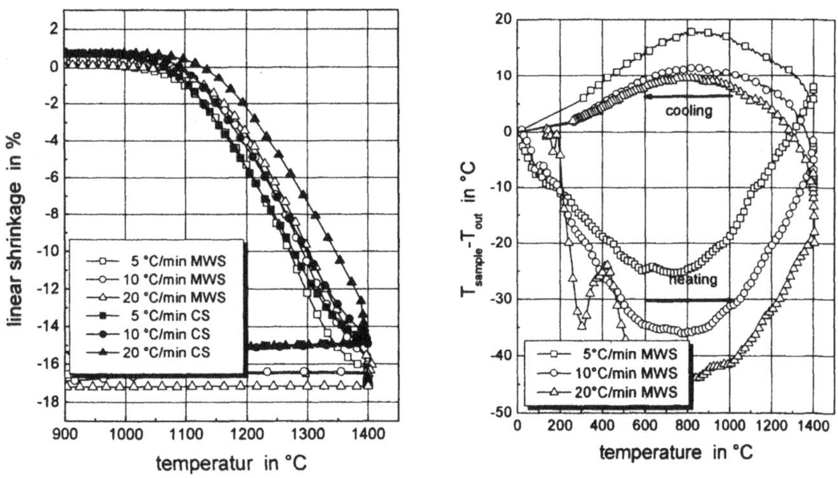

Figure 4. Linear shrinkage (left) and temperature gradients (right) during mm-wave (MWS) and conventional sintering (CS) of submicron high purity alumina.

But if we look at the results of identical experiments with low loss alumina samples, the effect is opposite (Fig. 4). With increased heating rates during mm-wave heating the onset of sintering, that means shrinkage is shifted to higher temperature values. This effect is similar but more pronounced with conventional heating. For conventional heating such effects are explained by temperature gradients within the sample and by a reduced annealing time at equal temperatures if heating rates are increased. If we look at the temperature gradients within the sample holder during heating that means at the temperature difference from the sample surface T_{sample} to the outer surface of the sample holder T_{out} (see Fig. 2), we can see that temperature gradients in the sample holder are changing sign if we change the material (see right graphs in Fig. 3 and 4). Furthermore this gradient increases with increasing heating rate. If we assume that there are similar temperature gradients within the sample as well, this has to be taken into account if one compares different heating methods.

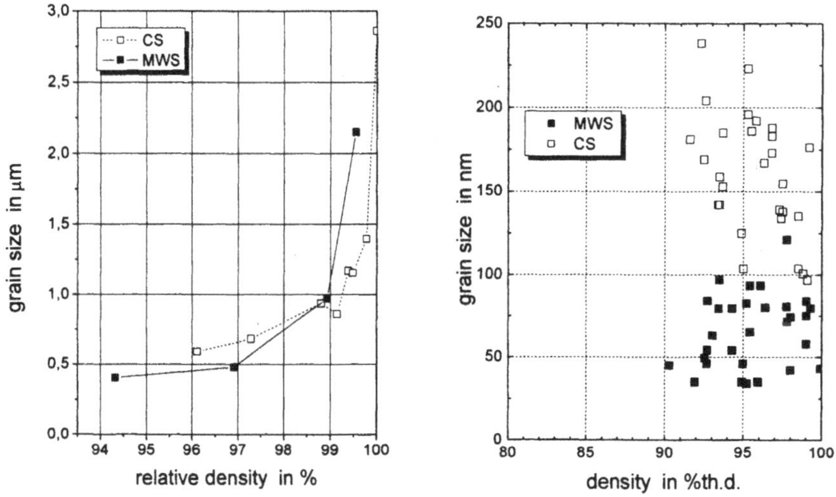

Figure 5. Grain size as a function of relative sintered density for zirconia (right) [5] and alumina (left).

Since temperatures can be measured at the sample surface only, the presence of temperature gradients within the sample unavoidably leads to systematic errors. In case of negative gradients as shown in the right graph of Fig. 4 with ultra-fine alumina, during mm-wave sintering as well as during conventional sintering, measured temperatures are higher than the average, that means the effective temperatures of the sample. And this error obviously increases with increasing heating rates. If temperature gradients are positive, as found with YSZ samples, which is typical for most dielectric materials during micro- or mm-wave heating, the systematic errors are opposite. This means that measured temperatures are too small and with increasing heating rates this error is increasing as well.

This clearly demonstrates that process parameters such as temperatures can hardly be used as the only argument to demonstrate microwave effects. Therefore further investigations have to be taken into account such as investigations on grain size evolution as a functions of sintered density. This correlation which is shown in Figure 5 for both materials, does not need the temperature parameter. This result reveals a clear difference in densification kinetics for the zirconia if we compare mm-wave and conventional heating. The grain size of mm-wave sintered samples is in average by a factor of about two smaller than the grain size of conventionally sintered samples. Such a remarkable difference could not be discovered for the alumina. A possible explanation for these different results might be the difference in mm-wave materials interaction. Dielectric absorption is much more pronounced with YSZ than with the high purity ultra-fine alumina. If there is only little mm-wave materials interaction, that means little mm-wave absorption, non thermal microwave-effects can obviously be expected to be small as well or even absent.

The combination of mm-wave heating and conventional resistant heating, the so called hybrid sintering, is the only way to solve the problem of temperature gradients within bulk materials. But this solution is restricted to materials with dielectric losses sufficiently high, that means higher than losses by surrounding materials such as the dilatometer sample holder and the material for thermal insulation, respectively. Otherwise mm-wave power is predominantly absorbed in surrounding materials, which are acting then as an additional radiation source for infrared heating.

Investigations of hybrid heating with a submicron YSZ from Toso, Japan in comparison with conventional heating as well as with pure mm-wave heating are shown in Figure 6. The graphs show the linear shrinkage with two different heating rates (5 and 20 °C/min) as well as the temperature gradients from the sample surface T_{sample} to the inner wall of the thermal insulation box T_{wall} heated with MoS_2 heating elements for hybrid and conventional heating. With mm-wave heating, temperature gradients are measured between the sample surface and the outer surface of sample holder (see Fig. 2). Conventional heating results in a shift of the onset of linear shrinkage to higher temperatures as well as in growing temperature gradients if the heating rate is increased as it has been observed with high purity alumina (Fig. 4). With hybrid heating where temperature gradients can be controlled to almost zero for all heating rates, the shift of the dilatometer curves is neglectable. With pure mm-wave heating we observe a big shift in the onset of shrinkage of more than 100 °C compared to conventional and hybrid sintering. But this difference obviously reduces with reducing temperature gradients to higher temperatures (open triangles in Figure 6).

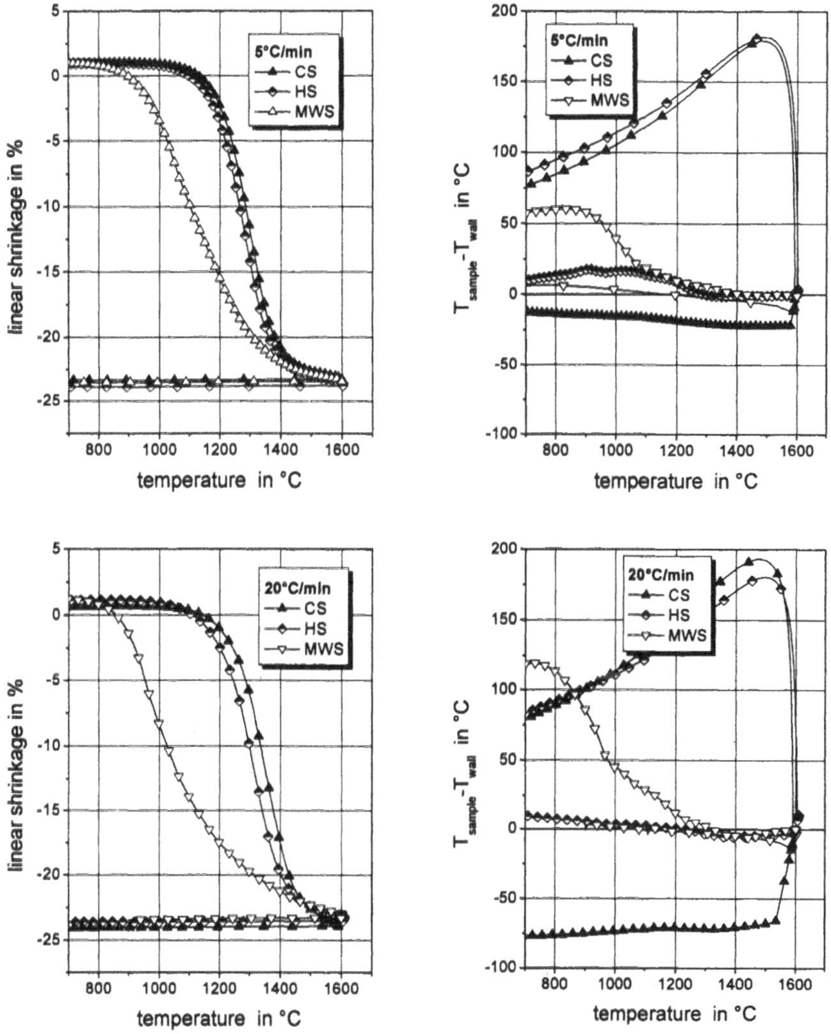

Figure 6. Linear shrinkage (left) and temperature gradients (right) during mm-wave (MWS), conventional (CS) and hybrid sintering (HS) of Toso YSZ at a heating rate of 5 °C/min (top) and 20 °C/min (bottom).

This dilatometer system can be used with changing heating conditions as well. That means one could start the sintering process with hybrid heating and then changing to conventional heating by switching off mm-wave power, as it has been reported by Wroe and Rowley with a comparable system at 2.45 GHz [6]. Equivalent experiments have been performed with nanocrystalline zirconia samples as shown in Figure 7. If the shift of the dilatometer curves to lower temperature values with mm-wave power on indicates an enhanced densification, one would not expect the curve of linear shrinkage going back to the conventional line (full squares) as we found. The slope of the dilatometer curves, after switching off mm-wave power, is much smaller than the slope with conventional heating at the same temperature or density. But if the shrinkage rate is temperature driven or if it is a function of density, it would be expected to be the same. This is probably another clear indication for temperature gradients been responsible for such a behavior. If we look at the measured temperature gradient within the sample holder of the dilatometer system it can be clearly seen, that temperature gradients present during hybrid heating (open squares in right graph of Fig. 7) are completely changing to temperature gradients present with conventional heating (full squares) if mm-wave power is switched of (open triangles).

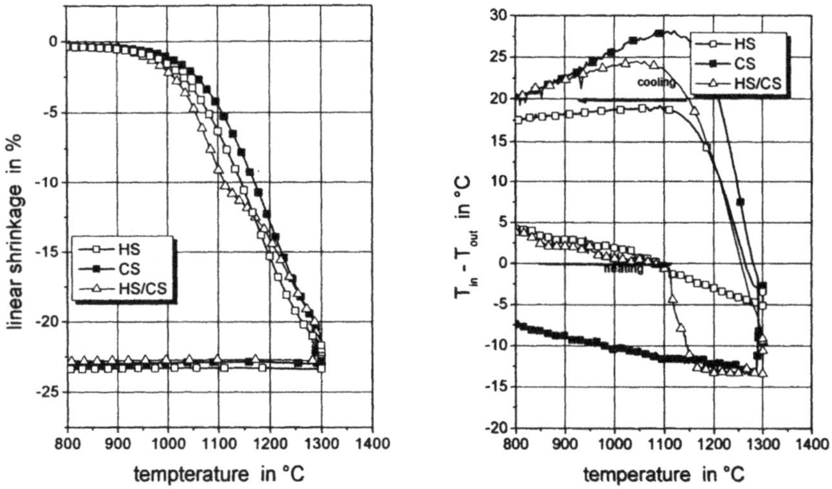

Figure 7. Linear shrinkage (left) of nano-crystalline YSZ and temperature gradients (right) under various heating conditions with a heating rate of 20 °C/min.

SUMMARY AND CONCLUSIONS

Dilatometry is a versatile technique for detailed inverstigations of ceramics sintering. The influence of the heating method on the sintering process can be very effectively investigated if different heating methods such as microwave heating, conventional heating and microwave assisted heating are unified in a single system. Systematic errors due to the comparison of results from different experimental systems and therefore different temperature measurement instruments can be neglected. Such a system was implemented in a 30 GHz millimeter-wave applicator and has been used for systematic investigation of different commercial oxide ceramics such as alumina and zirconia. The combination of linear shrinkage measurement and measurement of temperature gradients inside the sample holder of the dilatometer system gives a clear indication that a comparison of different heating methods has to be done carefully. Each heating method distinguishes by its specific type of temperature gradients which are opposite for volumetric microwave heating compared to conventional heating. So the temperatures usually measured at the sample surface are not reliable values as long as the sample is not in thermal equilibrium with its environment. In order to support conclusions on non-thermal microwave effects which are drawn from differences in process parameters, such as reduced sintering temperatures or processing time, additional materials characterization is indispensable.

REFERENCES

1. Yu.V. Bykov, K.I. Rybakov, V.E. Semenov; "High-temperature microwave processing of materials", *J. Phys. D: Appl. Phys.*; **34**, R55-R75, (2001).

2. G. Link, S. Rhee, M. Thumm; "Dilatometer System for Investigations of Sintering in a mm-Wave Oven"; *Proc. of the 36th Annual Microwave Symposium of the IMPI*; San Francisco, April 18-21, pp. 23-26, (2001).

3. G. Link, L. Feher, M. Thumm; "Hot wall 30 GHz cavity for homogeneous high temperature heating"; *Proc. Int. Conf. Microwave andHF Heating*, Valencia; ed. J.M. Catalá-Civera et alValencia, Sept. 13-17, pp.165-168, (1999).

4. W.H. Sutton, "Microwave processing of ceramics - an overview"; *Microwave Processing of Materials III, MRS Symp. Proc.*; **269**, ed. Janney M. A. et al Pittsburgh, PA, pp. 3-20, (1992).

5. S. Rhee; "Mikrowellenprozeßtechnik für keramische Werkstoffe der Mikrosystemtechnik"; *Wissenschaftliche Berichte, FZKA-6768, Forschungszentrum Karsrluhe* (November 2002); dissertation, University Freiburg, Germany 2002.

6. F.C.R. Wroe and A.T. Rowley, "Microwave enhanced sintering of ceramics"; *Microwaves: Theory and Application in Materials Processing, Ceramic*

Transactions; **59**, ed. Clarc D.E.; Amer. Cer. Soc. Westerville Ohio, pp. 69-76, (1995).

THERMOGRAPHY FOR OPTIMIZING HIGH FREQUENCY AND MICROWAVE INSTALLATIONS

W. M. Van Loock
Ghent University
St. Pietersnieuwstraat 41
B-9000 Ghent, Belgium

ABSTRACT

This paper reports on thermography applied to an existing high power microwave installation to improve its functioning in an industrial production process. The method is useful not only for locating weak spots and defective areas but also for determining the corresponding high frequency and microwave energy loss in power installations.

INTRODUCTION

A major problem with industrial high frequency and microwave installations is that they are often operated continuously for several weeks at a time. Conscious design and implementing intelligent protection systems against many forms of failure have now extended considerably the running time of such machines. The overall energy efficiency is also very important; it should be as high as possible, avoiding any energy wastage. Short periods of operation and intermittent operation due to all kinds of failures – arcing for example – also decrease overall efficiency considerably.

At 0.9 GHz, the generators are reliable magnetrons having an efficiency of more than 80%, which is interesting because of the concern of rational use of energy. However, overall energy efficiency is usually much lower.

Today, there are many possibilities for optimizing high frequency and microwave power installations - for example better process control and computerized programs for remodeling. Applying a modeling software turns out to be expensive because of problems such as long learning time and manpower involved for manipulating it regularly. This paper reports on applying infrared thermography to a high power microwave installation. The goal was to study the energy balance and to find out ways to improve the overall energy efficiency. Thermography uses a camera to display and digitally record the surface temperature of objects. Such cameras are extensively being used to locate hot spots or areas with an abnormally high temperature, because these are likely to be defective.

As explained in this paper, an infrared camera also allows the assessment of the microwave losses inside the installation to be assessed by determining the heat loss from the outer surface.

An infrared camera is relatively expensive equipment, but it can be rented and careful study in advance can reduce the recording time.

This is perhaps the first time that thermography has been used for to optimize a microwave installation. The method reported here can be applied to high frequency installations as well.

INFRARED AND THERMOGRAPHY

In thermography, an infrared camera is used for a pictorial recording of the natural infrared radiation of an object. The details of infrared and thermography can be found in many books - for example, in the work of Gaussorges[1]. Only essential details are presented here.

Infrared is part of the optical electromagnetic energy spectrum and covers conventionally the wavelength range from 0.76 μm to 1 mm. The range is divided into many sub-ranges with specific denominations. For infrared heating applications, the range of .76 to 10 μm is important, while thermography uses the range from 2 – 14 μm.

A beam of electromagnetic energy incident on an object is partly reflected, partly absorbed, partly diffracted, and partly transmitted. Provided no diffraction takes place - for example, in the case of a small beam on a large object - the general law of energy conservation gives

$$\rho + \alpha + \tau = 1 \qquad (1)$$

In this equation ρ, α and τ are the power reflection, absorption and transmission coefficient, respectively - or reflectivity, absorptivity, and transmissivity. The penetration depth of absorbed infrared is limited to fractions of microns and in some cases to about 1 mm. The result is that the power transmission coefficient is practically zero for most materials, simplifying Eq. 1. Only a few materials are transparent to infrared frequency and such materials are used for lenses in infrared cameras.

The picture of a video camera is based on reflected light from the objects that are illuminated by a source of visible light in the viewing field. An infrared camera uses the natural infrared energy that is emitted by the surface of all objects, provided their temperature is higher than absolute zero or –273°C. This is Planck's radiation law: at a given temperature, a material surface emits electromagnetic energy over a broad wavelength range. Maximum energy per unit of time is emitted at a wavelength given by Wien's law. Integration of Planck's Law over all wavelengths gives the Stefan-Boltzmann law: the emitted power per unit surface of a material is equal to a constant times its absolute temperature T to the fourth power:

$$S = \varepsilon \, s \, T^4 \tag{2}$$

S power per unit surface W/m^2
s Stefan-Bolzmann constant
T absolute temperature in degree kelvin ($T = 273 + °C$)
ε emission factor, emissivity coefficient

When a material emits infrared, there is a problem with its surface or in other words with the interface material free space. An ideal material surface called a black body surface has $\varepsilon = 1$, but practical materials have an ε smaller than one; ε is the emission coefficient of the material (other denominations are used, but do not confuse this with the dielectric constant). The emissivity equals the power absorption coefficient for incident infrared, this is the reciprocity law of Kirchhoff ($\varepsilon = \alpha$, see Eq. 1). For practical materials $\varepsilon = 0.7$ to 0.8 but for a mirror-like surface ε may be as low as 0.1.

Radiative heat exchange between two elementary surface areas 1 and 2 at temperatures T_1 and T_2 is proportional to

$$T_1^4 - T_2^4 \tag{3}$$

provided the reciprocity law is valid[2]. A temperature difference is necessary for radiated heat exchange: no temperature difference, no heat exchange.

An infrared camera detects the power flux emitted by the surface as a voltage that is converted to a temperature according the Stefan-Boltzmann law, provided ε is known. This can be measured or assessed for the surface areas viewed by the camera but it complicates the use of thermography for quantitative purposes; and so it is not detailed here. As a practical rule, a thermographical recording should not show "bright" shadows and if observed, they can be eliminated by proper displacements of the camera simplifying the problem of large differences in surface emissivity.

LEAKAGE

Prior to the use of an infrared camera, the high frequency and microwave leakage levels around the installation should be well below the immunity level specification of the camera. If not, a specially shielded camera housing must be used.

Current regulations allow, as a first approximation, a constant total leakage power of 1 to 10 W, independent of the high frequency or microwave power of the installation. Above 100 kW, the total leakage power is therefore negligible in overall energy efficiency determination, provided the installation complies.

MICROWAVE LOSS AND CORRESPONDING SURFACE HEAT LOSS

Thermography reveals localized hot spots due to bad joints, welds, flange connections, insulation defects and defective areas.

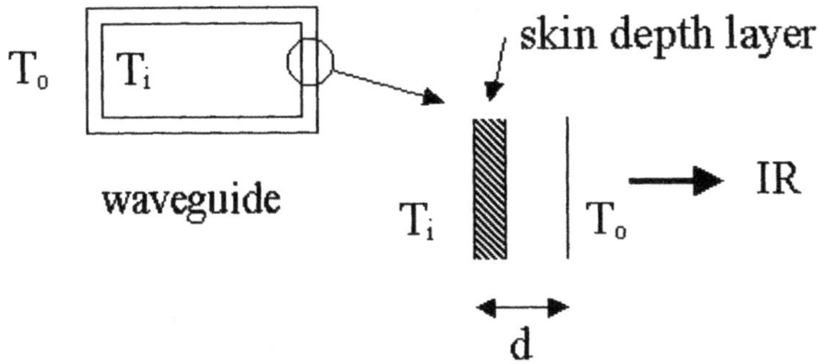

Fig. 1 Infrared radiated flux (IR) is due to microwave absorption in a surface layer (skin depth) inside the waveguide wall.

For assessment of microwave loss, the problem is its relationship with the measured infrared flux on the outside surface of the installation. This relationship is different for an applicator wall, a flange, a structure or a waveguide wall. As an example, consider Fig. 1. Microwaves are absorbed in the skin depth layer inside the waveguide wall. In abnormal conditions such as high conductivity of the waveguide metal, high standing wave ratios, rough surfaces or a bad design, the wall losses cannot be neglected. Conduction and convection heat loss is proportional to a temperature difference and infrared radiation loss predominates at high temperature differences, according to Eq. 3.

An installation running continuously is in steady state conditions and conduction heat flux Q through a wall of total thickness d (Fig. 1) can be calculated with the linear equation

$$Q = (Ti - To)\,(\lambda/d) \tag{4}$$

T_i being the temperature of the inner surface, T_o of the outer surface and λ is the effective thermal conductivity. The measured T_o corresponds with an outward radiated heat flux in W/m^2 according to Eq. 2. This heat originates in the skin depth layer due to microwave absorption. The temperature T_i of the layer can be calculated with Eq. 4 provided conduction loss along the waveguide can be neglected The radiated flux inside the waveguide falls back on the walls, Eq. 3. Conduction loss along the waveguide is seen as a temperature decrease and can be assessed with Eq. 4. The observed surface temperature distribution allows assessment of the microwave losses using Eq. 2, 3 and 4, provided parameters are known such as wall composition and surface areas. At lower surface temperature convection loss may become important and has to be taken into account.

APPLICATION

Thermography has been applied to a 0.9 GHz 300 kW microwave installation for sterilization. Sterilization is a temperature-time treatment (for example, at 125°C) requiring a pressurized applicator[3], which means that any temperature increase above 125°C is a hot spot due to microwave loss.

The layout of such an installation is standard. To cope with electromagnetic compatibility regulations, the generators are installed in a separate room. The waveguides connecting the water cooled microwave generators have different sections and components, such as a matching device, a window section and a transition section. Foodstuff enters the applicator at 20°C and microwaves provoke the temperature jump and it remains in the applicator during the time necessary for the sterilization treatment. A hot air pressure circuit maintains 125°C but water in the foodstuff may evaporate which is an endothermal reaction. The applicator is 95% thermally insulated. The determination of the overall energy efficiency as the ratio of required energy for treatment of the foodstuff to electricity consumption is complicated but straightforward.

RESULTS

The infrared camera revealed hot spots on the walls of the applicator due to hot spots inside the applicator wall. Connecting tubes and waveguides showed along their length away from the applicator, a progressive decrease in temperature down to the ambient temperature.

The surface temperature distribution in operation with no microwaves is the reference surface temperature distribution of the installation, but this is difficult to achieve (expensive due to waste of foodstuff). It turned out that different power settings of the individual magnetrons was more practical. The surface temperature distribution changes to a new steady state pattern allowing the determination of the origin of the detected heat loss: thermal or microwaves.

The infrared camera revealed high loss in a waveguide section of reduced height. The total surface area and the temperature distribution have been determined. Using the approximation as explained a total microwave loss of 480 W for a generator power setting of 50 kW was found. This revealed poor design of a section 1 m long. However, a better design would result in only a 1% improvement of the overall energy efficiency. The return on a new design with its high associated cost is not clear.

CONCLUSION

Infrared thermography is a cheap way to optimize high frequency and microwave power installations. It allows the location of hot spots and defective areas due to insufficient thermal insulation, abnormal microwave absorption in bad welds, poor joints, rough surfaces, etc. Thermography can also be used to determine microwave loss qualitatively.

This is perhaps the first time that thermography has been applied to optimize an existing microwave installation. The procedures outlined here are also applicable to high frequency power installations.

REFERENCES

[1]G. Gaussorgues, "La thermography infrarouge", Lavoisier, 1989.

[2]M. Orfeuil, "Electric Process Heating", Battelle Press, Ohio 1987.

[3]GW. Van Loock, "Electromagnetic energy for pasteurization, another viewpoint" Microwave World, 17 (1) 23-27 1996.

INSTRUMENTATION FOR *IN SITU* DIFFRACTION OF MATERIALS UNDER MICROWAVE RADIATION

Andrew Harrison, Graeme Robb and Gavin Whittaker
School of Chemistry,
The University of Edinburgh,
Kings Buildings,
West Mains Road,
Edinburgh, EH9 3JJ, UK

ABSTRACT

The development of novel methods for investigating reaction mechanisms will be crucially important in answering questions about "microwave-specific" and "non-thermal" effects in microwave-heated materials, particularly in the solid phase. In this paper, we describe the design and application of apparatus that enables X-Ray and neutron diffraction experiments to be performed on materials during microwave irradiation and thus follow changes in chemical and physical properties. The most efficient design in terms of controlled application of significant microwave power to the sample incorporates a conventional waveguide, and may be adapted to X-ray and neutron use with suitable choice of window material. Using these methods, it has proved possible to monitor solid-state reactions and phase changes, and to provide information on the temperature variations within solid samples.

INTRODUCTION

The mechanism of microwave activation of chemical processes in solution and liquid phases continues to be the subject of numerous theoretical and experimental studies[1-3]. In contrast, research on the nature of microwave activation of solid-state chemistry and materials processing is not extensive, despite an increasing appreciation of the advantages provided by this method of heating.[4-8] There are notable exceptions[7], particularly concerning the nature of enhanced rates of ion migration under the influence of high frequency electric fields – the so-called 'ponderomotive' effect that may occur at surfaces or interfaces in ionic materials during microwave irradiation.[9-11] In this paper we describe developments in apparatus that will enable the most incisive method of solid-state structure determination – diffraction of X-rays or neutrons – to be applied *in situ*, elucidating physical and chemical changes in materials during microwave irradiation. Such work may also provide insights into the way in which biological material may be influenced by microwave irradiation,[12, 13] and perhaps inform the current debate about whether microwave telecommunications equipment such as mobile phones are harmful to health.[14]

The nature of the two experimental probes – X-rays and neutrons - is quite different in terms of some of the difficulties posed by the specific requirements of microwave equipment. Conventional waveguide systems enclose the sample and link it to the microwave source with conducting material, generally a metal whose thickness should be at least 50 microns. This may attenuate conventional sources of X-rays[15], but neutrons penetrate most metals much more effectively[16], and are generally not constrained in this manner by such apparatus. However, in both cases there may be significant problems measuring the temperature accurately and precisely – indeed there are difficulties even invoking the concept of temperature in the conventional thermodynamic sense for a system pumped with microwave energy. Even if one assumes that the system is at or close to equilibrium, there are problems with homogeneity of temperature over a millimetre or even sub-millimetre scale.[17] The two basic methods for temperature measurement - contact probes such as thermocouples and thermography devices such as pyrometers – may be unreliable. The former may perturb the microwave field at the point of interest, while the latter only register surface temperatures in an environment where surface cooling is highly significant. We shall show how diffraction may be used to provide a microscopic probe of temperature and temperature distribution over the whole sample.

Given these experimental challenges, apparatus was initially designed and built for *in situ* neutron diffraction on a stirred or circulated liquid sample.[16] X-ray diffraction is a more widely used structural probe than neutron diffraction, and the designs for this application initially centred on a novel co-axial applicator system suitable for powder samples. However, this system also had drawbacks: geometric constraints made it unsuitable for single-crystal diffraction using 4-circle instruments, and we also encountered problems in matching the source to the load. In response to the first problem, we have designed and tested a parallel plate applicator that allows measurements to be made with very little impedence to the X-ray beam (reported elsewhere[18]); the latter problem was solved by returning to a modified form of the original waveguide design. We describe each of these pieces of apparatus with representative applications to structural problems.

IN SITU NEUTRON DIFFRACTION

The penetrating nature of neutrons towards most metallic elements meant that it was relatively straightforward to modify a waveguide for *in situ* diffraction studies. The specific problem we first wished to study with such apparatus concerned the growth of colloidal metal oxide particles by hydrolysis of a precursor metal salt heated in aqueous solution. This is a common process in the production of particulate materials of controlled size and shape, and the hydrolysis of iron (III) salts to make iron oxides and oxyhydroxides was selected as a first test case because it is a particularly well-studied and interesting system.[19] The measurement involved scattering neutrons from the particles to determine their shape distribution as a function of time, and test particular theoretical models of the growth processes. The angle through which the neutrons are scattered is very small because the scattering objects are large compared with the neutron wavelength, so this technique is known as small angle neutron scattering (SANS). These measurements were conducted at the ISIS Facility at the Rutherford Appleton Laboratories, UK. This is a spallation source that provides a beam with a width of 12 mm at a port in a fixed position in the SANS diffractometer LOQ.

The intensity of the scattered beam is measured with a detector whose position is also fixed, so our instrument had to comply with these geometric constraints. As a consequence, a compact waveguide system was devised that presented the sample to the neutron beam, as shown in Figure 1.

A 0-1 kW variable power microwave source (Microwave Heating Type GU14) delivered 2.45 GHz microwaves to the sample space through waveguide 9a (312) components. Although the neutron beam was able to penetrate the waveguide, two 17.5 mm diameter holes were incorporated as windows, reducing background scattering. The relatively small diameter of these holes means that there is no requirement for chokes or metal foil windows to be included in order to maintain microwave leakage within legal safety limits. The 50 cm^3 sample was held in a glass round-bottomed reaction vessel and continuously pumped through a sample cell *via* a short 3 mm i.d. tube. The temperature was monitored with a gas pressure thermometer to an accuracy of better than 0.1 K and the microwave output power used to control the temperature of the sample to within ±0.5 K for the period of the experiment.

Data obtained from heating the sample of $FeCl_3$ in D_2O (which was used in place of H_2O because H but not D scatters the neutron beam strongly and incoherently) at 95° C indicated that particle formation begins rapidly within the first 30 minute time frame. Subsequent measurements, taken in further 30 minute intervals (which is the shortest time in which data of reasonable quality could be obtained in this experiment) showed that the particles first formed then agglomerated to produce macroscopic particles.[16]

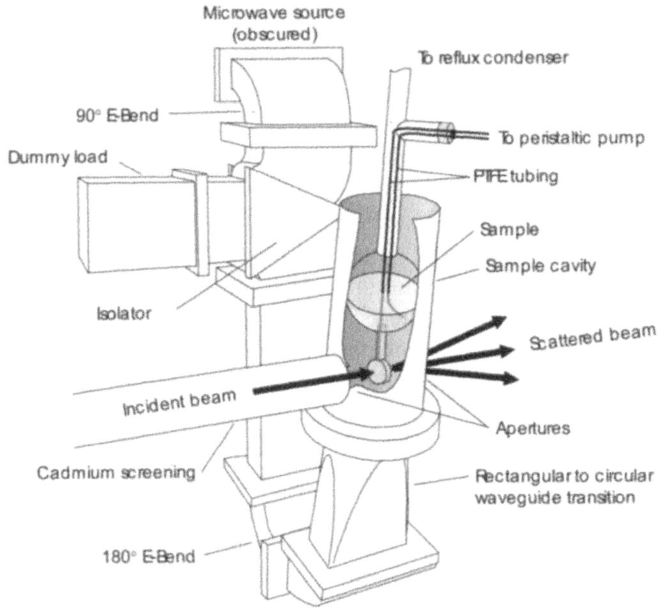

Figure 1. Apparatus for the *in situ* study of a microwave-driven reaction by small angle neutron scattering.

IN SITU X-RAY DIFFRACTION: COAXIAL APPLICATOR

Laboratory-based X-ray diffractometers are the most widely used means of studying the structure of solids, and so a microwave applicator that could be used with a commercially available, sealed-tube X-ray diffractometer was designed and built. The design used for the SANS measurements had to be adapted in a number of ways. In the particular diffractometer used, data are collected over a wide range of scattering angle 2θ, preventing the use of relatively small holes to bring the beam in and out. Instead, it was necessary to use some form of windows with a wide angle for the X-rays. The material with the best balance of electrical conductivity and X-ray transparency is beryllium, but it is a difficult material to use: it tends to form extremely toxic oxides and the machining involved in making the applicator would create harmful dust. However, aluminium foil (~100 μm) was found to be satisfactory with respect to beam attenuation. Health and safety constraints require the X-ray diffraction optics to be enclosed in radiation shielding. A way was therefore needed to bring the waveguide through this shielding, and this was accomplished initially *via* a 1 m length of coaxial cable. The design of the applicator is shown in Figure 2. Microwaves are generated with a commercial 0–1 kW variable power, ASTeX S-1000 operating at a frequency of 2.45 GHz and introduced to the sample cavity by an 'antenna' which reaches to within a few millimetres of the sample. Temperature is measured using a fluoro-optic probe. The equipment was designed to be used in conjunction with a Philips PW1730 diffractometer, using Cu K_α radiation (λ=1.542Å), in flat-plate (θ, 2θ) geometry.

To test the apparatus, the change in diffraction pattern of the microwave susceptor AgI was studied as it transformed on heating from the low-temperature

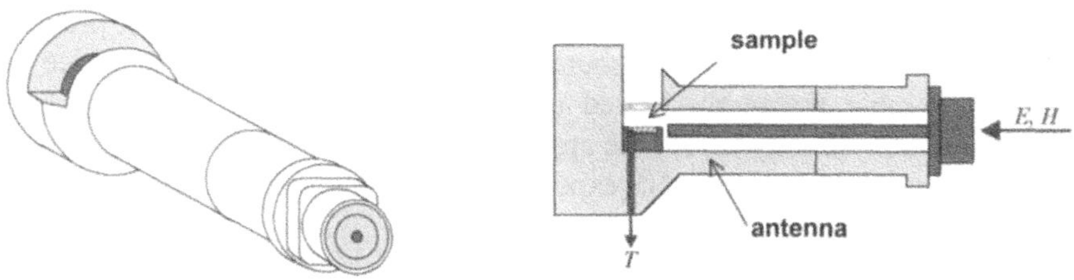

Figure 2. Prototype applicator for *in situ* X-ray diffraction during microwave heating. (a) Perspective. (b) Schematic, showing port for temperature probe (*T*), incoming microwave radiation (*E,H*) at the coaxial port and central antenna to re-emit microwaves.

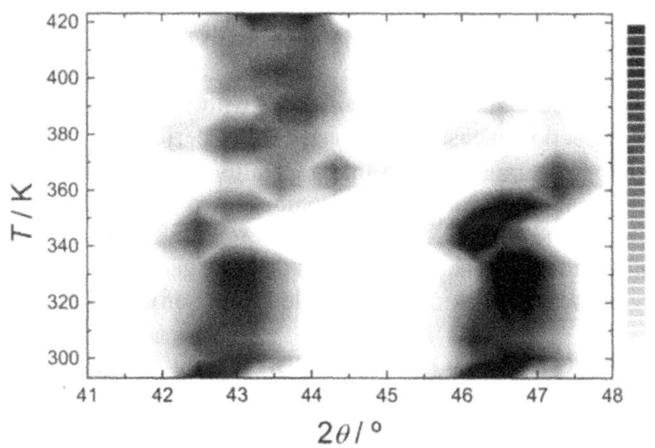

Figure 3. Contour map showing temperature variation of the intensity of Bragg reflections over the 2θ range 41°-48° for AgI heated by microwave radiation.

β-polymorph to the fast-ion conducting α-polymorph.[20-22] The temperature of this transition is generally taken to be 420 K, which is well below the maximum operating temperature of the fluoro-optic thermometer (~500 K). Characteristic Bragg reflections of both phases are found within the 2θ range 41° - 48° and Figure 3 illustrates the way in which this region of the diffraction pattern changes on warming. It is quite clear that the phase transition from β to α–AgI occurs at approximately 370 K, with an estimated error of 10 K which is about 50 K below the accepted literature value. We are currently investigating whether this is an experimental artefact, or some form of athermal microwave effect.

There are several ways in which this apparatus may be improved. First, a brighter X-ray source would enable data to be taken more quickly and in greater detail, and this could be achieved using synchrotron radiation. Second, problems were encountered in matching the source and the load, with the consequence that the co-axial cable frequently heated excessively. This could be circumvented using a circulator directly before the applicator, or using some form of waveguide similar to that used with the SANS experiment. Such apparatus is described next.

IN SITU SYNCHROTRON X-RAY DIFFRACTION

The apparatus used with the laboratory X-ray source was adapted for use at the UK Synchrotron Radiation Source (SRS) at the Daresbury Laboratories. Microwaves were still transmitted through a co-axial lead and antenna, but applied to the sample *via* a cylindrical waveguide illustrated in Figure 4.

The internal length of the applicator can be varied with a moveable short-circuit termination, allowing a degree of control over the field distribution. A slot was cut across the applicator then covered in aluminium foil to allow the passage of X-rays. The sample is placed as loose powder onto an insulating, ceramic post (Makor) at the centre of the applicator and aligned with the foil window and the incident X-rays. Diffraction measurements have been performed on samples of the fast-ion conductor β-aluminas using this apparatus on station 9.1 at the SRS Daresbury; these preliminary measurements demonstrated that high-quality X-ray data may be collected and refined, but to date the response towards microwave heating appears conventional. It should be noted that this geometry of cavity gives rise to poor tuning properties, making appreciable heating of the sample difficult.

IN SITU NEUTRON DIFFRACTION: WAVEGUIDE APPLICATOR

Problems of heat dissipation in the co-axial cable led to a reconsideration of waveguide-based applicators for *in situ* diffraction. A new applicator based on a cylindrical waveguide was constructed from stainless steel (Figure 5). The sample is positioned a quarter wavelength from the terminating short-circuit of the waveguide, where the electric field strength is at a maximum. The diameter of the applicator is such that only the fundamental TE_{11} mode propagates: the field intensity is a maximum at the centre of the circular cross-section. The waveguide fitted to the magnetron generator (ASTeX S-1000, variable power source up to 1 kW) is rectangular and therefore a transition piece is required between the two different types. The distance of the sample from the rectangular to circular transition is sufficient to quench any extra modes thus excited. This system was first designed for neutron diffraction experiments using the high resolution powder diffractometer (HRPD) at the ISIS Facility. As with the SANS experiment there are geometric constraints: the incident beam is fixed in position, but now there are several banks of detectors at small angles, as well as centred at 90°, and in a back-scattering position. The apparatus was therefore designed to have a wider angular range for its windows, and the material of choice for neutrons is vanadium, which has a very low coherent scattering cross section. The window is edged with neutron-absorbing gadolinium sheet to minimise scatter from other

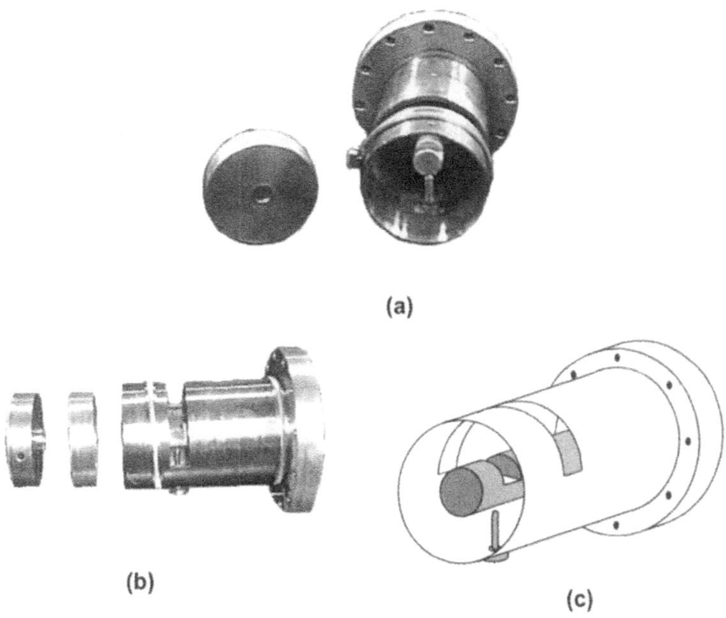

Figure 4. Apparatus for use with station 9.1 at the SRS, shown (a) end-on, (b), side-on and (c) schematic. The radial slot is covered with thin metal foil windows in use (see also Figure 6).

components of the microwave heating system.

This apparatus has been tested effectively on $BaTiO_3$, obtaining high quality diffraction patterns with and without microwave heating.[18] The lattice parameter of this material (indeed of most materials) can be used as a form of thermometer, though it should be noted that in the particular case of $BaTiO_3$ interpretation of the data is greatly complicated by the presence of minority phases. However, our experiments did provide some evidence for hot spots in the sample, with a spread of cell parameters that indicate a distribution of temperature.

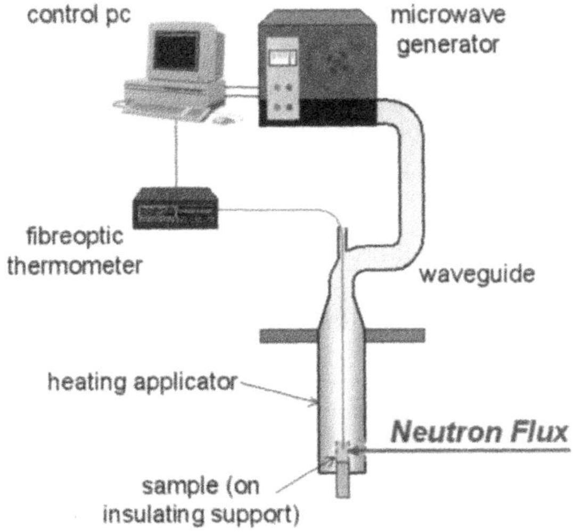

Figure 5. Apparatus for diffraction experiment on HRPD, shown schematically.

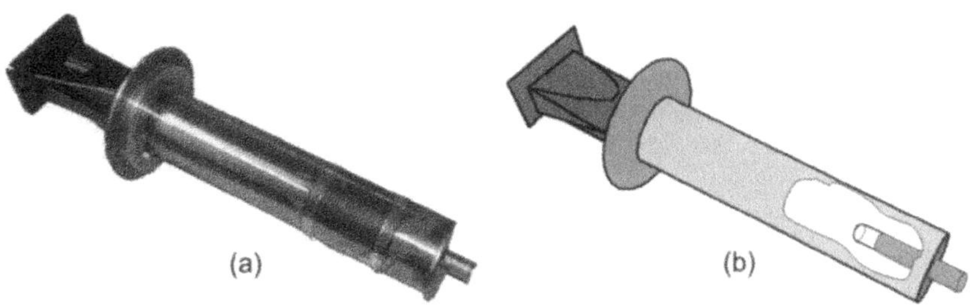

Figure 6. Microwave applicator for powder neutron diffraction: (a) photo showing vanadium foil windows; (b) schematic showing sample position.

IN SITU X-RAY DIFFRACTION: WAVEGUIDE APPLICATOR

The apparatus for *in situ* powder diffraction at a neutron source based on a waveguide applicator can be readily adapted for a conventional laboratory X-ray source. As the diffraction pattern is measured as a function of 2θ, the sample plate rotates by θ, so the spindle on which this plate is mounted must be able to rotate within the waveguide; it should also be located so that the sample is at or near a maximum in the electric field. The waveguide had a circular cross-section with a diameter sufficient to support the TE_{11} fundamental mode, and was provided with slots covered in aluminium foil to allow the passage of X-rays; it could be tuned by means of a sliding short-circuit termination as illustrated in Figure 7. A hole was cut in the safety shielding of the diffractometer, the waveguide was inserted and then gaps between the two made safe with lead tape. It proved most practical to place the microwave generator on top of the radiation shielding, more-or-less above the sample plate. This required a bent section of waveguide. Bends with efficient transmission are based on large radius, rectangular cross-section waveguides, with circular cross-section waveguide in the region of the sample plate – partly because this provided the optimum geometry for data acquisition in the. A mitre bend for circular waveguide tubing was designed and built for the applicator, and although overall matching was not perfect, it was adequate for the purpose. The new applicator has several advantages over that used previously: the power is limited by the magnetron, not the apparatus; the wavelength can be adjusted to focus maximum power at the sample position; and lengthy experiments are possible without the danger of overheating. Initial testing of this apparatus indicates that it can heat a moderately good microwave susceptor such as powdered silicon quickly, while collecting high quality powder diffraction data.

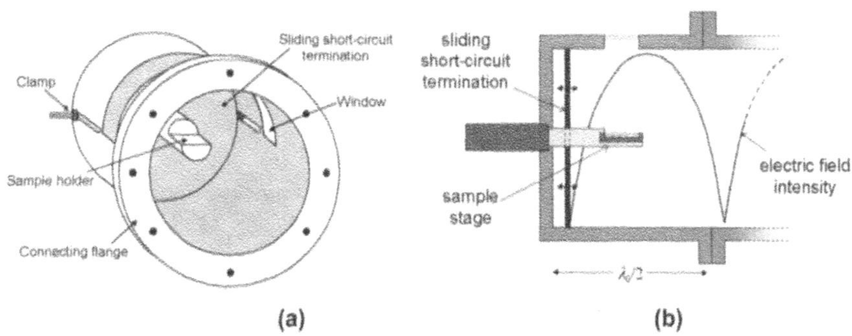

Figure 7. The improved microwave applicator for use with the Philips PW1730 instrument; (a) in perspective, (b) in cross-section.

CONCLUSIONS

We have illustrated the challenges that arise in bringing microwaves to a diffraction experiment through a description of the evolution of apparatus we designed and built for X-ray and neutron experiments. We are now in a position to perform more detailed experimental studies of chemical and physical changes in solid-state systems during microwave irradiation.

ACKNOWLEDGEMENTS

Construction of microwave apparatus was conducted at the University of Edinburgh with the help of S.Mains, D.Paden, J.Ashfield, K.Anderson and S.Johnson and also with input from workshops at the Rutherford Appleton Laboratories. The various diffraction experiments reported here and in greater detail in work in some of the references was performed with R.K.Heenan, R.M. Ibberson, S.M. King, G.S.Oakley, G.-H.Wang, C.C. Wilson, and I.D Youngson. Financial support was provided by EPSRC.

REFERENCES

1 D. Stuerga and P. Gaillard, Microwave athermal effects in chemistry: A myth's autopsy .2. Orienting effects and thermodynamic consequences of electric field, *J. Microw. Power Electromagn. Energy*, **31** [2] 101 (1996)

2 C.R. Strauss and R.W. Trainor, Invited Review - Developments in Microwave-Assisted Organic- Chemistry, *Aust. J. Chem.*, **48** [10] 1665 (1995)

3 A.G. Whittaker and D.M.P. Mingos, Microwave dielectric heating effects in chemical synthesis, in *Chemistry under extreme or non-classical conditions*, R. van Eldik and C. Hubbard, Editors. 1996, Wiley: New York. p. 479.

4 D.M.P. Mingos and D.R. Baghurst, Applications of Microwave Dielectric Heating Effects to Synthetic Problems in Chemistry, *Chem.Soc.Rev.*, **20**[1]1 (1991)

5 A.G. Whittaker and D.M.P. Mingos, The Application of Microwave Heating to Chemical Syntheses, *J. Microwave Power and Electromag. Energy*, **29** [4] 195 (1994)

6 S.A. Galema, *Chem. Soc. Rev.*, **26** 233 (1997)

7 Y.V. Bykov, K.I. Rybakov, and V.E. Semenov, High-temperature microwave processing of materials, *J. Phys. D: Appl. Phys.*, **34** R55 (2001)

8 Lidstrom P, Tierney J, and e.a. Wathey B, Microwave assisted organic synthesis - a review, *Tetrahedron*, **57** 9225 (2001)

9 J.H. Booske, R.F. Cooper, and I. Dobson, Mechanisms for Nonthermal Effects on Ionic Mobility During Microwave Processing of Crystalline Solids, *J. Mater. Res.*, **7** [2] 495 (1992)

10 J.H. Booske, R.F. Cooper, S.A. Freeman, K.I. Rybakov, and V.E. Semenov, Microwave ponderomotive forces in solid-state ionic plasmas, *Phys. Plasmas*, **5** [5] 1664 (1998)

11 S.A. Freeman, J.H. Booske, and R.F. Cooper, Modeling and numerical simulations of microwave-induced ionic transport, *J. Appl. Phys.*, **83** [11] 5761 (1998)

12 H. Bohr and J. Bohr, Microwave enhanced kinetics observed in ORD studies of a protein, *Bioelectromag.*, **21** 68 (2000)

13 J.A. Laurence, P.W. French, R.A. Lindner, and D.R. McKenzie, *J.Theor.Biol.*, **206** 291 (2000)

14 W. Stewart, Mobile Phones and Health: a report of the Independent Expert Group on Mobile Phones. 2000, Chilton: National Radiation Protection Board.

15 G.R. Robb, R. Djangazov, A.G. Whittaker, and A. Harrison. Microwave reactor for in-situ X-ray diffraction measurements - Evidence of non-thermal effects in silver iodide. in *Proceedings of the Second International Conference on Microwave Chemistry*. 2000. Antibes - Juan-les-Pins: AMPERE.

16 A.G. Whittaker, A. Harrison, G.S. Oakley, I.D. Youngson, R.K. Heenan, and S.M. King, Preliminary experiments on apparatus for *in situ* studies of microwave-driven reactions by small angle neutron scattering, *Rev. Sci. Instrum.*, **72** [1] 1 (2001)

17 C. Gerk and M. Willert-Porada. Application of microwaves (2.45 GHz) to processing of metallic, ceramic and composite materials. in *Proc. Int'l Conf. on Microwave Chemistry, AMPERE, Antibes*. 2000. Antibes - Juan-les-Pins: AMPERE.

18 A. Harrison, R. Ibberson, G.R. Robb, A.G. Whittaker, C. Wilson, and D. Youngson, In situ neutron diffraction studies of single crystals and powders during microwave irradiation, *Faraday Discuss.*, **122**, 363-379 (2002)

19 G.H. Wang, G. Whittaker, A. Harrison, and L.J. Song, Preparation and mechanism of formation of acicular goethite- magnetite particles by decomposition of ferric and ferrous salts in aqueous solution using microwave radiation, *Mater. Res. Bull.*, **33** [11] 1571 (1998)

20 J.E. Maskasky, High phase-transition temperature for beta-AgI to alpha-AgI and an explanation of the memory effect, *Phys. Rev. B: Condens. Matter*, **43** [7] 5769 (1991)

21 C.A. Rains, J.R. Ray, and P. Vashishta, Phase transformation and polytypism in silver iodide: a molecular dynamics study, *Phys. Rev. B: Condens. Matter*, **44** [17] 9228 (1991)

22 J.L. Tallon, Constant-stress molecular dynamics: the phase diagram of silver iodide, *Phys. Rev. B: Condens. Matter*, **38** [13] 9069 (1988)

Agricultural Products

MICROWAVE AND RADIO-FREQUENCY POWER APPLICATIONS IN AGRICULTURE

Stuart O. Nelson
U. S. Department of Agriculture, Agricultural Research Service
Richard B. Russell Agricultural Research Center
P. O. Box 5677, Athens, Georgia, 30604, U. S. A.

ABSTRACT

A brief review is presented of potential applications for radio-frequency and microwave power applications in agriculture. Included are applications for stored-product insect control, seed treatment to improve germination and seedling performance, conditioning of products to improve nutritional value or prolong keeping quality, soil treatment to control weeds, crop drying, and seed treatment for elimination of bacterial human pathogens. Limitations for practical application are also discussed.

INTRODUCTION

The potential for agricultural applications of radio-frequency (RF) energy for the solution of various problems in agricultural production, crop handling and storage, and product preservation and conditioning has been considered for many years [1, 2]. With the development of economical microwave power equipment and widespread use of microwave power for heating food, additional agricultural applications have been studied. This paper is concerned with RF and microwave dielectric heating applications in agriculture for purposes other than food processing, and consists of a short review, update, and discussion of some of those potential agricultural applications that may be of interest to the microwave and RF power industry.

INSECT CONTROL

One of the earliest records of interest in the use of RF energy for agricultural purposes involved the reactions of insects that infest stored products to high-frequency electric fields. A number of studies were conducted between 1930 and 1960 on the possibilities of controlling such insects that infest cereal grains, seeds, tobacco, wood and a few other commodities [3-6]. Analyses of dielectric heating revealed that relative dielectric properties of insects and the host materials, as well as geometrical relationships, were important in determining likelihood of differential heating of the insects and that selective heating of the insects was more probable in materials of relatively low moisture content [5, 6]. Subsequent research on stored-grain insects established the influence of physical factors, such as frequency, modulation, and intensity of the electric field and moisture content of the host materials, and of entomological factors such as insect species, developmental stage, and physiological responses to RF dielectric heating exposures [7-9]. Broad-frequency-range measurements of the dielectric properties of insects and grain revealed that prospects for selectively heating the insects were better at frequencies between 10 and 100 MHz than they were at microwave frequencies [10, 11], and experimental dielectric heating treatments of infested wheat at 39 and 2450 MHz confirmed this assessment [11]. New measurements of the dielectric properties of stored-grain insects extended the data to 20 GHz and showed that the dielectric properties of insects continued to decrease as frequency increased above the X-band [12, 13].

Interest continues in the possibility for insect control methods involving RF and microwave treatments, not only for grain and similar commodities [14, 15], but also for fruits and nuts [16-18]. Some optimism was expressed about controlling stored-grain insects with 10.6 GHz microwave treatments [14], but dielectric properties measurements show better selective heating available at lower frequencies in the 10 to 100 MHz range [9, 13]. Recent studies have included pulse modulated high-frequency (HF) electric fields in combination with vacuum and HF-induced plasma. Dielectric heating treatments of infested in-shell and shelled pecans at 40 MHz showed potential for providing high mortality in pecan weevils [16], and 27-MHz treatments were effective in killing codling moth larvae in in-shell walnuts [17]. Microwave treatments at 915 MHz, along with holding time at elevated temperatures and hydrocooling of cherries to preserve quality, were considered possible for quarantine purposes, but more research was needed before recommendation of the process [18].

SEED TREATMENT

Another early application for RF dielectric heating involved the treatment of crop seeds to improve or stimulate germination. Several reports of accelerated germination or beneficial influences of such treatments were noted, but consistency of such experiments appeared lacking [19]. However, in dielectric heating of alfalfa seed and some other small-seeded legumes, in which an impermeable seed coat is a natural phenomenon preventing the entry of moisture required for germination, consistently favorable responses were obtained [20, 21]. The germination of seed lots was increased by rendering the impermeable seed coats permeable to water [22, 23]. Favorable responses to RF and microwave treatments of such seed lots were also noted in greenhouse and field plantings

[24], and beneficial effects were retained in treated seeds after many years in storage [25, 26]. Traditional treatments for reduction of hard-seed percentages consist of an abrasive process that scratches the seed coat to make the seeds permeable to water. This process, called scarification, has a damaging effect if seed is not planted promptly after scarification, so scarified seed lots cannot be held over for use the following season without serious deterioration in quality. Thus, dielectric heating treatments offer a significant benefit, but the practice has not yet been commercialized. Germination responses to RF treatments of seeds of more than 80 plant species have been summarized [27]. Favorable responses were achieved with several small-seeded legumes as a result of reductions in percentages of hard seed, and increases in germination and acceleration of germination was noted in some, but for many species no improvement was obtained. Favorable results of microwave heating treatments on *Acacia* seeds through reduction of hard-seed percentages were also obtained [28]. Recent studies on treatment of alfalfa seed, artificially inoculated with human pathogenic bacteria, showed increased germination and moderate reductions in pathogen populations, which may provide an additional incentive for practical use in the seed sprouting industry.

RF and microwave heating treatments have been studied in connection with seed treatments to reduce seed infections with microorganisms. A seed-borne fungus infection in barley was eliminated by RF dielectric heating, but only when the seed was soaked in water before treatment, and the germination of the barley was seriously reduced [29]. Reductions in infection by the same organism were observed following microwave heating treatments, but again seed germination was reduced [30]. Another seed-borne fungus infection of wheat was reportedly eliminated by dielectric heating treatments of several minutes duration without reducing seed germination [31]. However, microwave heating of wheat seed infected with the same organism reduced infection to a low level but also lowered germination and seedling vigor [32], so further work seems necessary to confirm the usefulness of RF and microwave heating for controlling seedborne pathogens.

PRODUCT CONDITIONING

Conditioning of agricultural products by RF and microwave heating treatments for improvement in quality has also been explored. Raw soybeans have a natural antinutritional factor when fed to non-ruminant farm animals or when utilized by humans. This factor inhibits the action of the enzyme trypsin in the digestion process. Steam treatment or extrusion processes are used to inactivate the trypsin inhibitor, but dry air heating is ineffective, because the presence of water is required. RF and microwave heating treatments were found to produce the desired effects, because they are rapid and the necessary action is achieved with the water that is present in the soybeans [33, 34]. They also resulted in a more appealing appearance of the soybean meal, and the lipoxygenase enzyme responsible for the development of off-flavors was inactivated. Similar RF treatments were found to improve the storability of pecans, which become rancid when not properly refrigerated [35, 36].

SOIL TREATMENT

Treatment of soils with microwave energy has been proposed as a method of weed control by killing the weed seeds in the soil [37]. However, large amounts of energy are required to raise the soil temperature to the lethal level for the seeds, and attenuation of microwave energy by the soil reduces the field intensities rapidly to ineffective levels as the waves penetrate into the soil [38]. Thus, this technique seems unlikely to become practical. The reader is referred to an earlier review for an extensive treatment of this topic [38].

CROP DRYING

Use of microwave energy has often been suggested as a way to accelerate the drying of farm crops. However, the energy and equipment costs, in relation to the value of the products, has generally been discouraging. Use of dielectric or microwave heating to supplement conventional drying techniques has also been considered, whereby the processing time can be shortened with less costly energy requirements. Although microwave heating can accelerate the drying process, limits on speed of drying for corn have been observed in work at 915 MHz and 2.45 GHz due to reduction in product quality if dried too rapidly [39]. Soybeans were dried with 2.45-GHz microwave energy without damaging seed quality if absorbed energy did not exceed 0.13 W/g of moist material [40]. Experience with small scale microwave drying studies has also pointed out real engineering challenges in designing drying systems of sufficient scale for practical use [41]. More recently, different microwave energy application techniques have been studied for drying fresh fruit [42, 43]. If improved product quality and time and energy savings can be demonstrated for high-value products, RF and microwave dielectric heating may find practical uses.

STERILIZATION

Dielectric heating has been explored for the purpose of reducing the populations of bacterial human pathogens. Most of these studies have been conducted in the field of food technology, but a few have involved agricultural products before they are introduced into the food arena. One such study concerned the potential bacterial contamination of alfalfa seed used for production of sprouts for human consumption [44]. Studies of illness outbreaks associated with consumption of vegetable sprouts identified contamination of alfalfa seed by human pathogens as the likely source of the bacteria. Because RF and microwave dielectric heating treatments of alfalfa seed are known to be effective in increasing the germination of alfalfa seed containing high percentages of hard seed [45], and because the human pathogens responsible for the illness are known to be sensitive to heating treatments, it was logical to explore the use of dielectric heating to learn how effective such treatments might be in lowering the pathogen populations on alfalfa seed. Also, because the mechanical scarification process produces scratches in the seed coat, these injuries may provide attachment areas for bacteria that render them more resistant to control and removal by chemical washes. Studies showed that human pathogen populations on alfalfa seed were significantly reduced by 39-MHz exposures without damage to seed viability [46, 47]. However, the degree of reduction desired was not achieved without significant reductions in alfalfa seed germination. The combination of increased germination, with consequent improved sprout yield, and moderate bacterial population reduction, however, makes RF and microwave heating worthy of consideration for this application. Mechanical scarification of seed lots would no longer be necessary, and the consequent damage to seed viability if seed lots are held over could be avoided.

CONCLUSION

Many potential applications for microwave and lower frequency dielectric heating in agriculture have been considered, including insect control, seed treatment to improve seed germination and performance, conditioning of agricultural products to improve quality or shelf life, soil treatment for controlling weeds and other pests, crop drying, and sterilization of seeds used for sprout production. Although some of these applications were successful from a phenomenological viewpoint, most of them have not appeared practical from an economic viewpoint. Since many agricultural commodities have very low unit values, little additional expense can be tolerated in remaining competitive, and equipment for either RF or microwave dielectric heating and energy costs to operate such equipment are relatively expensive. Also since many agricultural operations are seasonal, it is difficult to amortize equipment costs effectively. Therefore, until the benefits provided by the application of RF and microwave power can be expected to exceed the costs involved, such applications are highly unlikely to be developed. On the other hand, If RF and microwave power can accomplish a necessary task that cannot be done in some other less expensive way, there may still be practical applications developed.

REFERENCES

[1] S. O. Nelson, "RF and microwave energy for potential agricultural applications," *Journal of Microwave Power*, vol. 20, no.2, pp. 65-70, 1985.

[2] S. O. Nelson, "Potential agricultural applications for RF and microwave energy," *Transactions of the ASAE*, vol. 30, no.3, pp. 818-822, 831, 1985.

[3] S. O. Nelson, "Electromagnetic and sonic energy for insect control," *Transactions of the ASAE*, vol. 9, no.3, pp. 398-405, 1966.

[4] S. O. Nelson and J. L. Seubert, "Electromagnetic and sonic energy for pest control," in *Scientific Aspects of Pest Control, Publication No. 1402*. Washington, DC: National Academy of Sciences--National Research Council, 1966.

[5] S. O. Nelson, "Electromagnetic energy," in *Pest Control -- Biological, Physical and Selected Chemical Methods*, W. W. Kilgore and R. L. Doutt, Eds. New York and London: Academic Press, 1967.

[6] A. M. Thomas, "Pest control by high-frequency electric fields -- critical resume," British Electrical and Allied Industries Research Association, Leatherhead, Surrey Technical Report W/T23, 1952.

[7] S. O. Nelson and W. K. Whitney, "Radio-frequency electric fields for stored-grain insect control," *Transactions of the ASAE*, vol. 3, no.2, pp. 133-137, 1960.

[8] S. O. Nelson, L. E. Stetson, and J. J. Rhine, "Factors influencing effectiveness of radio-frequency electric fields for stored-grain insect control," *Transactions of the ASAE*, vol. 9, no.6, pp. 809-815, 1966.

[9] S. O. Nelson, "Review and assessment of radio-frequency and microwave energy for stored-grain insect control," *Transactions of the ASAE*, vol. 39, no.4, pp. 1475-1484, 1996.

[10] S. O. Nelson and L. F. Charity, "Frequency dependence of energy absorption by insects and grain in electric fields," *Transactions of the ASAE*, vol. 15, no.6, pp. 1099-1102, 1972.

[11] S. O. Nelson and L. E. Stetson, "Comparative effectiveness of 39- and 2450-MHz electric fields for control of rice weevils in wheat," *Journal of Economic Entomology*, vol. 67, no.5, pp. 592-595, 1974.

[12] S. O. Nelson, P. G. Bartley, Jr., and K. C. Lawrence, "Measuring RF and microwave permittivities of adult rice weevils," *IEEE Transactions on Instrumentation and Measurement*, vol. 46, no.4, pp. 941-946, 1997.

[13] S. O. Nelson, P. G. Bartley, Jr., and K. C. Lawrence, "RF and microwave dielectric properties of stored-grain insects and their implications for potential insect control," *Transactions of the ASAE*, vol. 41, no.3, pp. 685-692, 1998.

[14] S. L. Halverson, W. E. Burkholder, T. S. Bigelow, E. V. Nordheim, and M. E. Misenheimer, "High-power microwave radiation as an alternative insect control method for stored products," *Journal of Economic Entomology*, vol. 89, no.6, pp. 1638-1648, 1996.

[15] A. A. Mishenko, O. A. Malinin, V. M. Rashkovan, A. V. Basteev, L. A. Basyma, Y. P. Mazalov, and V. A. Kutovoy, "Complex high-frequency technology for protection of grain against pests," *Journal of Microwave Power and Electromagnetic Energy*, vol. 35, no.3, pp. 179-184, 2000.
[16] S. O. Nelson and J. A. Payne, "Pecan weevil control by dielectric heating," *Journal of Micorwave Power*, vol. 17, no.1, pp. 51-55, 1982.
[17] S. Wang, J. N. Ikediala, J. Tang, J. D. Hansen, E. Mitcham, R. Mao, and B. Swanson, "Radio freqeuency treatments to control codling moth in in-shell walnuts," *Postharvest Biology and Technology*, vol. 22, pp. 29-38, 2001.
[18] J. N. Ikediala, J. Tang, L. G. Neven, and S. R. Drake, "Quarantine treatment of cherries using 915 MHz microwaves: temperature mapping, codling moth mortality, and fruit quality," *Postharvest Biology and Technology*, vol. 16, pp. 127-137, 1999.
[19] S. O. Nelson, "Electromagnetic Radiation Effects on Seeds," *Conference Proceedings - Electromagnetic Radiation in Agriculture*, pp. 60-63, 1965.
[20] S. O. Nelson and W. W. Wolf, "Reducing hard seed in alfalfa by radio-frequency electrical seed treatment," *Transactions of the ASAE*, vol. 7, no.2, pp. 116-119, 122, 1964.
[21] S. O. Nelson, "Use of microwave and lower frequency RF energy for improving alfalfa seed germination," *Journal of Microwave Power*, vol. 11, no.3, pp. 271-277, 1976.
[22] L. A. T. Ballard, S. O. Nelson, T. Buchwald, and L. E. Stetson, "Efects of radiofrequency electric fields on permeability to water of some legume seeds, with special reference to strophiolar conduction," *Seed Science and Technology*, vol. 4, pp. 257-274, 1976.
[23] S. O. Nelson, L. A. T. Ballard, L. E. Stetson, and T. Buchwald, "Increasing legume seed germination by VHF and microwave dielectric heating," *Transactions of the ASAE*, vol. 19, no.2, pp. 369-371, 1976.
[24] S. O. Nelson, L. E. Stetson, R. B. Stone, J. C. Webb, C. A. Pettibone, D. W. Works, W. R. Kehr, and G. E. VanRiper, "Comparison of infrared, radiofrequency, and gas-plasma treatments of alfalfa seed for hard-seed reduction," *Transactions of the ASAE*, vol. 7, no.3, pp. 276-280, 1964.
[25] S. O. Nelson and L. E. Stetson, "Long-term effects of radiofreqeuncy, infrared, and gas-plasma treatments on alfalfa seed," *Transactions of the ASAE*, vol. 26, no.5, pp. 1516-1521, 1983.
[26] S. O. Nelson, W. R. Kehr, L. E. Stetson, and W. W. Wolf, "Laboratory germination and sand emergence responses of alfalfa seed to radiofrequency electrical treatment," *Crop Science*, vol. 17, pp. 534-538, 1977.
[27] S. O. Nelson and L. E. Stetson, "Germination responses of selected plant species to RF electrical seed treatment," *Transactions of the ASAE*, vol. 28, no.6, pp. 2051-2058, 1985.

[28] V. N. Tran and A. K. Cavanagh, "Effects of microwave energy on *Acacia Longifolia*," *Journal of Microwave Power*, vol. 14, no.1, pp. 21-27, 1979.
[29] S. O. Nelson and E. R. Walker, "Effects of radio-frequency electrical seed treatment," *Agricultural Engineering*, vol. 42, no.12, pp. 688-691, 1961.
[30] M. M. P. Stephenson, A. C. Kushalappa, and G. S. V. Raghavan, "Effect of selected combinations of microwave treatment factors on inactivation of *Ustilago nuda* from barley seed," *Seed Science and Technology*, vol. 24, pp. 557-570, 1996.
[31] M. Cwiklinski and D. von Horsten, "Effect of exposure to radio-frequency electric fields on *Fusarium graminearum* in wheat seed," *ASAE Paper No. 016171*. American Society of Agricultural Engineers, St. Joseph, MI, 2001.
[32] M. V. B. Reddy, G. S. V. Raghavan, A. C. Kushalappa, and T. C. Paulitz, "Effect of microwave treatment on quality of wheat seeds infected with *Fusarium graminearum*," *Journal of Agricultural Engineering Research*, vol. 71, pp. 113-117, 1998.
[33] S. O. Nelson, "Effects of 42- and 2450-MHz dielectric heating on nutrition-related properties of soybeans," *Journal of Microwave Power*, vol. 16, no.3 & 4, pp. 313-318, 1981.
[34] A. Pour-El, S. O. Nelson, E. E. Peck, and B. Tjiho, "Biological properties of VHF- and microwave-heated soybeans," *Journal of Food Science*, vol. 46, no.3, pp. 880-885, 895, 1981.
[35] S. O. Nelson, S. D. Senter, and W. R. Forbus, Jr., "Dielectric and steam heating treatments for quality maintenance in stored pecans," *Journal of Microwave Power*, vol. 20, no.2, pp. 71-74, 1985.
[36] S. D. Senter, W. R. Forbus, Jr., S. O. Nelson, R. L. Wilson, Jr., and R. J. Horvat, "Effects of dielectric and steam heating treatments on the storage stability of pecan kernels," *Journal of Food Science*, vol. 49, no.3, pp. 893-895, 1984.
[37] F. S. Davis, J. R. Wayland, and M. G. Merkel, "Ultrahigh-frequency electromagnetic fields for weed control: phytotoxicity and selectivity," *Science*, vol. 173, pp. 535-537, 1971.
[38] S. O. Nelson, "A review and assessment of microwave energy for soil treatment to control pests," *Transactions of the ASAE*, vol. 39, no.1, pp. 281-289, 1996.
[39] G. E. Fanslow and R. A. Saul, "Drying field corn with microwave power and unheated air," *Journal of Microwave Power*, vol. 6, no.3, pp. 229-235, 1971.
[40] U. S. Shivhare, G. S. V. Raghavan, R. Bosisio, and M. Giroux, "Microwave drying of soybean at 2.45 GHz," *Journal of Microwave Power and Electromagnetic Energy*, vol. 28, no.1, pp. 11-17, 1993.

[41] G. S. V. Raghavan, P. Alvo, and U. S. Shivhare, "Microwave drying of cereal grain: advantages and limitations," *Postharvest News and Information*, vol. 4, no.3, pp. 79N-83N, 1993.

[42] H. Feng and T. Juming, "Microwave and spouted bed drying of diced apples: Effect of drying conditions on drying kinetics and product temperature," *Proceedings of the 33rd Microwave Symposium*, pp. 99-102, 1998.

[43] I. A. Ali, "Microwave drying of fresh dates with controlled power multi-slot waveguide radiators," *Proceedings of the 35th Annual Microwave Symposium*, pp. 85-88, 2000.

[44] L. R. Beuchat, "Pathogenic microorganisms associated with fresh produce," *Journal of Food Protection*, vol. 59, pp. 204-216, 1996.

[45] L. E. Stetson and S. O. Nelson, "Effectiveness of hot-air, 39-MHz dielectric, and 2450-MHz microwave heating for hard-seed reduction in alfalfa," *Transactions of the ASAE*, vol. 15, no.3, pp. 530-535, 1972.

[46] C.-Y. Lu, "Application of dielectric heating and ultrasound in combination with heat and chemical solutions to kill human pathogenic bacteria on alfalfa seeds. M. S. Thesis," in *Department of Food Science and Technology*: The University of Georgia, Athens, Georgia., 2001.

[47] S. O. Nelson, C.-Y. Lu, L. R. Beuchat, and M. A. Harrison, "Radio-frequency dielectric heating of alfalfa seed for reduction of human pathogens," *ASAE Paper No. 026001.* American Society of Agricultural Engineers, St. Joseph, MI, 2002.

MICROWAVE TECHNOLOGY FOR HEATING LARGE DIAMETER LOGS FOR SAPSTAIN TREATMENT AND STRESS RELIEF

Peter C S Carter
Fibre-Gen, Carter Holt Harvey
Manukau City,
New Zealand

Peter Hutchings
Industries Systems
333 Dandenong-Frankston
Dandenong South
Victoria, Australia, 3175

Nguyen Tran
Industrial Research Institute of Swinburne
Swinburne University of technology
533 Burwood Rd,
Hawthorn, Victoria, Australia, 3122

ABSTRACT

The paper discusses an important microwave solution for two urgent problems encountered in the timber industry. Microwave technology is a strong contender for both cases because of its selective and immediate heating. Experiments and extensive computer simulation demonstrate that it is feasible to heat a 100mm annulus of a big log uniformly. The remaining challenges to be overcome are (1) to achieve a throughput rate of 20m/min to be commercially viable and (2) to design a transport system, which can handle natural bends in logs.

Dielectric and physical properties of logs were measured. Laboratory microwave treatment and temperature profile results for both cases are presented and discussed.

INTRODUCTION

After harvesting, radiata pine logs are colonised almost immediately by several species of fungus [1]. This is a natural precursor to rotting, a process for decomposing fats, sugars and cellulose in wood. One serious undesirable aftermath of one particular type of fungus is the production of dark stains left by hyphae emanating from each germinated spore. This sapstain is not desirable commercially because it affects the timber quality. Fungus colonisation must be prevented within the first three days [2] after felling otherwise the commercial value of the timber decreases dramatically with every day of storage or transport

[1]. Short term preventive measures include ponding and water sprinkling and medium term preventive measures include chemical spraying and chemical immersion. For chemical immersion in dip tanks to be effective, baths must be regularly agitated and kept free of sawdust [2]. Chemical spraying spreads residues, which are undesirable and may become a serious public health issue [3]. Studies on the spread of sapstain show that the staining hyphae may propagate up to 10 mm per day. An immediately effective treatment is required and microwave technology is a potential solution.

Another unrelated problem that requires a similar microwave treatment is to rid off growth stresses in plantation logs [4,5], especially Eucalypt logs. After felling, the sap flow stops and drying begins; growth stresses and drying stresses cause cut logs to split. Serious splits develop to the extent that logs become unsuitable for high grade quality timber. Heat can relieve the stresses [4,5]. Hence microwaving is another potential solution.

Microwaves available for industrial applications in Australia and New Zealand centre on two frequencies: 922 MHz and 2450 MHz. The corresponding wavelengths are 325mm and 122mm, respectively. The ability of a material to absorb microwaves is manifested through its dielectric properties, which consist of the dielectric constant and the loss factor. The depth of penetration [6] of microwaves into the timber material is of interest in these applications and is determined mainly by its dielectric properties. In general, a large loss factor gives a shallow penetration depth. For lack of the proper dielectric properties, one can make an initial estimate of the penetration depth from those of the distilled water [6]. Its dielectric properties at 25^0C are 75 for the dielectric constant and 9.5 for the loss factor at the frequency of 2450 MHz. The resulting penetration depth for water is only 17.8mm. Given that a freshly cut log may have up to 80% water on the wet basis, this means that its dielectric properties are quite close to water and therefore microwaves at 2450 MHz are not an appropriate frequency to use. The dielectric properties [6] of water at 922 MHz are 77 for the dielectric constant and 3 for the loss factor. The resulting penetration depth of 151.4mm offers a more suitable option for heating a freshly cut log to a desirable depth of 100mm. Heat transfer accounts for the heat flow toward the pith and this will help improving the heating penetration with time. Keeping the temperature of the ambient around the log high by hot air will also help with the transfer of heat inwards.

The paper reports the results using microwave technology for the treatment of sapstain fungus in logs of radiata pine and growth stresses in logs of plantation Eucalypts. The concept of heating a large cylinder of dielectric material such as a log, which could be up to 800 mm in diameter, has been published in the Proceedings of the Ceramic Society by Tran [7].

STUDY OF THE SAPTAIN AND GROWTH STRESSES

Saptain [3,8] is caused by a number of species of fungus called *Ophiostoma* spp. *Ceratocystis* spp. and *Leptographium* spp., of which *O. piceae* is the most common. However the *Ophiostomatales* consisting of *O. minus, C. coerulescens* and *Leptographium* spp. are associated with deep stain in logs. A fungus is a primitive organism that produces 'seeds' called spores. Rodents, beetles and mites carry these spores to trees or they may be spread through splashing water [8]. Soon after a tree is felled, a germinating spore will be able to grow and extend its thread-like hyphae to seek out fats and sugars in special ray cells in the sapwood. Fortunately sapstain fungi do not cause serious structural damage but they do leave a stain, which reduces the timber quality. This has become a costly problem for the timber industry. In New Zealand, the loss of revenue is 100m$ [2]. The cost of treatment is estimated annually to be as high as 50 M$US world wide [1].

Radiata pine logs of different diameters must be heated to a sufficient depth to destroy all forms of sapstain fungi. Chemical spraying is the one method available at the moment to prevent fungus colonisation, but it is a surface treatment. Hence chemical treatment is most effective if it can be applied immediately after felling. Chemical fungicide faces registration restrictions and presents public hazards [3]. Logistics is another problem. It takes several days at least to move a large quantity of logs to a debarking plant. Debarking then takes several more days. At the same time, fungal growth continues at a significant rate of 10mm per day. By the time these logs reach a treating facility, the penetration of sapstain fungus to a depth of several centimetres has taken place. On the other hand, conventional heating is effective but too slow to handle 300,000 - 500,000tonnes/annum produced by Carter Holt and Harvey. Our laboratory experiments show that the target temperature for totally effective treatment is 60^0C to 70^0C.

Growth stresses in plantation timber [4,5] have attracted attention recently because of their adverse effect on timber quality. Internal stresses develop in all trees as they grow especially in an organised environment. In a plantation, due to organised planting, there is considerable inter-tree competition. Trees tend to have small crowns and grow tall and slender without many side branches [9]. They tend to develop a high stress level. When a tree is felled, splitting or radial cracking may start to develop almost immediately. Splitting can be delayed by water spraying or water immersion. But this is not satisfactory because it is not a cure.
Plantation Eucalypts of different diameters are smoked or steamed or heated [4,5] soon after debarking to relieve growth stresses. A more practical solution is required to improve the timber quality of logs. Earlier experiments indicate a stress relief is achieved in the temperature range from 80^0C to 100^0C and could take up to several days [4,5]. Recent investigation shows that pre-treatment with medium humidity at 50^0C begins to provide some stress relief [9].

Microwaves can penetrate some distance into a log. This augers well with the applications above because microwaves are not required to penetrate the entire log. The required depth depends on the extent of the fungus colonisation or the severity of the growth stresses. Microwave technology is the technology appropriate for both on the basis of volumetric heating and penetration depth. The final solutions and the final decision hinge on the amount of energy required to deliver a practical throughput at an economical cost.

MICROWAVE AND PHYSICAL CHARACTERISATION

Microwave Characterisation

The dielectric properties of samples of radiata pine were measured as soon as logs were delivered using an Automatic Network Analyser* fitted with an open-ended coaxial probe. The same procedure was applied to samples of plantation Eucalypts (*Eucalyptus globulus*).The method is described in detail by Tran [10]. It relies on the minute fringing field at the end of a coaxial cable to measure the dielectric properties. To obtain good results the surface of the material must be flat because the fringing field only extends to approximately one hundred microns from the open end. If the sample is not flat, air gaps will affect the accuracy. The density of each sample was measured by comparing its weight with that of the same volume of distilled water. The specific heat was measured with a DSC** (Differential Scanning Calorimeter) while the moisture content was measured by a TGA*** (Thermo-Gravimetric Analyser). The dielectric properties at 922 MHz and the physical properties are summarised in Table 1. All results were obtained under ambient conditions of 20^0 C and standard atmosphere.

From the dielectric properties, the depth of penetration for the sample of radiata pine is calculated to be 106mm and that for *E. globulus* is 113mm. These depths of penetration are within the expectation for achieving good treatment.

Table 1. Microwave and physical properties of radiata pine and E. globulus

	Dielectric constant	Loss factor	Density (g/cc)	Specific heat ($J/^0 C/g$)	Moisture content (%)
Radiata pine	38	3.0	0.97	1.6	65
E. globulus	30	2.5	1.1	1.5	50

*Agilent Technology 8722ES with 85070C dielectric probe kit
**DSC 2920 Modulated DSC by TA *Instruments*
***SDT 2960 simultaneous DSC-TGA by TA *Instruments*

MICROWAVE TREATMENT

Microwave equipment

The microwave equipment used consists of a microwave generator at 922 MHz, an isolator, a four stub tuner, forward and reflected power meters and a microwave chamber equipped with a mode stirrer and a carousel. The generator delivers 25 kW into a match load. The magnetron is fully protected by an isolator, which is capable of absorbing 30 kW of reflected power. The four stub tuner is capable of matching a reflection of 90% to less than 5%. The mode stirrer is used to improve the heating uniformity and it moves at 10 revolutions per minutes. The carousel is capable of carrying an 800mm diameter log and rotating at 2 revolutions per minute.

Experiment preparation

Radiata pine logs up to 800mm diameter were seeded with *O. floccosum*, a benign type of *Ophiostoma* fungus, which has the same growth rate as its stain bearing counterparts. After five days, hyphae were expected to reach a considerable depth and the logs were ready for microwave treatment. Each log was exposed to 20 kW of microwave for duration of 15 to 25 minutes. Thermocouples were inserted into 1mm diameter holes drilled into the surface to record the radial temperature distributions and their cooling curves. After treatment, sample wedges were taken to be sent to the University of Waikato [11] for culturing to determine the depth of colonisation of *O. floccosum* in control logs and the effectiveness of the microwave treatment in treated logs.

Samples of logs of *E. globulus* were exposed to microwave in the same fashion as radiata pine logs. After treatment, they were left in the laboratory for observation. Thermocouples were inserted into the treated logs to record the temperature distributions and their cooling curves.

RESULTS AND DISCUSSION

The culturing results [11] show that all fungi were destroyed in all microwave treated samples and the recorded temperatures varied between 60 and 70^0 C. The colonisation of *O. floccosum* showed an average penetration to a depth of between 60-70 and 70-80 mm after five days of incubation and two days in transit and storage. No live fungus was found at any sampled depth. Therefore it was concluded that microwave treatment was effective to at least a depth of 80 mm as determined by the depth of fungus penetration in control logs.

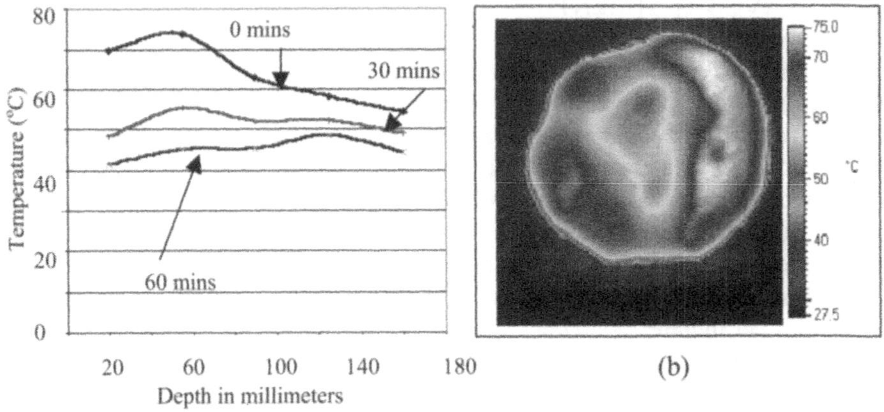

(a) (b)

Figure 1. (a) Radial temperature distributions and their variation with time.
(b) Photo shows thermographic temperature distribution

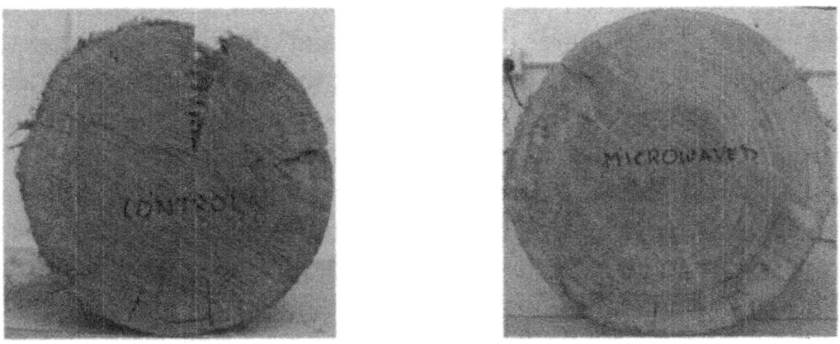

(a) (b)

Figure 2. (a) Growth stresses are evident in the control sample by the split
(b) Growth stresses are relieved in the microwave treated sample

The first test for the effectiveness of the microwave treatment was qualitative, based on visual inspection from making observations of radial cracking. The control samples showed a big crack which began a day after the timber arrived from a small surface slit. It gradually opened up to the final size as shown in Figure 2(a).

There were no serious splits in all the treated samples after several months as shown in Figure 2(b). When the cracking in the control sample was stabilised, this was determined by measuring the width of the split daily, two log samples of microwave treated Eucalypt, which never developed any serious cracks, were cut into boards. According to Mills [12] and Rosen [13] approximately 20mm thick boards should not bend or twist if all stresses had been relieved. This was the case for our boards. Another test used was the prong test. This was done by cutting prongs as shown in Figure 3 in each 20mm thick board, which should have a large slenderness ratio i.e. the ratio of length over width. Each prong was supposed to bend under the influence of stresses. The prongs should also be made along the grain and across the grain. Soon after they were cut, the prongs should remain straight for ever if all stresses were relieved.

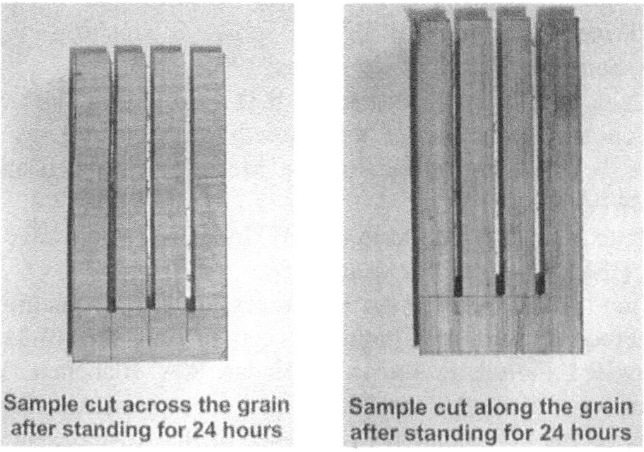

Figure 3. Prongs cut in a microwave treated board remain straight indicating that growth stresses have been relieved

The prongs were prepared as prescribed and all prongs remain straight soon after cutting. Observations were made after 24 hours and all prongs still remained straight. This proves that treated samples of *E. globulus* are relieved of the growth stresses.

CONCLUSION

Heating timber logs with microwave at 922MHz kills all seeded sapstain fungus in radiata pine logs and relieves growth stresses in *E. globulus*. The next stage is to develop a microwave machine for each application. The challenge to overcome is to create uniform heating and to transport logs, which are not straight naturally.

ACKNOWLEDGEMENT

The authors wish to thank Mr Michael Malcmann for preparing the experiments and taking all the data and Mr John Gillies for arranging the manufacture of Eucalypt boards and samples with prongs as required.

REFERENCES

[1]N.J. Strong, J.F. Webb, and R.A. Eaton "Sapstain Fungi in Pine and the Effect on Timber Ageing and Death" 7th International Congress of Plant Pathology, 9-16th August 1998, Edinburgh, paper 3.7.2S.

[2]New Zealand Radiata Pine Web Site "What are the common methods of protection" www.ronayne.co.nz/pine/faq7.html and /faq6.html.

[3]C. Breuil "Sapstain in Canadian Wood: the Causal Fungi, their Growth, Mechanism of Pigmentation, and Future Control Methods", www.wood.ubc.ca.

[4]T. Ohuyama "Assessment of Growth Stresses and Peripheral Strain in Standing Trees" Conferencia IURO sobre Silvicultura e Melhoramento deucaliptos, Salvador, Brasil, 24-29 August 1997.

[5]E.R.P. Jara, M.J. de A.C. Miranda, e R.D. Humphreys "Influence of Thermal Treatment on the Reduction of the Growth Stresses in Logs of *E. grandis*" Conferencia IURO sobre Silvicultura e Melhoramento deucaliptos, Salvador, Brasil, 24-29 August 1997.

[6]A.C. Metaxas, and R. Meridith (1983) "Industrial Microwave Heating", Peter Peregrinus, (IEE), Stevenage, p.60 and p.80.

[7]V.N. Tran "Applicator Design for Processing Large Quantities of Dielectric Materials" Proceedings of the Ceramic Society in 1994, pp 340-348.

[8]S.J. Kay, R.L.Farrell, E. Hadar, Y. Hadar, R.A. Blanchett, T.C. Harrington, "Sapstain in New Zealand – the Causes and a Potential Anti-Sapstain Solution" Proceedings of the 11th Biennial Conference of the Australasian Plant Pathology Society, p.21, 1997.

[9]G. Becker, and T. Beimgraben "Stresses in Beech", FAIR project CT-98 3606.

[10]V.N. Tran "Sandalwood Microwave Characterisation and Oil Extraction", Proceedings of Microwave and HF Heating, Bayreuth, Germany, September 2001, pp.35-37.

[11]J. Thaites, and R. Farrell "Analysis of Fungal growth for Microwave Experiment" Confidential report prepared for Carter Holt and Harvey, University of Waikato, New Zealand, 10 pages.

[12]R. Mills "Australian Timber Seasoning Manual" Australian Furniture Research and Development Institute Ltd, p.53, 1991.

[13]H.N. Rosen, "Stress Development During Wood Drying: an Overview", Proceedings of stress development and degrade during wood drying, pp.1-9, Sweden, September 1987.

INNOVATIVE MICROWAVE TECHNOLOGY FOR THE TIMBER INDUSTRY

G. Torgovnikov and P. Vinden
CRC Wood Innovations, University of Melbourne
Creswick, Victoria 3363, Australia

ABSTRACT

A number of wood species, particularly hardwoods, have very low permeability. This can cause problems during processing. The timber industry requires a technology which can provide an increase in wood permeability for vapours and liquids.

Intensive MW power applied to the wood generates steam pressure within the wood cells. Under high internal pressure the weak ray cells are ruptured to form pathways for easy transportation of liquids and vapours in the radial direction. Narrow voids are also formed in the radial-longitudinal planes.

It is possible to use microwave modification of wood to enhance existing processes and develop entirely new processing and product options for wood. It establishes opportunities for developing a number of new commercial applications.

INTRODUCTION

A number of wood species, particularly hardwoods, have very low permeability causes problems during timber processing. These include long drying times, timber degrade after drying, expensive drying processes, and difficulties in impregnating timber with preservatives and resins.

The main elements of wood structure include tracheids, libriform fibres, vessels and ray cells. Ray cells occupy 5 to 35% of the total wood volume, have thin walls and run in a radial direction from pith to cambium of the tree stem. Ray cells are the weakest cells in wood, and under high internal pressure can be ruptured preferentially, forming pathways for easy radial transportation of liquids and vapours. The net result is a substantial improvement in wood permeability.

The green wood moisture content varies between species and between sapwood and heartwood. It can range between 40 to 250%. Green wood readily absorbs microwave (MW) energy. This results in a high release of energy from within the material. Intensive MW power applied to the wood generates steam pressure within the wood cells. Under high internal pressure the weak ray cells are ruptured to form pathways for easy transportation of liquids and vapours in the radial direction.

An increase in the intensity of MW energy applied to the wood increases internal pressures, resulting in the formation of narrow voids in the radial-longitudinal planes. The number of cavities, their dimensions and distribution are controlled by the intensity of MW energy supplied. A several thousand-fold increase in wood permeability in the radial and longitudinal directions can be achieved in species previously found to be impermeable to liquids and gases (1, 2, 3). Other physical properties and technological attributes are also improved.

It is possible to use microwave modification to enhance existing processes and to develop entirely new processing and product options for wood. It establishes opportunities for developing a number of new commercial applications:

- rapid preservative treatment of heartwood of softwoods,
- the treatment of refractory wood species with preservatives,
- rapid drying of hardwoods,
- growth and drying stress relief in timber,
- manufacturing of MW modified wood "Torgvin",
- production of new wood-based material, for example "Vintorg".

MW TIMBER PROCESSING FOR PRESERVATIV TREATMENT

Biological agents are the major causes of wood deterioration. The greatest financial losses from bio-deterioration result from decay fungi. Different chemicals are used to prevent timber biological deterioration. A very low permeability of heartwood of softwoods and hardwoods of many species for gases and liquids does not permit the impregnation of wood with preservatives.

Heartwood of radiata pine (*Pinus radiata*) and Douglas Fir (*Pseudotsuga menziesii*) can not be impregnated with water-based preservatives. Experiments with Douglas fir heartwood timber with cross sections of 45x90, 70x70 and 90x90 mm showed that it is possible to substantially improve the permeability of this refractory species by intensive MW conditioning. After MW modification, timber can be impregnated with water or oil based preservatives or resins. These results provide new opportunities for increasing the life of Douglas fir heartwood using preservative treatment systems. The heartwood of Douglas fir increases in permeability such that impregnation with CCA (Copper-Chrome-Arsenic) solution results in high preservative uptakes 375 - 426 l/m3 compared to 60 - 90 l/m3 for controls.

Experiments with radiata pine heartwood timber showed increases in permeability such that impregnation with CCA solution results in high preservative uptakes 361 - 516 l/m3 compared to 120 -140 l/m3 for controls. The MW energy consumption needed to make Douglas fir and radiata pine heartwood permeable ranging from 110-120 kW-h/m3.

Experiments with green debarked Yellow Stringybark (*Eucalyptus muellerana*) posts with diameters ranging from 60 -100 mm showed that it is possible to make the refractory heartwood of this species very permeable by applying of intensive MW treatment. After MW modification timber can be impregnated with preservatives or resins to increase the service life of posts. Very high CCA uptake ranging from 340 - 400 l/m3, particularly in heartwood was indicative of much improved permeability. MW treatment of Messmate timber (*Eucalyptus oblique*) provides CCA uptakes of 160-225 l/m3 compared 19 l/min for control.

MW conditioning of impermeable species and subsequent impregnation with preservatives provides increasing timber durability and widening the treated timber usage.

RAPID DRYING OF HARDWOODS

Many commercial hardwood drying operations impose an extended period of slow air drying (up to 12 months) to reduce the occurrence of drying defects (checking and collapse) prior to kiln drying. MW conditioning of eucalypts provides an opportunity for fast kiln drying immediately after sawing, potentially without the need for steam conditioning. MW pre-drying treatment of hardwoods provides increase wood permeability . Depending on the MW intensity applied different degrees of wood permeability achieved.

The reduction in drying time for hardwoods provides a reduction in associated capital, space, energy and labour costs whilst a reduction in drying defects increases the yield of high quality timber by approximately 5%. Microwave pre-treatment of Victorian grown Messmate timber reduces drying time by a factor of 5-10 times that of unmodified wood. Intensive microwave conditioning of the green wood overcomes the propensity of the wood to collapse during kiln drying.

Commercial drying trials with MW modified Messmate sawn timber were undertaken to estimate the economic advantages of MW conditioning. Messmate timber measuring 28 x 90, 45 x 90 and 90 x 90 mm were processed using a 60 kW microwave installation (frequency 0.922 GHz). Moisture losses after MW conditioning ranged from 20 – 35% MC. The subsequent moisture content of timber before kiln drying was 60 – 70%. The results of kiln drying at a commercial installation (Black Forest Sawmills, Victoria) using conventional hardwood drying schedules showed a 5-10 times reduction of kiln drying time.

MW energy consumption for timber conditioning prior to drying ranges from 80 – 160 kW-h/m^3. The MW treatment costs for fast timber drying vary depending on plant output (3,000 to 15,000 m3/y) and electricity costs in the range: at 3 shifts per day from AU$18.6 to 37.2/m3 and at 2 shifts per day from AU$23.4 to 44.3/m3. In many instances the costs of MW timber treatment are acceptable to industry and provide an attractive option for the fast drying of refractory hardwoods.

GROWTH AND DRYING STRESS RELIEF IN TIMBER

Large volumes of timber are projected to become available from plantation timbers. Plantation grown blue gum (*Eucalyptus globulus*) and other eucalypt species are primarily destined to be chipped for low value exports. Extreme growth stresses inherent in young eucalypts limit the processing of these resources into higher value sawn timber. This problem costs the Australian timber industry more than 1.4 billion dollars each year. Therefore it is very important for the timber industry to find opportunities for reducing the negative effect of growth stresses in timber.

Preliminary experiments have shown that microwave processing of timber reduces internal stresses in wood by modification of the wood structure. Yellow Stringybark (80 –100 mm) posts were processed in a 60 kW microwave installation (frequency 0.922 GHz). Microwave irradiation and the formation of micro checks resulted in stress relief and the maintenance of post integrity. Control samples (receiving no microwave irradiation) eventually cracked and resulted in the total splitting and destruction of posts due to internal development of stress during drying.

The stress release induced in small diameter posts can be replicated in larger diameter logs. Economic assessment shows that the cost of MW modification of Yellow Stringybark posts for impregnation can be AU$ 29.9 - 43.8 per m3 (or AU$ 0.56 - 0.68 per post) if electricity costs are AU$ 0.077 per kW-h, then the share of electricity cost is about 50% of total cost of MW treatment. Further work is on going to determine microwave treatment parameters needed to achieve stress relief in large diameter logs and to provide consistently high standards of the processing.

MW MODIFIED WOOD "TORGVIN"

The application of very intensive microwaves can give rise to a novel wood product Torgvin, which has a multitude of cavities oriented in the radial - longitudinal planes. Torgvin has very high permeability, increased flexibility, altered shrinkage, heat conductivity, acoustic and mechanical properties and lower densities compared to unmodified wood (2).

Density and volume

The oven-dry wood density reduction and volume increase following Torgvin production varies between species in the range from 0 to 15.5 %.

Permeability

After MW wood treatment the permeability of wood increases in the radial direction with a smaller increase in the tangential direction. Impermeable Messmate (*E. oblique*) heartwood becomes permeable after MW conditioning and increases the air permeability at low pressures in the radial direction up to 1200 times. Higher degrees of MW modification provide a several thousand fold increase in wood permeability in the radial and longitudinal directions.

The uptake of CCA (copper-chrome-arsenic) solution after pressure impregnation was used as an index of increasing wood permeability to liquids. After MW conditioning and impregnation the liquid uptake increased 5 – 5.7 times for Douglas fir heartwood, 3 – 4.3 times for radiata pine heartwood, 11–14 times for Messmate and 7.4 – 8.7 times for Yellow Stringibark posts compared to their unmodified controls..

Mechanical properties of Torgvin

The MW modification of wood properties results in a loss of mechanical strength of wood which increases depending on the intensity of MW irradiation, the speed of conveyor, and time of exposure. An increase in MW power leads to a progressive reduction in both MOE (modulus of elasticity) and MOR (modulus of rupture). Investigations into the influence of MW wood modification on mechanical properties shows that intensive MW treatment can reduce MOE in the range 35 -90 % (of initial MOE) and MOR - in the range 20 - 94 % (of initial MOR). The minimum reductions in strength properties after MW wood modification are summarised in table 1.

Table 1. Minimum losses in MOE and MOR to effect changes in wood permeability

	MOE	MOE	MOR	MOR
	Tangential dir.	Radial dir.	Tangential dir.	Radial dir.
Radiata pine	26	4	23	6
Messmate	17	12	15	15

Electric energy consumption for Torgvin production ranges from 120 to 300 kW/m3 depending on the required degree of modification, species, timber density and moisture content.

Intensive MW treatment of different wood species (Radiata pine, Douglas fir English oak, Messmate, Mountain Ash, Jarrah, Manna Gum) shows that the wood can be transformed into a new material - Torgvin with altered properties: improved permeability, reduced density, reduced heat conductivity (better heat insulation), reduced shrinkage and swelling, improved acoustic properties (better sound insulation). These properties open up a number of new fields of application for this wood material. One application for Torgvin is the production of a new material Vintorg.

VINTORG

Vintorg is manufactured from MW modified wood (Torgvin) by impregnating the modified wood with resin and curing in a press (3,4). Vintorg can has strength properties similar to LVL (laminated veneer lumber). The process is particularly suitable for using the low value core wood of species like radiata pine and Douglas fir, but to date we have found the process suitable for any species.

Vintorg material has the following advantages:

- High strength properties (comparable with LVL)
- Natural appearance and structure
- Good dimensional stability
- The use of low grade material (for example heartwood of radiata pine or Douglas fir)
- Simple technology compared with LVL
- Cheap production

A comparison of the strength properties of Vintorg compared to other structural materials is given in table 2.

Table 2. Comparison of structural lumbers (based on softwoods)

	LVL	Stress-graded lumber (T30)	GLU-LAM L-40	Vintorg (estimated)
Allowable stresses (MPa):				
bending	18	11	14	11 -14
shear	1.7	1.0	1.2	1.2 - 1.4
MOE	10,000	7,000	8,400	7,000 - 10,000
Practical Max dimns. (mm):				
length	unlimited	6,000	30,000	6,500

thickness	75	150	150	250
width	1200	225	2000	250

Production process

Vintorg production is much simpler compared to LVL production.

The manufacturing stages of Vintorg production include:

1. Microwave timber modification: the MW installation is used for increasing the heartwood permeability of softwoods and reducing the wood moisture content from 30 - 40% to 10 %
2. Resin impregnation: used for timber impregnation with a resin in a cold bath with a cross conveyor.
3. Pre-pressing: for removing surplus resin after impregnation.
4. Hot pressing: used for resin curing and cross section formation.
5. Sawing and finishing of the material.

The success of this technology arises from microwave modification which renders any wood species totally permeable to resin impregnation. The microwaves have identified weak planes in the wood which are subsequently bonded with resin under pressure to provide stronger wood of the same dimensions as the unmodified wood. On-going research is investigating the improvements in strength that can be achieved by increasing the density of wood by compressing Torgvin to smaller cross sectional volumes than the original unmodified wood.

Resin consumption ranges from 6-8%. However, on going research is investigating the potential for reducing resin consumption and using alternative resins and treatments to modify the visual characteristics of Vintorg. Substantial economies are possible in the manufacture of Vintorg compared to LVL production. The process also lends itself to small-scale production.

CONCLUSIONS

The study of wood modification using MW treatment demonstrates selective ray cell modification, and the formation of a number of narrow cavities in the radial/longitudinal planes. The volume of the cavities and their dimensions can be controlled. Such wood treatment increases the permeability of wood for liquids and gases mainly in the radial and longitudinal directions and modifies the properties of the wood. MW treatment of different wood species (Radiata pine, Douglas fir, English oak, Messmate, Mountain Ash, Jarrah, Manna Gum) transforms the wood into a new material, Torgvin, which has altered properties.

It is possible to use microwave modification of wood to enhance existing processes and to develop entirely new processing and product options for wood.

It establishes new opportunities for developing a number of new industrial applications:

- Rapid preservative treatment of heartwood of softwoods
- The treatment of refractory wood species with preservatives
- Rapid drying of hardwoods
- Growth and drying stress relief in timber,
- New material – "Vintorg" production

A cost analysis of MW conditioning of different species of timber in Australia indicates costs of approximately AU$ 20 – 60/m^3 depending on electricity costs. These costs are reasonable for industry and can provide the commercialisation of MW method for wood modification.

REFERENCES

1. P. Vinden, J. Romero and G.Torgovnikov, "A Method for Increasing the Permeability of Wood", Patent Appl. Intern. Publication No WO 99/64213,1998.

2. P. Vinden and G. Torgovnikov, "The Physical Manipulation of Wood Properties Using Microwave." Proceedings (p.240–247) International Conference of IUFRO, *The Future of Eucalypts for Wood Production*. Tasmania, Australia, March 19–24, 2000.

3. G. Torgovnikov and P. Vinden, "New Wood-Based Materials "Torgvin" and "Vintorg". Proceedings (p.756-754) *5th Pacific Rim Bio-Based Composites Symposium*. Canberra, Australia, December 10-13, 2000.

4. P. Vinden and G. Torgovnikov. "Modified Wood Product and Process for the Preparation Thereof", International Patent Application No. PCT/AU02/00315 (in processing), 2002.

Applications

SUCCESSFUL MICROWAVE PROCESSING APPLICATIONS IN INDUSTRY

Nguyen Tran
Industrial Research Institute of Swinburne
Swinburne University of Technology
Hawthorn, Victoria, Australia

ABSTRACT

Microwave technology assumes a new dimension of importance in the 21st century. This is based on the widespread use of non-metallic materials in every day life. Plastic, ceramic and composite materials are thermal insulators, which present a problem for conventional heating. Microwaves, on the other hand, can penetrate these thermal insulators volumetrically and cause immediate heating. The paper discusses a new paradigm in high power microwave design involving accurate measurements of the dielectric properties and the use of simulation software such as HFSS or QW-3D or Microwave studio and CFD. Dielectric properties measurements and simulations of the actual applicator including the load are proposed as a standard procedure to arrive at designs instead of the traditional simplified mathematics and experimentations.

INTRODUCTION

The attractiveness of the microwave energy for industrial processing is demonstrated by the research activities in many disciplines on the use of the technology. There are four international conferences covering the field. There have been successful microwave processes in use throughout the world. But the technology remains semi-obscure. One reason may be due the perception that microwaves are not safe. The positive effect of this perception is that microwave safety has been the most scrutinised subject to date. The technology is much safer now because there is a strong scientific research interest on the issue. Another reason is that a microwave process is usually developed in house, with the active involvement of a champion [1]; therefore it remains a confidential process, obscure to the outsiders. The third reason is that there is limited high power microwave engineering teaching infrastructure at colleges and universities. The fourth reason is the cost of the microwave hardware. The one off manufacturing machine is expensive compared to large volume production. Microwave technology holds an ideal position for bridging applications within and between science, engineering and industry. A converging mechanism must be found to make the technology becoming more widespread [1].

A new paradigm is proposed in this paper to replace the common experimental approach in microwave process design. It will reduce the development cost and create successful processes required by industry.

The new paradigm proposed consists of (1) accurate measurement of the dielectric properties, (2) adequate simulation using electromagnetic and thermodynamic simulators to minimise the reliance on extensive experimentations, and (3) direct construction of the final working machine.

MEASURMENT OF THE DIELECTRIC PROPERTIES

There are well known people such as Von Hippel [2] from MIT or Stuart Nelson [3] from the US Department of Agriculture, who have devoted their entire working life to dielectric properties measurements.

Many techniques, and their variations, have been developed. Since the introduction of the Automatic Network Analyzer (ANA), there have been four fast methods, compared to the slotted line technique used by von Hippel [2]: (1) open ended coaxial line method; (2) open shielded coaxial line method; (3) transmission/reflection method; (4) and open and short circuit method.

Open ended coaxial line method

The open-ended probe is depicted in Figure 1. Basically it uses the broadband response of a coaxial line and the small fringing field capacitance at the end. The external fringing capacitance C_o is loaded with the dielectric properties of the material to be measured. A calibrated network analyzer records the reflected signal S_{11} to enable the calculation of the dielectric properties.

This method is most suitable for large dielectric constant and medium to high loss factor, and preferably the sample material should be liquid or semi-liquid. A solid sample must be optically flat to obtain accurate results, because the fringing fields extend only a minute distance, of the order of tens to hundreds of microns. Minute air gaps between the sample and the probe may cause significant errors. For low frequency measurements, the centre conductor may be extended to increase the terminal capacitance. Capacitance approximation provides a simple way of representing the open end, and this works well for frequencies up to 20 GHz. The technique can be extended to higher frequencies by modeling the field distribution as reported by Baker-Jarvis et al [4]. In the near future, automatic measurements using field modeling will become available. Tran et al [5] used the open ended probe to characterize vegetables and food from 0.1 to 10 GHz.

Referring to Figure1, the input admittance of the probe is given by:

$$Y/Y_o = j\omega C_f Z_o + j\omega\varepsilon C_o Zo + G_o Zo\varepsilon^{5/2} \qquad (1)$$

Where 2a = inner diameter, 2b = outer diameter, C_f = internal fringing capacitor, C_o = external fringing capacitor, G_o = radiation conductance and $Z_o = 1/Y_o$ = characteristic impedance of the coaxial line (usually 50Ω). A calibrated automatic network analyzer measures the complex reflection coefficient S_{11}. Since $Y/Y_o = (1\text{-}S_{11})/(1+S_{11})$ and from (1), ε is equal to:

$$\varepsilon = (j\omega C_t Z_o)^{-1}.(1\text{-}S_{11})/(1+S_{11}).(1+C_f/C_o) - C_f/C_o \qquad (2)$$

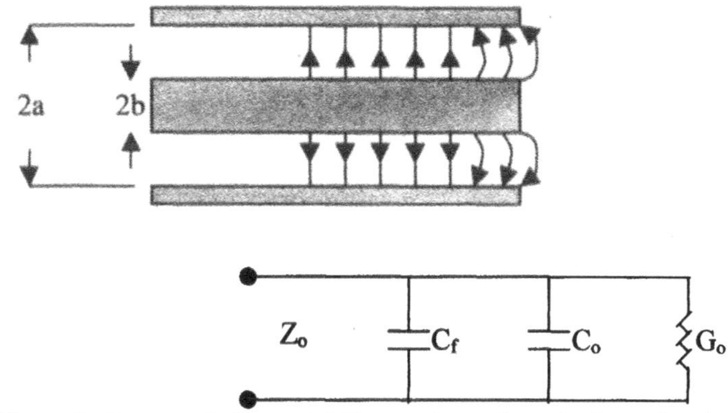

Figure 1. Open ended coaxial line probe and equivalent circuit

Where $C_t = C_f + C_o$, $\varepsilon = \varepsilon_o\, \varepsilon_r$ and $\varepsilon_r = \varepsilon_r'\text{-}j\varepsilon_r''$
ε_r' = dielectric constant and ε_r'' = loss factor

According to Hewlett Packard [6], a 3.6mm coaxial line open ended probe can be used to measure the dielectric constant up to 100, and the loss factor greater than 0.05, when the flatness is at least 25 microns, the sample diameter is greater than 20mm and sample thickness is greater than $20\text{mm}/\sqrt{\varepsilon_r}$. In general, the accuracy of this method is ± 2% for ε_r and > ± 0.01 for ε_r''.

Open shielded coaxial line method:

This method involves extending the outer conductor well past the inner conductor as reported by Bussey [7]. The inner conductor is shielded hence the radiation conductance becomes zero. Basically the probe is equivalent to a capacitor, which can be varied by varying the length of the inner conductor. Such a probe has been used to measure dielectric properties from 0.1 to 18 GHz [7]. The open termination must be correctly represented to achieve the desired accuracy. The open circuit admittance can be represented by at least one of three methods [7] and the method involving an extension of the centre conductor is often preferred due to its simplicity. An extension ΔL is added to the physical length of the inner conductor (Figure 2) proposed by Bussey [7]:

$$\Delta L = (b\text{-}a)(0.6034+0.9464x^2 + 18.19x^{5.127}) \quad (3)$$
$$\text{for } x = b/\lambda < 0.3$$

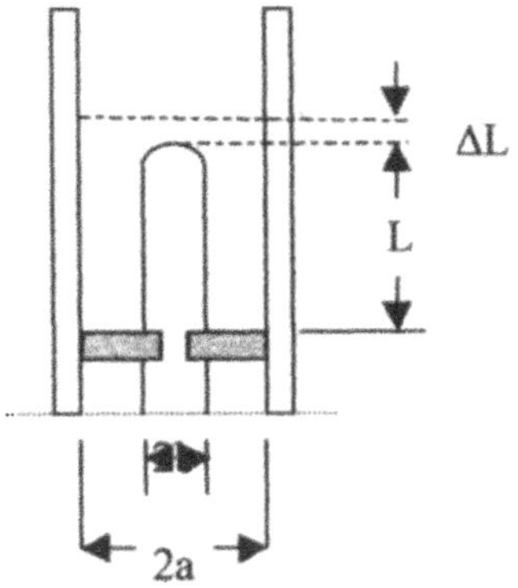

Figure 2. Open shielded coaxial line probe

S_{11} at the interface of the sample is measured to produce the admittance Y_m:

$$Y_m = (1\text{-}S_{11})/(1+S_{11}) = Y_o \tanh(\gamma d) \quad (4)$$

Where Y_m = calibrated admittance of the sample
Y_o = characteristic admittance of the coaxial line
$= 2\pi(\varepsilon/\mu)^{1/2}/\ln(b/a)$
d = corrected length of the inner conductor
$\gamma = (-\omega^2\mu\varepsilon)^{1/2}$, $\varepsilon = \varepsilon_o\, \varepsilon_r$ and $\varepsilon_r = \varepsilon_r'\text{-}j\varepsilon_r''$

Y_o and γ contain ε, and when the sample is non magnetic a simple solution is obtained because: $\mu = \mu o$. Equation (4) is transcendental, which has an infinite number of solutions. To determine a correct result, *a priori* knowledge of the dielectric properties must be available. Otherwise two samples having different lengths must be used [2]. The correct answer is the intersection of the two solution sets or curves. The length of the center conductor must be correctly selected to achieve minimum error.

Tran et al [8] used the open shielded coaxial probe to characterize the dielectric properties of granular materials. Typical accuracy for this method is ± 5% for ε'_r and ± 0.005 for ε_r''.

Transmission and reflection method

The transmission and reflection method uses both ports of the network analyzer. The method was discussed by Weir [9] and was further improved by Baker-Jarvis et al [10] and Boughriet et al [11]. The dielectric properties can be calculated by one of several expressions given by [9-11]. For example:

$$S_{21}S_{12}-S_{11}S_{22} = \exp[(-2\gamma_o)(L_{air}-L)](z^2-\Gamma^2)/(1-z^2\Gamma^2) \quad (5)$$

Where L_{air} is the length of the sample holder, L is the length of the sample and Γ is the reflection coefficient at the air-dielectric material interface assuming infinite material [10] and $z=\exp(-\gamma_o L)$ and (5) is transcendental.

Short circuit and open circuit method

The short circuit and open circuit method uses only one port with a short circuit then an open circuit placed behind the sample to measure the respective admittances, whose product is:

$$y_\varepsilon = Y_{sc}\,Y_{oc} = \sqrt{(\varepsilon_r - (\lambda/\lambda_c)^2)}/\sqrt{(1-(\lambda/\lambda_c)^2)} \quad (6)$$

Where Y_{sc} = admittance of the sample terminated by a short circuit.

Y_{oc} = admittance of the sample terminated by an open circuit.

λ_c is the cut off wavelength and λ is the free space wavelength.

Let $y_\varepsilon = g + jb$, ε'_r and ε''_r can then be calculated from (6) as:

$$\varepsilon'_r = (g+(\lambda_g/\lambda_c)^2)/(1+(\lambda_g/\lambda)^2)$$

$$\varepsilon''_r = -b/(1+(\lambda_g/\lambda)^2)$$

Where λ_g is the guide wavelength.

According to Altschuler [12], it is accurate for low to high loss materials.

When the loss is high and the sample length is sufficiently long, the dielectric properties are obtained from the reflection of an infinite sample [12].

APPLICATOR SIMULATION

The first requirement is to obtain the microwave heating pattern for a certain applicator design. Accurate knowledge of the dielectric properties is essential to achieve accurate simulation. The variation of the dielectric properties with temperature must also be known. The important information required is the heating pattern and the reflection coefficient of the process over the frequency range dictated by the magnetron of the microwave generator. Software such as HFSS, Microwave Studio or QW-3D [13] can be used. Accurate representation of an applicator keeps experimental verification to a minimum. When used properly, simulation removes any doubt imposed by the use of approximate mathematics, which has to be used in the past when digital computation is not yet appropriate.

The second requirement is to couple the electromagnetic solution to thermodynamic software such as CFD or Mathcad to determine the ongoing thermodynamic behavior of the process. Some software such as HFSS or QW-3D has a component of thermodynamics included.

There are about 70 software packages [13] available to analyze electromagnetic problems but not all are suitable for analyzing microwave heating problems. Indications are that with the advances in the PC technology, software simulation will become more sophisticated. Different methods of numerical analysis are used to find the solutions for the Maxwell's equations and the wave equation either in the integral equation form or in the differential form. The former leads to the boundary element method (BEM). The latter leads the finite difference method (FDM) and the finite element method (FEM). The FDM has two offshoots, viz. the finite difference frequency domain (FDFD) and the finite difference time domain (FDTD).

HFSS (High Frequency Structure Simulation) uses FEM and each element is a tetrahedron because it the best shape to model the 2nd order interpolation of the differential equations involved. If computer memory is available, there appears to be no limit to the accuracy of FEM using HFSS. Simulation over a broadband of frequency is possible by using the same mesh structure at different frequencies.

QW-3D on the other hand uses FDTD. It is a full wave conformal FDTD simulator. A structure is subdivided in a changeable regular rectangular grid and Maxwell's equations are approximated both in space and time. With FDTD, the energy propagates through the grid so that traveling waves and standing waves are shown in real time. Excitation is done with a sinusoidal or a pulse waveform.

A simulator provides the field solution at every point in a structure and the reflection coefficient. The ability of a post processor to animate the field solution helps visualize the heating pattern in real time.

The field solution must be coupled with the temperature variation in the dielectric properties and with a thermodynamic package to observe the dynamic behavior of the process. CFD is a software package which can be used for the thermodynamic coupling. Mathcad and Mathlab can also be used for computation.

Example of an applicator design

Several microwave machines have been successfully made for industry using the procedure described above. These applications vary from food, chemical, liquid, rubber, and ceramic to waste recycling.

As an example of a typical design, a network analyzer was used to obtain accurate dielectric properties. Simulation was then followed with an accurate drawing of the applicator and the load. Figure 3 shows that microwaves at 2450

MHz did not penetrate much into the load and the reflection was high. It was clear that matching was required. A tuning stub was added and the improved result is shown in Figure 4. The reflection was reduced. This design process was completed in five simulations; each took about 2 minutes on a Pentium 3 with 256 Mb of RAM. The applicator worked and no experimentation was required.

As a second example, let's consider a new design of a brick kiln having a load diameter of 2m and is heated with a 922MHz microwave source. The dielectric constant=10 and loss tangent=0.2. A tapered horn applicator is selected and Figure 5 shows the simulation results. The reflection is less than 5% and there is no tuning stub required. A deployment of multiple horn applicators around the kiln will provide adequate heating for the whole kiln. The simulation time could have been reduced to one quarter by applying the symmetrical E and H planes to reduce the computation requirement.

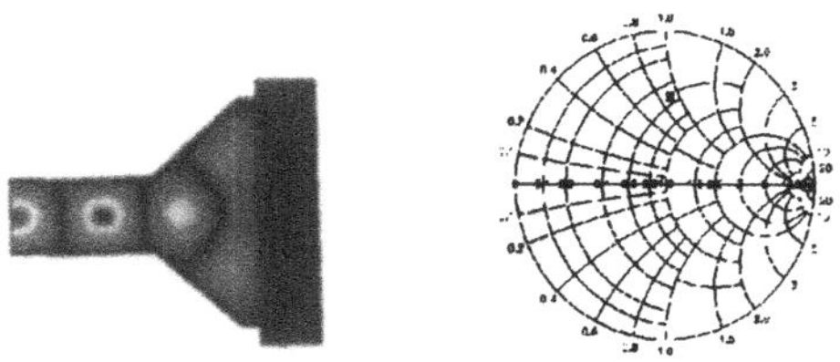

Figure 3. Field pattern and reflection when the horn is not matched

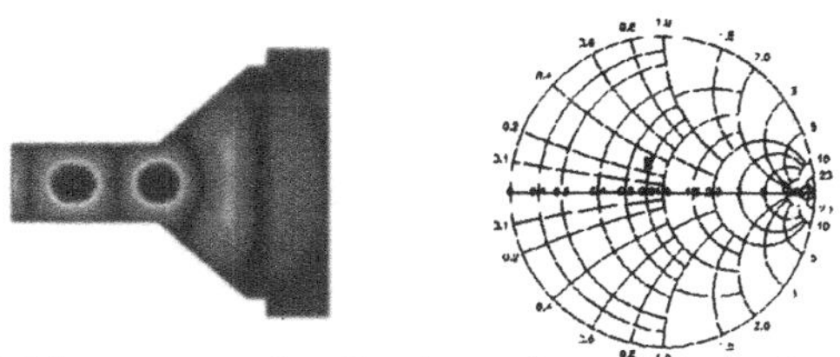

Figure 4. Field pattern and reflection when the horn is matched

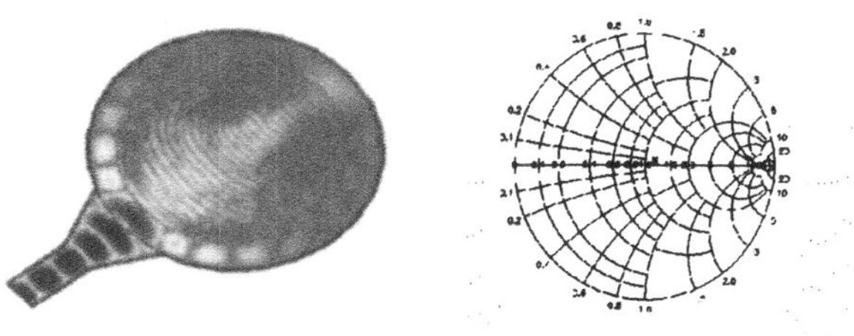

Figure 5. Field pattern and the reflection of a cylindrical brick kiln

CONCLUSION

Microwave heating with this new design approach will be practiced by a wider population and it will become a converging mechanism needed [1].

REFERENCES

[1]W.R. Tinga "Microwave and RF – application, engineering and social issues", Ceramic Transactions Volume 111, Microwaves: Theory and Applications in materials processing V, pp577-585, 2000

[2]A.R. von Hippel, (ed.) "Dielectric materials and applications", Wiley and Sons, New York, 1954, pp.63-122.

[3]S.O. Nelson "Comparison of measured and calculated powdered mixture permittivity", Ceramic Transactions Volume 111, Microwaves: Theory and Applications in materials processing V, pp101-105, 2000

[4]J. Baker-Jarvis, M.D. Janezic, P.D. Domich, and R.G. Geyer "Analysis of an open-ended coaxial probe with lift-off for non destructive testing", IEEE Trans. Instrum. Meas., vol. 43, Oct. 1994, pp. 711-718.

[5]V.N. Tran, S.S. Stuchly, and A. Kraszewski "Dielectric properties of selected vegetables and fruits, .1 - 10 GHz", J. Microwave Power, 19(4),1984 ,pp. 251

[6]Hewlett Packard Co. "Basics of measuring the dielectric properties of materials" HP application note 1217-1, Feb. 1992.

[7]H.E. Bussey, "Dielectric measurements in a shielded open circuit coaxial line" IEEE Trans. Instrum. Meas. vol. 29, 1980, pp120-124.

[8]V.N. Tran, and S.S. Stuchly "Dielectric properties of granular materials using an automatic network analyser", IMPI Symposium, Cincinnati, Ohio, USA, 30 August-2 September 1987.

[9]W.B. Weir "Automatic measurement of complex dielectric constant and permeability at microwave frequencies", Proc. IEEE, vol. 52, Jan. 1974, pp.33-36.

[10]J. Beker-Jarvis, E.J. Vanzura, and W. Kissick "Improved technique for determining complex permittivity with transmission/reflection method", IEEE Trans. MTT, vol.38, August 1990, pp.1096-1103.

[11]A.H. Boughriet, C. Legrand, and A. Chapoton "Non iterative stable transmission/reflection method for low loss material complex permittivity determination", IEEE Trans. MTT, vol. 45, 1997, pp.52-57.

[12]H. Altschuler "Dielectric constant", in handbook of microwave measurements, vol. 2, Sucher and Fox. Eds. Brooklyn, N.Y.: Polytechnic Press, 1963, ch. 9, p. 495.

[13]V.V. Yakovlev, "Commercial EM Codes Suitable for Modelling Microwave Heating- a Comparative Review", Scientific Computing in Electrical Engineering, Lecture notes in Computational Sciences & Engineering, 18, Springer, 2001, pp.87-96.

INNOVATIVE APPLICATION OF DIELECTRIC HEATING TECHNOLOGIES TO CONCURRENTLY REDUCE ENERGY USAGE AND EMISSIONS IN POLYMER MANUFACTURING PROCESSES

J.W. Cresko, Electrotechnology Applications Center, 3835 Green Pond Road, Bethlehem, PA, 18020, USA. E-mail: cresko@etctr.com; Phone: (610) 861-5572.

ABSTRACT

There is a great opportunity in mature chemical process industries (CPI) to modify unit operations to address pollution prevention and energy efficiency issues. Modifications can range from incremental optimization efforts within a unit operation, to extensive revamping of a sequence of unit operations. This paper describes changes to a critical separation process used in the manufacture of polyvinyl chloride, with important applications to volatile organic compound (VOC) and hazardous air pollutant (HAP) reductions. In this case, dielectric heating techniques are the enabling technologies.

A highly efficient separation and recovery process has been developed that utilizes electromagnetic (EM) energy as the driving force to significantly increase the efficiency of stripping unreacted monomer from select polymer product streams. During manufacture, direct steam injection is a standard industrial method of heating emulsions of polyvinyl chloride (PVC) to the stripping temperature. After stripping, the product is dried to a final, powdered form. Replacing direct steam injection with a highly controllable dielectric heating step is advantageous since significant energy savings are realized by reducing the amount of water to be removed in the final drying step. In addition, improved stripping efficiency results due to an increased bulk processing temperature as well as by preferential EM heating of the monomer, yielding lower subsequent volatile emissions.

A pilot–scale system is described that utilizes microwave (MW) and radio frequency (RF) technologies to enhance the thermal and mass transport of unreacted vinyl chloride monomer (VCM) from PVC emulsions. The goal is to integrate this EM induced unit operation into current polymer manufacturing operations in order to achieve a very efficient separation process.

BACKGROUND

Polyvinyl chloride (PVC) is an important polymer that is used in a wide variety of products, including pipes, paints, packaging, bottles, wire insulation, etc. The greatest growth has been seen in construction markets (new and replacement), where PVC siding, windows and flooring are replacing traditional wood products. Commercial production of PVC started in the US in the 1930's and production of PVC in the US alone has grown steadily from about 3 million tons in 1985 to 15 million tons in 2001. Current worldwide production exceeds 25 million tons per year.

The Standard Vinyl Chloride Monomer (VCM) Stripping Process

The companies that manufacture the bulk PVC resin used in these products polymerize vinyl chloride monomer (VCM) in a suspension or emulsion process (in water).[1,2] After polymerization, the PVC is typically dried to a powdered or pelletized form. This raw material is then shipped to various manufacturers to be processed into products. In the 1970's the VCM monomer from which PVC is polymerized was connected to serious health hazards, including hepatic angiosarcoma (a form of liver cancer). Consequently, the removal of unreacted VCM from polymerized PVC is an important step in the manufacture of PVC resin.[2]

The Electrotechnology Applications Center (ETAC) is working with a PVC resin manufacturer to reduce their emissions of VCM. The area of interest is a post-polymerization step where an emulsion of 50% PVC and 50% water is heated, and VCM (CH_2CHCl) stripping is performed in a large falling film unit operation (stripper). The conversion of the VCM reactant to PVC product is limited by the polymerization kinetics to a maximum of approximately 90%. As such, unreacted VCM must be reduced below state and federally allowed levels prior to further processing. The emulsion polymerization process in the United States is regulated to 2,000 ppm on a daily average.[3]

Figure 1 shows a basic process flow diagram (PFD) of the PVC emulsion heating and stripping operation that is typical in the industry. The emulsion is heated first by a shell & tube heat exchanger to ~60 °C, pumped to the top of a 10 meter tall stripper, and heated to ~70 °C by direct steam sparge just prior to injection into the stripper. This process is performed while under vacuum of ~20 inches Hg.

The electromagnetic (EM) enhanced stripping process that is being developed will be deployed on parallel production lines, each running a maximum of ~40 gpm (~300 lb/min.). Although this operation achieves VCM emissions below the current 2,000 ppm limit, it is anticipated that regulatory pressures will demand lower emissions in the future.

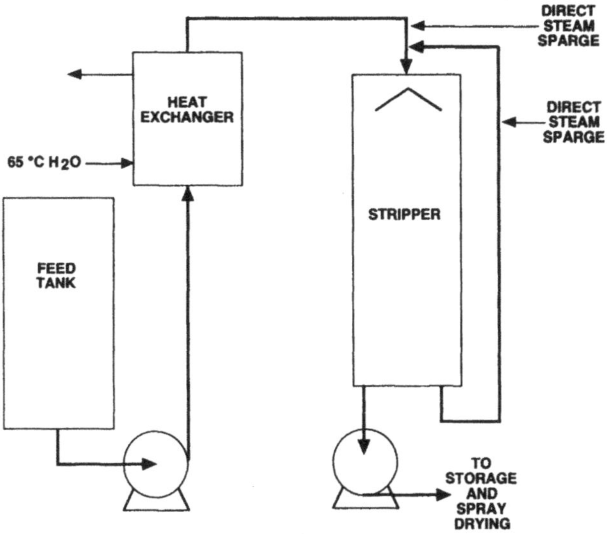

Figure 1 – Basic PFD of the VCM Stripping Process

Reinventing the VCM Stripping Process

The overall objective is to develop a highly efficient EM energy heating process, and the specific goal is to significantly increase the efficiency of stripping unreacted VCM from PVC emulsion while concurrently reducing energy usage. It was theorized that an EM heating step at the current direct steam sparge location would provide significant efficiency improvements in the stripping operation by two mechanisms. First, by allowing a *higher temperature rise of the emulsion*, and second by *selective excitation of the VCM molecule*. Each step enhances the diffusion of VCM out of the PVC particle prior to injection into the stripper.

It is not possible to simply increase the energy input at the heat exchanger or at the steam sparge to increase stripping efficiency (see Figure 1). Emulsions of PVC are generally unstable. Effective increases in temperature rise are limited when heating with traditional techniques; in particular, too much energy input to the emulsion (by thermal or mechanical means) can cause an undesirable semi-solid state (known as "creaming out"). By using EM technology that can volumetrically heat the emulsion prior to stripping, it was theorized that the emulsion could be heated to higher temperatures and significantly reduce VCM levels in subsequent unit operations.

Further, stripping efficiencies are temperature and diffusion rate limited.[4,5,6] Emulsion resins of PVC are comprised of very small (micron sized) PVC particles suspended in water. During polymerization the water acts as a carrier, as well as a heat transfer fluid, to move heat from the particles to the walls of the reactor vessel. After polymerization, residual VCM in the water phase is relatively easily separated from the water (and is largely temperature dependent). Efficiently removing VCM entrained within the PVC particles is more difficult. Further advances in stripping efficiency depend upon driving the residual VCM from the PVC particles to the water phase. Again, raising the temperature by traditional methods will cause the PVC emulsion resin to become unstable. The potential that EM heating could increase stripping efficiencies would be synergistically enhanced if another key component of the theory is realized: that enhanced mass diffusion of VCM from the PVC particles to the water phase results from preferentially coupling the EM energy to the VCM.

Both microwave (MW) and radio frequency (RF) technologies are used in certain industrial heating operations,[7,8,9] but it is not believed that these EM technologies have ever been used to drive this type of industrial separation process. Recent advances in MW and RF technologies,[10,11] as well as regulatory and economic concerns, now make the deployment of such technologies favorable.

Technical feasibility – Pilot-Scale Systems and Initial Performance Data

Based on the premise that EM heating techniques could enable significant advances in separation technology in general, and VCM reductions in particular, preliminary studies were initiated to assess the feasibility of MW and RF technologies to heat the PVC emulsion resin. The tests started by assessing the heating response of batch samples, and progressed to heating tests of flowing material. In order to understand the overall system dynamics and to provide a method of comparing MW versus RF directly, the heating and stripping processes were replicated with a pilot-scale system.

Heating tests with RF energy utilized a 27.12 MHz power, supplied via 50-ohm technology using an Advanced Energy (Voorhees, NJ; 609 627-6100) solid state power supply and matching network. The setup allows the precise and controlled application of energy at a fixed frequency from the power supply, which is connected by a 50-ohm cable to a matching network that ensures a constant forward power into the product by monitoring and actively controlling the reflected power to zero.

All MW heating tests were performed at 2,450 MHz; the static heating rate tests used a multi-mode cavity, and the dynamic arrangement became the test bed for various applicator designs. Important points of the testing program are summarized:

Static Testing: Initial work involved static testing (heating discrete, stationary samples of various volumes) to assess the EM heating response. It was demonstrated that very high temperatures (>87°C), measured by fiber optic temperature probes, could be attained without breaking the emulsion. Static testing indicates that the PVC dispersion resin responds well to both RF and MW electromagnetic heating as measured by *Heating Rate* (°C/m) for a given *Specific Power* (W/g). A *Measure of Efficiency* (MOE) can be calculated by dividing the *Heating Rates* by *Specific Power:*

$$MOE = HR/SP \quad (g\text{-}°C/W\text{-}s) \tag{1}$$

The MOE allows for a direct comparison of the heating efficiency between various EM heating techniques.

The RF power was applied to the product in a capacitive applicator arrangement (SRF-1) as well as a Strayfield applicator arrangement (SRF-2). Samples were heated by MW in a Cober LBM 1.2A laboratory microwave research oven (Cober Electronics Norwalk, CT 203 855-8755). The unit is capable of providing up to 1.2 kW of continuously variable output power. The static tests are summarized in Table I.

Table I – Summary of Static RF & MW Tests

Test Configuration*	***Specific Power (W/g)***	***Heating Rate (°C/sec)***	***% Wt. Loss***	***MOE (g-°C/W-s)***
SRF-1	7.4-7.7	0.5-1.2	1.0-11.0	118×10^{-3}
SRF-2	4.5-17.9	0.6-2.5	0.4-0.7	71×10^{-3}
SMW-1	4.5-11.2	0.5-1.4	0.2-1.4	132×10^{-3}

Test Configuration:
SRF-1 = Static RF-1 → 'Parallel plate' applicator
SRF-2 = Static RF-2 → 'Strayfield' applicator
SMW-1 = Static MW-1 → 'Multi-mode' cavity

Dynamic Testing: Since the ultimate goal is continuous heating of the PVC emulsion in an industrial system, a laboratory-scale reproduction of the process was constructed. By controlling input energy, emulsion flow rate and applicator configuration, it was demonstrated that the product can be heated in a continuous flow setup, with both MW and RF, up to approximately 88°C without deleterious effects to the emulsion. This is significantly greater than the temperature rise achievable by steam sparge.

The setup consisted of a variable speed peristaltic pump, tubing and valves to allow the emulsion to be exposed to either RF or MW energy in a one-pass or recycle configuration. In the setup, the key control variables affecting the temperature rise of the flowing product are:

- Input emulsion temperature
- Flow geometry
- Flow rate
- Applied power

Preheated PVC latex emulsion samples were heated by RF at flow rates in the range of 0.24 to 1.00 l/min through a straight section (DRF-1, DRF-2, DRF-3) or annular section (DRF-4, DRF-5, DRF-6) of low-loss piping. Power levels from 200 to 800 watts were used. No appreciable change was observed when a static mixer was inserted into the section of straight tubing (DRF-2, DRF-3). Indeed, all variations of the RF applicator showed similar heating rates. Figure 2 demonstrates that volumetric flow rate is the key variable, and that within the range of RF applicators that were tested there is comparatively little influence of the applicator to the heating temperature gain.

Analogous MW tests were performed. However, the residence time was varied by coiling the tubing to up to six turns on an approximate diameter of ~14 cm. The PVC emulsion was pumped through the system at volumetric flow rates of 0.24 to 1.00 l/m, with inlet and outlet temperatures measured by fiber optic probe. The dynamic RF and MW tests performed are summarized in Table II.

Stripping Tests: A stripping chamber was added to the laboratory-scale setup to study devolatilization. A basic PFD of the system is shown in Figure 3. The PVC emulsion processed by RF & MW energy and raised to high heat conditions resulted in a four-fold reduction of the VCM concentration based upon the United States Environmental Protection Agency (USEPA) Analytical Test Method 107. Quality control tests for rheology, heat stability and color showed no deleterious effects from the increased stripping temperature. Most notably, a solids increase of 4% was observed. This will reduce the energy consumed by the subsequent spray drying unit operation by 12%.

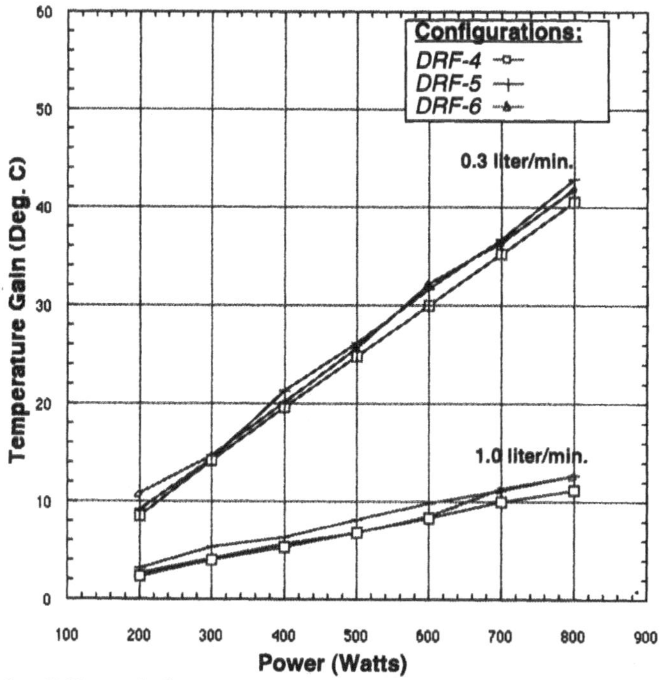

Figure 2 – Effect of Flow Rate and RF Applicator Variation on Heating

Table II – Summary of Dynamic RF & MW Tests

*Test Configuration**	*Specific Power (W/g))*	*Temperature Rise (°C)*	*Flow Rate (l/min)*
DRF-1	200-800	2-51	0.24-1.00
DRF-2	200-800	4-50	0.24-1.00
DRF-3	200-800	3-50	0.24-1.00
DRF-4	200-800	2-49	0.24-1.00
DRF-5	200-800	3-49	0.24-1.00
DRF-6	200-800	3-52	0.24-1.00
DMW-1	200-800	8-34	0.24-1.00

Test Configuration:

DRF-1 = Dynamic RF-1 → 1" OD, 7/16" ID, no static mixer
DRF-2 = Dynamic RF-2 → 3/4" OD, 1/2" ID, static mixer
DRF-3 = Dynamic RF-3 → 1" OD, 7/16" ID, static mixer
DRF-4 = Dynamic RF-4 → 3/16" annular gap
DRF-5 = Dynamic RF-5 → 1/4" annular gap
DRF-6 = Dynamic RF-6 → 1/8" annular gap
DMW-1 = Dynamic MW-1 → tubing coiled in multimode MW cavity

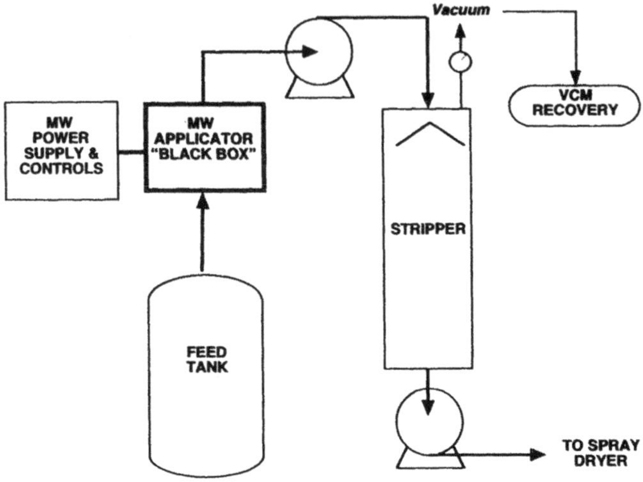

Figure 3 – Basic PFD of the Experimental VCM Stripping Process

MW versus RF - Both MW and RF heating technologies show viability in improving VCM stripping efficiency. The technology and economics of equivalent RF and MW heating applicators and systems at a pilot-scale level (at ~5 kW power level) were investigated to determine the preferred technology. The MW feasibility testing initially relied on a multi-mode applicator, which can essentially be described as an "*unfocused*" design. In order to perform a quantitative head-to-head analysis of RF versus MW at the 5 kW level, a focused RF applicator (based on DRF-1) and system as previously described was compared to an analogous 5 kW MW system with a single-mode (*"focused"*) applicator design.

The MW applicator consisted of a single pass or multi-pass (to vary the residence time) through a terminated waveguide, with tuning achieved by a four-stub tuner. The single pass setup was prone to spot overheating resulting in plugging and flow cessation. Multiple passes reduced this by eliminating the tendency to overheat at one spot, and by increasing the residence time which allowed a lower input power to achieve the same temperature rise.

Tests were performed starting at a baseline of ~60 °C and heated to a target temperature of ~87 °C = delta T of ~27 °C. The MW versus RF analysis considered a number of factors, including:

- Heating Efficiency
- VCM Stripping Efficiency

- In-plant Integration Advantages and Disadvantages
- Design/Process Robustness

Both technologies show promise for industrial deployment. Based upon technical and process integration concerns and upon analysis of the initial performance data, the decision was made to deploy MW as a pilot-scale operation at the manufacturing facility. However, the design and integration of an optimized MW applicator is critical to successful deployment.

Process Development – Refining the MW Heating and Devolatilization Process

Subsequent to gathering initial performance data, the small-scale pilot system used to heat and strip the PVC was used as a foundation for testing advanced design and process modification concepts. These include testing and refining the design of the applicator, the fluid flow of the emulsion through the applicator and the feedback control of the applied power based upon the individual unit operation as well as overall process conditions. An innovative MW applicator design has resulted, in which the emulsion makes two-passes in the gap between two slotted waveguides. The applicator allows for efficient power transfer and sufficient residence time to achieve a controllable and stable process.

A computer numerical model of the applicator was developed, and since the mechanisms of dielectric heating are quantified by the dielectric properties of the materials, these measurements were made (and utilized by the model). Important to the design process are the dielectric properties of the water/PVC emulsion and the temperature dependence of the emulsion to assess temperature stability criteria. These measurements were be made by vector network analysis techniques in the MW frequency range based upon standard Nicholson-Ross techniques.[12,13] The model has resulted in refinements to the optimal design.

Next Steps – Scale-up: A large-scale pilot unit is currently being designed that will deploy meritorious design ideas at elevated volumetric flow rates and applied power levels in the industrial setting. In comparison with the data gathered from the small-scale pilot unit, scaling relationships will be developed that will allow more accurate prediction of the full-scale requirements. Process data gathering activities will mirror those on the small-scale pilot unit, but will also include data gathered over long time intervals. Data gathered during the heating of different formulations will form a basis for process control algorithms. Inspection of the equipment between the long process runs will define maintenance requirements.

CONCLUSIONS

This work fundamentally addresses the growing concern for pollution prevention and energy efficiency. We set out to investigate creative and cutting-edge technologies to re-engineer critical unit operations to reduce volatile organic compound/hazardous air pollutant (VOC/HAP) emissions while concurrently emphasizing more efficient utilization of energy and materials. Our research indicates increased efficiency of separating and recovering VCM from the PVC resin manufacturing process, hence a significant reduction of that VOC/HAP can be attained. Concurrently, energy usage can be significantly reduced, since the industry standard steam sparge is replaced, resulting in less water that must be removed in a subsequent spray drying operation.

REFERENCES

1. L.I. Nass, C.A. Heiberger (eds.), "Encyclopedia of PVC, Volume 1: Resin Manufacture and Properties" 2nd Ed., Marcel Dekker Inc., New York 1986, ISBN: 0-8247-7427-2
2. R.J. Albalak (ed.), "Polymer Devolatilization," Marcel Dekker Inc., New York 1996. ISBN: 0-8247-9627-6; Chapter 16. H.M. Joseph, *Devolatilization of Polyvinylchoride*, pp. 495-506.
3. United States Code of Federal Regulations, Title 40, Chapter I, Part 61, Subpart 64, Emission Standards for Polyvinyl Chloride Plants.
4. F.W. Billmeyer, "Textbook of Polymer Science," J. Wiley, New York, 1984. ISBN: 0-471-03196-8
5. C.J. Geankoplis, "Transport Processes and Unit Operations," 2nd ed., Allyn and Bacon Inc., New York, 1983. ISBN: 0-205-07788-9
6. R.B. Bird, W.E. Stewart, and E.N. Lightfoot, "Transport Phenomena," J. Wiley & Sons, New York, 1960. ISBN: 0-471-07392-X
7. A.C. Metaxas,, "Foundations of Electroheat," John Wiley & Sons, Inc. Chichester, UK, 1996. ISBN: 0-471-95644-9
8. Roussy and J.A. Pearce "Foundations and Industrial Applications of Microwaves and Radio Frequency Fields," John Wiley & Sons, Inc. Chichester, UK, 1995. ISBN: 0-471-93849-1
9. D.E. Clark, W.H. Sutton and D.A. Lewis, eds., "Microwaves: Theory and Application in Materials Processing of IV," Ceramic Transactions 80, American Ceramic Society, Westerville, OH, 1997. ISBN: 1-57498-025-4
10. A. Mavretic, E. Gergin and J. Stach, "Materials Processing by RF Power," pp. 561-567 Proceedings of the 7th International Conference on Microwave and High Frequency Heating, Valencia, Spain Sept. 13-17, 1999. ISBN: 84-7721-781-5.
11. A. Mavretic and J. Allen., "Integration of Fifty Ohm Technology for Materials Processing," pp. 267-273 Proceedings of the 6th International Conference on MW & HF Heating, Fermo, Italy Sept. 9-13, 1997.
12. D. Blackham, "Use of Network Analyzer and Coaxial Probe to Determine Complex Permittivity," Material Research Society Proceedings, vol. 189, pp. 123-128, Pittsburgh, PA, 1991. ISBN: 1-55899-214-6
13. A.M. Nicholson and G.F. Ross, "Measurement of the Intrinsic Properties of Materials by Time Domain Techniques," IEEE vol. IM-TC, January, 1990.

ADVANCES IN MICROWAVE CURING OF CONCRETE

S.L. Mak, R. Banks, D. Ritchie and G. Shapiro
CSIRO Manufacturing & Infrastructure Technology
PO Box 56
Highett, Victoria 3190, Australia

ABSTRACT

A significant amount of concrete used in building and construction around the world is manufactured under factory conditions. These precast concrete products cover a wide range of types, applications and manufacturing processes. One common requirement for efficient production is obtaining a sufficiently high strength for formwork removal, product lifting or application of prestressing. Since the rate of hardening in concrete is temperature dependent, heating is sometimes used to accelerate the curing rate of concrete products. This is particularly important in cooler climates and when the casting equipment is capital intensive. Some common methods for heat curing include the use of steam or hot water, either at atmospheric pressure or in high-pressure autoclaves.

The drive for production efficiency and increasing awareness of the environmental impacts of manufacturing, has prompted research into alternative methods for rapid curing of concrete. There have been relatively few new innovations in rapid heat curing technologies over the last few decades. In recent years, the use of industrial microwave heating technology for rapid curing of precast concrete has been investigated by CSIRO at both benchtop and pilot scale. Our results show that microwave curing cycles of less than six hours can provide sufficient strength for formwork removal and prestressing without impairing concrete quality. This equates to a two- to threefold increase in production rate in selected applications, and a two- to fivefold reduction in process energy requirements.

INTRODUCTION

Current heat curing practice in the precast concrete products industry is generally based on the application of conductive heat through either direct contact with steam, hot water-/steam-heated casting beds or in high-pressure steam autoclaves. Different types of products are heated in different ways. Pipes and poles are typically steam cured in their steel molds in curing tunnels. Flat cast or slip-formed structural wall and slab elements are typically heat cured through steam or hot water-heated sub-beds. Structural beams and girders are typically also heat cured through their closed formwork via hot water or steam pipes welded onto the formwork. Autoclaved aerated concrete or fiber-cement boards are heat cured under high pressure in autoclaves.

The use of alternative techniques for heating concrete has been investigated sporadically in the last 20 years. Two of the main forms of alternative heating are ohmic (or direct electric) heating and microwave heating. Watson (1) reported research on microwave heating of concrete in 1968. However, very little additional work was reported over the next 20 years. Interest in microwave curing of concrete was again revived in the 1980s through to the present (2–9).

The use of microwave heating to process concrete potentially covers the entire spectrum of

- Preheating of materials prior to the product casting or primary curing stage.
- Rapid curing and/or drying as a primary process stage.
- Drying as a secondary post-curing stage.

These may be required individually or in combinations that are either applied sequentially or with some overlap.

Heat-accelerated curing of concrete aims to promote rapid cement hydration in order to achieve a sufficiently high strength at the end of the manufacturing stage, but without any detrimental impacts on product quality. Cement hydration is thermally sensitive and hydration reactions are exothermic. Therefore, it is possible to speed up cement hydration by increasing the temperature of a product, through either external heating or promoting the internal evolution of heat from the exothermic cement hydration. However, this needs to be done without excessive drying, which may deprive cement of water for hydration. Heating also needs to be applied at a rate that does not induce thermal stresses that may cause cracking, particularly when the product is beginning to achieve rigidity and stiffness. The temperature at which a cementitious material is cured may also have a profound influence on its microstructure and morphology. These have consequences on its long-term durability and engineering performance. Therefore, heat curing is typically applied according to a relatively strict regime, e.g.:

- Heating rate no greater than 20°C/hr.
- Maximum "soak" temperature of no more than 70°C.
- Cooling rate of no greater than 20°C/hr.

With the aid of heating, precast concrete production typically revolves around a 24-hour cycle. Therefore, a product needs to be strong enough to be moved, lifted or prestressed within a 20-hour time period, with the remaining time spent preparing for the next production cycle. Faster production cycles may be achieved with the aid of accelerating chemicals, rapid hardening cements or with the use of special production techniques.

Significant innovations in accelerated curing technology needs to deliver substantial improvements over existing techniques, in terms of curing speed, product quality and operational efficiency. Research at CSIRO set out to investigate the viability of using industrial microwave heating for rapid curing of precast concrete, with the following goals:

- Reduce process time by at least 2 times without sacrificing product quality.
- Reduce energy use by at least 2–3 times compared to existing processes.

The principal target was to deliver total curing cycles of less than 6 hours, given that existing processes require anywhere between 12 and 24 hours.

EARLY AGE HEATING CHARACTERISTICS

Microwave Heatability

Cement-based products typically show moderate lossiness depending on the actual composition of the material. Table I shows values of permittivity, loss tangent and penetration depth for materials measured at 20°C using a waveguide technique over a frequency band of 2.3–2.6 GHz. Typically, ε'' ranges between 2 and 3 for either mortar or concrete, whilst penetration depth D_p ranges between 25 and 30 mm. With no coarse aggregate content, sand/cement mortars typically show slightly greater potential for absorbing microwaves. For both mortar and concrete, an increase in the water/cement ratio typically increases ε'', accompanied by a slight reduction in D_p; the effect being more pronounced for mortars in the range of water/cement ratio considered.

Table I. Permittivity and loss factors for concrete and constituent materials at 20°C/2.45GHz

Material	Permittivity		Loss tangent tanδ	Wavelength in material λ (mm)	Penetration depth D_p (mm)
	ε'	ε''			
Water	77.5	12.0	0.15	14	14
Sand 2% mc*	4.4	0.1	0.03	58	313
Sand 6% mc	7.2	0.4	0.06	46	129
Mortar w/c† = 0.4	13.6	2.7	0.20	33	27
Mortar w/c = 0.5	19.7	5.8	0.29	27	17
Concrete w/c = 0.4	13.9	2.6	0.18	33	28
Concrete w/c = 0.5	15.1	2.9	0.19	31	26

* Moisture content calculated on basis of dry sample weight.

† Water/cement ratio.

Bulk Microwave Heating of Concrete

Microwave heating of concrete products requires consideration of

- *Through depth uniformity*, i.e. how far microwaves can penetrate the material and heat up a bulk material rapidly.
- *Thermal overruns*, particularly when aided by exothermic heat of hydration.
- *Spatial uniformity* – level of control over a wide area of product.

As a general principle, microwaves can provide a rapid thermal boost, bringing a product up to its required "soak" temperature. This is aided by the relatively deeper penetration of microwaves into a product when compared to conductive heating. This is evident in the plot of maximum thermal gradient versus bulk heating rate in Fig. 1, measured from 800 × 800 × 100 mm thick panel specimens subjected to either microwave heating or sub-bed steam heating. The microwave heating was carried out in an instrumented 10 kW pilot-scale oven with a movable specimen trolley, as shown in Fig. 2. The average temperature development within the concrete panels is shown in Fig. 3.

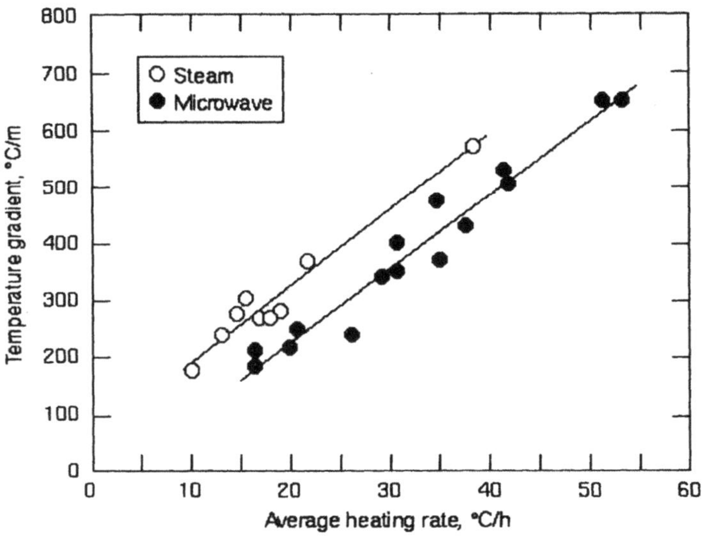

Fig. 1. Maximum temperature gradient across a 100 mm thick concrete panels for various bulk heating rates – steam or microwave heating at 2.45 GHz.

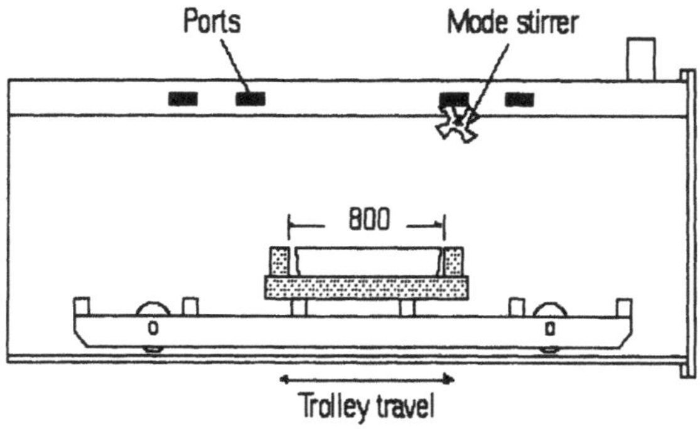

Fig. 2. Microwave heating layout in CSIRO's 10 kW pilot microwave facility.

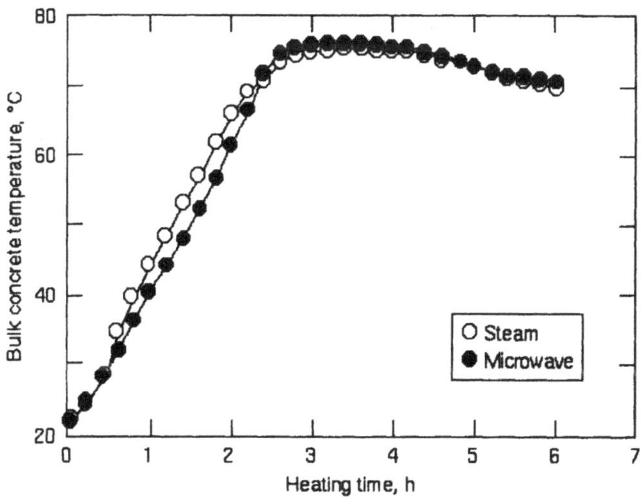

Fig. 3. Temperature development in 800 × 800 × 100 mm thick slabs subjected to steam or microwave heating from top face at a nominal bulk heating rate of 20°C/hr.

For average bulk heating rates in the range of 10–60°C/hr, thermal gradients are significantly lower in the microwave-heated slabs when compared to the conventionally heated slabs. This illustrates the ability to use microwaves to heat a product more quickly and more uniformly across its thickness. This also opens up the possibility that with the appropriate microwave frequency, significantly faster safe heating rates are possible when compared to conventional conductive heating. Our work on high slump 60 MPa structural precast concrete indicates that safe microwave heating rates up to 60°C/hr are feasible, i.e. up to three times the rate normally acceptable for conventional heating of concrete.

Acceleration of Early-Age Setting Behaviour

Controlled microwave heating reduces the setting time in a Portland cement mortar by up to 70%. The results in Fig. 4 show the relationship between penetration resistance and time from mixing for a mortar made with Type GP cement and with a water/cement ratio of 0.5. Under standard conditions, the initial and final setting times were 4.9 and 6.9 hours respectively.

Using a moderate microwave heating rate of 35°C/hr to a maximum temperature of less than 70°C, both the initial and final setting times for the mortar were more than halved to 1.9 and 2.4 hours respectively. The rate of setting reduced as the microwave heating rate was increased. When heated at 120°C/hr, the initial and final setting times reduced further to 1.3 and 1.8 hours respectively.

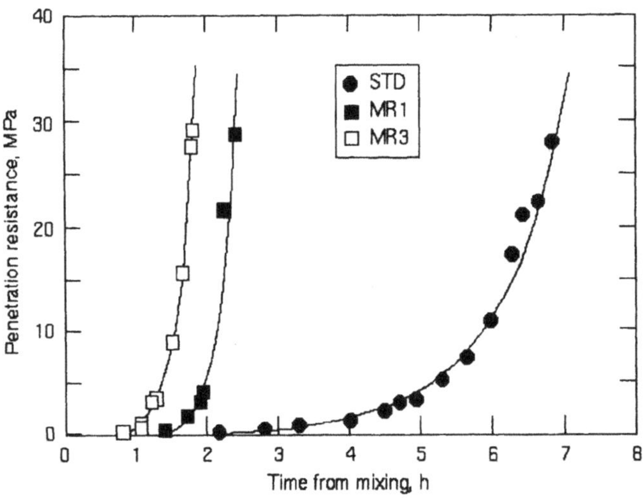

Fig. 4. Penetration resistance versus time for Portland cement mortar under standard conditions (STD), and microwave heating rates of MR1 30°C/hr and MR3 120°C/hr respectively.

PROPERTIES OF HARDENED CONCRETE

Early-Age and Medium-Term Strength Development

Rapid microwave heating can significantly increase early-age strength without undermining medium-term strength gain. Depending on the concrete mix design and curing regime, compressive strengths in excess of 30 MPa can be achieved within a 5-hour microwave curing cycle, as shown in Fig. 5. In this case, the concrete was a low slump concrete typical of the type used in slip-formed precasting, and the active microwave heating time was less than 10% of the total curing cycle time. The 28-day strength of the microwave-cured concrete was almost 70 MPa, or about double that achieved at the end of microwave curing. As shown in Fig. 5, the compressive strength of both microwave-cured and standard-cured concrete was almost identical at 28 days.

With controlled microwave curing, the accelerated early age strength gain was followed by sustained strength development even in the medium to long term. The results in Fig. 6 show two strength development curves for a 60 MPa concrete rapidly heated with microwaves with either 0 or 2 hours of delay after casting. It is evident that compressive strength remained stable for up to 3 years and the period of preheating delay had no impact on long-term strength.

Within sensible limits, it is possible to increase the heating rate with microwaves without adversely affecting concrete strength. When microwave heating was used, the results in Table II show that doubling the heating rate from 20°C/hr to 40°C/hr did not cause any reduction in 28-day strength. However, doing the same with conventional heating caused the 28-day strength to be reduced by almost 15%.

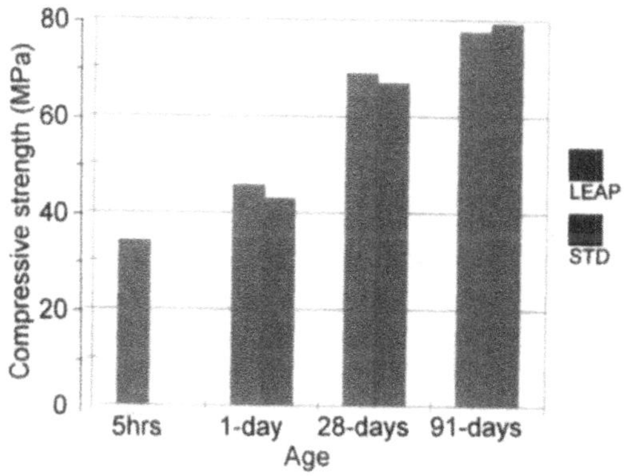

Fig. 5. Strength development in 60 MPa low slump – standard or microwave curing (LEAP).

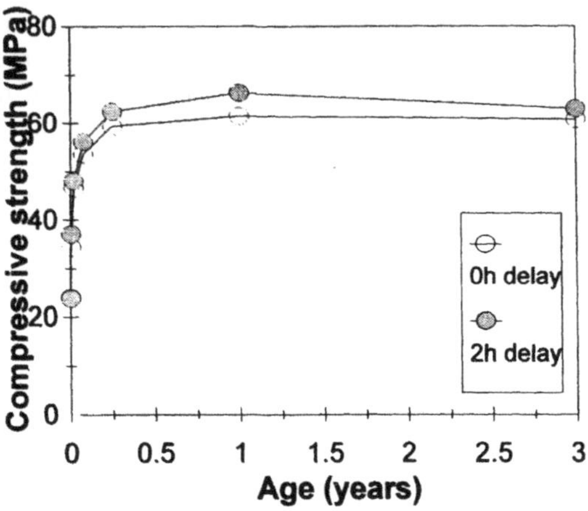

Fig. 6. Long-term strength development of 60 MPa microwave-cured concrete.

Table II. Compressive strength in standard specimens and for concrete cores extracted from 800 × 800 × 100 mm panels subjected to three curing regimes

Specimen type	Curing	1-day strength (MPa)		28-day strength (MPa)	
		GP	HES	GP	HES
100 × 200 mm ϕ cylinders	Standard fog 23°C			58.0	61.5
100 × 100 mm ϕ cores from 800 × 800 × 100 mm slabs	SLC*	22.5	31.5	51.5	56.0
	Steam STR20†	32.0	38.5	44.0	53.5
	Steam STR40†	27.5		38.5	
	Microwave MWR20‡	29.5	39.0	42.0	53.0
	Microwave MWR40‡	30.5	40.5	42.5	52.5

* SLC = limited moist curing for six days followed by laboratory air drying.
† Steam STR20/STR40 = bulk heating rates of 20 and 40°C/hr respectively.
‡ Microwave MWR20/MWR40 = bulk heating rates of 20 and 40°C/hr respectively.

Near-Surface Quality of Concrete

Accelerated heating can impair the near-surface quality of heat-cured concrete. The depth of water sorptivity within a 24-hour period, d_{24}, is one measure of near-surface quality, with a lower value indicating relatively improved durability potential. For either steam or microwave heating, the d_{24} values at 28 days became increasingly larger when heating rates increased. However, the relative change in near-surface quality due to increasing bulk heating rate was found to be larger in steam-cured concrete when compared to microwave-cured concrete. This is evident from the comparison of results for concrete containing Type GP cement shown in Fig. 7. As the bulk heating rate increased from approximately 20 to over 40°C/hr, the d_{24} value increased by 40%, i.e. from 27 to 38 mm. By contrast, the doubling of heating rate in the steam-cured concrete more than doubled the d_{24} value, increasing it from 21 to

47 mm. This is possibly due to the increasingly large thermal gradients that are generated when conventional steam heating rates are increased. These results are consistent with the compressive strength results discussed earlier, where, within limits, increasing the rate of heating using microwaves did not produce any deleterious effects when compared to conventional heating. These findings provide supporting evidence of the benefit of using a heating mechanism that has relatively deeper penetration and which is able to heat a bulk element more rapidly without introducing large thermal gradients.

Properties of Steel-Reinforced Concrete Elements

The performance of steel-reinforced concrete subjected to microwave curing was not dissimilar to that of steam curing. The results in Fig. 8 show load displacement curves for 100 mm thick panels tested in flexure. The panels were manufactured from a 60 MPa mix with 90 mm slump, and the panels were reinforced with one layer of RF-81 Ribmesh. The results indicate that both the ultimate load and post-peak behavior in slabs subjected to either steam or microwave curing were similar.

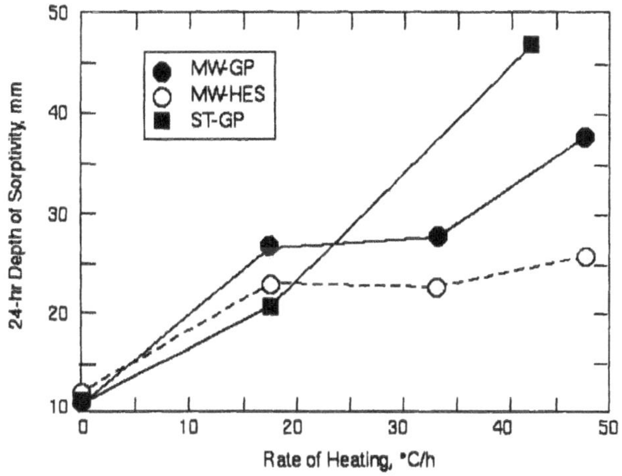

Fig. 7. Depth of water sorptivity at 24 hours versus heating rate – steam or microwave curing.

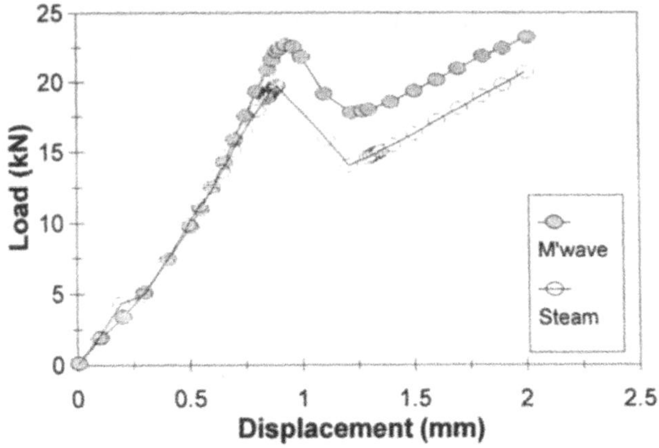

Fig. 8. Load versus displacement curves for 100 mm thick panels loaded in flexure.

CONCLUSIONS

Controlled microwave heating of medium-thick, bulk concrete elements is feasible on a pilot scale, providing an alternative rapid curing technique for precast concrete. With total curing cycles of less than 6 hours, compressive strengths in excess of 25 MPa can be achieved, with a very high level of residual strength development even without any further moist curing. Microwave heating provides relatively deeper penetration when compared to that achieved using conventional steam heating. The deeper penetration achieved with microwave heating also means that, unlike conductive heating with steam, compressive strength and near-surface quality are not adversely affected when bulk heating rates are doubled from 20 to 40°C/hr.

REFERENCES

[1]A. Watson, "Curing of Concrete," in *Microwave Power Engineering*. Edited by E.C. Okress. Academic Press, New York, 1968.

[2]X. Wu, J. Dong and M. Tang, "Microwave Curing Technique in Concrete Manufacture", *Cement & Concrete Research*, **17**, 205–10 (1987).

[3]J.T. Hutchison, H.M. Chang, H.M. Jennings and M.E. Brodwin, "Thermal Acceleration of Portland Cement Mortars With Microwave Energy," *Cement & Concrete Research*, **21**, 795–9 (1991).

[4]L. Dongxu and W. Xuequan, "A Study on the Application of Vacuum Microwave Composite Dewatering Technique in Concrete Engineering," *Cement & Concrete Research*, **24**, 159–164 (1993).

[5]S.B. Della, S. Lai and M. Pinna, "Microwaves for the Hyper-Accelerated Curing of Concretes," *Betonwerk + Fertigteil-Technik.*, **12** (1994).

[6]C.K. Leung and T. Pheeraphan, "Very High Early Strength of Microwave Cured Concrete," *Cement & Concrete Research*, **25**[1], 136–46 (1995).

[7]C.K.Y. Leung and T. Peeraphan, "Microwave Curing of Portland Cement Concrete: Experimental Results and Feasibility for Practical Applications," *Construction & Building Materials*, **9**[2], 67–73 (1995).

[8]S.L. Mak, K.K. Sagoe–Crentsil, A.H. Taylor and D. Ritchie, "Application of Microwave-Accelerated Processing in Cement and Concrete Technology," Proc. 20th Conf. on Our World in Concrete & Structures, Otani, Singapore, 23–25 March 1995. Edited by C.T. Tan, vol. XIV, pp. 117–20.

[9]S.L. Mak, G. Shapiro and T. Son, "Accelerated Heating of Concrete With Microwave Curing," Proc. 4th CANMET/ACI/JCI Int. Conf. on Recent Advances in Concrete Technology, Tokushima, Japan, 7–11 June 1998, pp. 531–42.

HYBRID MICROWAVE FIRING OF HEAVY CLAY PRODUCTS.

Garth V.A. Tayler
Acme Brick Company, Technical Department, 2821 W 7th Street, Fort Worth, Texas 76107, USA.

Mike Anderson and Mike Hamlyn
Staffordshire University, School of Engineering and Advanced Technology, P.O.Box 333, Beaconside, ST18 ODF, UK.

ABSTRACT

Thermal analyses and dielectric property measurements of numerous heavy clay bodies provided data for the selection of optimum body compositions. The addition of carbon increased the complex permittivity of the bodies tested, in such a way that it improved the efficiency of microwave coupling in the firing process. The Microwave Autogenous Firing (MWAF) mechanism was identified.

The key variables that influenced this microwave heating process were examined. The test firing methodology used varying techniques of forward power application. The surface area of the test pieces, the application of preheated secondary air supply to the cavity and the application of modulated-pulsed microwave power, were found to influence temperature differential across the section of the test pieces examined. In light of these research findings, the paper proposes some of the design features of a continuous and intermittent, hybrid microwave process for the firing of heavy clay products.

INTRODUCTION

The MWAF process, as applied to the firing of heavy clay products, has been found to be an effective mechanism for exploiting the efficiencies of the volumetric heating effect of microwave energy. It was found that controlled applications of microwave energy to carbon enriched heavy clay bodies, resulted in minimal amounts of power being required to achieve effective microwave heating of the load. The combination of microwave energy with conventional gas firing has been found to be an effective way of speeding up the firing process, as well as a means of reducing total energy consumption (1).

Figure 1 illustrates the reactions, which take place during the heating of a typical heavy clay material. Varying amounts of energy are required at the different stages of the heating cycle, to complete the reactions.

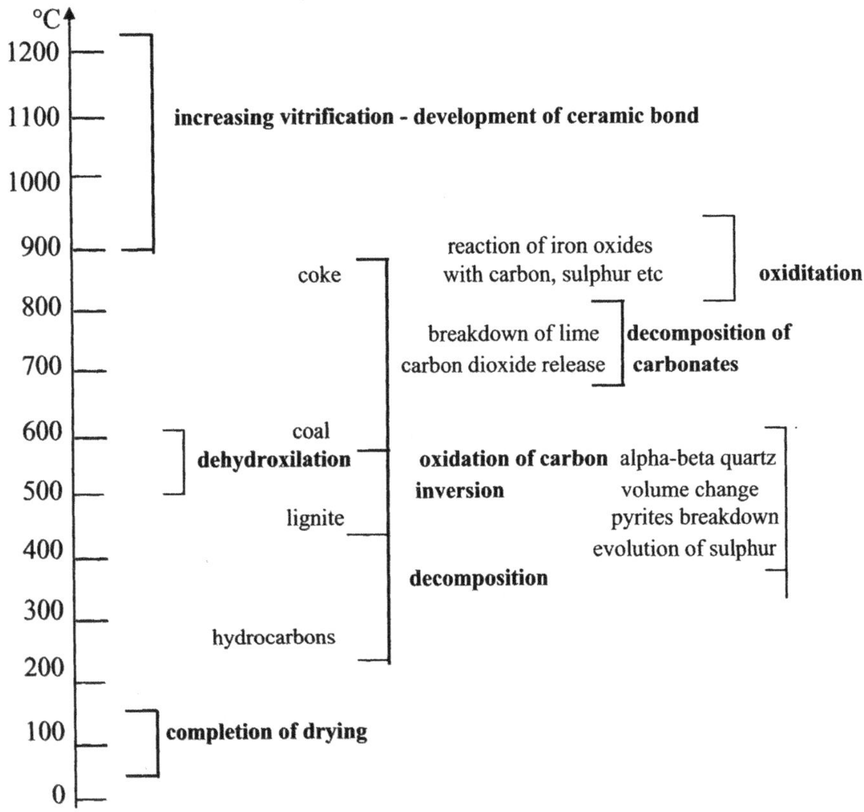

Figure 1: Reactions during heating of a typical heavy clay mineral

Experimental microwave firings have shown that it is feasible to design and develop a hybrid firing system utilizing the *autogenous combustion mechanism* that is initiated by a microwave energy source.

Autogenous combustion can be described as the heat transfer process developed from the ignition and combustion of the carbon additive within the heavy clay body (2). This technique of kiln firing is typically used in clamp kiln firing systems for clay brick. More recently, tunnel kilns have been designed and built with burner/fan circulation zones that re-circulate the sensible heat developed from the autogenous combustion in the preheat and furnace zones, thereby reducing gas usage. Figure 2 illustrates the typical heating cycle of a heavy clay tunnel kiln firing process. The temperature cycle is required to follow differing heating rates, to facilitate the effective completion of the different reaction stages, illustrated in Figure 1. As a result of this, the firing cycles are lengthy and sluggish, due to the poor thermal conductivity of the clay body, as well as the poor efficiency of the convection and the radiation heat transfer mechanisms.

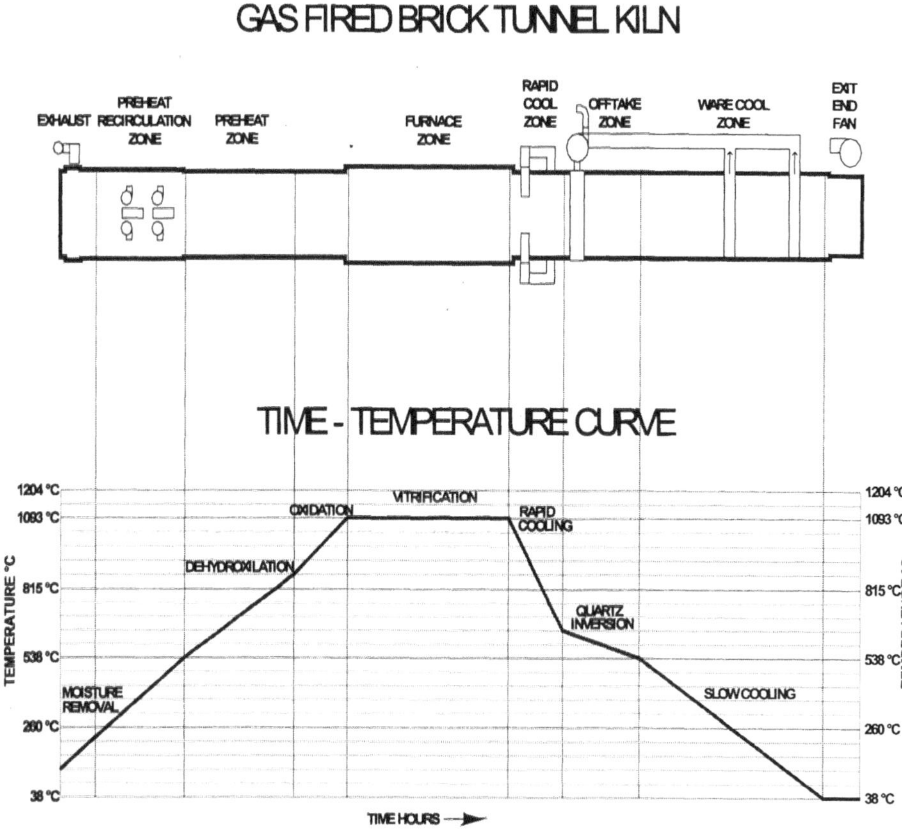

Figure 2: Layout and heating cycle of a typical gas fired heavy clay tunnel kiln.

The introduction of the more efficient volumetric heating (3) of microwave energy, at selected portions of the heating cycle, provides the opportunity to reduce the overall cycle time and improve the specific energy usage.

Research efforts have also shown that with this technology, the microwave energy can be applied as a minor proportion of the total energy used in the system.

The hybrid gas/microwave firing technology, using the *autogenous combustion microwave firing* (MWAF) mechanism, could be developed and adapted for continuously fired brick kilns as well as intermittently fired kilns.

EXPERIMENTAL

Research on the MWAF mechanism was conducted at the TRI laboratories of ESCOM, South Africa, in collaboration with Staffordshire University, UK. The hybrid firing system, used Kanthal resistance elements in combination with a 6kW, 2450 Mhz, microwave system. An insulated stainless steel-lined cavity was used. The cavity was lined on the sides and bottom with the Kanthal resistance elements rated at 12kW. The wave-guide was fitted with a circulator and stub tuner. A quartz window was installed in the wave-guide to prevent heat and moisture from causing magnetron over-heating and possible arching in the wave-guide. The control equipment was designed to measure sample temperature using optical fiber thermometry. Forward power and reflected power were monitored to develop an understanding of the firing behavior of the materials during the autogenous firing phase of the cycle. The method of application of the microwave power was adjusted and optimized over numerous firings, in order to achieve the best possible heat transfer mechanism. During the microwave stages of the cycle, the heat transfer mechanism was enhanced greatly by the provision of a convective heat dissipation technique. The measurements taken during the many test firings and the observations of the fired results, enabled the researcher to draw some important conclusions on the key variables of this hybrid microwave heating process. The research findings provided guidelines for the design of a continuous or intermittent firing hybrid microwave process.

Clay Bodies

Typical heavy clay bodies were studied. They consisted primarily of a kaolinitic fireclay combined with silica sand, pulverized fly ash (PFA) additive and carbon fly ash (CFA) additive. The CFA, a by-product from an industrial processing industry, contained approximately 40% fixed carbon and served as a "doping" agent to increase the complex permittivity of the bodies (4). Figure 3 illustrates how the carbon additions increased complex permittivity of the body. This increased carbon content, resulted in more efficient coupling (5) with the microwave field. The control of this key variable facilitated the development of the MWAF process.

Test firings

The objective of the test program was to use controlled microwave energy applications to ignite (6) carbon within the body. This autogenous combustion of the carbon was sustained by pulsed microwave power applications during firing cycle. Network analyzer measurements were taken of both an empty and loaded

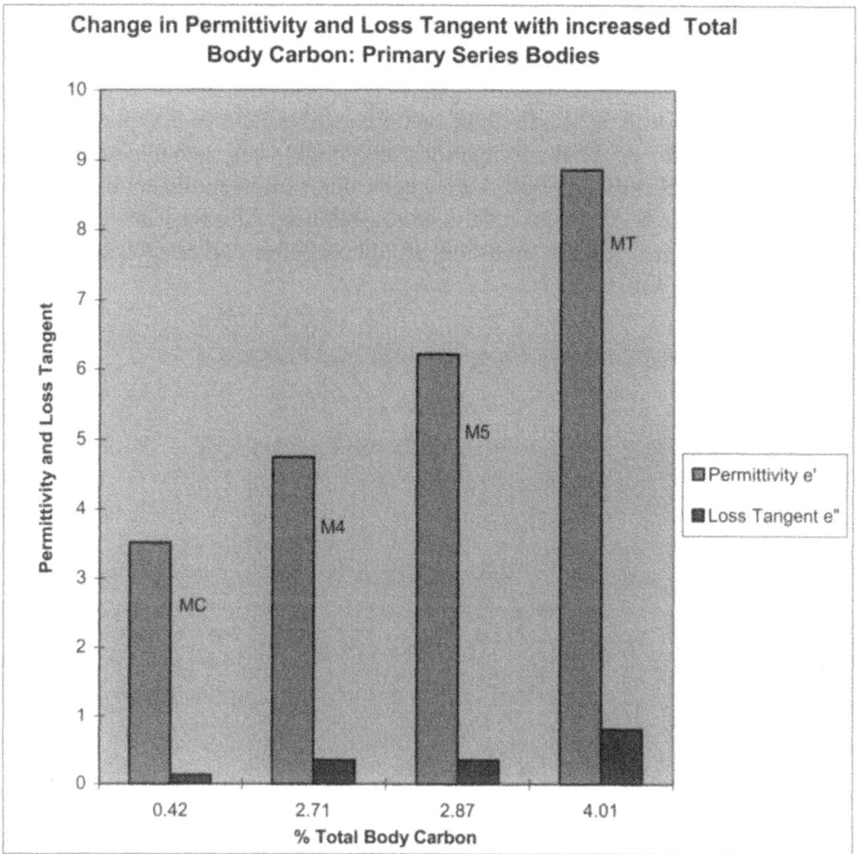

Figure 3: Increased complex permittivity with increased carbon content.

cavity. Adjustments were made to optimize the field distribution within the cavity. The loaded cavity, containing a sample mass of approximately 1.5 kg, indicated good microwave power absorption with minimal reflected power. Increased sample-hack setting density further improved field distribution in the load.

The preheat cycle (from 20°C to 600°C) was programmed to finish over a time cycle at a pre-determined ramp rate. The transition from the radiant heat, preheat cycle, to the microwave autogenous firing cycle, was operated manually. The energy usage during the preheat cycle was recorded using an energy data tracker. The microwave firing cycle was programmed to be either a "full forward power cycle", after the 600°C dwell was completed, or a "modulated-pulsed forward power cycle" using the automatic control loop facility available in the Microsoft Lab View software used with the test furnace. Figure 4 illustrates a firing cycle using the modulated-pulsed method of applying the forward microwave power. Firings were carried out in a "single high" sample setting pattern. Results of each firing were analyzed. A heat dissipation technique was achieved by the introduction of air, preheated to 500°C. This technique simulated the conditions of an industrial, furnace (7). The best results were obtained with perforated sample pieces that had a greater surface area and produced better carbon burnout with more uniform vitrification.

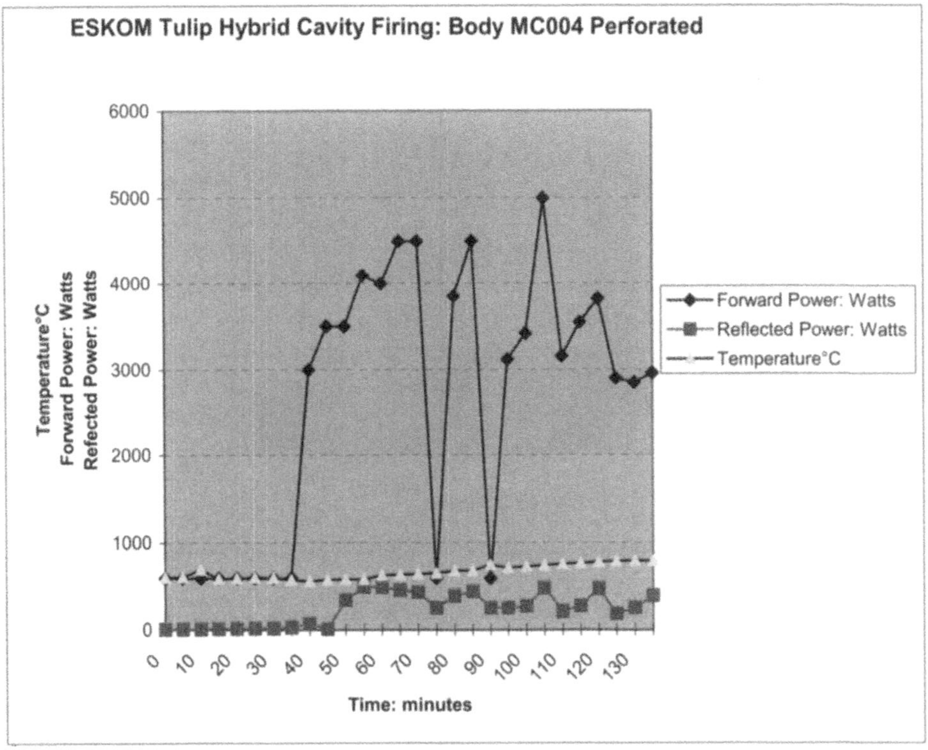

Figure 4: Firing cycle using the pulsed-modulated forward power application.

An optimum carbon content of 2% was found to produce the most effective carbon burnout, resulting in a uniformly vitrified body. This carbon content also produced efficient microwave coupling and hence facilitated effective autogenous combustion. These findings provided the guidelines for a proposed hybrid MWAF system.

PROPOSED CONTINUOUS HYBRID MWAF PROCESS.

Applying these research findings, the design of a continuously firing, roller-bat kiln is proposed and illustrated in Figure 5. The counter-flow of the preheated gasses i.e. the gasses that flow in the opposite direction to the ware passing through the kiln, is an important feature of the Hybrid MWAF process. Controlled volumes of this preheated air, at a suitably controlled pressure, in combination with the recirculation of gasses from high velocity gas burners in the preheat and firing zones, ensures that the ware is uniformly pre-heated before entering the microwave ignition and autogenous combustion zones. The 915 MHz microwave ignition and autogenous combustion zones are hybrid gas/microwave zones. The choice of 915 Mhz was made due to better coupling efficiency and depth penetration achievable with this frequency. Pulsed application of microwave power, in these three separate zones, facilitates the autogenous combustion of the carbon in the body and an efficient and rapid heat transfer mechanism. Conventional thermo-couple and infrared temperature measurement techniques are used to control gas firing settings and the forward power microwave levels in such a way that the programmed set point temperatures in the zones are maintained automatically. The heat dissipation effect of the counter flow gasses counteracts the reversed-thermal gradient effects (8) developed by the microwave heating systems. This important feature prevents thermal runaway from developing, a problem seen in the microwave-only test firings.

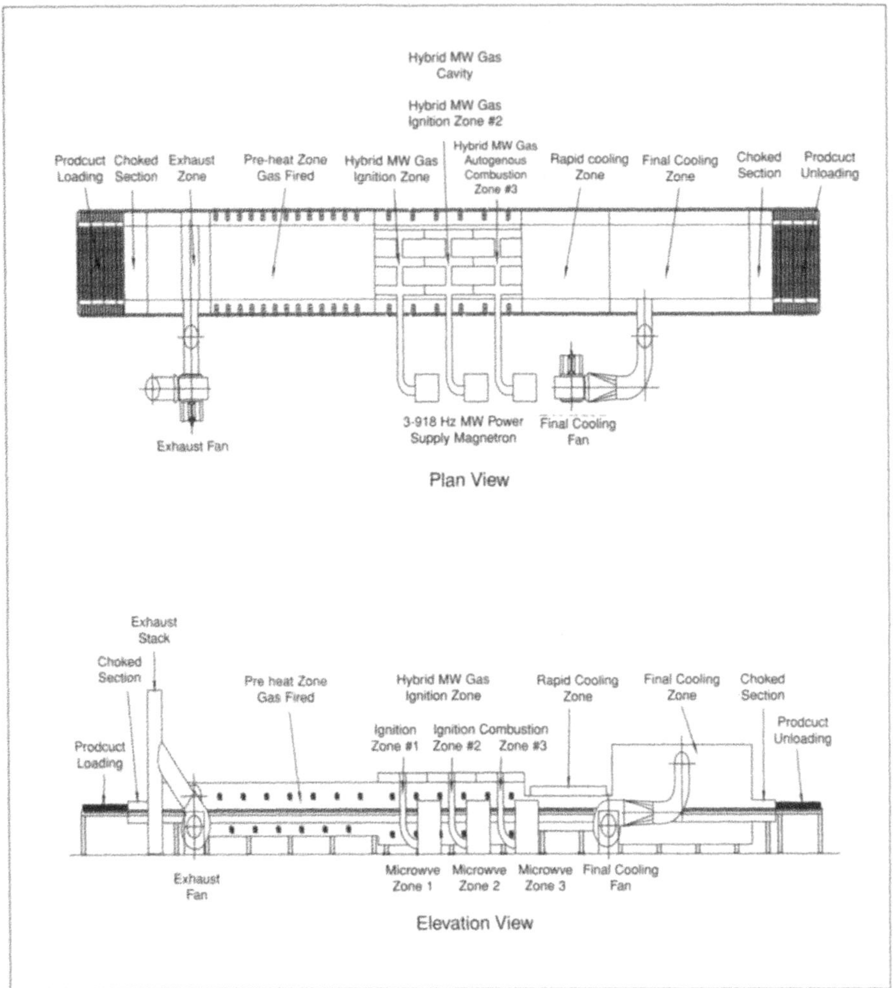

Figure 5: Proposed continuously firing, hybrid MWAF roller-bat kiln.

PROPOSED INTERMITTENT HYBRID MWAF PROCESS

The proposed hybrid MWAF intermittent brick kiln consists of an insulated fire brick (IFB) lined steel shell, constructed as a 13.8 meter diameter circular structure, which supports a domed crown, manufactured from IFB masonry or Z-block refractory fiber. The entire outer surface of the IFB lining is sealed, using a Teflon bonded, stainless steel foil layer, to render the structure completely gas-

tight. Firing is carried out in 50 hours, using 6 large high velocity, pulse fired gas burners. The kiln pressure is automatically controlled.
In the proposed hybrid intermittent kiln, the microwave energy is launched into the large cavity, using 915 Mhz generator-magnetron systems. Four wave guide entry points are positioned symmetrically at the base of the crown. This system is designed to introduce the microwave energy after the dehydroxilation stage of the firing cycle. The pulsed, forward microwave power, follows the programmed temperature set point schedule, to achieve the following conditions:

- Maximum uniformity of heat dissipation within the load.
- Minimal "reversed thermal gradient" effects.
- Minimized thermal runaway conditions.
- Maximum degree of carbon burnout.
- Reduced firing cycle time.

For an effective, continuous or intermittent MWAF firing process, the key operational controls of the process are:

- the speed of the product traveling through the unit (continuous firing).
- the composition of the material i.e. total carbon content.
- the level and method of forward power application in the ignition and autogenous combustion stages of the cycle.
- the volume and pressure of combustion air supplied to the cavity.

CONCLUSIONS

The work has indicated that the use of a hybrid microwave firing system to initiate autogenous combustion is highly feasible. The addition of a suitable carbon component, improves the complex permittivity of the material being fired. The use of a controlled pre-heated secondary air, is the key to the development of a heating process which effectively minimizes thermal runaway. The effective control the conventional gas firing in combination with the modulated-pulsed microwave portions in the overall energy balance, enables the more expensive microwave portion of the total energy requirement, to be minimized.

The numerous test firings and the analyses of the results yielded the following:

- The use of perforated test pieces produced improved results due to better circulation and greater surface area, which facilitated a more effective gas diffusion process.
- Body carbon contents, which exceeded 2%, displayed efficient microwave coupling, but detracted from the final properties of the fired body due to the "carbon-trap" phenomenon.
- The best carbon burnout and vitrification results were achieved using an air supply preheated to 500°C and introduced into the cavity through a perforated base on which the samples were mounted. This preheated air

supply produced improved heat dissipation and reduced the "reversed thermal gradient" effect.

- A modulated-pulsed forward power technique resulted in better heat distribution within the load.
- An excessive supply of air to the cavity resulted in a higher microwave power demand. An optimum combination of air and modulated-pulsed microwave power is required.

An industrial scale MWAF system, utilizing hybrid gas/microwave firing, would need to adhere to most of the above design requirements to be effective. The research has shown that using this hybrid microwave/gas technology (9), firing cycles can be reduced by as much as 50%, with resultant energy reductions of as much as 35%.

REFERENCES

(1) S.M Bond, "Microwave-Assisted Processing - from Laboratory to Production". Paper presented at the 2nd World Microwave Congress, Florida, USA (2000).

(2) Tayler G.V.A., "The use of Microwave Energy to Initiate Autogenous Combustion for the Firing of Heavy Clay Products". PhD thesis, Staffordshire University, England (April 1998).

(3) I. Ahmad, R.Dalton. and D. Clark, "Unique Application of Microwave Energy to the Measurements on Ceramics". Material Research Society Symposium Proceedings, Vol. 269, 611-616 (1992).

(4) Clark D.E., Ahmad I. and Dalton R.C. "Microwave Ignition and Combustion Synthesis of Composites." Mater. Sci. Eng. A144, 91-97 (1991).

(5) Evans N.G. "The Dielectric Properties of Ceramics at 2.45 GHz and Their Influence on Microwave Firing". PhD thesis, Staffordshire University, England (April 1995).

(6) Hamlyn M.G., Bowen A.L., Jones G. and Jackson S.M. "Microwave Firing Process for Earthenware Ceramics". High Frequency and Microwave Processing and Heating Conference, Arnhem (1989).

(7) Kulbekov M.K. and Saibulatov S.Z. "Mechanism and Kinetics of Carbon Burnout During Calcinations of Resource-conserving Ash-based Ceramic Materials." Izvestiya Akademi Nauk SSSR Neorganicheskie Materialy vol. 27(5), 1074-1078 (1991)

(8)Mataxas A.C. and Meredith R.J. "Industrial Microwave Heating". Peter Peregrinus, London (1983).

(9)"The Microwave-Assisted Gas Firing of Ceramics". Future Practice Report 57, EA Technology, UK (1999).

RECENT RESEARCH AND DEVELOPMENT ON MICROWAVE POWER APPLICATIONS AT THE UNIVERSITY OF ELECTRONIC SCIENCE AND TECHNOLOGY OF CHINA

Tian-Ren Ji
College of Physics and Electronics
University of Electronic Science and Technology of China (UESTC)
Chengdu, Sichuan 610054
PR China

ABSTRACT

A brief review is given on some recent research and development of microwave power applications (MPA) at the Microwave Plasma Laboratory of UESTC. It covers mostly the application of research done in recent years in the fields of industrial chemistry and particularly new materials such as: diamond film and many other chemical vapor deposition (CVD) films; microelectronics; surface modification and treatment of materials, etc. We also introduce the specific products developed during this period, including: a series of low ripple computer controlled microwave power sources, with the power range from 0.2kW to 10kW at 2450MHz, and 5kW and 20kW at 915MHz; fine quality microwave transmitting systems. Also introduced are various microwave applicators and systems: 8kW Microwave Rubber Vulcanizing System; 1kW Microwave Polymer Processing System; 20kW/915MHz Industrial Microwave Oven; 12kW/2450MHz Microwave Industrial Drying System; 5kW/2450MHz Microwave Muffler Furnace; 200W/2450MHz compact ECR system; 1kW/2450MHz Multicusp ECR System; 5kW/2450MHz TM_{013} mode downstream system; 10kW/2450MHz Ellipsoid Cavity MPCVD System; coaxial and waveguide cavity for atmospheric MP torches. Compared with some advanced western countries, we are behind in processing technology and hardware equipment. But this shows the great opportunities, technological challenges and a better perspective for us to follow.

INTRODUCTION

With the rapid development and potential application of microwave power applications (MPA) in the fields of industrial chemistry and particularly new materials, such as diamond, GaN and other chemical vapor deposition (CVD) films, microelectronics, surface modification and treatment of materials, we have taken the key project named "Basic and Applying Study on Microwave Plasma" from China National Science Foundation Committee. In order to carry on this significant research project, the Microwave Plasma Laboratory (MPL) was organized in 1994 with a strong team of researchers. Since then we have been carrying on not only the systematic basic theoretical study on magnetic microwave plasma headed by the member of Chinese Academy of Sciences, Professor Liu Shenggang, but also building up our Lab furnished with advanced microwave cold and hot testing instruments and CAD computer system, which have provided us with a strong ability on theoretical study, design and digital simulation. We also have a series of Microwave Plasma Systems and equipment with diagnostic measuring instrumentation (Figure 1), and we are doing a lot of related experimental research and development work in this field (Figure 2).

Figure 1 – Microwave plasma ECR equipment in our laboratory

Figure 2 – Microwave plasma experiments using the ECR system

All of these studies show that microwave power and microwave plasma have excellent performance and great potential application perspectives in many interdisciplinary science fields. Now, our MPL has been appointed as the provincial key laboratory.

RESEARCH AND DEVELOPMENT ON MICROWAVE POWER APPLICATIONS (MPA)

On the basis of our Lab's set, we have developed a series of 0.2kW through 10kW at 2450MHz, and 5kW and 20kW at 915MHz low ripple, high stability computer controlled microwave sources. We have also developed fine microwave transmission systems with high performance circulator, high power dummy load, motorized "smart" tuner, various kinds of microwave reaction cavities and all other related microwave components. This equipment has not only met our research needs but provided nearly sixty universities and institutes with their different MPA, such as microwave rubber vulcanizing, polymer and other materials modification, industrial heating and drying, diamond and other MPCVD films, high quality optical fiber manufacturing, microwave ceramics sintering, long life microwave ion source for ion implantation and neutron accelerator, microwave plasma torch and related materials treatment (Figures 3 thru 6).

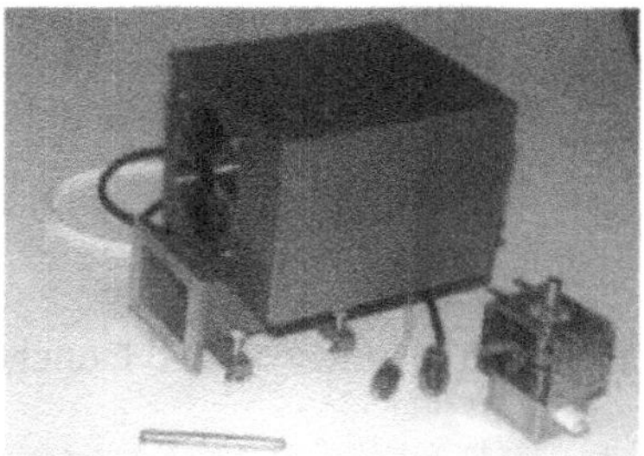

Figure 3 – 1kW water cooled, variable magnet Magnetron Assembly

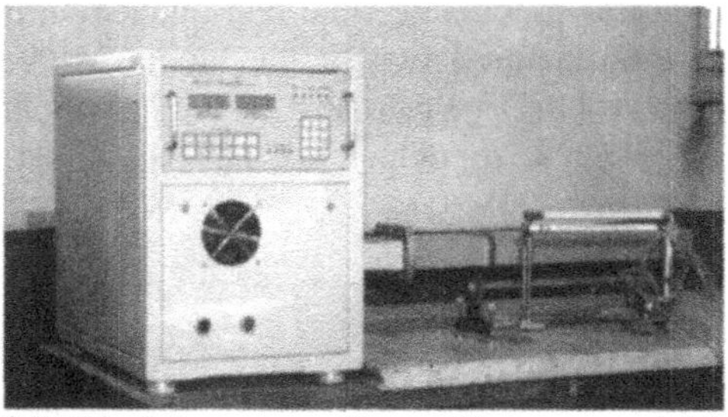

Figure 4 – 1kW, 2450MHz Microwave Power Source, model MPS-2010C

MPA in industrial fields has developed in China for many years but has still not form a big business for it faces the challenge of traditional operating ways with its comparatively higher cost. In view of this situation, we have improved our microwave power sources, transmission system and applicators, and try all means to lower their costs, choose proper customers, provide good services and achieve good results in following application fields.

Figure 5 – Some cavities and waveguide components

Figure 6 – 10kW Magnetron with magnet coil and launcher

Rubber Vulcanizing

Microwave rubber vulcanizing is a good field of MPA for it has high quality of rubber strips, much shorter processing time and quick return of investment. We have developed and manufactured two of the model MRVS-2080c (8 x 1kW, 2450MHz) microwave rubber vulcanizing system and full-set product line (Figures 7 and 8) since 1998, for which we have received favorable customer comment.

Figure 7 – 8kW, 2450MHz Microwave Rubber Strips Vulcanizing System

Figure 8 – The complete Rubber Strips Vulcanizing Production Line

Chemical Synthesis

Microwave chemical synthesis, decomposition material modification and assisted extraction technology is a new front joint science, which has been rapidly developed within the past ten years and has shown great prospects for the future. Our recent research and development work has resulted in the MPS-2010C series, 1kW/2450MHz microwave polymer material processing system. We are now developing a popular multi-function microwave extracting system and wish to do more in the near future.

Industrial Applications

In the field of microwave industrial application, we have simply followed customers' requests and developed different equipment for their various applications. The following are some of the equipment we have developed and delivered to customers during these eight years.

1). In 1994, a model MIDS-9200 (20kW, 915MHz) versatile industrial microwave drying system was produced and exported to Korea (Figure 9).

2) In 1998, a model MPPS-2010C (1kW, 2450MHz) microwave polymer material processing system was built for Sichuan University to do middle scale experiment.

3) Model MPOS-2008C (800W, 2450MHz) high precision optic fiber temperature controlled microwave pathology analysis system.

4) Model MDDS-2008-2024 series, (8kW-24kW, 2450MHz) modular assembled style multipurpose microwave drying and bacterium extinguishing equipment (Figure 10).

5) Model MDO-2005 (5 x 1kW, 2450MHz) large volume, multi-functional microwave drying oven is designed for drying various kinds of special materials (Figure 11).

6) We are currently developing model MHDS-2004-2032 series (4kW-32kW, 2450MHz) modular assembled new type of microwave low temperature

combined drying equipment for drying special valuable susceptor materials in request of many customers.

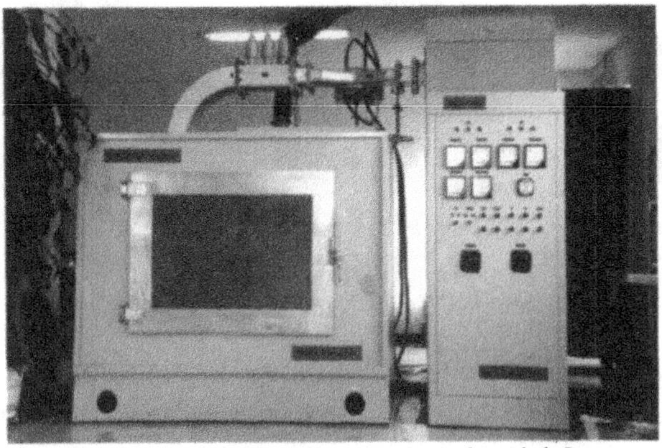

Figure 9 – 20kW, 915MHz Versatile Microwave Industrial Oven

Figure 10 – 12 x 1kW, 2450MHz Microwave Drying Conveyer System

Figure 11 – 11 x 1kW, 2450MHz Industrial Microwave Oven

Microwave Plasma Chemical Vapor Deposition (MPCVD)

Microwave power combined with plasma is the most attractive application field. On the basis of the above results, we have begun to develop two complete MPCVD systems. One is a new multi-function, computer controlled electron-cyclotron resonance (ECR) system. It is furnished with an advanced low ripple 1kW, 2450MHz microwave power source, a high performance microwave transition system, the multicusp permanent magnets ECR cavity, a large volume plasma processing cavity, four gas line mass flow controllers, a high vacuum turbo pumping and measuring system, and all parameters controlled by a computer system (Figure 12).

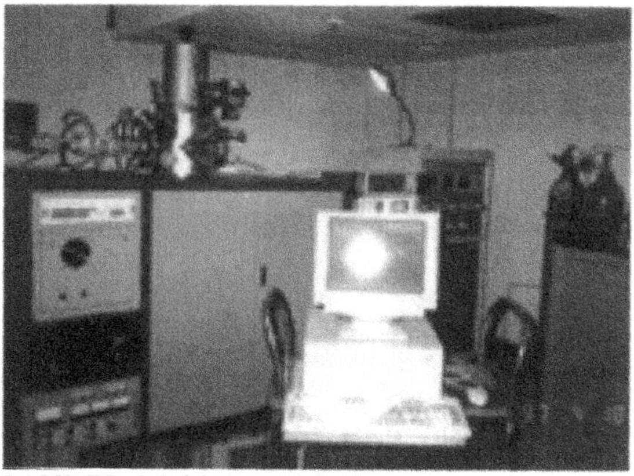

Figure 12 – 1kW, 2450MHz Multicusp ECR System

The other is a 5kW, 2450MHz system (Figure 13) which is a standard down-stream TM_{013} mode cylindrical cavity including high vacuum, four gas lines, substrate DC bias and temperature control, and complete system computer control. This is mostly suitable for MPCVD at the middle and higher gas pressures, which has already provided to six Institutes for deposition of diamond, GaN and other films.

Among the MPCVD systems is the recently developed 10kW, 2450MHz MPCVD system with a special rotational ellipsoidal cavity as the reacting applicator which provides outstanding performance (Figure 14). The simulation revealed that the electric field strength in an ellipsoid cavity exhibits two pronounced maximum at the two focal points. By coupling microwave energy into one of the maximum focal points, it generates a strong electric field at the other maximum focal point, and thus to excite an intense spatially stable plasma where the substrate be placed. It is very promising that by using this system, we can realize a large area, optical quality diamond film MPCVD with stable operation and higher deposition speed.

Figure 13 – 5kW, 2450MHz Downstream CVD System

Figure 14 – 10 kW, 2450MHz Ellipsoid Cavity MPCVD System

Microwave Ceramic Sintering

Microwave ceramic sintering (MCS) is another focal point in our R&D work. Compared with traditional furnace sintering, MCS has obvious advantages of whole material heating, and high thermal efficiency, fast temperature rise, high density and fine mixture. These benefits are especially suited for producing fine ceramics, functional and biological ceramics, and applications in the electronics industry and other fields. The first model MMFS-2050C, 5kWmicrowave high temperature sintering system we developed in 1998 (Figure 15) had passed the province level evaluation. Five systems have been delivered to our customers for sintering fluorescence powder , fine ceramic and functional ceramic components, which have all demonstrated satisfactory results. We are now going to cooperate with customers to develop the high power continuous MCS system and put it into industrialized application soon.

Microwave Plasma Torch

Microwave plasma torch (MPT) is a new technology developed in western countries in recent years. It is stimulated by microwave discharge, operated at 100

Torr to atmospheric pressure and has features of high temperature, high plasma density and easily controlled chemical reaction with materials. This technology is widely used in CVD synthesizing of SiO_2 and diamond films, surface modification of metal materials, fine processing of limestone slices, etc. We have designed two kinds of MPT experimental systems, one of which is a coaxial cavity type and the other is waveguide cavity type (Figure 16), which can work with 1kW or 5kW, 2450MHz microwave power source and different gases. At atmospheric pressure and above, we can achieve a high temperature, adjustable torch. In cooperation with several institutes, various MPT reactors and jet heads have been developed and experiments studied with good results.

Figure 15 – 5kW, 2450MHz Microwave Ceramics Sintering System

Figure 16 – Waveguide Cavity Microwave Plasma Torch System

Ion Implantation

Ion implantation is an important process in microelectronics and material modification field. Microwave ion source substituting the traditional hot filament ion source will further improve ion consistence (1~2 times higher than old methods) and life time (from several tens or hundreds of hours to thousands, even ten thousands hours), and also will update the technology greatly. Upon request of customers, we have developed two kinds of microwave ion source systems. One is model MIS-2010C (1kW, 2450MHz) used in the ion implantation equipment and medical neutron accelerator and the other is model MEIS-2002C (200W, 2450MHz) used in Molecular Beam Epitaxy (MBE) equipment (Figure 17). Comparing to normal ion sources they all have shown fine results and greatly increase the ion source's life.

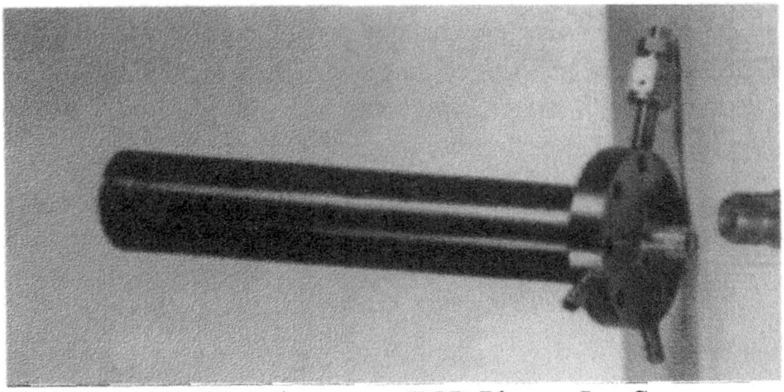

Figure 17 – 200W, 2450MHz Microwave ECR Plasma Ion Source

SUMMARY

MPA in industrial fields has developed for over thirty years both in western countries and China and seemed quite difficult to form a big business, but if we pay special attention to reliable operation, high performance and low cost, customers real requests and good service, there will certainly be great potential for market development. MPA in chemistry is a new field and now just in the research market. Think of those traditional huge chemical engineering projects: if microwave energy would be accepted, there will be a lot of good work for us. Microwave energy stimulated plasma associated with processing of new materials and many other applications are great opportunities, technological challenges and even show better perspective. All these science areas in China now have achieved rapid development, but we are still behind compared to advanced western countries, especially in processing technology and equipment. We are confident to catch up soon through our great efforts and very good situation in China.

REFERENCES

[1] Jes Asmussen, Handbook of Plasma Processing Technology, 1994

[2] Ronard D. Taylor, Francis Perkins, Plasma Processing of Materials, National Academy Press, Washington D.C., 1991

[3] D. R. Wur, J. L. Davidson, Proceedings of the 7th International Conference on Solid State Physics and Actuators, 1993

[4] Seitaro Matsuo, and Mikiho Kiuchi, Japanese Journal of Applied Physics, Vol. 22, No. 4, April 1983

Electronic Materials

ELECTROMAGNETIC INDUCTION HEATING FOR ULTRA SHALLOW JUNCTION FORMATION

K. Thompson, J.H. Booske*, R.F. Cooper,
University of Wisconsin
1415 Engineering Drive
Madison, WI 53706

Y.B. Gianchandani
University of Michigan
2417A EECS
Ann Arbor, MI

ABSTRACT

The rapid heating, >125°C/sec, of silicon wafers to very high temperatures, > 1000°C, with microwave radiation, 2.45 GHz, has been applied as a rapid anneal technology for advanced CMOS processing. This heating technique repairs the damage introduced during the implantation process, activating a large fraction of the implanted dopants without causing substantial diffusion of the junction. Microwave spike anneals, with zero soak time at the target temperature, of boron implanted silicon have created p-type junctions with sheet resistances of 550 ohms/square at a depth of only 25 nm. These junctions satisfy the 90-nm CMOS technology node as established by SEMATECH. Further advances in microwave spike anneal technology, including ramp rates on millisecond time scales, should continue to push the limits of this technology.

INTRODUCTION

Advances in the microelectronics industry are driven by a continuous scaling down of the physical dimensions of the features that comprise silicon technology devices. Smaller transistors switch faster, require less power, and consume less space on a silicon wafer. This allows a greater number of more complex chips to be built on a single wafer. After twenty years of continuous advances in silicon processing technology, traditional methods are so mature that future improvements are increasingly difficult to achieve. To maintain the current pace of technology development, innovative and novel processing techniques must be developed.

One specific process that is considered fully mature is the implant and anneal process [1]. In a typical transistor structure, Fig. 1, as the lateral gate dimension shrinks, the depth of the implanted source drain regions must become shallower and contain a higher concentration of implanted dopants. This ensures that good isolation exists between the contact regions and the underlying channel, that the contact resistance is low, and that device degradation during operation is kept to a minimum. To aid in achieving these goals, a highly doped (average concentration of dopants should exceed $10^{20}/cm^3$), extremely shallow (less than 20 nm by 2005) extension region is added. Because the dopants are introduced through ion implantation, a process that leaves the surface of the wafer heavily damaged, a high temperature anneal must be performed to repair the damage and to move the implanted boron atoms onto silicon sites, where they become electrically active.

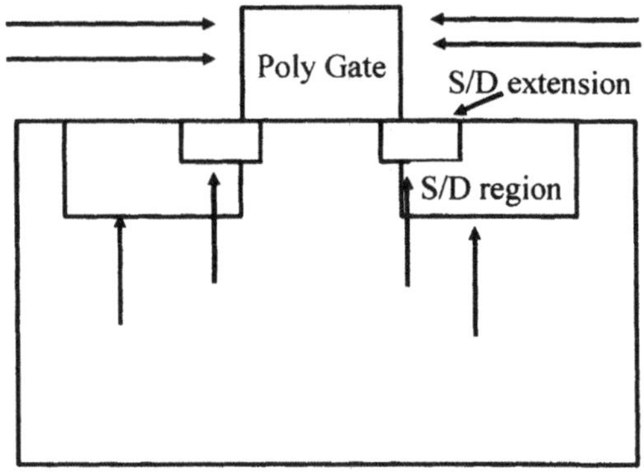

Fig. 1. Typical transistor structure. The arrows indicate the shrink direction.

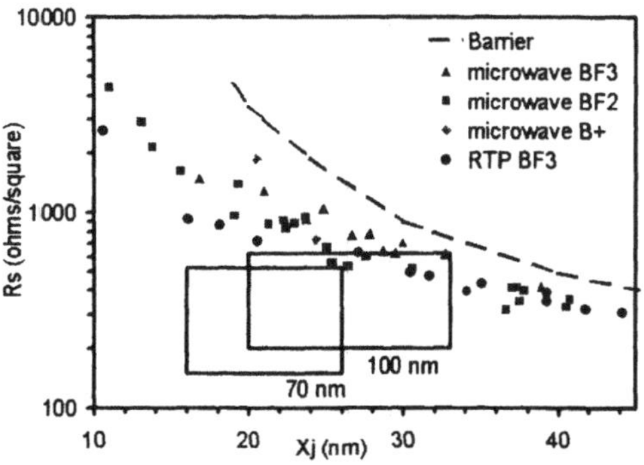

Fig. 2. A classic engineering trade-off is exhibited by the sheet resistance and junction depth requirements for source/drain extension formation in CMOS technology devices. For a given anneal technology, improvements in dopant activation are met by increases in dopant diffusion.

The inherent drawback to this scheme is that high temperature processing drives the dopants deep into the wafer, diluting the concentration and eliminating any benefit generated by the source/drain extensions. This creates a classic engineering trade-off between diffusion and activation as seen in Fig. 2, which plots the sheet resistance, Rs, a measure of active dopant concentration, versus junction depth, Xj, a measure of diffusion. For a given anneal technology any improvement in dopant activation is met by an equivalent increase in diffusion, and vice-versa. As a consequence, small changes in key operating parameters – e.g. final temperature, implant dose, or implant energy – stimulate movement of the Rs-Xj data point up and down an imaginary line, as shown in Fig. 2 by the dashed line. This line, known as the SEMATECH technology barrier [2], represents the limit in conventional anneal technology. Moving this curve to the south, by enhancing activation without increasing diffusion, or to the west, by restricting diffusion without impeding activation, can only occur through the introduction of a new anneal technology.

Current rapid thermal processing, RTP, strategies utilize lamp-based heating systems. Microwave RTP offers several advantages over lamp-based annealing. First, while the heating efficiency of silicon with lamp-based RTP systems decreases with temperature, it increases during microwave annealing [4]. The result is a sharper spike profile near the target temperature and the potential for higher ramp rates as more efficient microwave furnaces are constructed. This combination of faster ramp rates and minimized dwell time at the target temperature is known to reduce diffusion while maintaining a given level of activation [8]; effectively moving the sheet resistance-junction depth technology curve straight west. Another advantage is the greatly reduced design complexity for microwave RTP systems. Lamp-based systems struggle to maintain uniform heating patterns and to eliminate "hot" and "cold" spots on the wafer. It has been shown for microwave heating that the creation of a uniform electric flux in the cavity, through the introduction of hundreds of modes [3], directly results in a similarly uniform heating pattern across the wafer. Finally, the efficiency of lamp-based systems is limited by the amount of optical radiation that a silicon wafer can absorb at a given temperature. This limit has so far proven inadequate for advancing RTP technology beyond the 90 nm node [1]. While the absorption efficiency of microwave radiation also faces physical limits, the potential to heat silicon to temperatures in excess of 1200°C on time scales of hundreds of microseconds is currently under investigation. Such heating rates will open the door for zero-diffusion, high temperature anneal strategies, which are critical to achieving the most advanced technology nodes listed in the semiconductor technology road map.

One such novel technique, electromagnetic induction heating, EMIH, utilizes RF and microwave radiation to rapidly (125°C/sec) heat silicon wafers to temperatures in excess of 1050°C. The heating experiments discussed in this paper were performed with microwave radiation in a resonant cavity of radius 17 cm, Fig. 3 [4]. Up to 3000 watts at a frequency of 2.45 GHz was available from a magnetron source, but to date no more than 1500 watts has been necessary. The height of the cavity was left adjustable to tune in specific modes, and a radial tuning stub helped minimize reflected power. Although several modes resonate at 2.45 GHz, the dominant TM111 was used for all experiments. Modifications allow for specific control over the chamber ambient, especially the concentration of oxygen during the anneal process. The wafer was supported by four quartz supports, one at each corner, with a height of 1 mm above the cavity bottom to ensure that it was in a magnetic field maximum. Temperature was measured by an optical pyrometer, fed through the bottom of the chamber.

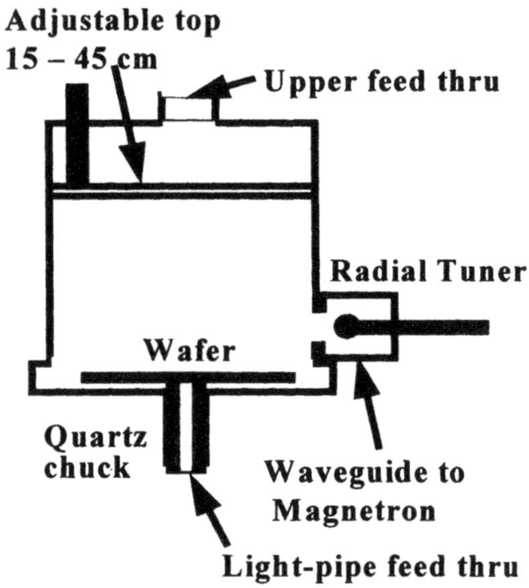

Fig. 3. Resonant cavity used to heat the silicon.

A plot of the temperature transient during microwave "spike" annealing is shown in Fig. 4. These rapid heating rates allow the silicon to be heated to the target anneal temperature, where the implant damage can be

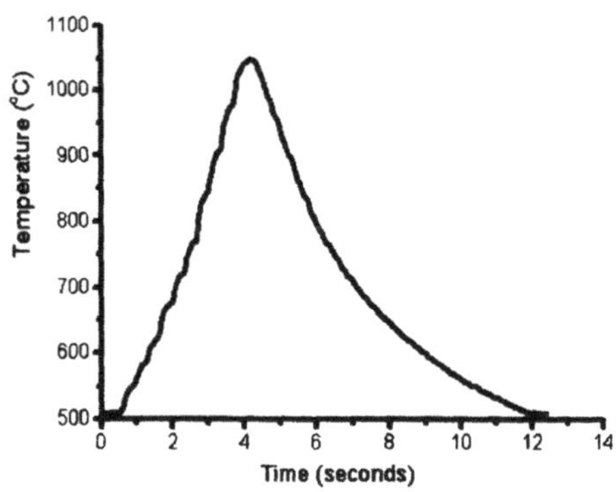

Fig. 4. Typical heating transient for a microwave spike anneal.

repaired and the dopants activated, and then cooled to room temperature in a very short amount of time. This severely restricts the amount of diffusion that can occur, maintaining the integrity of the source/drain extension, while still activating a large concentration of the implanted dopants.

ULTRA SHALLOW JUNCTION RESULTS

To test the effectiveness of microwave annealing for ultra-shallow junction formation, several samples were shallow implanted with boron and spike annealed. The results were then compared to present day state of the art annealing techniques using lamp-based RTP. The B^+ and BF_2^+ implants were performed by Varian Semiconductor Equipment Associates (VSEA) using a VIISTA80 beam-line implanter. The BF_3 P^2LAD samples were also implanted by VSEA using a technique called pulsed plasma doping [5] with a BF_3 source gas. Due to the nature of plasma doping, the implanted species is composed of a variety of ionic species, i.e. BF_3^+, BF_2^+, BF^+, B^+, with the majority species being BF^+.

It has been shown, for lamp-based RTP, that annealing in an ambient whose trace oxygen concentration, $[O_2]$, is maintained at 100 ppm in a nitrogen purge restricts the diffusion of boron, through elimination of oxygen-enhanced-diffusion effects [6]. Preventing the oxidation of the silicon surface also lowers the sheet resistance by preventing significant dose consumption by the newly grown oxide. As a result, the microwave anneals were performed both in an ambient whose $[O_2]$ was limited to 100 ppm and whose $[O_2]$ was uncontrolled (~21% oxygen).

Results from the spike anneal process were quantified through sheet resistance and SIMS measurements. Sheet resistance, which is a direct indication of active dopant concentration, was determined with a 4-pt. probe. The active dose, or number of electrically active boron atoms per square centimeter, was then determined by dividing the measured sheet resistance into $1.25*10^{17}$ [7], a number that accounts for the mobility and electronic charge of boron in silicon. The SIMS measurements provide the average volumetric dopant concentration as a function of depth in the wafer. This data is critical for quantifying the amount of diffusion that has occurred during the anneal and the "boxiness" of the final profile.

Boron concentration profiles for microwave spike annealed implants of 500 and 1100 eV ($10^{15}/cm^2$ of BF_2) are shown in Fig. 5, and 6. The 500 eV implant resulted in a fairly box-shaped junction whose final depth (measured at a boron concentration of $10^{18}/cm^3$) was 27 nm. The sheet resistance of this sample was 842 ohms/square, and the total integrated dose of boron contained within the silicon was $3.73*10^{14}/cm^2$. Compared to the total active dose (calculated as $1.25*10^{17}$ divided 842) of $1.5*10^{14}/cm^2$, approximately 40% of the implanted boron is active. The 1100 eV implant, Fig. 8, had a junction depth of 41 nm and a sheet resistance of 396 ohms/square. The total integrated dose was $4.14*10^{14}/cm^2$ and 76% of the implanted dose was activated. Shown on the same plot is a microwave spike anneal to 1050°C in an uncontrolled ambient. Although the slightly lower sheet resistance of 308 ohms/square results in an active boron concentration of 89%, the junction is significantly deeper. As was discussed earlier, the excellent activation comes only at a diffusion cost that is unacceptably high for advanced CMOS devices.

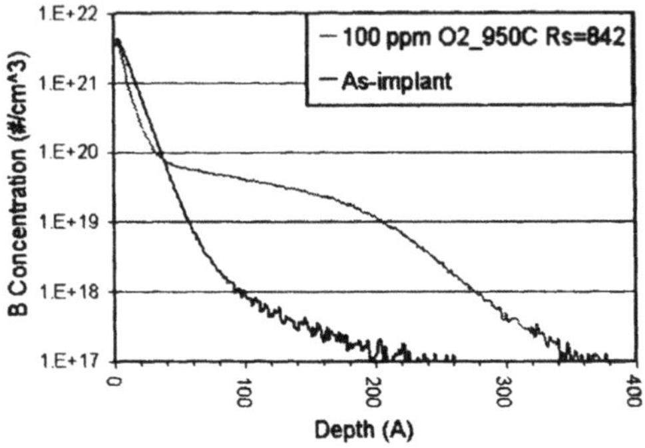

Fig. 5. SIMS results for a microwave spike anneals to 950°C for a 500 eV, $10^{15}/cm^2$ BF_2 implant. The sheet resistance is 842 ohms/square and the junction depth is 29 nm.

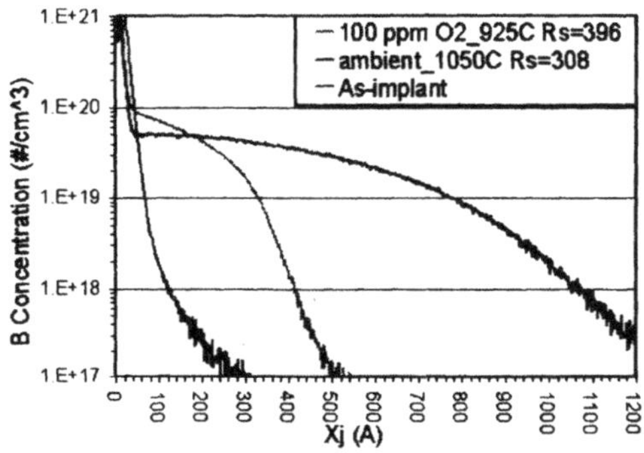

Fig. 6. SIMS results for microwave spike anneals to 925°C and 1050°C for a controlled ambient of 100-ppm $[O_2]$ and an uncontrolled ambient. The implant was 1100 eV, $10^{15}/cm^2$ BF_2.

While much of this junction diffusion can be attributed to the higher anneal temperature, the oxygen enhanced diffusion played a significant role as well. To study this in more detail, a series of 4500 eV, $10^{15}/cm^2$ BF_2 implants were microwave spike annealed to 1000°C in controlled ambients whose background oxygen concentration was varied from 33 ppm up to 210000 ppm. The results, Fig. 7, indicate that the $[O_2]$ in the anneal ambient does play a role in boron diffusion during microwave annealing. Reducing the oxygen concentration limits diffusion, and the junction depth is reduced by 6 nm as the amount of oxygen present during the anneal is purged down to 33 ppm. The sheet resistance initially decreases with lower oxygen concentration until it saturates and rises slightly below 33 ppm. The drop in sheet resistance is attributed to the minimization of dose loss in the extreme near surface of the silicon.

DISCUSSION

These anneals, as well as several other microwave spike anneals, performed in an ambient whose [O_2] was held to 100 ppm, form a new technology line that falls below the SEMATECH technology curve that was shown in Fig. 2. Also shown for comparison are several flash-lamp RTP anneals, performed in a controlled low ppm oxygen environment by Lerch et. al. using an advanced flash lamp RTP system recently developed at Mattson [8]. Both the microwave and the advanced lamp-based sheet resistance-junction depth data points form a very similar curve, indicating that while the heating mechanisms are very different, the final result is very similar. This clearly illustrates that the activation and diffusion mechanisms of boron in silicon are driven almost entirely by thermal forces.

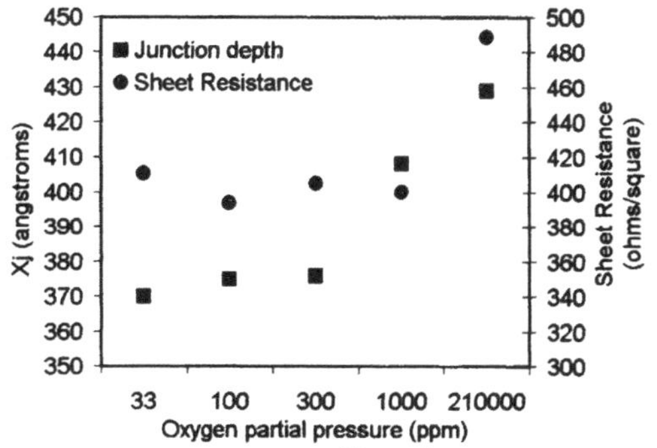

Fig. 7. 4500 eV, $10^{15}/cm^2$ BF_2 implant samples microwave spike annealed to 1000^oC at different background oxygen concentrations to illustrate the effect of oxygen in the ambient on the final profile obtained.

There are three identifiable physical differences between microwave and lamp-based RTP. The most obvious is that the electromagnetic frequency for microwave radiation is 3 – 5 orders of magnitude lower than optical radiation. Second, the electric and magnetic fields during microwave RTP penetrate into the wafer (especially at lower frequencies), while the optical radiation of lamp-based RTP has an extremely small skin depth, and couples only to the wafer surface. The third difference is a corollary of the first: optical photon energies are high enough to photo-activate chemical processes (compound formation and dissociation, defect production, etc.), whereas microwave photons are much too low in energy to cause such effects. In this last sense, microwave annealing is performed "in the dark", while lamp-based RTP illuminates the wafer with high intensity radiation in the infrared, optical, and ultraviolet regimes. It has been speculated by some that the photo-generation process, which is stimulated by this high intensity optical illumination, acts to decrease the diffusion of boron in silicon [8]. The effect was studied for spike anneal technology by performing comparison anneals using both microwave and lamp-based anneal techniques.

Silicon samples were implanted with either BF_2 at a dose of $10^{15}/cm^2$ and 4 keV energy, or with B^+ at a dose of $10^{15}/cm^2$ and 0.75 keV energy. The B^+ wafers were pre-amorphized with a dose of $2*10^{15}/cm^2$ Ge to avoid channeling effects. These implant conditions were specifically chosen to study the influence that F may have on the anneal process. Samples taken from each wafer were either microwave or lamp annealed, in a controlled ambient of 100-ppm oxygen, and direct comparisons between the two techniques were made.

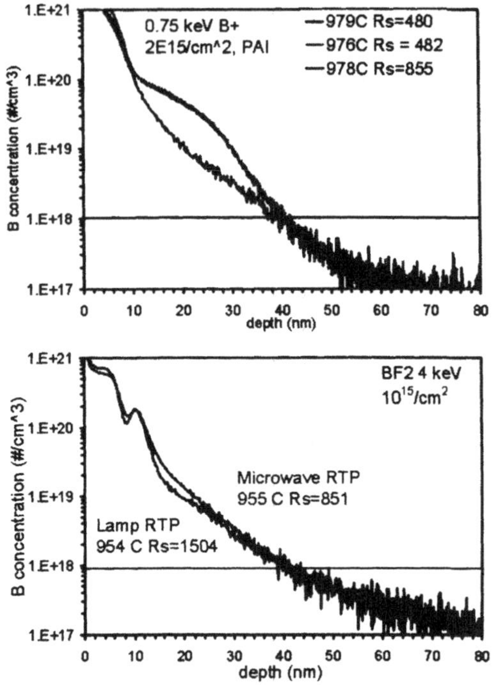

Fig. 8, 9. Boron concentration profiles from direct comparison anneals of microwave and lamp-based RTP systems.

The results, Figs. 8 and 9, show very intriguing phenomena. In Fig. 8, the B^+ implants annealed by microwaves suggest an enhanced near-surface diffusion compared with lamp-based RTP. While the final junction depth, measured at $10^{18}/cm^3$, is the same for both anneal techniques, the $10^{19}/cm^3$ concentration is much deeper for microwave RTP. The result is a more box-shaped profile. The microwave sheet resistance (480 ohms/square) is half that of the lamp-based RTP sample, a result of more activated boron atoms associated with the box-shaped profile. It should be mentioned that a similar box-type profile and activation (sheet resistance) *could* be obtained with lamp-based RTP, but at approximately 100°C higher temperature than with microwave annealing. The BF_2 implants in Fig. 9, however, show very little difference in the boron concentration profiles between the two anneal techniques. While there is a slight amount of increased diffusion in the extreme near surface layer, it is much less than that seen for the B^+ only implant. It is clear, therefore, that fluorine plays an important role in affecting B diffusion. At the same time, however, the microwave-annealed sample's sheet resistance is one-half that of the lamp-annealed sample, showing the microwave annealing results in much greater activation than lamp-based RTP at the indicated temperature. With F co-implantation, lamp annealing *can* achieve the same activate as microwave annealing, but, again, at approximately 100°C higher temperature than with microwave annealing.

The impact of fluorine on the diffusion of boron has been studied for some time. Downey clearly shows the usefulness of fluorine in restricting the diffusion of boron in silicon [9], and Robertson et. al. have shown that the impact of fluorine is a purely chemical effect and not related to more traditional damage effects [10]. Finally, Fair, in his studies of boron diffusion through thin gate dielectrics, details how fluorine prevents the optically enhanced diffusion of boron through silicon dioxide [11]. The latter effect is of greatest interest in this case. Through his work, Fair has conclusively shown that the diffusion of B through SiO_2 is significantly more rapid when the anneal is performed in a lamp-based RTP system instead of in a conventional furnace. He has further shown that the presence of F causes the diffusivity of boron in SiO_2 to be the same regardless of whether the anneals are performed in a furnace or through lamp-based RTP. His work indicates that F eliminates the kinetic mechanism introduced by high intensity optical radiation, which enhances the movement of boron through SiO_2. It is foreseeable that F has the same effect on the motion of B in silicon. While the enhancement effect of the illumination on diffusion would be reversed in the two cases, the general trend should hold true. This is what is indicated in Figs. 8 and 9.

CONCLUSION

A novel rapid thermal processing technique, RTP, microwave annealing, is capable of creating ultra shallow junctions or CMOS technology that satisfy the 90 nm technology node. These junctions are on a level equal to those formed using the most state of the art lamp-based RTP systems. The most significant differences between the two techniques include: (1) microwave annealing results in higher boron activation than lamp annealing at the same temperature, (2) without F co-implantation, microwave annealing produces more near-surface diffusion, a more "box-shaped" profile, and more boron activation than lamp-annealing, (3) with F co-implantation, the two methods produce similar diffusion profiles but the microwave annealing still provides more boron activation for identical temperatures.

It remains to be seen whether these physical differences in the reaction chemistry are sufficient to motivate commercial adoption of microwave RTPP over lamp RTP for next generation technology objectives. To that end, microwave annealing offers several other potential advantages over lamp-based RTP. First, while the heating efficiency of silicon with lamp-based RTP systems decreases with temperature, it increases during microwave annealing [3]. The result is a sharper spike profile near the target temperature and the potential for higher ramp rates as more efficient microwave furnaces are constructed. This combination of faster ramp rates and minimized dwell time at the target temperature is know to reduce diffusion while maintaining a given level of activation [7], effectively moving the Rs-Xj technology curve straight west. Another advantage is the greatly reduced design complexity for microwave RTP systems. Lamp-based RTP systems struggle to maintain uniform heating patterns and to eliminate "hot" and "cold" spots on the wafer. These "hot" and "cold" spots are created by uncontrolled reflections f high intensity radiation which bound around the non-absorbing chamber walls until finally begin absorbed by the wafer. It has been shown for microwave heating that the creation of a uniform electric flux in the cavity, through the introduction of hundreds of modes [13], directly results in a similarly uniform heating pattern across the wafer. Finally, the efficiency of lamp-based systems is limited by the amount of optical radiation that a silicon wafer can absorb at a given temperature. This limit has so far proven inadequate for advancing RTP technology beyond the 90 nm node [1]. While the absorption efficiency of microwave radiation also faces physical limits, the potential to heat silicon to temperatures in excess of 1200°C on time scales of hundreds of microseconds is currently investigation. Such heating rates will open the door for zero-diffusion, high temperature anneal strategies, which are critical to achieving the most advanced technology nodes listed in the semiconductor technology road map.

ACKNOWLEDGMENTS

The authors would like to acknowledge Dan Downey, Sue Felch, and Edwin Arevalo of Varian Semiconductor Equipment Associates for their generous help providing state of the art PLAD BF_3 and beamline BF_2 implanted samples along with SIMS analysis and numerous lengthy discussions concerning anneal technology. Finally we are grateful to Intel for supporting Keith Thompson through an Intel Fellowship, and to the Wisconsin Alumni Research Foundation, the Civilian Research Development Foundation and the National Science Foundation for funding this work.

REFERENCES

1. J. O. Borland, T. Matsuda, K. Sakamoto, "Shallow and Abrupt Junction Formation: Paradigm shift at 65-70 nm," Solid State Technology, p. 83 – 94. (June 2002)

2. www.sematech.org

3. Y. Bykov, A. Eremeev, V. Holoptsev, I. Plotnikov, N. Zharova "Spike annealing of silicon wafers using millimeter-wave power," presented at the 9th International Conference on Advanced Thermal Processing of Semiconductors, IEEE RTP 2001, pp. 82, Anchorage, Alaska, 25-29 Sept. 2001.

4. K. Thompson, Y.B. Gianchandani, J.H. Booske, R.F. Cooper, "Direct silicon-silicon bonding by electromagnetic induction heating," Journal of Microelectromechanical Systems, V. 11, p.285 – 292. (August 2002)

5. D. Lenoble, M. J. Goeckner, S. B. Felch, Z. Fang, J. Galvier, A. Grouillet, "Evaluation of plasma doping for sub-0.18 micron devices," presented at the 1998 International Conference on Ion Implantation Technology, p. 1276. (1998)

6. D. F. Downey, J. W. Chow, W. Lerch, J. Niess, S. D. Marcus, "The effects of small concentrations of oxygen in RTP annealing of low energy boron, BF2 and arsenic ion implants," presented at the Materials Research Society. (1999)

7. Sheng. S. Li, "Semiconductor Physical Electronics," Plenum Press, New York, p.135. (1993)

8. W. Lerch, B. Bayha, D.F. Downey, E.A. Arevalo, "State of the art techniques for ultra-shallow junction formation," in Rapid Thermal and Other Short-Time Processing Technologies II, p. 321 of Proceedings of the 199th meeting of the Electrochemical Society, Washington DC. (25 March 2001)

9. D.F. Downey, J.W. Chow, E. Ishida, K.S. Jones, "Effect of Fluorine on the diffusion of boron in ion implnated Si," Applied Physics Letters, Vol. 73, p. 1263-1265 (31 August 1998)

10. L. S. Robertson, P. N. Warnes, K. S. Jones, S. K. Earles, M. E. Law, D. F. Downey, S. Falk, J. Liu, "Junction depth reduction of ion implanted boron in silicon through fluorine ion implantation," in Si Front-end Processing – Physics and Technology of Dopant-Defect Interactions II, Proceedings of the Materials Research Society. Vol. 610, p. B4.2.1 (2000)

11. R.B. Fair, "Anomalous B Penetration Through Ultra thin Gate Oxide During Rapid Thermal Annealing," IEEE Electron Device Letters, p. 466-468. (Sept. 1999)

MICROWAVE SINTERING OF INTEGRATED PASSIVE DEVICES

B. Vaidhyanathan, K. Annapoorani and J. Binner
Institute of Polymer Technology and
Materials Engineering
Loughborough University
Loughborough, Leicestershire
LE11 3TU, United Kingdom

R. Raghavendra
Littelfuse Ireland Ltd
Ecco Road
Dundalk, Co. Louth
Ireland

ABSTRACT

Microwave sintering of Integrated Passive Devices (IPDs) has been carried out for the first time. The sintered samples were characterised for density, microstructure, composition and electrical performance. It was found that IPDs with varistor/capacitor formulations could be microwave sintered to more than 98% densification within 3 hours of total cycle time, which is one tenth of the time required by conventional methods. Microwave sintering resulted in better products with finer grain structure and the electrodes in the multilayer devices were kept intact without delamination. There was no significant inter-diffusion observed between the varistor/capacitor and ceramic/electrode interfaces. The microwave technique also completely eliminated the separate binder burnt-out step. The electrical properties of the sintered samples were also found to be better or matching with the required values.

INTRODUCTION

Multilayer varistors and capacitors constitute the most important class of dielectric devices and correspond to a billion dollar market. Demands for internet capability, multiband access, video and more increasingly miniaturised mobile products are forcing electronic manufacturers to reassess the ways in which products are designed and assembled. The present world trend in the electronic industry is towards miniaturisation. Although the size reduction decreases overall product size, a finite board space is still needed per component for assembly considerations and the number of components remains the same. Hence recently multilayer structures incorporating varistors and capacitors are fabricated together[1] to form a single component that provides the advantages of both. Such components are known as Integrated Passive Devices (IPDs) and offer significant benefits in terms of space reduction and assembly cost.

The key step that determines the electrical properties of the final component during IPD manufacture is the sintering process. However the conventional procedures of processing these multilayer devices, involving separate multi-step binder-burnout and long sintering schedules and are often of lengthy duration[1]. Also conventional co-firing methods of integrated passive devices suffer from electrode de-lamination and unwanted inter-diffusion of matrix components at high processing temperatures. The higher sintering temperatures required to achieve reasonable densification, and the associated high energy consumption, necessitates strategies based on alternative processing methodologies. IPDs have a considerable commercial market and any improvements made in the processing methodology will have an impact. The present study aims at using microwave methodology for the efficient processing of these devices. Though microwave technique has been used for the processing of a number of inorganic materials[2-5], it has not been employed for the sintering of IPDs to our knowledge to date.

EXPERIMENTAL

IPD samples with a capacitor/varistor formulation namely, v18mla0805, which has capacitor on one side and varistor on the other side, was provided by Littelfuse Ireland Ltd. The capacitor formulation is bismuth niobate doped with magnesium oxide and the varistor formulation consists of 98% of zinc oxide and 2% addition of antimony oxide, bismuth oxide, cobalt oxide, etc. The microwave cavity used is capable of providing a maximum output power of 6 kW and operating at 2.45 GHz. Inconel sheathed, shielded K-type Cr/Al thermocouples and an infrared pyrometer (LAND instruments, UK) were used for temperature measurements. The thermocouples were calibrated using the boiling point of water and melting point of vanadium oxide in the microwave cavity. Microwave sintering experiments were performed at 950°, 1000°, 1025°, 1050° and 1100° C. The soaking time has been varied between 20-60 minutes. Fiberfrax high purity insulation was used as a casket for heat containment.

Density measurements were performed using the Archimedes method and 15 representative samples were taken for every set of conditions and the average density values are taken. Comparative conventional heating experiments were performed using an electric furnace. Sintered samples were investigated using optical microscopy (Olympus Microscopes, Japan) for electrode delamination and interface cracking. Scanning electron microscopic measurements were carried out using SEM; Leica Cambridge Stereoscan S360 and the extent of diffusion of niobium and bismuth at the interface were studied using energy dispersive X-ray analysis (EDX). The electrical tests were carried out at Littelfuse (Keithley instrument: Model 2410), Ireland and the results have been used as a pointer for optimising the experimental parameters in the microwave sintering trials.

RESULTS AND DISCUSSION

The time-temperature profile for the microwave and conventional processing of IPDs is given in figure 1. The time and energy saving efficiency of the microwave methodology is evident and the microwave heating requires only one tenth of the time compared conventional methods for the processing of IPDs Also the microwave procedure does not necessitate a separate binder burn out stage (a typical conventional procedure involves a slow binder removal stage), which is an important advantage for the efficient processing of electroceramics[6].

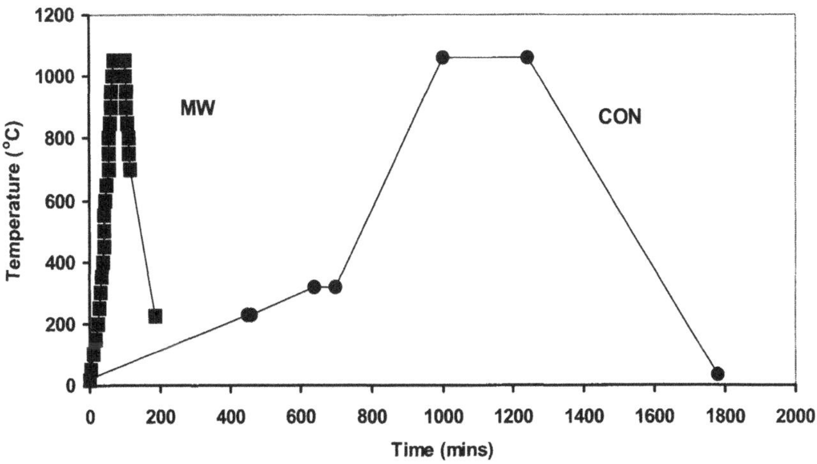

Figure 1: Comparison of microwave and conventional t-T profiles for the processing of integrated devices.

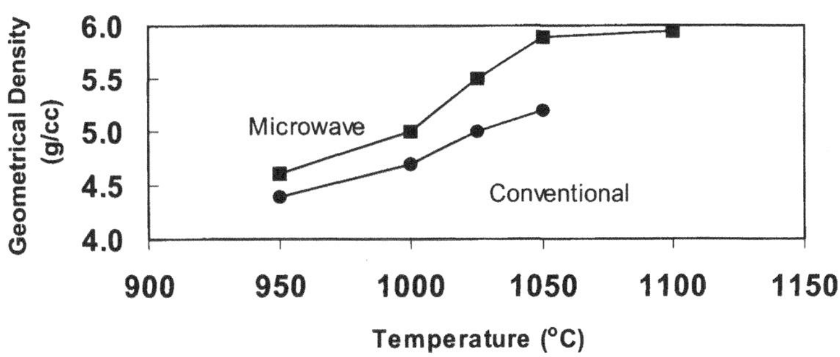

Figure 2: Densification of IPDs in microwave and conventional sintering

a) Density measurements

Figure 2 provides the densification achieved in microwave and conventional processing of IPDs. It is clear that at any given temperature the microwave process provides much higher densification. The figure also suggests that the density values of the microwave sintered samples tend to saturate around 1050°C. At higher temperatures (>1050°C), the capacitor region sintered more than the varistor part, leading non-uniform shrinkage and de-lamination. The microwave sintering experiments were also performed at various soaking times ranging from 20-60 minutes and the densification increased with increasing sintering time.

b) Microstructural characterisation and composition analysis

Optical microscopic investigations (figure 3) revealed that the metal electrodes in the microwave sintered multilayer devices were intact without de-lamination and no electrode 'buckling' was observed, a common problem due to non-uniform (conventional) sintering in these devices.

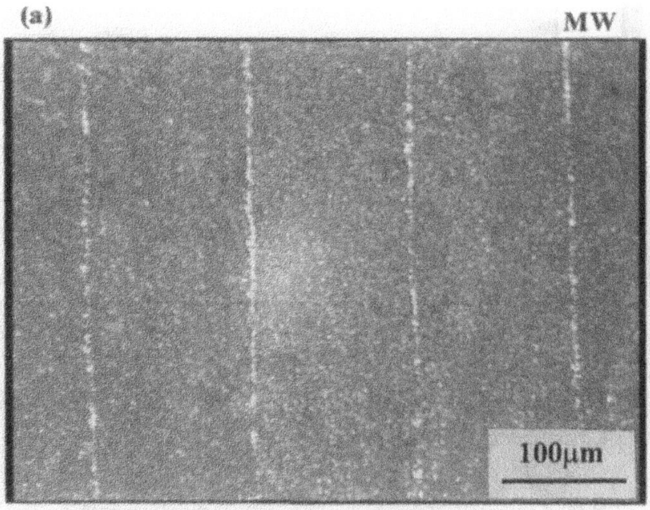

Figure 3. Optical micrograph of the microwave sintered IPD sample- the varistor side; indicating that the electrodes are intact without de-lamination.

The SEM micrographs in figure 4 provide the comparison of the interfaces of the microwave and conventionally sintered IPDs at low and high magnifications. It is evident that the varistor/capacitor interface is clear, distinctly preserved and that no inter-diffusion can be seen in the microwave-processed samples. The average grain size was found to be 4 μm on the varistor side and 7 μm on the capacitor side.

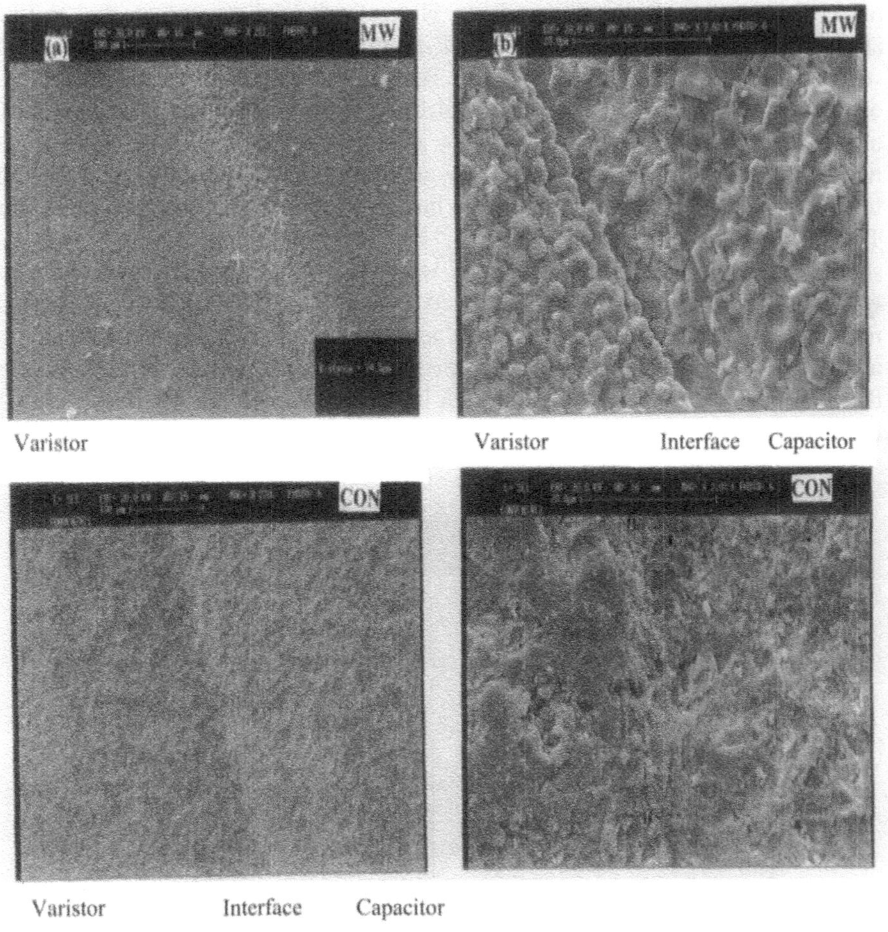

Figure 4. Comparison of the interfaces of the microwave sintered and conventionally sintered IPD samples at a) low and b) high magnification. Note that the interface is clear, distinctly preserved and that no inter-diffusion is seen in the MW sample.

The microwave sintered samples were further characterised using EDX for Bi and Nb concentrations as well as for overall composition, since inter-diffusion is found to be the major problem in long duration conventional processing of these devices. Figure 5 gives the Nb line analysis of the sample microwave sintered for 45 mins. As seen, there is no appreciable amount of Nb observed in the varistor region. Since the Nb lines overlap with that of the electrode material, sharp peaks are seen at the electrode positions (this also indicates the well-preserved nature of the electrodes). This clearly indicates that the rapidity associated with the microwave method diminishes the diffusion of Nb into the varistor region from the capacitor side. Recently it was observed that in the case of conventionally cofired composite MLCC's there is an appreciable amount of electrode material migration into the ceramic layers[7]. In this context, the microwave sintering appears to be very attractive since it not only minimises the inter-diffusion of ceramic components but also reduces the electrode migration. The Bi line analysis on the 45 mins MW sintered sample (figure 6), reveals significant Bi in the capacitor side and wherever the electrode was present there was a drop in the intensity observed, since there is no Bi found in the electrode region. This reveals that the electrodes are well preserved in the capacitor region also. The base line for Bi in the varistor region was not quite smooth, indicating the presence of very small amount of Bi in the varistor region, which is generally expected.

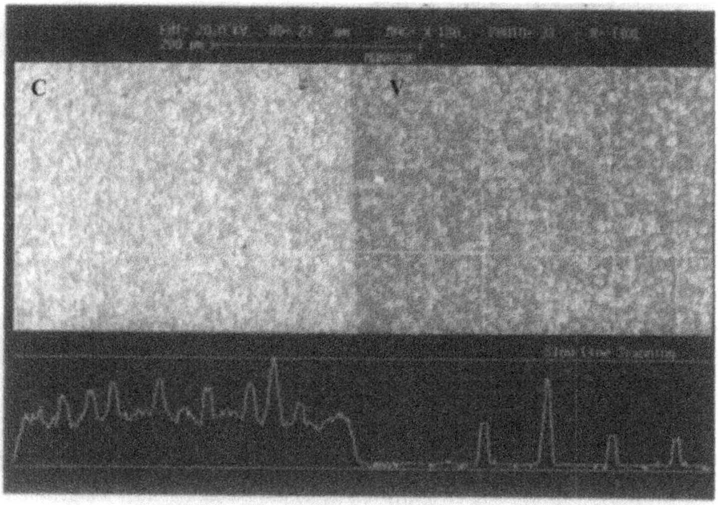

Figure 5: Line analysis of Nb on the sample sintered at 1050°C /45 mins

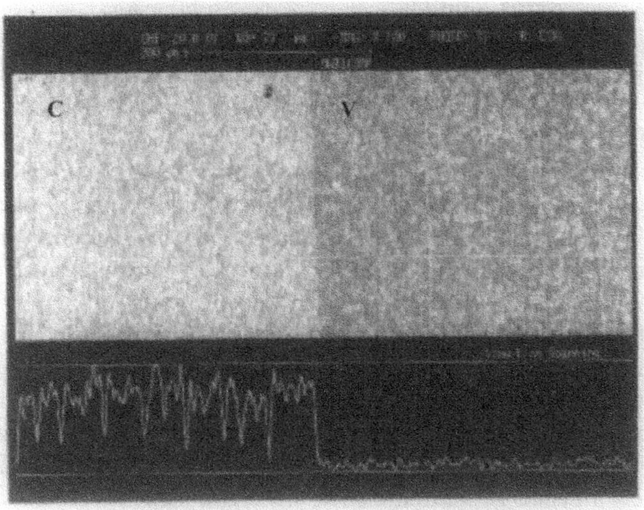

Figure 6. Line analysis of Bi on the sample sintered at 1050° C / 45 mins.

c) Electrical Performance

Table 1. Electrical characteristics of mw sintered IPD v18mla0805

Sintering conditions	**Vnom**	**Variance**	**Clamp ratio**	**Leakage**
1050°C / 30 mins	45.23	6.60	1.290	0.927
1050°C / 45 mins	38.53	6.93	1.343	0.925
1050°C / 30 mins (Flowing air + better casket design)	43.57	3.93	1.323	0.930
Required values	**>25**	**~3.5**	**<1.5**	**>0.8**

The electrical properties of the sintered IPD samples were measured at Littelfuse, Ireland and the observed values are tabulated (see Table 1). It is clear that the electrical performance of the microwave sintered IPDs were found to match, and in most cases better, the required values. Especially high values of Vnom, very good values of clamp ratio and better leakage characteristics were observed. However initially the parameter 'Variance', which is the deviation in the Vnom values, was found to be higher than the expected level and could be due to the lack of thermal homogeneity during the processing. This was circumvented by adopting a better casket design for heat containment and passing a burst of air to achieve thermal equilibrium during the microwave processing of these multilayer devices (see Table 1). Use of secondary susceptors was also found to be beneficial.

CONCLUSIONS

It has been found that Integrated Passive Devices can be sintered using microwaves to more than 98% densification within 3 hours of total cycle time. This is one tenth of the time required by conventional methods. The microstructural investigations revealed that the microwave sintering results in better products with finer grain structure. The electrodes in the multilayer devices were kept intact without delamination. There was no significant inter-diffusion noticed between the varistor/capacitor and ceramic/electrode interfaces. The microwave technique also completely eliminated the separate binder burnt-out step. The electrical properties of the sintered samples were also found to be better or matching the required values. In general the simplicity, rapidity and better product performance of the microwave technique makes it an attractive sintering methodology for the processing of IPDs.

AKNOWLEDGEMENTS

The authors thank Littelfuse Ireland Ltd for their valuable support in carrying out this work. One of the authors (BV) also thanks the EPSRC for financial assistance.

REFERENCES

[1]F.J. Toal, J.P. Dougherty and C.A. Randall, "Processing and Electrical Characterization of a Varistor-Capacitor Multilayer Device", *J. Amer. Cer. Soc.*, **81** (9), 2371-80 (1998).

[2]J.G.P. Binner, T.E. Cross and Y.Yin. "Microwave assisted CVI for ceramic matrix composites", *Ceram. Trans.*, **80**, 349-356 (1997).

[3]L.P. Martin, D. Dadon, M. Rosen, D. Jershon, K. Rybakov, A. Birman, J. Calame, B. Levush, Y. Carmel and R. Hutcheon, "Effects of anomalous permittivity on the microwave heating of zinc oxide", *J. Appl. Phys.*, **83**, 432-437, (1998).

[4]B. Vaidhyanathan, P. Raizada and K.J. Rao, " Microwave assisted fast solid state synthesis of niobates and titanates", *J. Mater. Sci. Lett.*, **16**, 2022-2025 (1997).

[5]Schulz, R.L., Folz, D.C., Clark, D.E. and Wicks, G.G., "Microwave Destruction/ Vitrification of Electronic Components", *Ceram. Trans.* **36**, 81-88 (1993).

[6]D.K. Agrawal, R. Raghavendra and B. Vaidhyanathan, " Production of Passive Devices", **US. Patent. 6,399,012,** (2002).

[7]R. Zuo, L. Li, Y. Tang and Z. Gui, " Mutual diffusion at hetrogeneous interfaces in co-fired composite MLCCs", *British Ceram. Trans.*, **100** (1), 38-40, (2001).

Waste Treatment and Recycling

MICROWAVE WASTE REMEDIATION: AN UPDATE ON APPLICATIONS AND CURRENT RESEARCH

R.L. Schulz
Corning, Inc.
SP-PR-01-3
Corning, NY 14830
schulzrl@corning.com

D.E. Clark and D.C. Folz
Virginia Polytechnic Institute and State University
Dept. Materials Science and Engineering
231 Holden Hall
Blacksburg, VA 24061
dfolz@vt.edu

G.G. Wicks
Westinghouse Savannah River Co.
PO Box 616
Aiken, SC 29803

ABSTRACT

As the population of the world increases, society is faced with the challenges associated with responsible waste management. Many landfills are at or near capacity, businesses are faced with stricter regulations regarding disposal of chemical and solid wastes. Air pollution abatement is essential for private and commercial vehicles along with the emissions that result from manufacturing processes. Microwave energy has often been examined as a means for waste remediation efforts. Over the past three decades, microwave processing has been investigated for high, intermediate and low-level radioactive waste encapsulation, decomposition of emissions from combustion and energy production, sterilization and volume reduction of medical/biological waste, soil decontamination, tire recycling, dewatering of sludges, as well as other industrial wastes. This paper provides an overview of recent studies that utilize microwave energy as a waste reduction or remediation tool.

BACKGROUND

As our population increases and the amount of waste to be managed expands, new and less-offensive methods are sought to recycle or dispose of large volumes of solid, gaseous or liquid wastes. Typical methods include landfill, incineration, thermal oxidizers, after burners, and filters. In some cases, production facilities are re-located to new locations or other countries, where environmental regulations are less stringent. None of these methods are without adverse consequences. While microwave processing is not a panacea for all disposal concerns, there are many reasons to consider microwave energy for waste remediation applications. Some of these include[1,2]:

- equipment flexibility, cost, maintenance, portability, ease of process control,
- possible reduced floor space requirements (smaller "footprint"),
- remote and mobile capabilities,
- significant volume/weight reductions,
- immobilization of hazardous materials to meet regulatory requirements,
- efficient processing times,
- enhanced chemical reactivity,
- in-situ processing,
- ability to handle several waste streams simultaneously,
- variable temperature processing within same system.

Several previous papers in these and other microwave symposia have provided overviews of microwave waste remediation applications[3,4,5,6,7,8,9,10,11,12]. Some of the topics of discussion have included the following:

- infectious/medical wastes,
- solvents and sludges,
- used rubber, tires, plastics,
- contaminated soils,
- contaminated concrete,
- chemical wastes,
- radioactive wastes,
- gaseous emissions,
- sewage, waste water,
- filter regeneration,
- separation processes (petroleum, coal, other minerals) and,
- electronic circuitry.

Some of these areas remain in research and development stages, while several have been put into practice or are successfully commercialized. The primary impediments for most scale-up operations are financial and/or the reluctance of authorities to accept microwave energy as an acceptable alternative to more conventional means of processing (i.e., vitrification of radioactive wastes). The goals of this overview are to highlight commercially developed uses of microwave technology for waste processing, new research, and advancements in the use of microwave energy for environmentally related enterprises.

COMMERCIAL SUCCESSES – PILOT SCALE FACILITIES

Medical Waste

One of the most notable commercial successes using microwave energy for waste processing has been sterilization of medical wastes. Several companies (Sanitec, Environmental Waste International) have developed processes and equipment capable of handling the tons of infectious and biological wastes generated each year by the health care industry[13,14]. The Sanitec systems have been installed worldwide with over 70 systems currently in operation in the US, Canada, Brazil, Japan, United Kingdom, Saudi Arabia, Korea and the Philippines. Sanitec currently has 2 systems available with capacities ranging from 220-900 pounds (100-400kg) of waste per hour. In addition, both units are available as mobile systems, allowing facilities to share expenses and maximize utilization of the equipment. The systems do not use incinerator technology, as it has been proposed that disinfection is sufficient remediation for ~95% of the medical waste generated. The waste from medical facilities is placed in a cart that is hydraulically loaded into the unit where it is fed through the first of two shredding devices. Volume reductions on the order of 80-90% are realized after shredding. A screw conveyor moves the shredded waste through the system where the sterilization process begins with injecting steam into the chamber. The moistened waste is then subjected to microwave energy through 4-6 sources (1.2kW each at 2.45GHz) in order to uniformly and completely disinfect the waste. The waste material is held at temperatures of 95-100°C for a minimum of 30 minutes. Upon completion of the process, treated materials can be transported to municipal landfills. Separation of wastes is not necessary (sharps can be processed), odor is controlled through HEPA filters, and due to the volume that can be processed, time and labor are significantly reduced. Lower disposal and hauling fees are also a benefit due to the volume reductions.

Environmental Waste International (EWI) also has developed a system for processing medical waste[9]. This system uses 14 magnetrons with outputs of up to 3kW each operating at 2.45GHz. The EWI system can process at 1230kg of waste per day using 2000kWh of energy. There is an operational unit in Liverpool (Speke) in the UK, with orders pending for additional units, another in the UK and one in Taiwan. In EWI's model MD1000 system, waste is loaded into the unit either in bags or cardboard trays. A conveyor moves the waste to the processing chamber. The chamber is purged with nitrogen, the magnetrons are activated and the waste is subjected to a 50-80 minute controlled irradiation cycle that EWI refers to as reverse polymerization, whereby chemical bonds are broken in an inert atmosphere. The resulting material is a sterilized carbon residue. The residue is allowed to cool and is then passed through a grinder to ensure complete destruction of the sharps. Sterilization is guaranteed as the waste is reduced to ash, significant volume reductions are realized, as are expenses related to time, labor, hauling and disposal.

Tires

Much research has been conducted on disposing or re-using scrap tires (see Table 1)[15]. As can be seen from these data, each year the US alone generates over 281 million scrap tires. Microwave energy has been investigated as a method for breaking down used tires and creating a de-vulcanized material for use in new tires[6] and as a method for pyrolyzing the tires and recovering fuel oil, carbon and steel[16].

Table I. Scrap Tire Statistics for the US as of 2001[10].

Number of scrap tires generated annually	281 million (5.58 million tons)
Stockpiled scrap tires	300 million
Number of scrap tires used as fuel source	115 million
Industries using tires as fuel source (number of facilities) (BTU's per pound = 15,000)	Cement kilns (36), Pulp/paper (18), Industrial boilers (17), Utility boilers (11)
Scrap tires used in civil engineering applications	40 million
Scrap tires processed into ground rubber	33 million
Number of scrap tires going to a market	218 million
Number of tires used in a pyrolysis process	0
States with scrap tire legislation	48
States that ban whole tires from landfills	38
Sources of scrap tires	Passenger car: 84%; trucks: 15%; aircraft, off-road, heavy equipment: 1%

Environmental Waste International proposes use of its Reverse Polymerization equipment for remediating scrap tires. A pilot scale facility was operated for years (1994-1998) at EWI in Ajax, Ontario[10]. The microwave system was capable of processing 300 scrap tires per day and recovered hydrocarbons in the form of oil and gases, carbon black and steel. Data gathered during operation of the facility was used to develop the larger scale TR1500 model. A full-scale facility in California is under construction.

Sharma et al[11], mentions a facility designed by BRC Environmental capable of processing 5000 tons of used tires per year using microwave energy. Recovery of 36% of the carbon material was possible and the system uses a condenser/distillation system to capture and separate gases and oils. However, particulars on the facility (location, current production, inception date, etc.) were not disclosed in the review article.

Sewage Applications

WASTECH International, Inc., a company based in New Hampshire, has developed two systems that use microwave energy to treat sewage waste[17]. Both systems are capable of handling up 600gal/day of waste and are designed to replace or upgrade home septic tank systems. The units can also be used as part of a municipal waste treatment program. The STM-2000 and CS-4000 models use 240VAC (30A) and have a battery backup to keep the system operational in the event of a power failure. The waste material is channeled to the microwave chamber where the solid wastes are reduced to ash. In the STM-2000, liquid wastes are filtered through ceramic media and are passed on to an existing infiltration field or municipal sewer. The CS-4000 provides additional treatment of the liquid waste, including ultraviolet light to destroy human pathogens. In new or replacement construction, the size of the infiltration field may be reduced due to the quality of the effluent. In some cases, the effluent from the CS-4000 may be used as irrigation water. Both systems require routine maintenance by service technicians (removal of ash, replacement of filters) every six months and performance is monitored via telephone links. Currently, the systems are available only in the New England area.

RECENT RESEARCH INITIATIVES

Gas/Emissions Decomposition

Several different research groups have examined the use of microwave energy for the decomposition of different gases. The following paragraphs summarize some recent research results.

NO_x: Baeva et al[18] used a microwave plasma reactor to decompose NO_x emissions. Potential applications for this research include flue gases and automotive exhaust gases. In this set-up, a 3GHz-pulsed microwave source (1.4MW) was used to generate the plasma at atmospheric pressures. To create a symmetric distribution of the electric field, a cylindrical high-Q cavity fitted with a quartz tube and micrometer adjustable plungers was used. The pulsed microwave energy was applied to the cavity through coupling holes in the cylinder wall. Synthetic NO_x emissions were pumped through the chamber with a vortex flow system. The gases were mixed in several stages in ensure good mixing and care was taken to avoid oxidation of NO to NO_2. A Fourier

transform infrared spectrometer was used to determine gas concentrations. The results indicated that reduction rates for N_2/NO_x of up to 90% were possible. The addition of oxygen was shown to decrease decomposition rates but additions of up to 9% were possible. The change in frequency was considered to have a negligible effect, but increased power, pulse length, repetition and optimal flow rates were all considered the important parameters for optimal decomposition. The authors believe that reduction improvements are possible with a moderate increase in repetition pulse rate and a shorter pulse time.

Tang et al[19] also discuss the use of microwaves for NO_x decomposition from engine exhaust. Their approach was to use microwave energy and an Fe/NaZSM-5 catalyst (ZSM-5 is a high-silica, low-aluminum content zeolite). In the study, a comparison of microwave energy to conventional reaction was made along with the extent of conversion of NO_x based on the amount of Fe loading, oxygen and NO_x concentrations, flow rates, as well as the effect of water in the gas stream. Results indicated that the effects of the iron addition level off after ~ 10% loading and the addition of microwave energy (~55W) resulted in a ~60% conversion of NO_x to N_2. With the same test parameters in the conventional reactor, no conversion was observed. Optimal conversion was obtained with <2% O_2, however, conversion was stable at levels of O_2 of up to 10%. Conversion rates decreased linearly with increasing flow rates and addition of water to the gases reduced conversion rates.

Zhenming et al[20] propose the use of microwave energy to improve the efficiency of exhaust conversion at cold-start conditions. When the engine is cold and the automobile is started, the catalytic converter is almost completely ineffective in decomposing the emissions. This remains true until the exhaust gas temperatures are over 300°C. With microwave energy to preheat the converter, and a gas inlet temperature of 60°C, the decomposition rate was 50%. At inlet temperatures of 110°C, conversion rates ranged from 92-100% for the various components in the simulated exhaust. To achieve similar efficiencies without the microwave assist, inlet gas temperatures need to be greater than 450°C. For these experiments, the authors used a SiC foam ceramic as a substrate that was immersed in a slurry containing CeO_2, ZrO_2, $Pd(NO_3)_2$, $La(NO_3)_2$, and γ-Al_2O_3. The final catalyst contained 2g Pd, 5g La_2O_3, 6g ZrO_2 and 44g CeO_2 per liter. The coated substrate was subjected to several heat treatments and was used for both microwave and conventional experiments. The microwave system used during the study was a 25W unit operating at 2.45GHz.

Fluorocompounds: Wei and Fang[21] propose the use of microwave plasma to destroy fluorocompounds such as CF_4, C_2F_6 and CHF_3. These materials are regularly used in the semiconductor industry for cleaning chemical vapor deposition chambers and plasma etching. As these gases have been shown to contribute to the greenhouse effect, their release to the environment must be limited. In this study, individual gases were exposed to a microwave-generated surface wave plasma using oxygen has the primary additive gas. By manipulating the microwave power level, the oxygen concentration and the pressure, 99% destruction of the gases was achieved.

Hydrogen Sulfide: Researchers at Argonne National Laboratory have developed a system using microwave energy to treat hydrogen sulfide waste[22]. This process, referred to as the Microwave-Induced Sulfur and Hydrogen Autolysis (MISHA), recovers both the hydrogen and the sulfur from the extraction process used to remove sulfur from oil. A pressurized microwave reactor is used to split the hydrogen sulfide gas into its constituent components. The sulfur produced is a high-purity liquid, and the gas is a mixture of hydrogen and hydrogen sulfide. The unreacted hydrogen sulfide is recycled back through the process and the hydrogen and sulfur are recovered for re-use (hydrogen) or sale. Some of the primary advantages of this process over other current technologies are reduced capital costs (~33% less), reduced operating costs, lower sulfur emissions and lower energy usage. The process is currently at research level.

Solids/Liquids

Electric Arc Furnace Dust: Xia and Pickles[23] describe a laboratory scale process to recover metals from electric arc furnace dust (EAF). The dust is a waste product created from the melting of scrap metal in electric arc furnaces. The composition of the dust varies depending on the composition of the scrap metal, the type of metal being produced and operating conditions. Primary constituents include zinc, iron, lead, chromium, nickel, and manganese. It has been estimated that approximately 700,000 tons of EAF dust are produced each year in North America and, as would be expected, landfill costs for disposing of this type of waste containing heavy metals are high. Several options for treating this waste have been investigated, but no practice is widely commercialized. The process described in this study uses hydrometallurgy (leaching using acidic or caustic alkaline solutions) and microwave energy to enhance the dissociation of metallic ions. The experimental set-up consisted of a small, 900W 2.45GHz microwave, Teflon® leaching vessel and a condenser. Molarity of the leachate was varied as were solids loading, process time and power level. The authors concluded that microwave-assisted leaching resulted in a rapid dissolution of zinc oxide. The recovery process was completed in minutes as opposed to hours for conventional leaching. Higher power levels increased dissolution rates and an 8M solution appeared optimal for the recovery of zinc. The authors also showed that zinc was preferentially recovered over lead, leading them to conclude that since the zinc oxides and zinc ferrites absorbed more microwave energy than lead oxides or sulfates, localized heating at the zinc solid/liquid interface was greater than the solution as a whole, increasing leach rate.

Flyash: The EMR Microwave Technology Corp. in Canada has developed a process for removing organic carbon from flyash. Flyash, generated as a by-product of energy production has applications in the construction industry as an additive to cement, providing residual carbon is <3%. Laboratory tests at EMR indicate reduction of residual carbon to <1% with their process. A pilot-scale facility is planned with an industrial partner, Bateman Engineering[24].

Plastic Wastes: Researchers at Cambridge University have investigated microwave energy for the destruction of high- and low-density polyethylene (HDPE, LDPE) used primarily in food and beverage containers designed for domestic use[25]. Approximately 70% of these wastes end up in Britain's municipal solid waste stream and are landfilled. In their experiments, a 5kW modified commercial microwave oven was used. The microwave was fitted with a 180cm-diameter quartz reaction vessel and was surrounded with microwave-transparent refractory. A carbon bed was used as a susceptor and thermocouples were used to monitor temperature. The carbon material was stirred using 2 stainless steel impellers with a maximum speed of 6rpm. The carbon bed was stirred to help maintain a stable consistent temperature. Nitrogen was used to purge the reactor of oxygen. Emissions were passed through a series of condensers and cold traps followed by a filter to trap any remaining aerosols. For most experiments, HDPE pellets were used to simulate the plastics wastes, but some experiments were carried out on toothpaste tubes (laminated PE with aluminum). Characterization of liquids and waxes collected in the condensing system, as well as the gas samples collected, were analyzed using gas chromatography with mass spectroscopy (GC-MS). For each experiment, 50g of pellets were added to the pre-heated carbon bed. Optimum temperatures for decomposition were found to be between 500 and 600°C. Total time for decomposition was approximately 120 seconds at 600°C. Analysis of the condensates revealed a mixture of alkenes, alkanes and dialekenes (81-93%). Gases contained a mix of linear alkenes and alkanes with trace amounts of aromatics (i.e., benzene, toluene, and xylene). The process was found to be successful in decomposing the simulated plastic wastes as well as a laminated waste (toothpaste tube), an advantage over conventional pyrolysis methods.

Industrial Sludges: Researchers at the Applied Advanced Technologies group at Carnegie Mellon have developed a process to treat steel mill sludges[26]. The sludge consists of steel particles, oil and water. In this process, the waste as it is produced is passed through a microwave chamber where water-based cleaners separate the oil and metal into recyclable components. The oil is skimmed from the top and the clean metal solids are filtered out. The water-based cleaners are continuously recycled. The process is energy and cost efficient, taking only seconds to treat the waste is a low-temperature process and operates at atmospheric pressures. The cost is estimated at $20/ton, about 10% of current landfill charges. A pilot facility was scheduled to be on-line in 1998 at Las Cruces, NM. Commercial development by Thermal Wave International is planned following successful pilot trials. It is envisioned that a similar process can be applied to other industrial sludges as well.

Oilfield Applications: The Petroleum Technology Research Center and Saskatchewan Research Council are testing the feasibility of using variable frequency microwave energy to separate oil-solids-water mixtures and to upgrade heavy oil[27]. Previous experiments have been conducted using 915MHz and 2.45GHz with minimal success. The research is preliminary in scope with plans to continue research only if successful.

OTHER

For the most part, the term "waste" in this and other publications, has been interpreted to mean an undesirable by-product of some industrial or commercial application, or an article that is obsolete and has been discarded. However, often due to certain conditions, a material may also become a waste prior to the end of its expected lifetime, or before it is used for its intended application. In that light, the following are some applications of microwave energy to prevent loss of commodities such as building materials, food, and textiles.

- Microwave treatment of termite/beetle infested wood (Ecola Services, Inc., and Hydrex Pest Control) – eliminates need for pesticides
- Microwave treatment of textiles to eliminate mildew
- Microwave treatment of wheat grain to eliminate granary weevils
- Sterilization of food in packages (including E-coli2)

SUMMARY

Microwave energy can offer efficient and effective methods for reducing hazardous and non-hazardous waste streams. Evidence of significant reductions in volume of primary and secondary waste steams have been recorded by many researchers in the field. There is no doubt that commercialization of microwave waste remediation processes and systems shows great potential and that additional applications, as well as the companies to exploit these microwave technologies, will increase significantly over the next 10 to 20 years.

Much work has been performed leading up to the achievements outlined in this paper. These research summaries are available in published works of the 1st and 2nd World Congresses on Microwave and RF Processing as well as in publications of the Materials Research Society, AMPERE proceedings and others.

REFERENCES

[1] R.L. Schulz, Microwave Remediation of Electronic Circuitry Waste and the Resulting Gaseous Emissions, Ph.D. Dissertation, University of Florida (1998).
[2] D.A. Jones, T.P Lelyveld, S.D. Mavrofidis and S.W. Kingman, *Resources, Convservation and Recycling,* Vol. 34, pp. 75-90 (2002).
[3] T.R. Krause and J.E. Helt, in, Microwaves: Theory and Application in Materials Processing II (D.E. Clark, W.R. Tinga, J.R. Laia, eds.), *Ceramic Transactions*, Vol. 36, The American Ceramic Society, Westerville, OH, pp. 53-60 (1993).
[4] S.J. Oda in, Microwaves: Theory and Application in Materials Processing II (D.E. Clark, W.R. Tinga, J.R. Laia, eds.), *Ceramic Transactions*, Vol. 36, The American Ceramic Society, Westerville, OH, pp. 73-80 (1993).
[5] W.M. Van Loock, in, Microwaves: Theory and Application in Materials Processing IV (D.E. Clark, W.H. Sutton, D.A. Lewis eds.), *Ceramic Transactions*, Vol. 80, The American Ceramic Society, Westerville, OH, pp. 619-626 (1997).
[6] G.G. Wicks, R.L. Schulz and D.E. Clark in, Microwaves: Theory and Application in Materials Processing V (D.E. Clark, J.G.P. Binner, D.A. Lewis eds.), *Ceramic Transactions*, Vol. 111, The American Ceramic Society, Westerville, OH, pp. 411-418 (2001).
[7] S.J. Oda in, Microwave Processing of Materials III (R.L. Beatty, W.H. Sutton and M.F Iskander, eds.), *MRS Symposium Proceedings*, Vol. 269, Materials Research Society, Pittsburgh, PA, pp. 453-464 (1992).
[8] G.G. Wicks, R.L. Schulz, D.E. Clark and D.C. Folz, **Microwave Treatment of Vulcanized Rubber,** U.S. Patent # 6,420,457. Issued: July 16, 2002.
[9] G.G. Wicks, R.L. Schulz and D.E. Clark, **Medical Waste Treatment and Decontamination System,"** U.S Patent # 6,262,405. Issued: July 17, 2001.
[10] G.G. Wicks, D.E. Clark, R.L. Schulz, **Method for Recovering Metals from Waste,** U.S. Patent #'s 6,143,139 Issued: November 7, 2000 and # 5,843,287 Issued: December 1, 1998.
[11] G.G. Wicks, D.E. Clark, R.L. Schulz,**Tandem Microwave Waste Remediation and Decontamination System,** U.S. Patent # 5,968,400. Issued: October 19, 1999.
[12] R.L. Schulz, G.G. Wicks, D.E. Clark and D.C. Folz **Microwave Off-Gas System,** in prosecution.

[13] Sanitec, Inc. http://www.sanitec-inc.com. Accessed July (2002).
[14] Environmental Waste International. http://www.ewmc.com. Accessed July (2002).

[15] Rubber Manufacturers Association. http://www.rma.org/scraptires. Accessed July (2002).

[16] V.K. Sharma, M. Mincarini, F. Fortuna, F. Cognini and G. Cornacchia, *Energy Convers. Mgmt.,* Vol. 39, No. 5/6, pp. 511-528 (1998).

[17] WASTECH International, Inc. http://www.wastech.com. Accessed August (2002).

[18] M. Baeva, H. Gier, A. Pott, J. Uhlenbusch, J Hoschele and J. Steinwandel, *Plasma Sources Sci. Technol.*, Vol. 11, pp. 1-9 (2002).

[19] J. Tang, T. Zhang, D. Liang, H. Yang, N. Li and L. Lin *Applied Catalysis B: Environmental* Vol. 36, pp. 1-7 (2002).

[20] Y. Zhenming, Z. Jinsong, C. Xiaoming, L. Qiang, X. Zhigun and Z. Shimin, *Applied Catalysis B: Environmental* Vol. 34, pp. 129-135 (2001).

[21] T-C. Wei and Y.S. Fang. http://www.ee.ualberta.ca/icops2002. Accessed July (2002).

[22] Argonne National Laboratory. http://www.techtransfer.anl.gov/highlights/9-1/misha.html. Accessed July (2002).

[23] D.K. Xia and C.A. Pickles, *Minerals Engineering*, Vol. 13, No. 1, pp. 79-94 (2000).

[24] EMR Microwave Technology Corp., http://emrmicrowave.com/press/FEB9-00.html. Accessed August (2002).

[25] C. Ludlow-Palafox and H.W. Chase, *Ind. Eng. Chem. Res.,* Vol. 40, pp. 4749-4756 (2001).

[26] Carnegie Mellon Research Institute. http://cmu.edu/cmri/adm_waste.html. Accessed August (2002).

[27] Petroleum Technology Research Centre. http://www.prtc.ca/projects. Accessed August (2002).

MICROWAVE IRRADIATION OF ASBESTOS CONTAINING MATERIALS

A. Corradi, C. Siligardi, P. Veronesi
Dept. of Environmental and Materials Engineering, University of Modena and Reggio Emilia, Via Vignolese 905, 41100 Modena, Italy
paolov@unimore.it

G. Marucci, M. Annibali
ENEA- ERG-SIEG, Centro Ricerche Casaccia, Roma, Italy

G. Ragazzo
ITREC – Centro Ricerche ENEA Trisaia, Rotondella (Mt), Italy

ABSTRACT

It is possible to reach thermal decomposition of serpentine (main crystalline phases in asbestos commercial items) by microwave irradiation in situ, within short period of time. Preliminary experiments were performed in closed multimode cavity on small samples, and their dielectric properties were measured. Computer modelling of a 3-waveguide open applicator helped in designing a remote controllable dedicated system, able to heat large asbestos samples directly on site. Depending on the output power, on sample positioning and lining, it was possible to lead the samples to complete inertisation, or, prolonging the thermal treatment, vitrification. The treated samples were composed of forsterite (magnesium silicate), which were used in the body composition of other ceramics, such as tiles, refractories and bricks.

INTRODUCTION

Asbestos is a collective name given to minerals that occur naturally as fiber bundles which may be several centimeters long, having a diameter in the millimeter range. In industrial practice asbestos minerals are not classified on a mineralogical basis, but rather they are known under their commercial name. However, asbestos can be classified in two basic mineral groups, serpentine (chrysotile) and amphibole, which contain different percentages of the six minerals of particular interest reported in table I.

Table I. Selected chemical and physical properties of common asbestos minerals [2]

Characteristic	Chrysotile	Crocidolite	Amosite	Anthophyllite	Tremolite	Actinolite
Chem analys (%wt)						
SiO_2	38-42	49-56	49-52	53-60	55-60	51-56
Al_2O_3	0-2	0-1	0-1	0-3	0-3	0-3
Fe_2O_3	0-5	13-18	0-5	0-5	0-5	0-5
FeO	0-3	3-21	35-40	3-20	0-5	5-15
MgO	38-42	0-13	5-7	17-31	20-25	12-20
CaO	0-2	0-2	0-2	0-3	10-15	10-13
Na_2O	0-1	4-8	0-1	0-1	0-2	0-2
others	11.5-13	1.7-2.8	1.8-2.4	1.5-3.0	1.5-2.5	1.8-2.3
Decomposition temperature (°C)	450-700	400-600	600-800	600-850	950-1040	620-960
Fusion temperature of residual material (°C)	1500	1200	1400	1450	1315	1400

In the past, asbestos was massively used due to its outstanding properties, such as very low thermal conductivity, good resistance to fire and reaction with biological and chemical agents. Moreover, it exhibits a remarkable mechanical resistance and sound absorption capabilities, which added to the success of asbestos as a building material. Its low cost and low/medium density suggested its use as an insulator for the transportation industry. Recent statistics estimate that world asbestos production in the 80's reached more than 4 million tons/year [1].

EXPERIMENTAL

Materials and Characterization

The investigated samples were almost pure serpentine asbestos, containing a small amount of an organic binder, used for building insulation and provided by ENEA (Ente per le Nuove tecnologie, l'Energia e l'Ambiente). The thermal behavior of the samples was determined by means of differential thermal analysis (DTA, Netzsch, STA 409) on as-received samples. X-ray powder diffraction (XRD) was used to determine crystalline phase modification due to thermal treatments. Powders were prepared by sampling different regions of each specimen and manually grinding the material in an agate mortar. The samples were mounted on a automated diffractometer (Philips, PW 3710), which used Cu-

Kα radiation in the 2θ range of 5°-70°. The microstructures were examined using scanning electron microscopy, (SEM, Philips, PSEM 505).

Dielectric Properties Measurement and Computer Modelling

Dielectric properties of asbestos samples were measured using a network analyzer (HP-85107B), and a dielectric probe (HP-85070B). Measurements were performed in the range 2-10 GHz, by slightly pressing the probe onto the sample surface, kept at room temperature.

Commercially available modelling FDTD software (Concerto 2.0 - Vector Fields) was used to simulate the behavior of the high-power open applicator. Load characteristics input was derived from measured data and literature results [3].

Microwave Heat-Treatments at 2.45 GHz

Preliminary tests were conducted in a multimode applicator (Radatherm VPMS – Sydney [4]) using two B-type thermocouples to monitor temperature changes during microwave irradiation. A Mikron M680 fiber optic system was used in the temperature range of 600-1600°C in order to validate thermocouple measurements. Heat treatments were conducted at various power level (900, 1000, 1200W) for 10-15 minutes. The microwave cavity was lined with alumina fiber refractories.

On-site heat treatments were conducted using a remote controlled translating open applicator, fed by 3 magnetrons, for an overall power output of 18 kW. Fixed power runs were conducted for times ranging from 5 to 15 minutes, depending on the load dimensions. Temperature was monitored by means of a 2-channel fiber optic system (ENEA, adapted), from 300 to 1200°C.

RESULTS

Preliminary Tests and Dielectric Properties Measurement

Preliminary tests showed that, after 13 minutes of microwave irradiation in a multimode applicator, 150g asbestos-samples reached 1200°C. Visual inspection, as well as optical microscopy, however, evidenced poor homogeneity of the treatment; the samples presented morphologically altered regions facing along with apparently unaltered portions of the material [5].

Dielectric property measurements at room temperature of the as-received samples provided an average value of the complex permittivity of 2.068-j0.053.

Computer Modelling

The geometry of the open applicator and the relative position of the load is shown in Figure 1, together with the electric field x component resulting from the calculations. Load position and distance from the applicator was varied during successive simulation runs, in order to determine a satisfactory geometry able to provide a homogenous treatment.

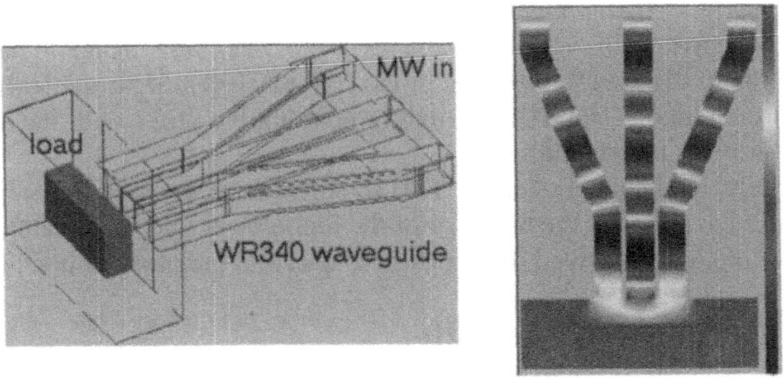

Figure 1. Open applicator geometry and electric field x component in the middle xy section

Dissipated power thermal maps evidenced the presence of inner regions presenting higher values of dissipated power. These regions were corresponded with the vertical projection of the open applicator end section, as shown in Figure 2, with a variable and fixed scale for intensity. It is worthwhile noting that the asbestos samples, after the inertization treatment, showed the presence of a thermally altered region which strictly corresponded with the modelling results, as depicted in Figure 3.

Open Applicator: Heat Treatments

The idea behind the remote controllable open applicator for in situ asbestos inertization was the possibility of treating horizontal and vertical surfaces without the need for a human operator to attend the process on site, thus coming in contact with dangerous asbestos particles. The movable applicator, optimized during the present study represents only the first step, being able to translate on wheels and thus treating horizontal surfaces. Acquiring the second degree of freedom, however, requires only minor modification of the existing apparatus. Figure 3b represents the open applicator during the heat treatment of asbestos, positioned on bricks.

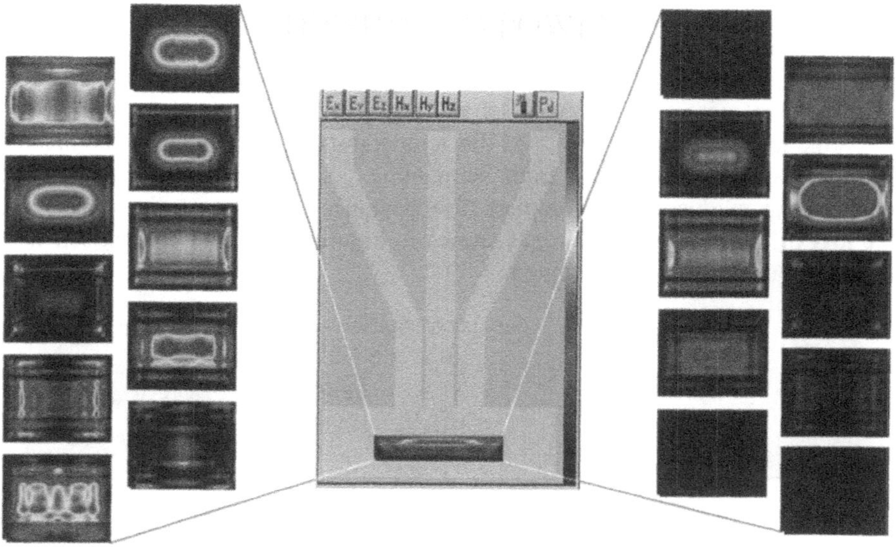

Figure 2. Dissipated power: xz sections (left and right) and xy section (center)

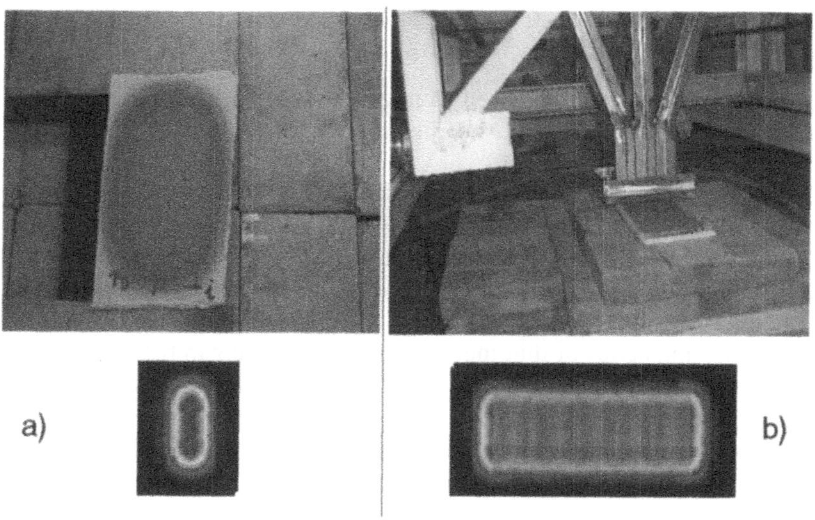

Figure 3. a) microwave inertization in stationary conditions and dissipated power modelling results; b) microwave inertization translating the applicator and dissipated power modelling results (superimposition).

By varying the output power or the duration of the treatment, two different results could be achieved: inertization and vitrification. In the first case, previously shown in Figure 3b, the heating process of 3 layers of asbestos, equivalent to more than 500 g at 18 kW for 9 minutes allowed the samples to reach the temperature of 900 °C. At this temperature a phase change is completed [6,7], producing forsterite from serpentine. X-ray diffraction and SEM observations evidenced the complete disappearance of the serpentine phase, and the loss of the fibrous structure, as shown in Figure 4, where a comparison with the as-received sample is presented.

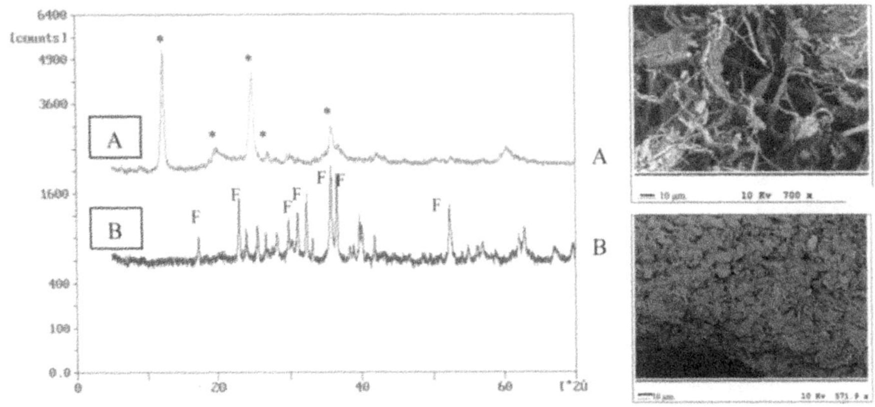

Figure 4. XRD pattern and SEM micrograph of the samples A) as received; B) after microwave inertization. *=serpentine; F= forsterite.

When processing time was increased to 12 minutes, temperature rose well above 1200°C and vitrification of the samples was achieved. Experimental conditions during heating are shown in Figure 5, where the protective shielding was removed to allow photographs to be taken. The shielding consisted of a metal wire net vertically surrounding the applicator end at a distance of 500 millimetres, and presenting series of thin metal chains closing the gap between the applicator-end Z-level and the ground level. The effect of this experimental setup was to reduce significantly microwave leakage, but the limit of 5 mW/cm^2 (FDA Part 1030.10c) was reached only at about half a meter from the applicator end, in case on non-absorbing load, or at a distance of less than 100 millimeters, when the asbestos samples, or other absorbing loads, were heat-treated. It is worthwhile noting that the whole system was designed to be remote-controlled, and it effectively was during the experimental runs, always being the human operator at a distance higher than 10 meters.

Figure 5. Open applicator during asbestos vitrification.

Either in the inertized form, or in the vitrified one, the resulting material can be regarded as a secondary raw material, which can be used for introducing silicates and magnesium oxides in ceramic bodies compositions[8-11]. Thus as a conclusion of the present work, different percentages (1, 3, 5%) of inertized or vitrified asbestos were successfully introduced as substitutes in bricks, refractory bricks, single firing tiles and porcelain stoneware tiles, following the typical industrial cycles.

CONCLUSIONS

The possibility to inertize or to vitrify asbestos layers directly on site, by means of a microwave open applicator was demonstrated [12]. Dielectric properties measurement and computer modelling helped in optimising the microwave-matter interaction conditions, allowing better uniformity of the microwave heat treatment as compared to commercially available multimode applicators.

Inertized asbestos no longer contained the serpentine phase and the peculiar acicular microstructure, but transformed into forsterite, a harmless magnesium silicate. Moreover, forsterite derived from asbestos inertization proved to be a suitable secondary raw material for the ceramic, bricks and refractory industry. The authors are still investigating the possibility to include inertized or vitrified asbestos in percentages higher than 5% weight.

ACKNOWLEDGEMENTS

The present study was granted by CNR (gr. n° 0200172.PF37), in the framework of the "Gruppo Nazionale per la Difesa dei Rischi Chimico-Industriali ed Ecologici". Authors would like to thank Dr. Eng. V. Fiumara and Dr. Eng. A.A. Barba, University of Salerno. They would like to show their appreciation to Prof M. Romagnoli and Dr. M.R. Rivasi, University of Modena and Reggio Emilia.

REFERENCES

[1] BGS, "British Geological Survey world mineral statistics 1979-83: 70th anniversary issue", (1983)

[2] World Health Organisation, "Asbestos and other natural mineral fibres", *Environmental Health Criteria*, **53**, 17-27 (1986)

[3] World Health Organisation, "Chrysotile asbestos", *Environmental Health Criteria*, **203**, 10-26 (1998)

[4] Veronesi P.,Leonelli C., Corradi A.B., Annibali M., "Variable power microwave system for high temperature materials treatment", *Materials Engineering Monograph*, 49-54 (2000)

[5] Marucci G., Annibali M., Carboni G., Gherardi G., Ragazzo G., Siligardi C., Veronesi P., Lusvarghi L., Rivasi M.R., "Characterization of microwave inertized asbestos containing materials", *Materials Engineering Monograph*, 115-125 (2000)

[6] Evans B.W., Johannes W., Oterdoom H.V.T., "Stability of chrysotile and antigorite in the serpentinite multisystem", *Schweiz. mineral. petrogr. Mitt.*, **56**, 79-93 (1976)

[7] Chernosky J.V., "The stability of chrysotile, $Mg_3Si_2O_5(OH)_4$, and the free energy of formation of talc, $Mg_3Si_4O_{10}(OH)_2$", *Geol. Soc. Am. Abs. Prog.*, **5**, 575 (1973)

[8] M. Nagamori, A.J. Plumpton and R.Lo Houillier, "The activation and chemical utilization of asbestos tailings - a review", *Ind. Mineral*, 144–156(1980)

[9] Y.M. Chiang, D.P. Birnie, W.D. Kingery, "Physical Ceramics — Principles for Ceramic Science and Engineering", 3rd ed. John Wiley & Sons Inc., (1997)

[10] Hassan H. M. Darweesh, "Building materials from siliceous clay and low grade dolomite rocks", *Ceramics International*, **27**[1], 45-50 (2001)

[11] T. W. Cheng, Y. C. Ding, J. P. Chiu, "A study of synthetic forsterite refractory materials using waste serpentine cutting", *Minerals Engineering*, **15**[4] , 271-275 (2002)

[12] C. Leonelli, C. Siligardi, G.C. Pellacani, G. Gherardi, G. Marucci, "Inertization process of asbestos containing materials with microwave energy", Italian Patent, IT1302348, Sept.1998

MICROWAVE-ASSISTED RECOVERY OF PLATINUM

Thorsten Gerdes and Monika Willert-Porada
University Bayreuth,
D-95448 Bayreuth, Germany

Rudolf Burghardt'
DEGUSSA AG
D-45764 Marl, Germany

Philip Bahke
InVerTec, Innovative Verfahrenstechnik e.V.
D-95448 Bayreuth, Germany

Ingo Kleinwächter
dmc2 Division, OM Group,
D-63403 Hanau, Germany

ABSTRACT

To achieve a sufficient purity of recycled platinum multiple cycles of dissolving, precipitation and calcining are necessary. In the paper a microwave heated calcining step of platinum sal ammoniac and a microwave assisted dissolving process is discussed. The benchmark for the microwave assisted processes are conventional heated processes performed in the same applicators.

INTRODUCTION

Platinum metals are used in various areas of industry as catalysts. The exhaust gas catalysts require approximately 10 %, shortly probably 20 %, of the total consumption of platinum group metals [1]. Other important fields of application are the reforming of crude oil or the catalytic oxidation of ammonia to nitrogen oxides or nitric acid. In the future the fuel cell industry might develop into an important consumer of platinum.

Traditionally the quote of recycling is very high and the primary production is in the same scale as the secondary production. With view of the constantly rising platinum return flows the cost structure of the secondary production becomes more relevant.

The separation of the platinum group metals are the last steps in the processing of the ores and even in the case of recycling material.

The individual metals are separated first roughly and refined afterwards to the commercial purity of 99.9 %. Platinum metals are transferred by special oxidation and reduction steps and addition of ammonium chloride into a scarcely soluble salt, that crystallizes in the solution and can be filtered. [2]. A scheme of the separation process is shown in figure 1

After the first precipitation the platinum salt has insufficient purity and special cleaning steps are always required. In the case of platinum sal ammoniac the salt is calcined at 800°C developing a platinum sponge. The platinum sponge is dissolved again with aqua regia or chlorine-saturated hydrochloric acid in general at t 80°C. This time consuming cycle is passed through, until the platinum has the necessary purity.

The decomposition effected from platinum sal ammoniac occurs according to the following reaction equation:

$$3\ (NH_4)_2[PtCl_6] \rightarrow 3\ Pt + 2\ NH_4Cl + 16\ HCl + 2\ N_2$$

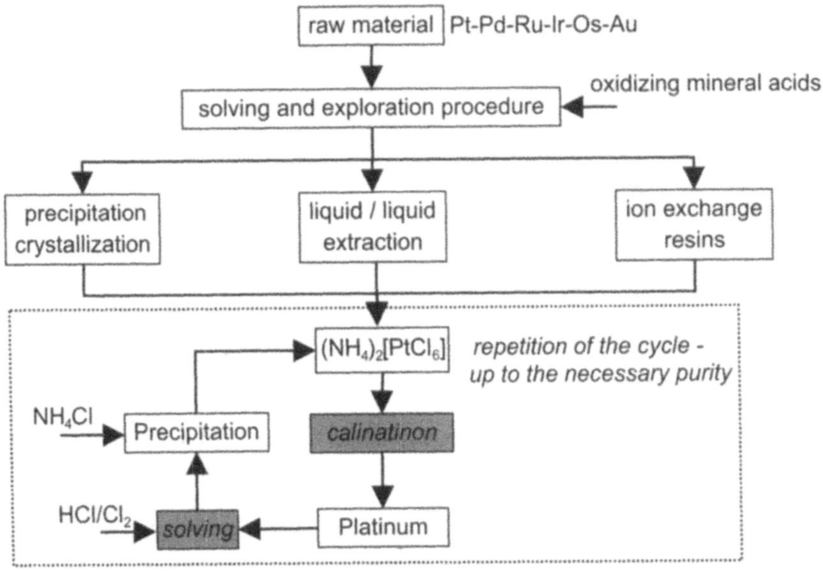

Figure 1: Scheme of the separation courses during the production of platinum

AIM OF INVESTIGATIONS

To achieve a sufficient purity of recycled platinum multiple cycles of dissolving, precipitation and calcining are necessary. The commercial calcining process of platinum sal ammoniac is operated in resistance heated tube furnaces with a batch size of 50kg for 24 h at 800°C. The basic process requirements are total decomposition of the Pt-salt to a homogeneous platinum sponge and the removal of impurities like ammonium chloride or tellurides.

In the case of an additional purification cycle a very fine grained sponge with good dissolving properties is preferred. If a melting step is required, the calcining process should end up with a coarse and compact material with good melting behavior. In the conventional process the heating rate can not be increased because of the formation of a partial densified skin that suppresses the evaporation of decomposition products.

The main goals of the process development were higher heating rates and shorter holding times by using microwaves with 2.45 GHz as energy source. An additional goal was a more flexible process that allows a control of the platinum sponge morphology.

To achieve high dissolution rates of the platinum sponge the temperature of the platinum has to be as high as possible. On the other hand the solubility of chorine in the hydrochloric acid has to be as high as possible to increase the platinum conversion. But with increasing temperature of the hydrochloric acid the solubility of the chorine strongly decreases.

These opposite trends in solubility could be overcome by microwave heating the suspension and by a local overheating of the platinum particles.

RESULTS AND DISCUSSION: *CALCINATION OF THE PLATINUM SALT*

To quantify the decomposition of platinum sal ammoniac thermogravimetric analysis (TGA) from 200 to 600°C has been performed (see figure 2). Most of the weight loss (about 55%) occurs in the temperature range between 300°C and 400°C. The decomposition of the platinum salt complex leads to the emergence of ammonium chloride and the further decomposition of this compound to NH_3 and HCl.

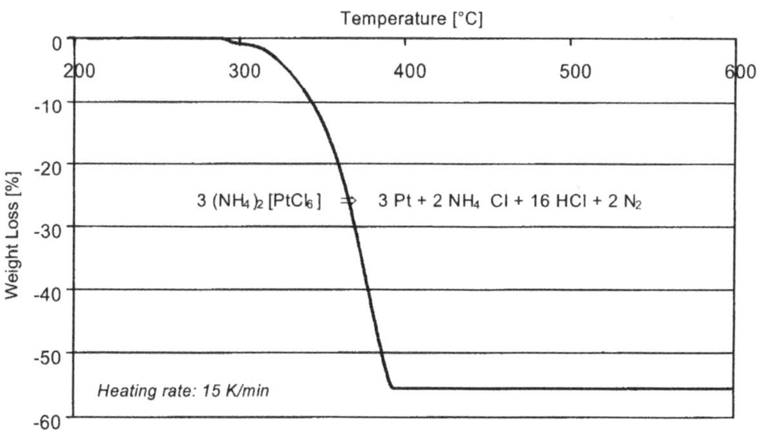

Figure 2: Thermogravimetric analysis of platinum sal ammoniac

The experimental setup for microwave calcining of platinum sal ammoniac is shown in figure 3. The cavity operates at 2.45 GHz with a maximum power of 2.7 kW. The platinum salt is arranged in an alumina crucible and surrounded by a thermal insulated quartz tube. During the heat treatment the tube is flowed through to enhance the transport of decomposition products.

For a comparison of the microwave calcination, the quartz tube was heated by a conventional electrical heating module.

The absorption of microwave energy during the calcining can be divided into three steps. In the temperature range of 20-100°C the absorbing behavior is dominated by the heating of remaining water and HCl. The heating behavior of the totally dried platinum sal ammoniac is moderate, but with the formation of the first platinum metal during the decomposition of the salt the absorbing rate strongly increases, because of the increase of dielectric losses.

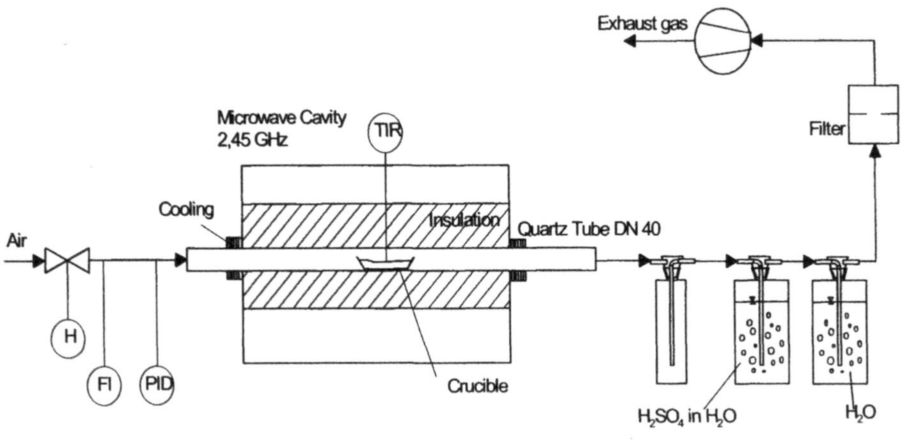

Figure 3: Experimental setup for calcining platinum sal ammoniac in a microwave field

To ensure the comparability with the commercial process, the calcination has been performed between 800°C and 1000°C. A temperature and power profile for microwave calcining of platinum is shown in figure 4

The most important benchmark for the success of the heat treatment is a analysis of weight loss of the generated platinum sponge by an additional glowing process. The weight loss should be lower than 0.10%.

In the case of microwave heating the weight loss is in the range of 0.03%, that fulfills the quality standard. In the case of material conventionally heated to 800°C followed by direct cooling down a weight loss during TGA of about 0.13%

is measured. Only by increasing the holding time at process temperature the weight loss can be reduced to the benchmark.

In the conventional process with a heat flux from the surface of the bulk to inside, the decomposition starts at the surface and leads to a less porous layer that suppresses the transport of decomposition products from the interior to the surface. In the case of microwave heating formation of a more densified top layer does not occur. This observation corresponds with the observation of an enhanced decomposition.

Because of the moderate dielectric loss of the platinum sal ammoniac and adequate penetration depth of the microwave radiation, the decomposition can occur in the volume of the material. And even if the penetration depth strongly decreases with the formation of metallic platinum, the sponge morphology with the high porosity seems to be sufficient for a homogeneous decomposition of the salt.

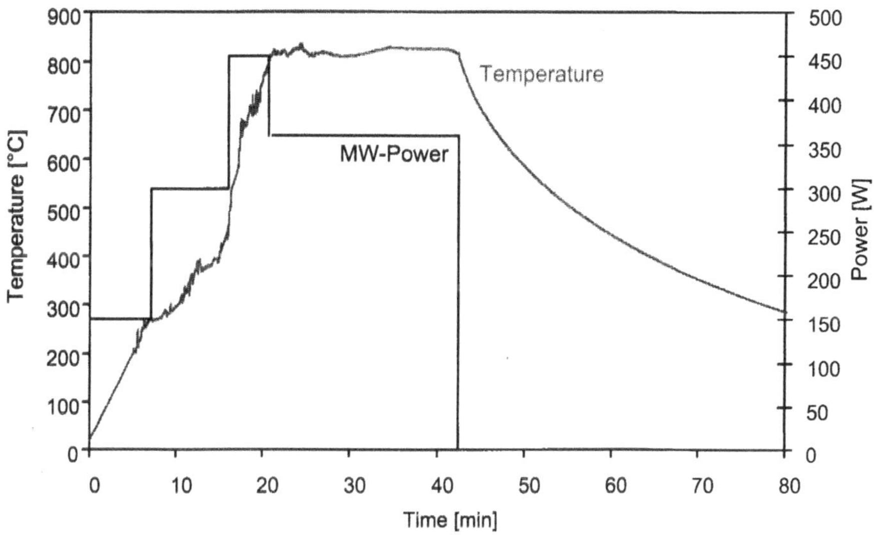

Figure 4: Temperature and power profile for microwave calcining of platinum (800°C/ 30 min)

The microstructure of a sponge calcined by microwave and conventional heating at 900°C is shown in figure 5. The overview shows for both cases a homogeneous high porous sponge. At higher magnification it is observed that platinum grain boundaries are more pronounced in the case of microwave heating. Even if the difference in the primary grain size is not changed significantly, in the case of microwave calcining the grain boundaries appears as if they have been etched. The difference in the surfaces is also visible at higher and lower calcining temperatures.

In the case of microwave heating a homogeneous platinum sponge as shown in figure 5 is only obtained if the magnetron is operated in an continuos mode. If the same average power is operated in a pulsed mode, local plasma ignition and sparking is observed (see figure 6). This effect of high field strength on the microstructure of the platinum sponge is visible in figure 7. The pulses with a length 0.5 – 1 sec cause local melting of platinum. To achieve such a local melting at a macroscopic temperature of 800°C a local superheating of about 900K is necessary.

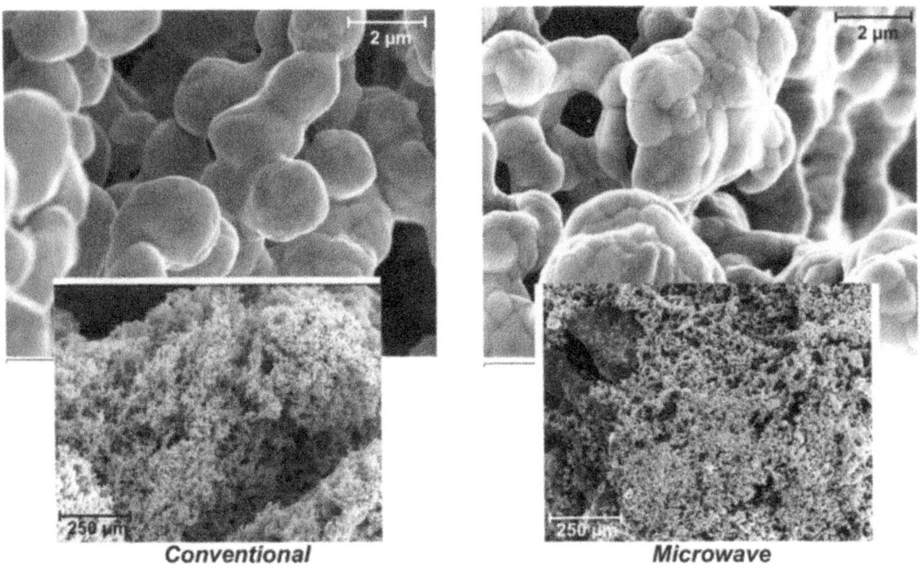

Figure 5: Overview and microstructure of microwave and conventional calcined Pt-Sponge (900°C / no Holding Time)

At calcining temperatures of nominal 1000°C in the case of pulsed microwave power larger areas of the platinum sponge are densified (figure 7, right microstructure). This microstructure is not desirable for a dissolving process but in the case of a subsequent melting process the melting behavior is better than this of a fine grained platinum sponge.

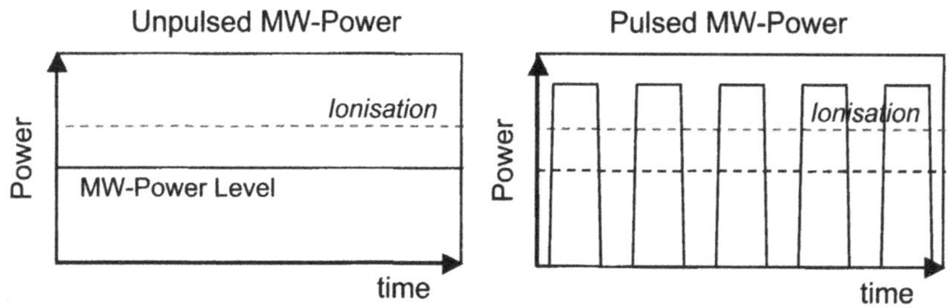

Figure 6: Relation between microwave operating mode and break down conditions for the same average power

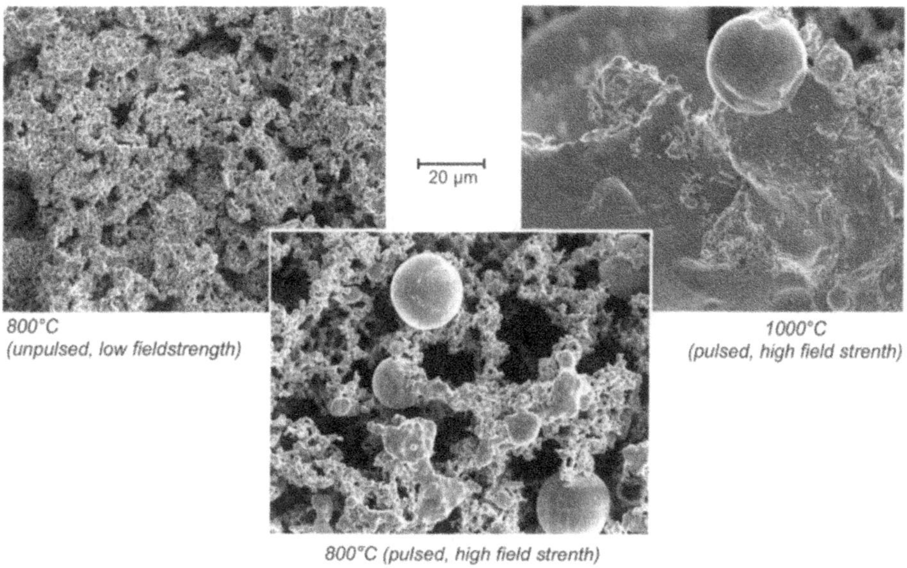

Figure 7: Adjustment of platinum sponge morphology by Microwave assisted calcining

A comparison of the resulting BET-surface for different calcining temperatures is shown in figure 8.

Even at temperatures of 500°C the specific surface area of the microwave calcined material is lower than in the case of conventional heating. This is also true in the case of an unpulsed microwave power. In the case of microwave heating the decrease of surface area strongly depends on temperature than in the case of conventional heating. The stronger decrease of surface area corresponds with thc progress of sintering. These observation of enhanced sintering by microwaves agrees the results in other materials systems [3-4]

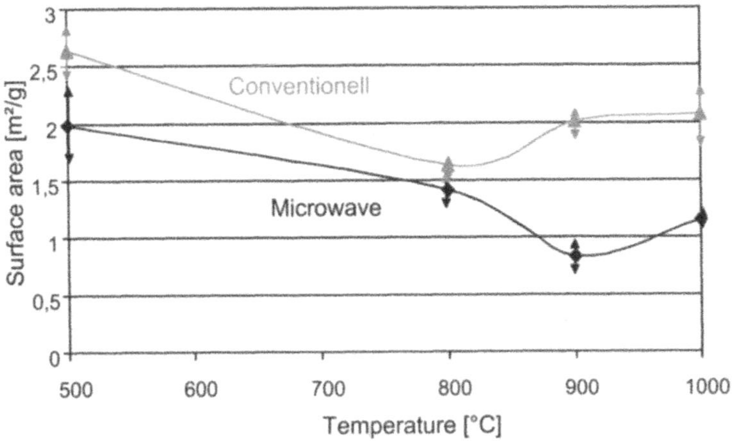

Figure 8: Development of BET-Surface area for conventional and microwave calcined platinum sponge (unpulsed microwave power).

RESULTS AND DISCUSSION: *DISSOLVING OF PLATINUM SPONGE*

The re-solving of the platinum sponge is performed in 6m HCl flowed through by chlorine bubbles to enhance the dissolving rate by solved chlorine. To achieve high chlorine concentrations corresponding with high dissolving rates of the platinum, the temperature should be low (figure 9, left side).

To accelerate the kinetics of the solving process the temperature should be as high as possible. These contrary demands on the process conditions lead to a strong maximum in the dissolving rate-temperature curve known from literature [2] (figure 9, right side).

By microwave heating with low average power but high power short pulses the macroscopic temperature of the hydrochloric acid can be held at moderate level. Because of the much higher dielectric losses of the fine grained platinum, the high energy microwave pulses should overheat the platinum particles and the resulting temperature gradient between platinum and solution should enhance the dissolving rate of the platinum.

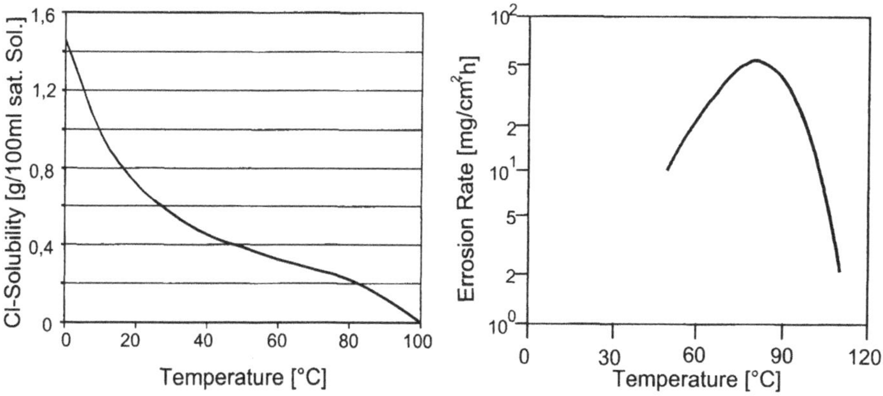

Figure 9: Solubility of chlorine in hydrochloric acid (left) and erosion rate of Platinum in 6m HCl/Cl_2 (right) as a function of temperature [2]

The experimental setup for the microwave assisted dissolving of platinum is shown in figure 10. The conventional experiments has been performed in the same apparatus but heated through the reactor wall.

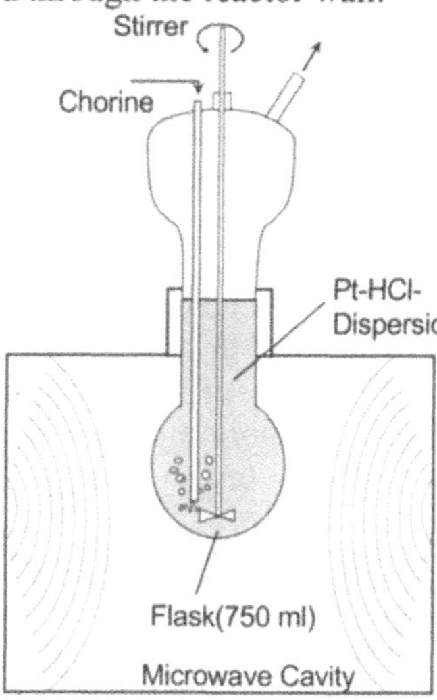

Figure 10: Experimental setup for microwave assisted dissolving of platinum

Although the penetration depth of the microwave radiation into the hydrochloric acid is strongly limited and the absorbing behavior is good, a superheating of the platinum seems to be possible because a considerable increase of the dissolving rate is observed (figure 11).

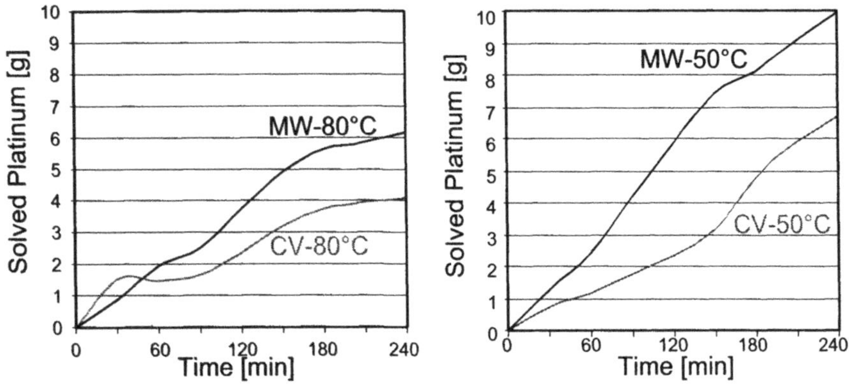

Figure 11: Comparison of conventional and microwave re-solving of platinum at 50°C and 80°C hydrochloric acid temperature

From literature data (see figure 9, right) the maximum of erosion rate should be in the range of 80°C. By superheating the platinum higher dissolution rates are observed at 50°C because of the higher chlorine solubility.

Depending on the reaction conditions the erosion rate of platinum is 50% to 100% higher than in the case of conventional heating.

SUMMARY

Platinum sal Ammoniac and Pt-sponge can be heated effectively with microwaves at 2.45 GHz. Contrary to the conventional calcining, no skin development is observed in the case of MW-heating. As a consequence the decomposition of the Pt-salt is more homogeneous than in the case of conventional heating, that leads to lower glowing losses of the microwave calcined platinum sponge.

The microwave power control mode has a strong influence on the obtained material. In the case of continuous MW-power a homogeneous Pt-sponge is obtained with comparable properties to the conventional calcined sponge. In the case of pulsed MW-power local melting of the Platinum is observed.

During the microwave assisted re-solving of platinum it was possible to overheat the Pt-particles in spite of the high dielectric losses of hydrochloric acid. Depending on the reaction conditions the re-solving speed of platinum can nearly be doubled by microwave heating.

A sufficient overheating of the platinum can only be obtained by pulsing the microwave power.

These investigations have shown that advantages during the separation cycle of platinum strongly depend on the kind of microwave power control. While continuous microwave heating lead to the desired homogeneous temperature in the case of dissolving the overheating of the platinum can only be achieved by high energy pulses at a moderate average power.

LITERATURE

[1]Hagelüken, *"Recycling von Autoabgaskatalysatoren- Stand und Perspektive für Europa", Metallwirtschaft und Metallmarkt.* 49 Band 7-8, 1995

[2]Renner, "*Platinmetalle und –Verbindungen*", Ullmanns Enzyklopädie der technischen Chemie Band 18. 4. Auflage, Weinheim

[3]W.H. Sutton "*Microwave processing of ceramics - an overview*", Mat. Res. Soc. Proc. Vol. 269 (1992) 3-20

[4]T. Gerdes, M. Willert-Porada, „*Microwave Sintering of Metal-Ceramic and Ceramic-Ceramic Composites*“, MRS Symp. Proc. Vol. 347 (1994) 531-537

USING $La_{1-x}Ce_xMnO_3$ PEROVSKITES AS THE ACTIVE COMPONENTS FOR SOOT FILTER REGENERATION BY DIELECTRIC FIELDS

Y. Zhang-Steenwinkel, J. Beckers, H.L. Castricum and A. Bliek
Department of Chemical Engineering, University of Amsterdam, Nieuwe Achtergracht 166, 1018 WV Amsterdam, The Netherlands
E-mail: YeZhang@science.uva.nl

ABSTRACT

Diesel engines increasingly tend to be equipped with filters to reduce the emission of potentially carcinogenic soot particulate matter. Whereas such (ceramic) filters are state-of-the-art, their in situ regeneration is not. As will be demonstrated, the application of perovskite coatings on these filter elements, and their subsequent periodic exposure to a dielectric field, may provide an elegant and energy efficient regeneration procedure. Perovskites are active oxidation catalysts and they possess favorable dielectric properties. Presently, the dielectric heating of $La_{1-x}Ce_xMnO_3$ perovskites is studied, showing that these materials can be rapidly heated in a dielectric field. Further, catalytic studies reveal that these materials have a good activity in CH_4 full oxidation, presently used as a convenient model system. Further structural analysis of these materials is presented using ICP, XRD, SEM and XPS. Finally, results are given for preliminary work into the SO_2 resistance of this class of perovskites.

INTRODUCTION

Interest in catalytic combustion is growing fast, especially in metropolitan and heavily industrialized areas, due to the strong environmental legislation on gaseous and particulate emission, both from gasoline and diesel engines.[1] This is particularly true in view of the soot emission problem related to diesel engines which, to an increasing extent, is viewed upon as a serious health problem. Several techniques are developed to eliminate the soot particulate emission, such as the use of homogenous fuel additives or the use of soot filters. Ceramic filters are used in a wide variety of applications for exhaust gas and waste gas cleanup. Whereas ceramic soot traps are state of the art technology,[2] the continuous in situ regeneration of these filter elements is not. While filters coated with oxidation catalysts are presently available, filter or exhaust heating is still required due to the high ignition temperature of diesel soot.[1] Currently, the use of a microwave sensitive catalytic material as a filter coating has been studied. For a filter fabricated in this way, the soot ignition temperature can be reached easily by exposing it periodically to a dielectric field, thereby allowing further subsequent self-sustained carbon burn-off, leading to full regeneration.

Ideal susceptor materials for the present purpose are also oxidation catalysts, thereby eliminating the need for additional full oxidation of carbon monoxide formed during soot combustion. Potential candidates are perovskite-type oxides, which can be described with the general structural formula ABO_3. The cations in the perovskite lattice can be partially substituted by foreign ones without inducing large changes in the crystalline structure, but changing systematically the catalytic and probably the dielectric properties.[3] The absorption of microwaves by perovskites has been reported before, e.g. in sintering studies.[4-6] Presently, the use of La-M-Mn (M = Ce) based perovskites has been opted as a class of potentially suitable coating materials.

The $La_{1-x}Ce_xMnO_3$ perovskites series were synthesized by co-precipitation followed by calcination at high temperature. The elemental composition, bulk structure, phase segregation and surface composition have been analyzed by inductively coupled plasma atomic emission spectroscopy (ICP-AES), X-ray diffraction (XRD), scanning electron microscopy (SEM) and X-ray photoelectron spectroscopy (XPS), respectively. The dielectric properties were tested in a dielectric field and the catalytic activity of these perovskites was assessed using CH_4 oxidation as a model system. Finally, the resistance to SO_2 was tested.

EXPERIMENTAL

Catalyst Preparation

A series of perovskites-type oxides $La_{1-x}Ce_xMnO_3$ ($x = 0$, 0.05, 0.1, 0.2 and 0.3) was prepared by co-precipitation, according to the method described by Kieβling et al.[7] A 0.5 N aqueous solution of corresponding nitrates was used with sodium hydroxide and hydrogen peroxide as the precipitating agent. The precipitate was filtered off, washed with distilled water, and dried overnight at 393 K in an air. The ground precipitate (<200 μm) was calcined at 1073 K in a air flow for 6 hours.

Characterization

The composition of these perovskites was determined by ICP on a multichannel Thermo Jarrel Ash ICAP 957 spectrometer, upgraded to ICAP 61. The XRD data were obtained on a CPS120 using a primary monchromator with 2θ from 0 to 120°, using CuKα radiation. The obtained SEM images were made using a Jeol High Resolution Scanning Electron Microscope, type JSM 6000F with a Field Emission Gun (FEG). The core levels and valence electronic structure of the $La_{1-x}Ce_xMnO_3$ perovskites (x = 0, 0.05, 0.1 and 0.2) were studied by using XPS. The spectra were measured in a vacuum ($5 \cdot 10^{-9}$ mbar) using an Omicron UHV system with a Mg Kα (1253.6 eV) X-ray source. A power of 25 W and a pass energy of 20 eV were used during the experiment. The core levels of La 3d, Mn 2p and O1s were recorded.

Dielectric Properties

Apparatus: The dielectric heating was carried out in a microwave system operated in travelling wave mode (TE 100 - monomode cavity, 2.45 GHz, using optical fiber), see Figure 1. The circulator is placed to protect the microwave source from the reflected radiation. The stub tuners are used to minimize the reflected radiation. A quartz tubular reactor (ID = 18 mm) is placed in the applicator section, which contains two power sensors. The water load at the end of the wave-guide absorbs any microwave radiation after passing though the reactor. The bulk temperatures were assessed using an optical fiber (Luxtron). In order to estimate the bulk temperature in the presence of microwave energy, a calibration was performed. The detection limit of this optical fiber is 373 K.

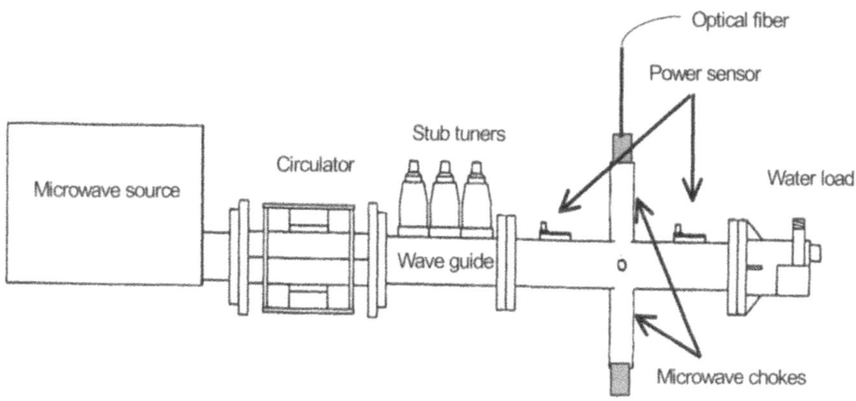

Figure 1: Schematic drawing of 2.45 GHz microwave system.

Dielectric behavior of $La_{1-x}Ce_xMnO_3$ (x = 0, 0.05, 0.2 and 0.3) in He flow: A sample with a constant volume of 10ml was placed in the reactor. The sample was dried in a conventional oven to 423 K in a He flow (GHSV = $6 \cdot 10^2$ hr^{-1}) with a heating rate of 5 K min^{-1}, followed by stabilization at 423 K for one hour. After cooling down to room temperature in a He flow, the sample was transported to the microwave cavity. The sample was heated up to 873 K in a He flow and kept there for 24 hours. Subsequently, the sample was rapidly and repeatedly heated up at 200 W with the same gas hourly space velocity.

Oxidation of CH_4 in conventional and microwave ovens: The catalytic activity of $La_{0.8}Ce_{0.2}MnO_3$ was tested using CH_4 full oxidation as model reaction, using both dielectric heating and conventional heating. About 3.35g (10ml) $La_{0.8}Ce_{0.2}MnO_3$ was first dried (see above). The CH_4 oxidation was carried out in the microwave oven as well as in the conventional oven in a $CH_4/O_2/He$ gas mixture (v/v/v 5/12.5/82.5, GHSV = $12\cdot10^2$ hr^{-1}). The resistance of this perovskite to SO_2 poisoning was studied for CH_4 full oxidation (see above), in the presence of SO_2 (200ppm).

RESULTS AND DISCUSSION

Catalyst Characterization

Elemental chemical analysis: The elemental composition, expressed as atomic ratios relative to Mn (Table 1), demonstrates that, for some perovskites, this parameter deviates from the theoretical value (La+Ce : Mn = 1:1). These results suggest that these perovskites are non-stoichiometric, as has been discussed in the literature.[8-10]

Table I. Elemental chemical composition of 1073 K calcined $La_{1-x}Ce_xMnO_3$ catalysts.

X	Elemental Chemical Composition (mol/mol Mn)		
	La	Ce	Mn
0.0	0.91	0.00	1.00
0.05	0.85	0.04	1.00
0.1	0.80	0.08	1.00
0.2	0.70	0.19	1.00
0.3	0.61	0.27	1.00

Bulk structure: The XRD patterns demonstrate that Ce is not totally incorporated into the perovskite lattice. Beyond 20 mol% Ce substitution, a separate CeO_2 phase is formed, as shown in Figure 2. These results suggest that a phase change has taken place, which is also confirmed by SEM image, see Figure 3. Cerium substitution has strong influence on the morphology and particle size of the perovskites. $LaMnO_3$ perovskite has a uniform particle size with an average diameter of 120 nm, whereas the morphology of $La_{0.8}Ce_{0.2}MnO_3$ is considered to be a poorly crystallized, layered structure with smooth surfaces. Since separate phases of neither Mn_2O_3 nor of La_2O_3 were detected, the perovskite consisted of a single phase with a bulk composition corresponding to $La_{1-x}\phi_yCe_{x-y}MnO_3$ (1>x>y; ϕ = A site cation vacancy).

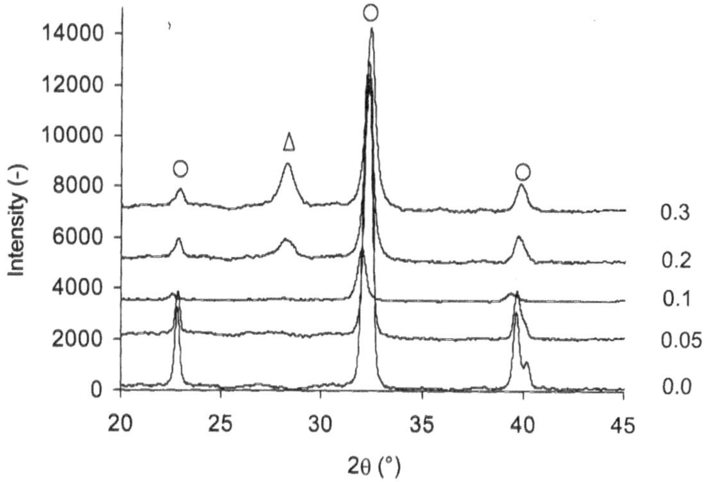

Figure 2: XRD patterns of perovskite $La_{1-x}Ce_xMnO_3$ (x = 0, 0.05, 0.1, 0.2 and 0.3) calcined at 1073 K: (○) perovskite phase; (△) CeO_2 phase.

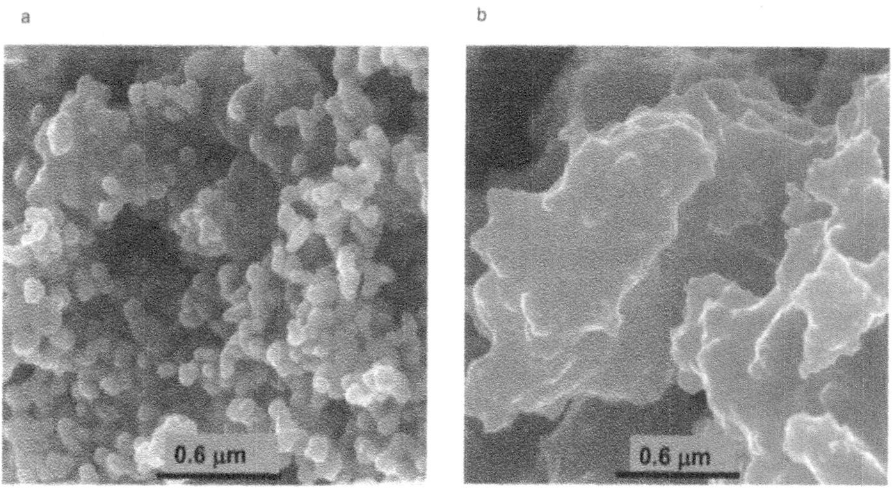

Figure 3: SEM image of: (a) $LaMnO_3$ and (b) $La_{0.8}Ce_{0.2}MnO_3$, bar corresponds to 0.6 µm.

Surface analysis: The XPS spectra showed that two types of oxygen existed on the perovskite surface: a peak at 529 eV accepted in general as lattice oxygen in the form of O^{2-} (metal oxide bond) and a peak at 531 eV assigned to the surface oxygen in the form of CO_3^{2-} or OH^{-1} (table 2).[11-13] The mean oxidation state of the Mn was determined by applying a peak synthesis procedure according to Ponce et al.[3] Since the peak positions of La $3d_{3/2}$ and La $3d_{½}$ did not change with Ce substitution and were close to those values of pure lanthanum oxide, lanthanum ions were present in the trivalent form throughout the sample.

Table II. La 3d, Mn 2p and O 1s binding energies for $La_{1-x}Ce_xMnO_3$ perovskites.

Degree of substitution (x)	La 3d (eV)	Mn 2p (eV)	O 1s (eV)
0	834.3	641.6	529.3
	838.4	642.9	531.2
		644.4	
0.05	834.6	641.6	529.5
	838.6	642.9	531.3
		644.6	
0.1	834.6	641.6	529.6
	838.7	642.9	531.5
		644.9	
0.2	834.3	641.9	529.6
	838.3	642.9	531.5
		644.7	

Dielectric heating: The dielectric heating experiments carried out in a He flow demonstrated that all La-Ce-Mn based perovskites are high loss dielectric materials. These materials could be heated without any irreversible change in bulk structure to a temperature of at least 1073 K, and kept for at least 24 hours (figure 4a). Figure 4b demonstrates that these perovskites can been rapidly heated up in a repeatable way to a stable temperature.

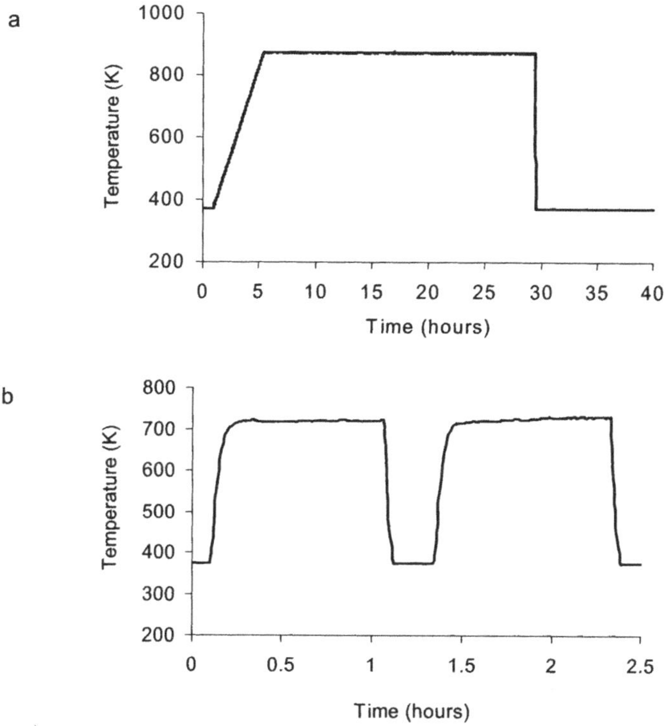

Figure 4: (a) Dielectric heating (2 K min^{-1}) under a He flow (GHSV = $6 \cdot 10^2$ hr^{-1}) of $La_{0.8}Ce_{0.2}MnO_3$ perovskite; (b) Two subsequent dielectric heating cycles in He for $La_{0.8}Ce_{0.2}MnO_3$.

Catalytic oxidation of CH_4: Full oxidation of CH_4 was performed on $La_{0.8}Ce_{0.2}MnO_3$ perovskite using conventional heating, as well as dielectric heating. The results show that the catalytic activity of this perovskite in CH_4 oxidation increases substantially using dielectric heating as compared to the conventional heating (see Figure 5a). A possible explanation of this difference may be found in the selective heating in dielectric fields, leading to the formation of "hot spots". This explanation will be critically addressed in follow-up studies.

Resistance to SO_2: The resistance to SO_2 has been studied for this perovskite in both ovens. It is known that SO_2 has a high electronic affinity. It can adsorb at room temperature in the form of SO_2^- or SO^- anions on metal oxides.[14,15] Base metal perovskites may, as reported, be subject to SO_2 poisoning. The performance of $La_{0.8}Ce_{0.2}MnO_3$ perovskite in CH_4 oxidation in the presence of SO_2 has been studied. To this end, SO_2 was either introduced at high reaction temperature (723 K), or SO_2 was pre-adsorbed at room temperature (298 K), followed by heating from 423 K to 723 K.

When SO_2 is introduced at 723 K, no significant deactivation occurs both for dielectric and conventional heating. This indicates that the influence of SO_2 at high reaction temperature is limited. However, when SO_2 is added at 298 K, loss of activity is observed for both heating modes (Figure 5a). The catalytic activity in the presence of SO_2 using dielectric heating appears to be even higher than that using conventional heating in the absence of SO_2. Similar results have been found for three-way automotive catalysts poisoned by SO_2 comparing a combination of dielectric and conventional heating to heating only by means of a conventional oven.[16] After prolonged exposure to SO_2 at 298 K (Figure 5b), the activity returns to its original level at 723 K, for dielectric heating. For conventional heating, on the other hand, a further irreversible loss in activity is observed. Clearly, the catalyst is highly resistant to SO_2, especially in case of dielectric heating.

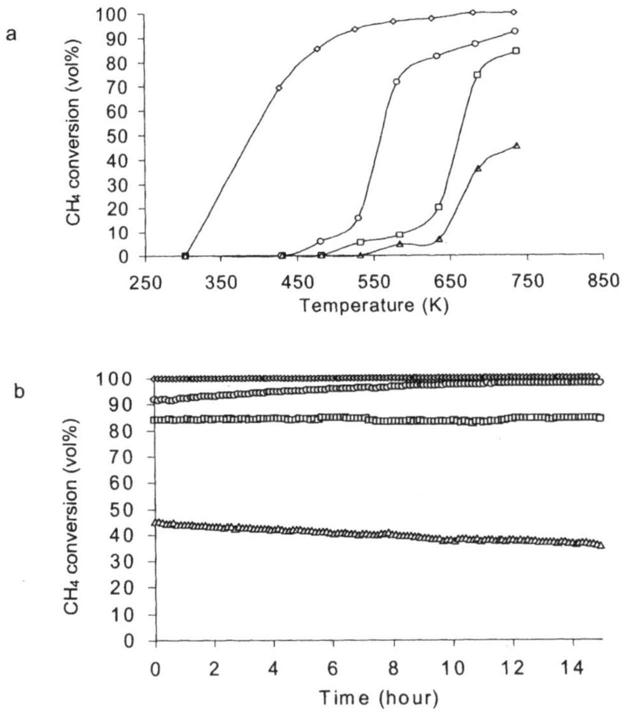

Figure 5: (a) CH_4 conversion over $La_{0.8}Ce_{0.2}MnO_3$ perovskite versus bulk temperature using dielectric and conventional heating in the presence and absence of 200ppm SO_2, GHSV = $12 \cdot 10^2$ hr^{-1}; (b) CH_4 conversion versus time-on-stream at 723 K when SO_2 is added at 298 K; (◇) dielectric heating, (□) conventional heating, (○) dielectric heating in the presence of SO_2 and (△) conventional heating in the presence of SO_2.

After the reaction, the samples that had been heated in the microwave and in the conventional oven contained a differently colored separate phase, which suggested that inhomogeneous poisoning had taken place. Fourier transform infrared spectroscopy (FTIR) and XPS measurements indicate that both samples contain an S-containing species, which may either be present in the bulk or on the surface. Further study is required to further assess the nature of these S-containing species.

CONCLUSIONS

Cerium substituted La-Mn perovskites are high loss dielectric materials, allowing rapid heating to temperatures in excess of 800 K with no irreversible change in the bulk structure, suggesting that these materials have high thermal stability. For catalytic CH_4 oxidation, dielectric heating results in much higher catalytic activity as compared to conventional heating. This is probably due to the "hot spot" effect. High resistance to SO_2 is observed for dielectric heating at high reaction temperatures, whereas an irreversible loss in activity occurs when conventional heating is used. Different poisoning mechanisms have taken place, as based on an inhomogeneous color change observed after the reaction. Our findings indicate that these perovskites can be applied as suitable coating materials for application in situ soot filter regeneration.

ACKNOWLEDGEMENT

The financial support of the Netherlands Research Council for this project is gratefully acknowledged. Thanks are due to Mr. P. Corbet and Mr. R. Haswell from Shell Research & Technology Center for performing SEM, and to Mr. Elgersma and Mrs. Mittelmeijer-Hazeleger for performing ICP and BET.

REFERENCE

[1]J.X. Ma, M. Fang, P. Li, B. Zhu, X.H. Lu and N.T. Lau, "Microwave-assisted Catalytic Combustion of Diesel Soot", *Applied Catalysis A: General*, **159** [1-2] 211-228 (1997).

[2]Y. Teraoka, K. Nakano, W.S. Ghangguan and S. Kagawa, "Simultaneous Catalytic Removal of Nitrogen Oxides and Diesel Soot Particulate over Perovskite-related Oxides", *Catalysis Today*, **27** [1-2] 107-113 (1996).

[3]S. Ponce, M.A. Peña and J.L.G. Fierro, "Surface Properties and Catalytic Performance in Methane Combustion of Sr-substituted Lanthanum Manganites", *Applied Catalysis B: Environmental*, **24** [3-4] 193-205 (2000).

[4]C.C. Chou, P.H. Chen and N. Lin, "Microstructural Characteristics of Microwave-sintered Semi-conductive $Pb_{0.6}Sr_{0.4}TiO_3$ Ceramics", *Ferroelectrics*, **231** [1-4] 625-630 (1999).

[5]B. Vaidhyanathan. D.K. Agrawal, T.R. Shrout and Y. Fang, "Microwave Synthesis and Sintering of $Ba(Mg_{1/3}Ta_{2/3})O_3$", *Materials Letters*, **42** [3] 207-211 (2000).

[6]M. Alliouat, L. Mazo, G. Desgardin and B. Raveau, "Microwave Sintering of Lead-based Perovskite Ceramics for Capacitors", *Materials Letters*, **5** [9] 328-332 (1987).

[7]D. Kieβling, R. Schneider, P. Kraak, M. Haftendorn and G. Wendt, "Perovskite-type Oxides – Catalysts for the Total Oxidation of Chlorinated Hydrocarbons", *Applied Catalysis B: Environmental*, 19 **[2]** 143-151 (1998).

[8] D.M. Smyth, "Oxidative Nonstoichiometry in Perovskite Oxides"; pp. 47-72 in *Properties and Applications of Perovskite-type oxides*, Edited by L.G. Tejuca and J.L.G. Fierro. Marcel Dekker, INC, New York, 1993.

[9] Y.N. Lee, R.M. Lago, J.L.G. Fierro, V. Cortés, F. Sapiña and E. Martínez, "Surface Properties and Catalytic Performance for Ethane Combustion of $La_{1-x}K_xMnO_{3+\delta}$ perovskites", *Applied Catalysis A: General*, **207** [1-2] 17-24 (2001).

[10] K.S. Chan, J. Ma, S. Jaenicke and G.K. Chuah, "Catalytic Carbon Monoxide Oxidation over Strontium, Cerium and Copper-substituted Lanthanum Manganates and Cobaltates", *Applied Catalysis A: General*, **107** [2] 201-227 (1994).

[11]K. Tabata, Y. Hirano and E. Suzuki, "XPS Studies on the Oxygen Species of $LaMn_{1-x}Cu_XO_{3+\lambda}$", *Applied Catalysis A: General*, **170** [2] 245-254 (1998).

[12]M. Imamura, N. Matsubayashi and H. Shimada, "Catalytically Active Oxygen Species in $La_{1-x}Sr_xCoO_{3-\delta}$ Studied by XPS and XAFS Spectroscopy", *Journal of Physical Chemistry: B*, **140** 7348-7353 (2000).

[13]A. Kowalczyk, A. Ślebarski, A. Szajek, J. Baszyński and A. Winiarski, "X-ray Photoemission Spectra of $La_{0.7}Sr_{0.3}MnO_3$ perovskite", *Journal of Magnetism and Magnetic Materials*, **212** [1-2] 107-111 (2000).

[14]R.J.H. Voorheve, D.W. Johnson, Jr., J.P. Remeika and P.K. Gallagher, "Perovskite Oxides: Materials Science in Catalysis", *Science*, **195** [4281] 827-833 (1977).

[15]W. Li, "Poisoning of Perovskite Oxides by Sulfur dioxide"; pp. 145-170 in *Properties and Applications of Perovskite-type oxides*, Edited by L.G. Tejuca and J.L.G. Fierro. Marcel Dekker, INC, New York, 1993.

[16]M.D. Turner, R.L. Laurence, K.S. Yngvesson and W.C. Conner, "The Effect of Microwave Energy on Three-way Automotive Catalysts Poisoned by SO_2", *Catalysis Letters*, **71** [3-4] 133-138 (2000).

Biomedical Research and Health Issues

CONTINUOUS AND PULSED MEDICAL MICROWAVE DEVICES

Fred Sterzer
MMTC, Inc.
12 Roszel Road, Princeton, NJ 08540 USA
sterzer@mmtc.com

ABSTRACT

Microwave technology can make important contributions to both diagnostic and therapeutic medicine. A number of medical devices that use microwaves are today in clinical use, and many new medical microwave devices are being developed. All microwave medical devices depend on the ability of microwaves to deeply penetrate into living tissues. The depth to which microwaves can penetrate tissues is primarily a function of the dielectric properties of the tissues and of the frequency of the microwaves. In general, the lower the water content of the tissue, the deeper a wave at a given frequency can penetrate. Also at the frequencies of interest, the lower the frequency, the deeper is the depth of penetration into tissues with given water content.

When microwaves penetrate into tissues they give up energy to the tissues. Thus microwaves can be used to non-invasively heat subsurface tissues (dielectric heating). Part of the thermal microwave noise power generated by any heated subsurface tissues passes through overlying tissues and reaches the surface. From measurements of this noise power it is possible to non-invasively estimate subsurface tissue temperatures. Also, high-power, low-duty cycle pulsed microwaves that do not significantly increase average tissue temperatures can produce beneficial medical effects that cannot be produced by heating with continuous microwaves.

Several medical devices that take advantage of the unique properties of microwaves will be described. They include devices for treating cancer and benign prostatic hypertrophy, accelerating wound healing, and for detecting subsurface inflammations and malignant lesions.

MICROWAVE HYPERTHERMIA TREATMENTS OF CANCER

In hyperthermia treatments of cancer malignant tumors are heated to temperatures of typically 42.5° Celsius, and maintained at these temperatures for about 30-45 minutes. The procedure is usually repeated several times at intervals of two or more days. Hyperthermia treatments can increase the efficacy of conventional radiation and chemotherapy because:

- Hyperthermia is often lethal to weakened malignant cells, including cells that were only sub-lethally injured by radiation or chemotherapy;
- Hyperthermia can increase the efficacy of those chemotherapeutic drugs that are more active at elevated tissue temperatures than at core temperatures;
- Hyperthermia can increase tumor blood flow thereby improving access of chemotherapeutic drugs to tumors and helping to oxygenate radiation-resistant hypoxic cells.
- Following exposure to excessive heat, cells produce heat shock proteins that can contribute to tumor regression.
- Cells that are killed by excessive heat activate T lymphocytes thus stimulating the body's own anti-cancer immunologic responses.

The benefits of adjuvant hyperthermia treatments are considerable. For example, in 2,234 human malignant tumors of various histological types that were treated at different institutions, 35% showed complete remission after radiation alone. When hyperthermia was added to radiation, the percentage of complete remissions increased to 65% [1].

Microwaves can be used to non-invasively produce hyperthermia in many important cancer sites, particularly in cutaneous and subcutaneous sites, and in sites that are accessible via natural openings of the body. Table I shows the depth of penetration of microwaves into tissues with high water content such as muscle and skin and tissues with low water content such as fat and bone for two microwave frequencies that are used in medical applications.

Table I. Depth of penetration of two microwave frequencies into tissues. The depth of penetration is defined at the distance into tissues when the power density of a plane wave is reduced by a factor of 0.14 [2].

Frequency	Depth of Penetration	
	Tissues with High Water Content (cm)	Tissues with Low Water Content (cm)
915 MHz	3.04	17.7
2.45 GHz	1.7	11.2

Numerous hyperthermia applicators (antennas) for broadcasting microwaves into tumors have been described in the literature, many of them in the International Journal of Hyperthermia. These applicators range from complex phase array applicators to simple waveguide and coaxial applicators. Deep-seated sites can be effectively heated with microwaves using interstitial applicators or microwave balloon catheters [3].

TISSUE ABLATION WITH MICROWAVES

Microwaves can be used to ablate (necrose) tissues. The range of temperature-duration thresholds for histologic damage at depth for mammalian tissues by localized heating in vivo are listed for three different times in Table II. The applicators used for microwave ablation are similar to those used for producing microwave hyperthermia.

Table II. Temperature-duration thresholds for histologic damage by localized heating at depth of mammalian tissues in vivo [4].

Time	Temperature	
	No Detectable Damage	Damage
10 seconds	52 °C	62 °C
1 minute	48 °C	56 °C
10 minutes	42 °C	51 °C

One of the most promising applications of microwave tissue ablation is its use as a stand-alone therapy for necrosing malignant tumors, particularly those tumors that are unresectable. However, even when surgery is possible, it may prove sometimes to be advantageous to use heat ablation instead of surgery because of the potential benefits of the stimulation of the immune system by the necrosed malignant cells. Remissions of cancer by fever caused by infections or the administration of bacterial toxins indicates that the immune system can sometimes be a powerful weapon against cancer [5,6]. The anti-tumor effects of malignant cells killed by local heating on distant metastatic sites has been repeatedly demonstrated in animal models, and has also been observed in human cancer patients [7,8]. In animal models it has been demonstrated that the immune response produced by killing cells with heat can be further enhanced by the administration of systemic immunostimulants [9].

Ablation may also offer benefits in certain instances over chemotherapy and radiation. As mentioned above, when cells are killed by excessive heat they not only produce heat shock proteins, but also activate T lymphocytes. In contrast to excessive heat that necroses malignant cells, chemotherapy and radiation therapy in general only sub-lethally injure malignant cells. These sub-lethally injured cells often kill themselves (apoptosis), but cells that die by apoptosis not caused by either viruses or bacteria do not elicit an immune response (no T cell activation) [10-12]. Furthermore, unlike cell ablation with heat, chemotherapy as well as radiation therapy over large areas depress the immunologic system.

MICROWAVE BALLOON CATHETERS FOR TREATING BENIGN PROSTATIC HYPERTROPHY (BPH).

The prostate is a small chestnut-shaped gland that surrounds the urethra of males immediately below the bladder. BPH is caused by the narrowing of the urethra by an enlarged prostate gland and is the most common cause of bladder obstruction in men. There are several microwave devices on the market that are designed to non-invasively treat BPH by modifying or ablating prostate tissues by heating the prostate gland with microwaves via the urethra. With microwave balloon catheters heating of the prostate gland with 915 MHz to moderately high temperatures is accompanied by simultaneously dilation with an expansion balloon of the part of the urethra that surrounds the prostate. By this simultaneous application of moderate pressure and heating it is possible to safely create long lasting enlarged lumens of the prostatic urethras (biological stents) in a single treatment session.

In Phase I and Phase II clinical trials approved by the Federal Drug Administration of the USA, Rockville, MD and funded by the Celsion Corporation, Columbia, MD, 140 patients with BPH have so far been treated with microwave balloon catheters.

MEDICAL APPLICATIONS OF PULSED MICROWAVES

High-power, low-duty cycle pulsed microwaves that do not significantly increase average tissue temperatures can produce significant forces at the interfaces between media of differing dielectric constants such as the interfaces between cell membranes and their surrounding, and also produce forces due to pulsed heating.

The direction of the forces at the interfaces between media of different dielectric constants due to oscillating electric fields such as those of microwaves are always from the medium with high dielectric constant to the medium with low dielectric constant as is illustrated in Figure 1. The magnitudes of these forces are proportional to the square of the electric fields [13]. Note that the local electric field in tissues can be much greater than the average electric field because of local field enhancement due to the inhomogeneity of tissues. Also the electric fields produced by high-power, low duty cycle pulsed microwaves that do not significantly increase tissue temperatures can be much greater than the fields produced by lower power continuous heating. For cells the force due to the oscillating electric fields can tend to expand the aqueous pores in cell membranes increasing the permeability of cells as is illustrated in Figure 2.

Other forces produced by pulsed high-power, low duty cycle microwaves are due to the periodic heating of the tissues during each pulse followed by cooling of the tissues during the inter pulse periods. This results in periodic expansion and contraction in the tissues.

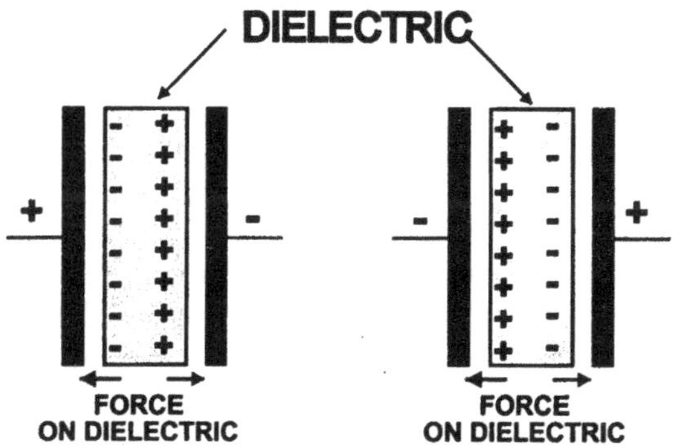

Figure 1. Force on a dielectric placed between two parallel plates in air. Note that this force is always in the same direction independent of the direction of the external electric field.

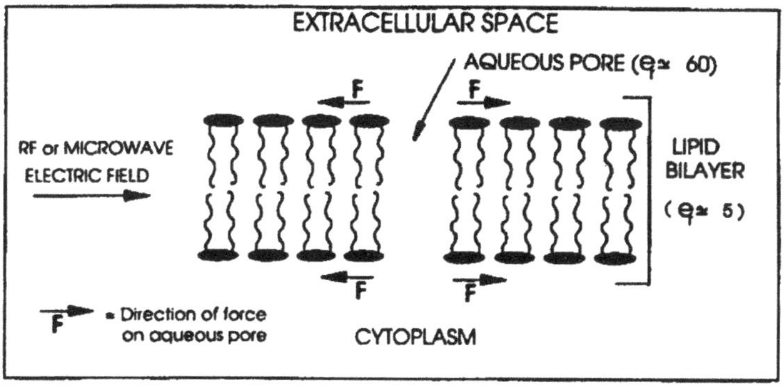

Figure 2. Direction of forces on aqueous pores of cell membranes due to RF or microwave fields. Note that the direction of the forces tends to widen the pores in the membranes, a phenomenon consistent with experimental observations of increased permeability of cells caused by pulsed RF and microwaves.

Pulsed electromagnetic fields that do not significantly increase average tissue temperatures can promote healing in both delayed and non-union fractures, accelerate the healing of soft tissue wounds, increase the permeability of cells, etc [14-16]. The biological effects responsible for these actions of pulsed fields are not well understood, but the forces on cells and on their membranes produced by pulsed fields likely contribute to the observed beneficial medical results.

While most of the studies on healing with pulsed electromagnetic fields used low RF frequencies, we have demonstrated in pilot studies using Copenhagen rats that wounds that are exposed daily to pulsed 915 MHz heal much faster than wounds that heal naturally without any intervention [17].

MICROWAVE RADIOMETERS FOR NON-INVASIVE MEASUREMENTS OF SUBSURFACE TISSUE TEMPERATURES

Medical microwave radiometers are instruments that measure the thermally generated microwave emissions from subsurface tissues, these emissions being proportional to the absolute temperature of the tissues. [18]. With such radiometers one can non-invasively estimate subsurface tissue temperatures to depth of several centimeters. Compared to other approaches for non-invasive measurements of subsurface tissue temperatures such Magnetic Resonance Imaging and Magnetic Resonance Spectroscopy, microwave radiometers are lower cost and simpler to use in the clinic.

Significant progress has been made in recent years in improving the accuracy and stability of microwave radiometers [19]. Digital microwave radiometers can be automatically calibrated for changes in ambient temperatures and drift in the electronics of the radiometer, and with the aid of software can remove the effect of skin temperature on displayed subsurface tissue temperatures [20]. Such radiometers have the potential of making important contributions to medical care in applications ranging from early detection of subsurface inflammations such as appendicitis to the detection of malignant tumors and early evaluation of the efficacy of anti-cancer treatments. These applications rely on the fact that inflamed tissues as well as growing malignancies tend to be hotter than surrounding normal tissues.

CONCLUSION

Microwaves have several characteristics that make them useful in important medical devices. These characteristics include:

- the ability to penetrate and heat tissues to a depth that is a function of the microwave frequency;
- the ability of pulsed microwaves to create significant forces in tissues while causing only minimal average heating;

• the fact that the thermal microwave noise power emitted by tissues is proportional to their absolute temperature.

ACKNOWLEDGEMENT

The author wishes to thank several of his colleagues, all of whom have made important contributions to the advancement of microwave medical device technology, for many helpful discussions. They are Dr. Augustine Chang, Celsion Corporation, Dr. Ester Friedenthal, Consultant, James R, Lawson and Daniel D. Mawhinney, MMTC, Inc., Drs. A. Melman and Joseph Mendecki,, Montefiore Medical Center, , and Paul R. Stauffer, University of California at San Francisco. The National Institutes of Health of the United States supported some of the work mentioned in this article.

REFERENCES

1. O. Dahl et al, "Status of Clinical Hyperthermia", Acta Oncologica, **38**, pp. 863-873, 1999.

2. C. C. Johnson and A. W. Guy, "Nonionizing Electromagnetic Wave Effects in Biological Materials and Systems", Proc. IEEE, **60**, 692-718, June 1972.

3. F. Sterzer, "Localized heating of deep-seated tissues using microwave balloon catheters", Chapter 4 in A. Rosen and H. D. Rosen "New Frontiers in Medical Device Technology", John Wiley & Sons, Inc. NY 1995.

4. P. P. Lele, "Physical Aspects and Clinical Studies with Ultrasonic Hyperthermia", Chapter 16 in F. K. Storm, "Hyperthermia in Cancer Therapy", G. K. Hall Medical Publishers, Boston, MA, 1983.

5. U. Hobohm, "Fever and Cancer in Perspective", Cancer Immunol Immunother. **50**, Oct. 2001.

6. G. Parmiani and M. T. Lotze, "Tumor Immunology", Taylor and Francis, 2002.

7. J. A. Dickson and S. A. Sha, "Immunologic Aspects of Hyperthermia", Chapter 23 in F. K. Storm, "Hyperthermia in Cancer Therapy", G. K. Hall Medical Publishers, Boston, MA, 1983.

8. B. Stawarz et al, "Transrectal Hyperthernia as Palliative Treatment for Advanced Carcinoma of Prostate and Studies of Cell-mediates Immunity", Urolog

9. Y. Takada et al, "Granulocyte-colony stimulating factor enhances anti-tumor effect of hyperthermia", Int. J. Hyperethermia, **16**, pp. 275-286, May-June 2000.

10. B. Sauter et al, "Consequences of cell death: exposure to necrotic tumor cells, but not primary tissue cells or apoptopic cells, induces the maturation of immunostimulatory dendritic cells", Journal of Experimental Medicine, **191**:423-33, February 7, 2000.

11. S. Basu et al, 'Necrotic but not apoptotic cell death releases heat shock proteins, which deliver a partial maturation signal to dendritic cells and activate the NF-kB pathway", International Immunology, **12**:1539-46, 2000.

12. S. Somersan et al, "Primary tumor tissue lysates are enriched in heat shock proteins and induce the maturation of human dendritic cells", Journal of Immunology, **167**:4844-52, 2001.

13. J. A. Stratton, "Electromagnetic Theory", Chapter II, McGraw Hill, 1941.

14. CAL Basset, "Fundamental and practical aspects of therapeutic use of pulsed electromagnetic fields (PEMFs), Crit rev Biomed Eng **17**, 451-529, 1989.

15. M. Itoh, et al, Accelerated wound healing of pressure ulcers by pulsed peak power electromagnetic energy (Diapulse), Decubitus, 24-25, 29-34, Feb. 1991.

16. F. Sterzer, Microwave medical devices, IEEE Microwave Magazine 3:65-70. 2002.

17. Dr. J. Mendecki, Montefiore Medical Center, Bronx, NY Private Communication.

18. F. Sterzer Microwave radiometers for non-invasive measurements of subsurface tissue temperatures. Automedica **8**: 203-11, 1987.

19. D. V. Land, "An efficient, accurate and robust radiometer configuration for microwave temperature measurements for industrial and medical applications", J. Microwave Power Electromagnetic Energy, **36** (3), 139-53, 2001.

20. J. R. Lawson and D. D. Mawhinney, MMTC, Inc. Princeton, NJ. Private communication.

IMAGE FORMATION ALGORITHMS WITH AUTOMATIC SKIN DETECTION FOR CONFOCAL MICROWAVE IMAGING

E.C. Fear and M. Okoniewski
Electrical and Computer Engineering
University of Calgary
2500 University Drive NW
Calgary, AB Canada T2N 1N4

ABSTRACT

The anticipated large contrast in electrical properties of normal and malignant breast tissues has sparked interest in microwave breast cancer detection. This paper describes both measurement probes required for characterization of an extensive database of breast tissues, as well as image formation algorithms that take advantage of skin reflections in the data. These algorithms are demonstrated on simulated and experimental data, demonstrating the feasibility of microwave tumor detection.

INTRODUCTION

The basis of tumor detection with microwave imaging is the significant contrast in dielectric properties of normal tissues and malignancies. This contrast is expected to persist over a wide frequency range, and is anticipated to be larger than the few percent contrast that may be imaged by x-rays. Several research groups have been investigating different approaches to imaging the contrast, as summarized in [1]. As the properties of breast tissues are not well defined over the frequency range of interest, part of our research program is directed towards compiling a database of the electrical properties of these tissues. This involves development of specialized probes capable of accurately measuring reflections from the tissue over the frequency range from 1 to 20 GHz. The probes and preliminary measurement results are described in this paper.

Confocal microwave imaging (CMI) was proposed by Hagness et al [1-4] and is a promising approach to breast tumor detection. To form an image with CMI techniques, reflections from the breast recorded at an array of antennas are synthetically focused. The coherent addition of reflections from scattering objects (tumors) results in a much larger response when the focal point is located at a tumor compared to normal tissue. CMI images indicate locations of strongly scattering objects in the breast and have the advantage of a simple image reconstruction algorithm.

Most work to date on CMI has involved computer simulations to determine the feasibility of this and similar approaches (e.g. [1-5]). Initial experimental work with tissue sensing adaptive array (TSAR) imaging and simple phantoms is ongoing [6]. This work has resulted in modification of image formation algorithms (compared to CMI) to automatically detect the location of the skin and window the data. In this paper, we outline the improved algorithm and summarize results obtained for a simple experimental phantom and simulations of a hemispherical breast model.

PART I: DIELECTRIC SENSORS

Small-aperture open-ended coaxial probes are the basis for dielectric spectroscopy of breast tissue specimens, due to their broadband response and simplicity, and lack of extensive sample preparation. Complete characterization of the probe response and the quality of the contact with the specimen are critical to the reliability and accuracy of the measurements. Very high frequencies contained in the spectrum of interest (up to 20 GHz) and considerable variation in size of the breast tissue specimens prompted a pilot study into the volume sensing properties of these probes using tissue equivalent liquids to ensure the highest possible accuracy in the dielectric-properties estimates [7]. The sensing volume depends mainly on the probe aperture, and is only slightly dependant on the material being measured. In addition, field energy focuses close to the probe aperture, with penetration depth very weakly dependent on frequency. During the extensive numerical and experimental analysis of simple teflon-filled probes made of sections of a semi-rigid cable, it became clear that small geometrical changes in the probe aperture region, arising e.g. from handling and inaccurate manufacturing, have profound effects on the observed reflection coefficient. Even small inaccuracies of the order of 0.1 mm can lead to significant changes in the probe properties, manifested by errors up to 25% for higher permittivity liquids at higher frequencies (15-20 GHz).

The above results show the need for an improved probe design and superior manufacturing processes to minimize the possibility of geometrical imperfections at the aperture. A new probe with a 10μm-flatness accuracy at the aperture has been built out of 5 sections of high quality coaxial cable filled with dielectrics whose coefficients of thermal expansion are matched to those of metal conductors (Fig. 1a). The probe is hermetically sealed at the aperture.

Due to the complex probe construction, standard calibration procedures cannot be used. However, accurate measurements of the probe shape have been made with the help of X-rays, allowing for numerical generation of the de-embedding S-matrix for the probe. This procedure yields measurement results that are consistently within 1-2% of the results obtained from simulations of an ideal homogeneous borosilicate probe. This indicates that the new probe is suitable as a sensor for high fidelity tissue measurements and characterization. It will be used in a switched, multi-probe configuration for fast and reliable tissue measurements.

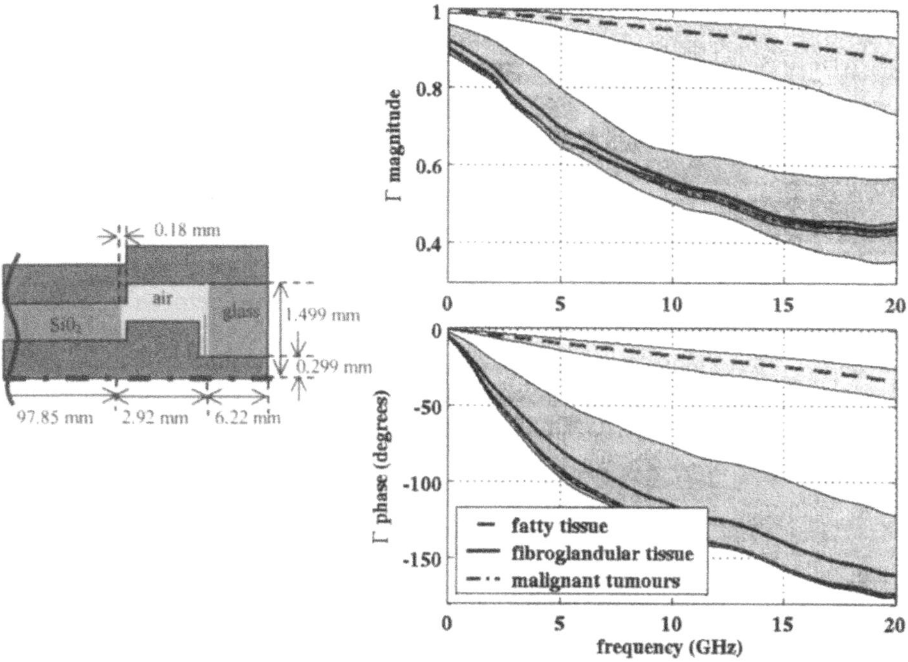

a) b)

Figure 1. a) Geometry of the new probe. b) Preliminary results on dielectric spectroscopy of breast tissue: shaded areas show the variation in the reflection coefficient measurements for fatty, fibroglandular, and malignant tissue samples.

Measurements of freshly excised breast tissue samples started in July 2002. Tissues from the breast reduction surgeries arrive in Pathology, where a pathologist cuts out samples of mainly fatty and mainly fibroglandular tissues from each breast. The samples are measured with the coaxial sensor at room temperature (23°C), and the measurement position is marked with dark ink. A vertical cross-section of the tissue sample is prepared and used to correlate the measurement results with the tissue morphology. Preliminary results (Fig. 1b) show a significant difference in the measured reflection coefficient of fatty and fibroglandular tissue of a healthy breast in the frequency range of interest. Results from few measurements on cancerous samples show significant difference between cancerous and healthy fatty tissue, however the differences between cancerous and healthy fibroglandular tissue are smaller. More measurements are required to determine the statistically valid dielectric properties of breast tissue.

PART II: IMAGING

Methods

Experimental Setup: The experimental setup is shown in Figs. 2 and 3, and consists of a PVC pipe in which a wooden hemisphere representing the tumor is placed. The experiment is designed to be extremely simple, as it serves only to verify concepts related to TSAR imaging. While the electrical properties of the materials are not similar to those of breast tissues, the contrasts between the materials are analogous to the contrasts between various breast tissues. Two configurations of the phantom are examined in order to image two orthogonal planes. In the first configuration (Fig. 2), the pipe and tumor are rotated about the center to represent scanning the antenna around the pipe. Data are recorded after each of 16 rotations, and used to create an image of the plane perpendicular to the pipe axis. In the second configuration (Fig. 3), the pipe is scanned past the antenna in 2.5 cm increments. Data are recorded at 21 locations, and an image of the plane parallel to the pipe axis is created. Imaging two orthogonal planes provides insight into detection and localization of objects in three dimensions.

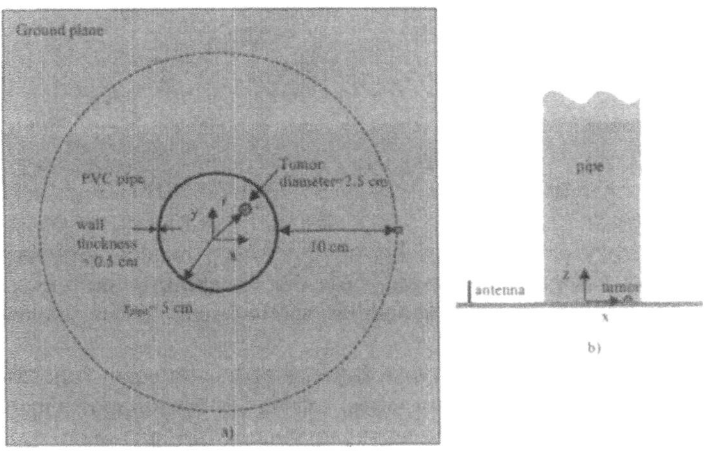

Figure 2. Configuration 1 a) top-down view and b) side-on view.

With both configurations, the phantom is illuminated with a resistively loaded monopole antenna [6]. Reflections are recorded using an 8720C vector network analyzer (Agilent Technologies, Palo Alto, CA, USA). Measurements are taken at 401 frequencies from 0.5 to 20 GHz, and 128 measurements are averaged at each frequency. Time domain signals are required for the image reconstruction algorithm, and conversion of the frequency domain measurements requires appropriate weighting, followed by an inverse transform. In this case, the weighting corresponds to a differentiated Gaussian signal with full-width half-maximum frequency content between 1.7 and 10.1 GHz. Inverse chirp-z transforms are used to convert to the time domain.

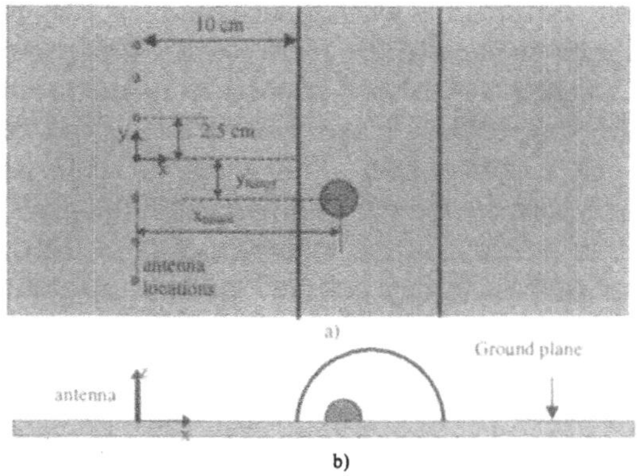

Figure 3. Configuration 2 a) top-down view and b) side-on view.

Simulations: Automatic skin detection algorithms are developed for the simple experimental models, so application of these algorithms to more complex models is required. The hemispherical breast model, shown in Fig. 4, provides a more challenging environment for the skin and tumor detection algorithms. The model and a resistively loaded antenna are immersed in a low-loss substance with electrical properties similar to breast tissue. The interior of the hemispherical breast contains objects representing glandular tissue (ε_r ranging from 11 to 15, σ from 0.4 to 0.5 S/m). The 6 mm diameter tumor has ε_r=50, σ=4 S/m. The remainder of the space is filled with fatty breast tissue (ε_r=9, σ=0.4 S/m). The glandular tissue has more structure and greater contrast with fatty tissue than the previously examined cylindrical models [5], thus increasing the clutter. Simulations are performed with the finite difference time domain method (FDTD), and involve exciting the antenna with a differentiated Gaussian pulse (frequency content greater than 10% of the maximum power between 1 and 8 GHz and center frequency near 4 GHz). Reflections are recorded at the illuminating antenna, the antenna is scanned to a new location, and this process is repeated. A synthetic conical array is formed by scanning the antenna to 10 locations per row and 12 rows spanning from the nipple to the chest wall. For automatic skin detection, an antenna is placed parallel to the chest wall and scanned past the breast (Fig. 4).

Image reconstruction algorithms: Similar image formation algorithms are applied to experimental and simulated data. The first step is calibration, or subtraction of reflections recorded without an object present in order to reduce the initial pulse. The dominant component of the resulting signal is the reflection from the layer of skin or PVC pipe. This reflection must be reduced in order to detect small tumors, which may have responses 50 dB or more below the skin reflection. The next step is an addition to the previously reported algorithm for simulated data [4]. An image is formed with the calibrated data by identifying the travel time from each antenna to the focal point, then time-shifting and summing the signals. The image is thresholded in order to isolate the skin or pipe response. The edges in the thresholded image are detected, and the largest group of connected edge pixels is identified. This group of pixels represents the pipe or skin, and is used to window the data. Specifically, the pixel closest to a selected antenna provides an estimate of the minimum distance between the antenna and object. The pixel furthest from a selected antenna suggests maximum extent of the object. Data arriving before the minimum distance and after the maximum distance are set to zero. This helps to reduce clutter in the image, as only reflections from phantom are retained (e.g. reflections from absorbers are time gated out).

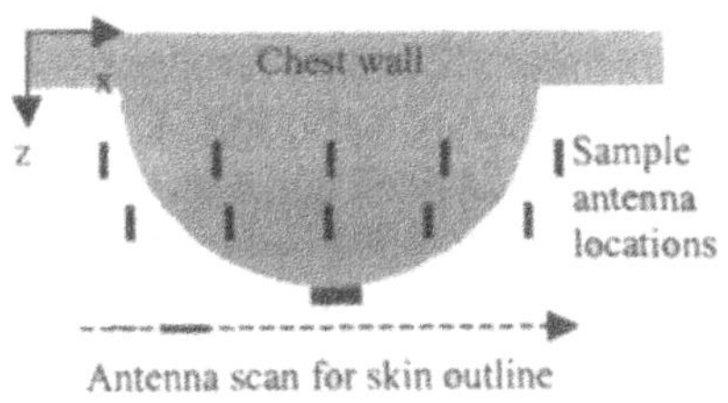

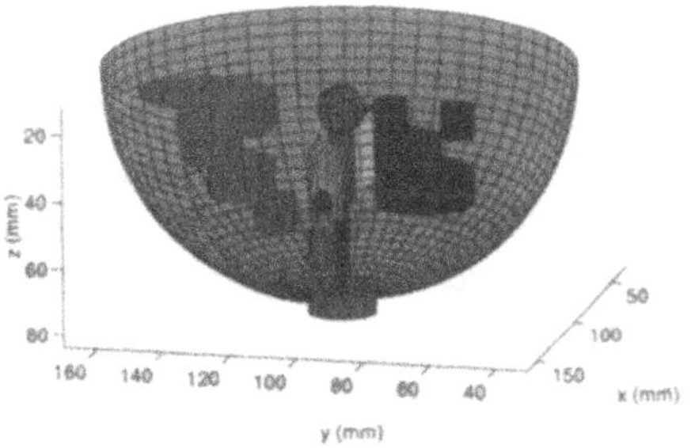

a) b)

Figure 4. Hemispherical breast model a) outline of breast model showing several antenna locations b) model interior.

Following the edge detection and windowing, an estimate of the skin reflection is subtracted from each signal and the data are focused to form an image. For the simulated data, an additional subtraction step is added. For each antenna, the skin-subtracted signal is estimated with the 4 nearest neighbour antennas, and this estimate is subtracted. This step accounts for local variations in the model. Focusing the data includes only a weighting based on relative distance between the focal point and antennas. Path loss compensation is not required, however the image created with simulated data is filtered to increase the importance of pixels nearer the center of the model. For experimental and simulated data, the success of tumor detection is evaluated with the signal-to-clutter ratio, which compares the maximum tumor response to the maximum clutter (non-tumor) response in the image.

Results and Discussion

Figs. 5 and 6 show results obtained for the PVC pipe phantoms. For both phantom orientations, the skin subtraction algorithm successfully reduces the reflections from the PVC pipe, allowing for detection of the 2.5 cm diameter hemispherical tumor. The SC ratio for Fig. 5a is 3.9 dB, and the maximum tumor response is located 2.38 cm from the center, which compares well with the physical location of 2.4 cm. The SC ratio for Fig. 5b is 3.01 dB, and the maximum tumor response is at x=14.0 cm, y=-1 cm. The tumor is physically located at x=-13.43 cm, y=-0.85 cm.

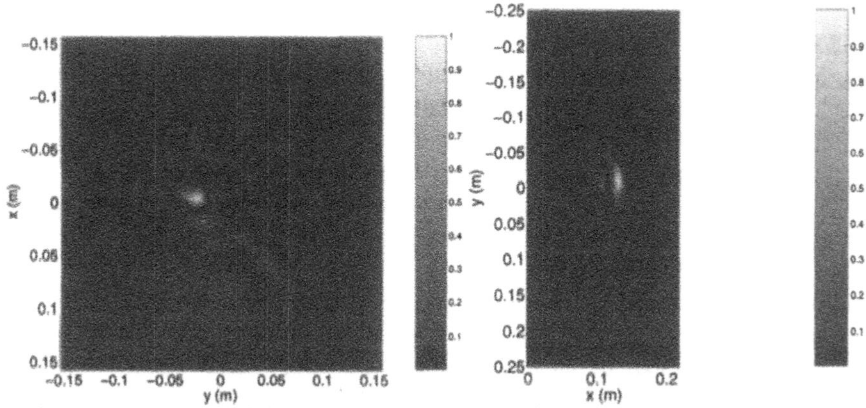

Figure 5. Images of 2.5 cm wood hemispheres in PVC pipes reconstructed with a) setup corresponding to Fig. 2, b) setup corresponding to Fig. 3.

Results for the hemispherical breast model are shown in Fig. 6. In this case, the tumor response is 70% of the maximum clutter response. While this does not provide reliable tumor detection, the tumor is evident in the image. When the previously introduced algorithms were applied to this model, the tumor was not evident in the images. With the automatic skin detection approach, the tumor response is well localized and is not similar (spatially) to the clutter response. This suggests that the algorithms form a basis for further improvements with the goal of providing reliable and accurate tumor detection.

CONCLUSIONS

Microwave breast tumor detection with the simple TSAR approach appears promising. Further data from the measurement program is crucial for evaluation of the system with realistic breast tissue properties.

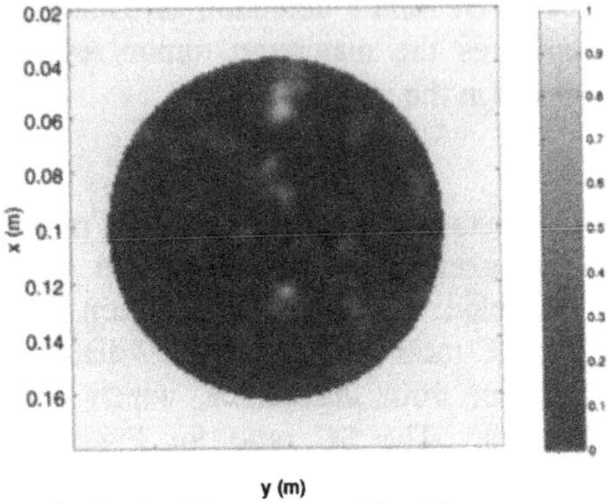

Figure 6. Image of hemispherical breast model with tumor at x=0.128m, y=0.1 m.

REFERENCES

1. E.C. Fear, S.C. Hagness, P.M. Meaney, M. Okoniewski and M.A. Stuchly, "Near-field imaging for breast tumor detection," *IEEE Microw. Mag.*, 3 [2] 48-56 (2002).
2. S. C. Hagness, A. Taflove, and J. E. Bridges, "Two-dimensional FDTD analysis of a pulsed microwave confocal system for breast cancer detection: Fixed-focus and antenna-array sensors," *IEEE Trans. Biomed. Eng.*, 45 [12] 1470-1479 (1998).
3. S. C. Hagness, A. Taflove, and J. E. Bridges, "Three-dimensional FDTD analysis of a pulsed microwave confocal system for breast cancer detection: Design of an antenna-array element," *IEEE Trans. Antennas Propag.*, 47 [5] 783-791 (1999).
4. X. Li and S. C. Hagness, "A confocal microwave imaging algorithm for breast cancer detection," *IEEE Microwave Wireless Components Lett.*, 11 [3] 130-132 (2001).
5. E.C. Fear, X. Li, S.C. Hagness, and M.A. Stuchly, "Confocal microwave imaging for breast cancer detection: Localization of tumors in three dimensions," *IEEE Trans. Biomed. Eng.*, 49 [8] (2002).
6. J. Sill, E.C. Fear, and M.A. Stuchly, "Microwave system for breast tumor detection: experimental concept evaluation", *IEEE Microw. Theory Tech.*, accepted, Sept. 2002.
7. D. Hagl, D. Popovic, S.C. Hagness, J.H. Booske, and M. Okoniewski, "Sensing Volume of Open-Ended Coaxial Probes for Dielectric Characterization of Breast Tissue at Microwave Frequencies", *IEEE Trans. Microw. Theory and Techniques*, accepted, Sept. 2002.

FDTD METHOD FOR MODELLING THE ELECTROMAGNETIC FIELD EFFECT ON THE HUMAN BODY IN MRI

Huawei Zhao
Centre for Magnetic Resonance,
The University of Queensland,
St. Lucia, Brisbane, Queensland 4072, Australia.

ABSTRACT

Peripheral nerve stimulation presents a fundamental limit to the gradient fields that can be safely used in a clinical MRI environment. The interaction between the human body and electromagnetic fields generated by gradient switches in an MRI examination is both spatially and temporally complex. Numerical modelling of the eddy currents induced in the human body by the pulsed field gradients in MRI presents a difficult computational problem. It requires an efficient and accurate computational method for high spatial resolution analyses with a relatively low input frequency. In this presentation, a new technique is described, which allows the FDTD method to be efficiently applied over a very large frequency range, including low frequencies. This is not the case in conventional FDTD based methods. A method of implementing streamline gradients in FDTD is presented, as well as comparative analyses which show that the correct source injection in the FDTD simulation plays a crucial rule in obtaining accurate solutions. In the method, no alterations to the properties of either the source or the transmission media are required. The method is essentially frequency independent and the source injection method has been verified against an example with an analytical solution. Results are presented showing the spatial distribution of gradient-induced electric fields and eddy currents in a complete body model.

INTRODUCTION

Peripheral nerve stimulation presents a fundamental limit to the gradient fields that can be safely used in a clinical MRI environment. The interaction between the human body and electromagnetic fields generated by gradient switches in an MRI examination is both spatially and temporally complex. The finite difference time domain (FDTD) [1,2] technique is a well established numerical technique for computationally modeling electromagnetic (EM) problems. However, there are two major difficulties involved in modeling the eddy currents that are induced in the human body due to gradient switching. The first problem is that the switching frequency of gradient coil current is very low relative to those normally modeled using FDTD methods; the second is the method used to inject the source in the computational domain. Recently, the numerical technique HR-FDTD (high resolution FDTD) [3] has been developed to overcome the first issue. In terms of source injection, the EM field source is the current in the gradient coil, a slow changing current source will produce a strong static magnetic field and a weak coupling between magnetic and electric fields. The FDTD method is normally used to

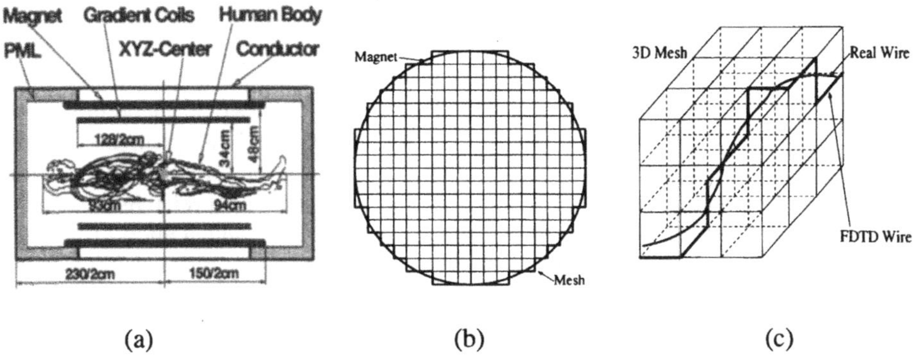

Figure 1. (a) The FDTD computational domain. (b) The magnet bore as a Perfect Conductor. (c) Edge approximation of an arc of the gradient coil.

solve problems for strongly coupled fields. The background static magnetic field also complicates the issue. In this paper, a method for the implementation of the current source in the coil is described and the details of the numerical model are given. The biological eddy currents induced by a gradient coil system in MRI scanner are highlighted in the final section of the paper.

FDTD MODELING WITH PULSED CURRENT SOURCES

The mathematical model is based on Faraday and Ampere's Laws which provide a system of equations that form an independent set of coupled relationships between time-varying electric and magnetic fields. The time-varying EM fields are described in vector form as:

$$\begin{cases} \nabla \times \mathbf{E} = -\partial \mathbf{B} / \partial t, \\ \nabla \times \mathbf{B} = \partial \mathbf{D} / \partial t + \mathbf{J}. \end{cases} \tag{1}$$

Where: $\mathbf{D}$ is the electric displacement, $\mathbf{E}$ is the electric field intensity, and $\mathbf{D} = \varepsilon \mathbf{E}, \varepsilon$ is the permittivity; $\mathbf{B}$ is the magnetic flux density, $\mathbf{H}$ is the magnetic field intensity and $\mathbf{B} = \varepsilon \mathbf{H}$, μ is the permeability. The electric current density is given by $\mathbf{J} = \sigma \mathbf{E}$.

In the FDTD model, the main objective is to investigate the interaction between the switched gradient coil and the human body, the main magnet inner bore is therefore simply treated as a conducting cylinder, with the two open ends truncated by absorbing boundaries, 'so-called' perfectly match layers (PML) [4]. A frequency dependent tissue model, developed by the United States Air Force Research Laboratory (www.brooks.af.mil/AFRL/HED/hedr), was used to calculate the exact human body dielectric and conductivity data. This real human body (see figure 1a) has its length 187cm, the relative permittivity are in the range from 1.0 to 2.14×10^5 and conductivities are in the range from 0.0 to 1.8913. A staircase mesh was used to approximate the magnet's inner bore, see figure 1b. For the gradient coil, the wire pattern was mapped into a regular grid, see figure 1c. The current only flows along the edge of each mesh element. Since the FDTD algorithm is second order in space, in general, the error introduced from the staircase boundary and discreted current rapidly becomes small as the cell size decreases. This assures that an accurate solution will be obtained when the grid size is small enough.

For example, if the current source is approximated by a path along the edge of the Yee cells, the finite difference form is taken as (firstly considering $\mathbf{J}$ to be z-directed)

$$E^{\tau+1}_{z(l,m,n)} = Ce_{1(l,m,n)} E^{\tau}_{z(l,m,n)} + Ce_{2(l,m,n)} \left(\frac{H^{\tau+1/2}_{y(l,m,n)} - H^{\tau+1/2}_{y(l-1,m,n)}}{\delta_x} - \frac{H^{\tau+1/2}_{x(l,m,n)} - H^{\tau+1/2}_{x(l,m-1,n)}}{\delta_y} \right), \quad (2)$$
$$- Ce_{3(l,m,n)} J^{\tau+1/2}_{z(l,m,n)}$$

where $Ce_{3(l,m,n)} = \delta_t / \varepsilon_{(l,m,n)}$. In the final analyses all components of $\mathbf{J}$ are considered. This equation represents that the curl of total magnetic field intensity minus the source current density is equated to temporal electric field changes. This is a good approximation in FDTD when $\mathbf{J}$ is changing rapidly (high frequency). However, when $\mathbf{J}$ is changing very slowly, as in this case, the magnetic field is a half time step behind the electric field in the FDTD scheme. Therefore, the electric field is immediately generated by the current density at the start point of a simulation. Moreover, the total magnetic field in FDTD simulation involves both zero and higher order fields. In the ideal case, the curl of the zero order magnetic field should be zero in free space. However, the curl operator could produce a large error when small numerical errors occur in the magnetic field. Consequently, the computed electric field solution could be larger than the true solution. This is illustrated when the frequency of current source goes to zero; at this time, according to the equation (2), electric field is still being generated in the system. It follows that it is preferable to avoid introducing the zero order magnetic field terms in these simulations.

In order to overcome this problem, we firstly review the nature of the low frequency current source. For static field problems, there is no coupling between the electric and magnetic fields, except in a medium of finite conductivity such as the human body, where

$$\nabla \times \mathbf{H} = \mathbf{J} = \sigma \mathbf{E}. \quad (3)$$

Thus, in a non-conducting medium, the static $\mathbf{E}$ and $\mathbf{H}$ fields may be calculated independently of each other. The coupling between the dynamic fields is due to the terms $\partial \mathbf{B} / \partial t$ and $\partial \mathbf{D} / \partial t$, and $\mathbf{J} = \sigma \mathbf{E}$ in the equation (1). This coupling becomes negligible as the time rate of change of the fields diminish. For such fields, it is possible to obtain series solutions by initially omitting the $\partial \mathbf{B} / \partial t$ and $\partial \mathbf{D} / \partial t$ terms from Maxwell's equations and then introducing their effects as correction terms.

To analyse the behaviour of dynamic fields, Maxwell's equations could be written in perturbed form, where the series solutions for $\mathbf{E}$ and $\mathbf{H}$ have the relations

$$\begin{cases} \nabla\times\mathbf{E}_0 = 0; & \nabla\cdot\mathbf{D}_0 = \rho_0; \\ \nabla\times\mathbf{E}_n = -\dfrac{\partial\mathbf{B}_{n-1}}{\partial t}; & \nabla\cdot\mathbf{D}_n = \rho_n; \qquad n\neq 0. \end{cases} \tag{4}$$

$$\begin{cases} \nabla\times\mathbf{H}_0 = \mathbf{J}_0; & \nabla\cdot\mathbf{B}_0 = 0; \\ \nabla\times\mathbf{H}_n = \mathbf{J}_n + \dfrac{\partial\mathbf{D}_{n-1}}{\partial t}; & \nabla\cdot\mathbf{B}_n = 0; \qquad n\neq 0. \end{cases} \tag{5}$$

The relations in (4) and (5) show that the time changes of each term of electric field will produce higher order terms of magnetic field, and vice versa. These relations have to be solved subject to boundary conditions, which is not a simple task. However, if we know the $\mathbf{H}_0$ term, then we can calculate the $\mathbf{E}_1$ term and the rest of the terms in the series are only dependent on the $\mathbf{E}_1$ term. This analysis gives us an indication of how to structure a FDTD method to produce solutions for dynamic EM fields. So, instead of using the perturbed forms of (4) and (5), we have used an FDTD method to solve for all of the 'higher order' fields, by using $\mathbf{E}_1$ as the source.

From Maxwell's equations, the field at source point can be specified as

$$\nabla^2\mathbf{E} - \mu\varepsilon\frac{\partial^2\mathbf{E}}{\partial t^2} = \mu\frac{\partial\mathbf{J}}{\partial t} + \frac{1}{\varepsilon}\nabla\rho. \tag{6}$$

If there is no free charge ($\rho = 0$), the electric field is only a function of $\partial\mathbf{J}/\partial t$. Consider a line current source, the exact solution can be expressed in terms of a Hankel function of the second kind [5] as

$$E_S = -\frac{1}{4}\mu H_0^{(2)}(kr)\frac{dI}{dt}. \tag{7}$$

Equation (7) is singular at $r = 0$. However, due to the finite size of the grid cell in the FDTD simulation, the field may be inaccurate close to the singularity. A modification of the source term for FDTD models can be achieved by using the effective radius r_0 of wire [6] to match the analytical solution. At the source point, in the discrete form, the electric field can be approximated as

$$E_{S(l,m,n)} = -\frac{1}{4}\mu \left\| H_0^{(2)}(kr_0) \right\| \frac{dI_{(l,m,n)}}{dt} . \tag{8}$$

Where $\left\| H_0^{(2)}(kr_0) \right\|$ is a scale factor.

If, for example, a source wire is positioned along the y-edge of the Yee-cell, the FDTD scheme could be written as

$$H_{z(l,m,n)}^{\tau+1/2} = H_{z(l,m,n)}^{\tau-1/2} - C_{h(l,m,n)} \left(\frac{E_{y(l+1,m,n)}^{\tau} - E_{yS(l,m,n)}^{\tau}}{\delta_x} - \frac{E_{x(l,m+1,n)}^{\tau} - E_{x(l,m,n)}^{\tau}}{\delta_y} \right), \tag{9}$$

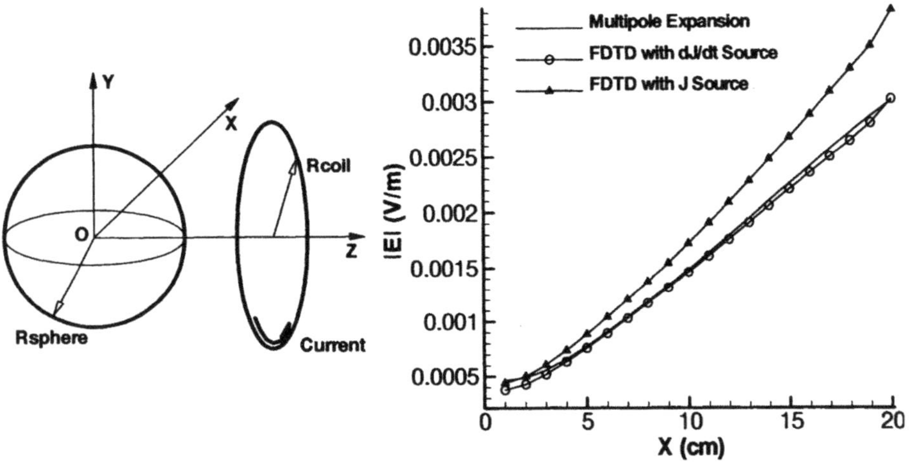

(a) (b)

Figure 2. (a) Schematic arrangement of a homogeneous conductive sphere near a circular current loop. (b) Comparison of the numerical analytic results for the magnitude of the electric field in the sphere.

where $E^{\tau}_{yS(l,m,n)}$ is the source term, which can be directly specified as a function of $d\mathbf{J}/dt$.

SIMULATION RESULTS

Firstly, we now consider the test case used by Keltner [7]: a loop coil of radius R_{coil} (26 cm) carrying uniform current I is placed adjacent to a lossy dielectric sphere of radius R_{sphere} (20 cm), the distance between the coil centre and the sphere centre is 10 cm, see figure 2a. The frequency of the current is 2.5 kHz and the amplitude is 1A. For the sphere, the conductivity is 0.2 S/m, the relative permittivity is 1. The reference calculation of electric and magnetic fields induced inside the sphere is made using the multipole expansion method (MEM). Both J and $\partial J/\partial t$ source FDTD schemes were simulated; the comparison with MEM results is illustrated in the figure 2b. These results demonstrate that, for this type of problem, accurate solutions for dynamic fields can be obtained by the injection of the current-change as a source.

If the current source were sinusoidal ($J = J_0 \sin(\omega t)$), then the derivative source would be a cosinusoid ($\partial J/\partial t = \omega J_0 \cos(\omega t)$). The peak electric field would then be proportional to ωJ_0. Therefore, higher frequency current sources generate stronger electric fields. Since we are mostly concerned with the peak induced electric field, and the spatial dimension is only a small fraction of the wavelength, it is (thankfully) not necessary to run the simulation for a long period of time. A ramp function, $E = E_0(1-\cos(\pi t/t_0))/2$, can be used to smoothly reach the maximum field at the source point, where the t is the simulation time and the t_0 is the time for the E to reach the peak E_0. The variable t_0 is selected to be large enough such that ct_0 (c is speed of light) is far greater than the space dimension and small enough to make the simulations efficient. The total simulation time should be

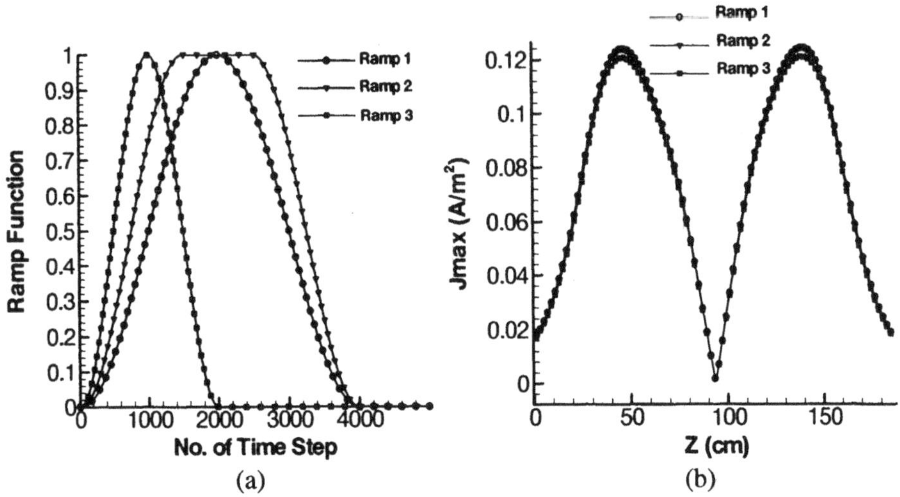

Figure 3. Test source function waveforms for a gradient coil. (a) the source function form, each time step represents 1.9×10^{-11} seconds; (b) the peak eddy current in a homogeneous cylindrical body.

long enough to let the peak pass through every point in the human body. To illustrate this point, we have run the simulation with three different source functions, see figure 3a, and they all give the similar eddy current distributions in a uniform cylindrical body 'excited' by a pulsed, streamline z-gradient coil, see figure 3b.

An MRIfdtd computer code with parallel computation has been developed. The gradient coil and the conductor boundaries are automatically mapped to the grid system. Complete details of the FDTD computational domain are illustrated in figure 1a. In this paper, a cell size of 10 mm was used, so the entire computational domain is divided into $N_x \times N_y \times N_z = 129\times129\times231 \approx 4.0\times10^6$ cells. A human body model containing 357918 cells was placed in the centre of the MRI scanner model. A trapezoidal current wave form, $I = AI_0t$, is used as the source with $A = 10^4$ in the gradient coils, and the gradient strength is $dB/dz \approx 40\ mT/m$ per coil. The numerical simulation ran for 5000 iterations with a ramp-up peak at 2000 iterations. The total computational time was about 8 hours using 4 CPU processors on a SUN Enterprise 450.

Figure 4a shows the maximum eddy current density induced along the human body for switched z, y and xyz gradient coil combinations. As expected, the xyz gradient coil combination produces the largest eddy current density and z-gradient coil produces the smallest. For comparative purposes, we have also simulated a sinusoidal gradient waveform often used in EPI, with a principal frequency of 833 Hz. The maximum eddy currents have a similar spatial distribution to those induced by the trapezoidal switch but have much smaller amplitudes. This demonstrates an advantage of using sinusoidal gradient waveforms where practicable. The two dimensional eddy current density distributions for the trapezoidal pulses are illustrated in figure 4b and clearly show the eddy current patterns in the body.

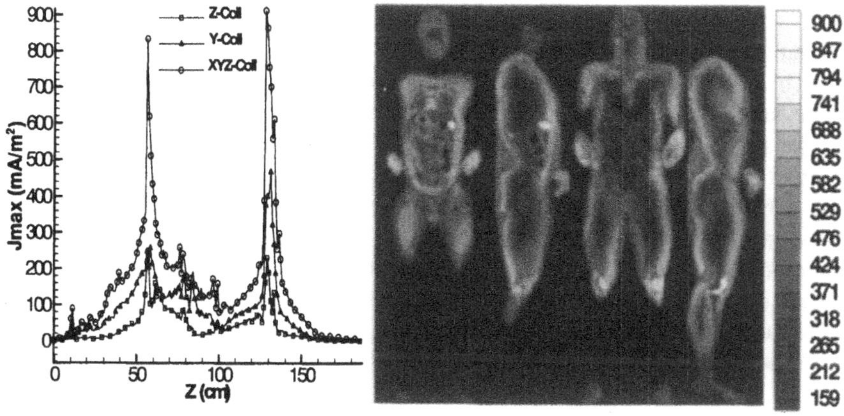

(a) (b)

Figure 4. (a) The maximum human body eddy current density induced by gradient coil in the MRI scanner. Each longitudinal point represents the maximum eddy current through the cross-section. (b) Eddy current density distribution generated by the xyz gradient coil combination in mA/m^2: (a) cross-sections at the largest induced current level in the chest; (b) cross-sections at the level of the knees.

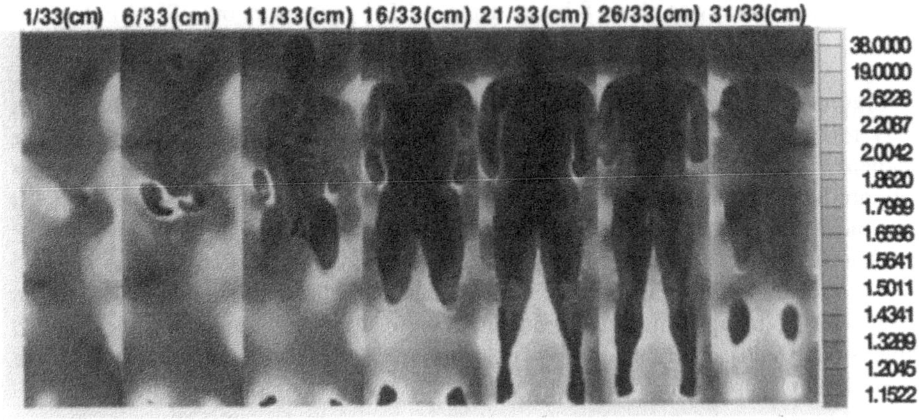

Figure 5. Amplitude Electric field distribution (V/m) at y=0: (a) a mid-coronal plane; (b) a mid-sagittal plane; and (c) a transverse plane 64cm from the origin.

The amplitude of the electric field distributions generated by the xyz gradient coils are shown in figure 5. These electric fields only exist during the transient state of current flow in the gradient coils. When the current reaches a constant value, these fields subside. The timing of this decay depends on the "electrical" relaxation behaviour of the tissues and will be the subject of future studies. From these simulations, it is clear that the electromagnetic field penetrates deeply into the body and that the induced eddy currents have complex distributions. In future work we plan to complete simulations of different gradient sets, patient positions and pulse sequences and attempt to correlate these with experimental reports of PNS.

CONCLUSION

A new method for the analysis of gradient induced effects in the human body has been presented. The method is a variant on the FDTD approach and essentially allows the interaction of electromagnetic fields with tissue at any frequency. It is therefore capable of modelling the effects of gradient and RF fields in MRI scanning. The analysis can accept a full imaging pulse sequence and analyse the induced fields at almost any point in time and space within the body model. The distribution of gradient induced E-fields and eddy currents in the body is important in predicting the safety of a pulse sequence and the biological consequences of particular gradient coil designs.

REFERENCES

[1] Furse CM, Gandhi OP. Calculation of electric fields and currents induced in a millimeter-resolution human model at 60 Hz using the FDTD method. Bioelectromagnetics 1998; 19: 293-299.

[2] S. Yee, Numerical Solution of Initial Boundary Value Problems Involving Maxwell's Equations in Isotropic Media IEEE Trans. Antennas Propagat 1966; 14: 302-307.

[3] Zhao H, Crozier S, Liu F. A high definition, finite difference time domain (HD-FDTD) method. submitted to Applied Mathematical Modelling, 2001.

[4] Berenger JP. A perfectly matched layer for the absorption of electromagnetic waves. Journal of Computational Physics 1994; 114:185-200.

[5] Dudley DG. Mathematical foundations for electromagnetic theory. IEEE Press Series on Electromagnetic Waves, IEEE, New York; 1994.

[6] Waldschmidt G, Taflove A. The Determination of the effective radius of a filamentary source in the FDTD mesh. IEEE microwave and guided wave letters 2000; 10(6): 217-219.

[7] Keltner, J.R., Carlson, J.W., Roos, M.S., et. al. Electromagnetic fields of surface coil in vivo NMR at high frequencies. Magn Reson Med 1991; 22: 467-480.

EPIDEMIOLOGICAL STUDIES OF CANCERS AND EXPOSURES TO RADIO FREQUENCIES

J. Mark Elwood
1 Rathdowne St, Carlton 3053 (Melbourne) Australia
Fax 61-3-9635-5320 melwood@ncci.org.au

ABSTRACT

This review will summarize the epidemiological studies assessing radiofrequency exposures and human cancer. Five types of epidemiological study contribute to this question: general descriptive studies, cluster studies, ecological studies, cohort, and case-control studies. The epidemiological evidence does not give clear or consistent results and requires careful consideration with regard to accepted criteria relevant to assessing causality. As with other science, the results cannot establish the absence of any hazard, other than to indicate that, for some situations, any undetected health effects must be small. All the studies have limitations in terms of the information on individual levels of exposure. However, the general population studies in the UK and Australia of subjects living close to large television and radio transmitters, the Motorola employees' study, the large population based study of mobile phone subscribers in Denmark, and the study of U.S. navy servicemen are particularly important. There are now four case-control studies published on brain cancer in relationship to personal use of mobile phones. Other studies are in progress, including those in the World Health Organization International EMF project.

CANCER IN PEOPLE LIVING NEAR RADIO AND TELEVISION TRANSMITTERS

While some trends have been seen, they have been inconsistent. In the largest study, of residents near 20 radio and TV transmitters in the UK, the results were generally negative, but a weak trend of a decrease in rates of adult leukemia with increasing distance from the transmitters was seen, of borderline statistical significance [1]. This followed a cluster study based on the Sutton Coldfield transmitter; no causal inference can be drawn from a cluster investigation alone [2]. In a further study of this transmitter [3] using cancer data for a later period, in children a marginally significant trend in leukemia incidence was seen for boys, based on small numbers, but there was no trend for girls, or in adults. An earlier study of childhood cancer in San Francisco showed no geographical association with a large transmitter [4].

In Sydney, Australia [5], increased incidence and mortality rates of childhood leukemia in areas close to VHF-TV transmitters were shown, but a further analysis showed that the excess applied only to one of the three areas close to the transmitter [6]. In Rome residents living near the Vatican radio station [7], which has no VHF or television transmissions, there was a significant trend in leukemia mortality for men, but no association for women; for childhood leukemia, based on eight cases, there was a significant trend of risk decreasing with increasing distance.

STUDIES OF PEOPLE EXPOSED TO RADIOFREQUENCIES THROUGH THEIR OCCUPATION

A cohort study [8] of 195,775 US Motorola employees showed that, compared to a general U.S. population, their death rates from all causes, from cancer, and from cancers of the lymphatic/hemopoetic system and the central nervous system (CNS) were all significantly reduced, characteristic of a 'healthy worker effect'. Comparing workers with higher radio frequency (RF) exposures with other workers in the same cohort, there were no increased risks of brain cancer, lymphatic and hemopoetic cancers, or Hodgkin's disease. The exposure information is limited; if the maximum levels of exposure were within current standards, the negative result is less informative. The results show no evidence of an increase in any specific cancer. Although a small increase or decrease cannot be excluded, the study is sufficiently powerful to reasonably exclude a substantial excess of leukemia or lymphoma in about ten years from first occupational RF exposure in these workers.

A study of U.S. navy servicemen showed no clear increase in cancer in personnel likely to have had substantial exposure to radar [9; 10]. These men served on U.S. navy ships at the time of the Korean War. Three groups likely to have had relatively high exposure to radar were identified, with the potential for exposures exceeding 100 mW/cm^2, although their usual exposures were well below 1 mW/cm^2. All the subjects were exposed to 60 Hz fields from electrical equipment, and some would have had considerable exposure to solder fumes, chlorinated solvents, and oils and greases. For the entire cohort compared to U.S. white male death rates the all-cause and cancer death rates were significantly reduced.

Compared to other men in the cohort, those in the high radar exposure group had significant reductions in total mortality and in deaths from cancer, but there was a significant excess in all leukemias (relative risk, RR, 1.48, 95% confidence limits of 1.01-2.17). Deaths from brain cancer were non-significantly reduced. The reductions in cancers were seen in each of the three high exposure groups, but the increase in leukemias was restricted to one of the groups. The actual exposure levels are unknown, and radio frequency exposures may have been substantial in all groups. There was no information available on lifestyle factors such as smoking, or on the post-Navy exposures or occupations of these men.

In earlier studies, a cohort study in the Polish military showed substantial excesses of total cancer and of several types of cancer [11], but there are design faults in the study likely to give bias in the exposure information [12-14]; the results are inconsistent with those of other studies. Studies of U.S. amateur radio operators showed an excess in one of nine types of leukemia assessed, although they may have had other relevant exposures [15]. A study of female radio and telegraph operators working at sea showed an excess of breast cancer and uterine cancer, but other confounding factors are relevant [16]. A detailed study of electrical workers in Quebec and France showed an excess of lung cancer, but their exposures were not primarily to RF [17].

STUDIES OF PARTICULAR TYPES OF CANCER

In case-control studies of particular types of cancer, RF has been one of a number of exposure factors assessed, and these studies do not show consistent results. While radio frequency exposures were related to brain cancers in U.S. Air Force personnel, higher rank had a stronger association [18]. In U.S. civilians, an excess risk was seen only with the combination of radio frequency exposures and other electrical/electronic job exposures, not for RF alone [19]. Other studies show results which vary with the method of collecting the information, are non-significant, or are open to the problem of multiple testing.

STUDIES OF CANCER IN ASSOCIATION WITH THE USE OF CELLULAR TELEPHONES

General population cohort study of cell phone users in Denmark.

Johansen *et al* [20] performed a prospective cohort study in Denmark of 420,095 cell phone subscribers since 1991, 58% with digital Global System for Mobile Communications (GSM) and the rest with analogue Nordic Mobile Telephone (NMT) systems. Incidence of all cancers was reduced, RR 0.86, in men, but not in women. For men, the incidence ratios of most smoking related cancers were reduced, while testicular cancer was non-significantly elevated. There were no increased risks seen for central nervous system (CNS) tumors or for leukemia; no trends were seen with latency up to five or more years or with age at first subscription, and no differences were seen between analogue and digital phones. The study was not able to assess intensity of use and the average period of follow-up was only 3.1 years. However, the study provides considerable evidence against any large increase in risk within a few years of use [20].

Case-control studies of brain tumors and cell phone use

In a Swedish study [21; 22] of 209 subjects with brain tumors (malignant or benign), ever-use of a cell phone showed no association, odds ratio 0.98, 95% confidence limits 0.69 to 1.41. In a later paper based on the same study [22], a marginally significant increased risk for tumors in the temporal, occipital, or temporoparietal regions was shown where cell phone use was on the same side. This result was based on one of many comparisons, and on only 13 of the 209 subjects, and the data on laterality of use may be open to recall bias.

In the U.S., Muscat *et al* [23] studied 469 patients with primary malignant brain cancers. Analyses by years of phone use (up to 4 or more), hours of use per month (up to 10 or more), and cumulative hours (up to 480 or more), showed no excess risks and no significant trends. There were no significant associations with any brain site; sub-division by pathological type showed no significant associations, although the risk for neuroepitheliomatous tumors was 2.1 (95% limits 0.9 - 4.7), based on 35 cases. This study shows no excess risks, even for the specific locations of tumors that were highlighted in the previous Swedish case-control study.

A larger U.S. case-control study of 782 patients with malignant or benign brain tumors [24] also showed no association with phone use. There was no association with laterality, no increased risk for temporal, parietal or frontal lobe tumors, and no increased risk with specific subtypes of tumors; the risk for neuroepitheliomatous tumors was 0.5 (95% limits 0.1 – 2.0), based on 25 cases.

Hardell *et al* [25] performed a further case-control study of 1429 patients with malignant or benign cranial tumors diagnosed in Sweden. A small but significant increased risk was seen with use of analogue phones, odds ratio 1.3 (95% confidence limits 1.02-1.6), increasing to 1.4 with more than 5 years latency, and 1.8 with over 10 years latency. The highest risks were seen for temporal lobe tumors, and with use of the phone on the same side as the tumor, although this was seen also for brain sites outside the area of phone exposure. There was no increased risk seen with digital or cordless phones. Risks were increased for acoustic neuromas and meningiomas, both benign tumors; there was no association with malignant tumors. In two U.S. studies of acoustic neuroma [26] [24], and in the Danish study [20], no associations with phone use were seen.

In a case-control study of 398 subjects with brain tumors in Finland, using computerized data linkage, there was a non-significant increase in risk (odds ratio 1.3, 95% limits 0.9 to 1.8) for ever having held a cell phone subscription. Gliomas showed a higher and statistically significant association with analogue phone subscription (odds ratio 2.1, limits 1.3 to 3.4).

DISCUSSION

The epidemiological evidence does not give clear or consistent results which indicate a causal role of radio frequency exposures in human cancer. On the other hand, the results cannot establish the absence of any hazard, other than to indicate that, for some situations, any undetected health effects must be small [12]. None of these studies give good information on individual levels of exposure. The studies of general populations living near radio or television transmitters relate to radio frequency exposures likely to be below currently accepted standards, while the studies of military personnel and occupational groups may include some higher exposures.

There are now six published studies of brain tumors and use of mobile phones, with mixed results. The main limitation of these studies is the short time frame. The second Swedish study [25] shows an overall increase in risk of brain tumors which is maximal for analogue phone use, temporal lobe tumors, and (within the temporal area) acoustic neuromas, and with phone use on the same side, which are all consistent with exposure intensity considerations. However all these associations are based on small numbers. The data on laterality of use are based on a single question about predominant use, which does not exclude some amount of phone use on the other side, and is open to biased responses. The only other study showing an overall increase in risk, although non-significant, is the Finnish study, which lacks any precise exposure information. The two U.S. studies that have the most detailed exposure information show no overall increase, and neither does the Danish cohort study that has the advantage of a prospective design. Neither of the U.S. studies nor the Danish prospective study show any higher risk with temporal tumors, or with acoustic neuroma, nor any association with phone use on the same side; in fact for acoustic neuromas there is a trend towards a stronger association with contralateral phone use.

The epidemiological studies of leukemia and radiofrequency exposures are difficult to interpret. The results show little consistency, but several show some weak associations.

CONCLUSION

The epidemiological results fall short of the strength and consistency of evidence which is required to come to a reasonable conclusion that radio frequency emissions are a cause of human cancer. The studies individually are limited and so their results cannot be clearly interpreted in terms of cause and effect. The evidence is weak in regard to its inconsistency, limitations in the design of many of the studies, the lack of detail on actual exposures, the limitations of the studies in their ability to deal with other likely factors, and in some studies there are likely to be biases in the data used.

REFERENCES

1 H.Dolk, P.Elliott, G.Shaddick, P.Walls, and B.Thakrar, "Cancer incidence near radio and television transmitters in Great Britain 2: All high power transmitters," *Am J Epidemiol*, 145, 10-17 (1997).

2 H.Dolk, G.Shaddick, P.Walls, C.Grundy, B.Thakrar, L.Kleinschmidt, and P.Elliott, "Cancer incidence near radio and television transmitters in Great Britain 1. Sutton Coldfield transmitter," *Am J Epidemiol*, 145, 1-9 (1997).

3 D.Cooper, K.Hemmings, and P.Saunders, "Re: "Cancer incidence near radio and television transmitters in Great Britain I. Sutton Coldfield transmitter II. All high power transmitters"," *Am J Epidemiol*, 153, 202-204 (2001).

4 S.Selvin, J.Schulman, and D.W.Merrill, "Distance and risk measures for the analysis of spatial data: a study of childhood cancers," *Soc Sci Med*, 34, 769-777 (1992).

5 B.Hocking, I.R.Gordon, H.L.Grain, and G.E.Hatfield, "Cancer incidence and mortality and proximity to TV towers," *Med J Aust*, 165, 601-605 (1996).

6 D.R.McKenzie, Y.Yin, and S.Morrell, "Childhood incidence of acute lymphoblastic leukaemia and exposure to broadcast radiation in Sydney - a second look," *Aust N Z J Public Health*, 22, 360-367 (1998).

7 P.Michelozzi, A.Capon, U.Kirchmayer, F.Forastiere, A.Biggeri, A.Barca, and C.A.Perucci, "Adult and childhood leukemia near a high-power radio station in Rome, Italy," *Am J Epidemiol*, 155, 1096-1103 (2002).

8 R.W.Morgan, M.A.Kelsh, K.Zhao, K.A.Exuzides, S.Heringer, and W.Negrete, "Radiofrequency exposure and mortality from cancer of the brain and lymphatic/hematopoietic systems," *Epidemiology*, 11, 118-127 (2000).

9 C.D.Robinette, C.Silverman, and S.Jablon, "Effects upon health of occupational exposure to microwave radiation (radar)," *Am J Epidemiol*, 112, 39-53 (1980).

10 F.D.Groves, W.F.Page, G.Gridley, L.Lisimaque, P.A.Stewart, R.E.Tarone, M.H.Gail, J.D.Boice, Jr., and G.W.Beebe, "Cancer in Korean war navy technicians: mortality survey after 40 years," *Am J Epidemiol*, 155, 810-818 (2002).

11 S.Szmigielski, "Cancer morbidity in subjects occupationally exposed to high frequency (radiofrequency and microwave) electromagnetic radiation," *Sci.Total.Environ*, 180, 9-17 (1996).

12 J.M.Elwood, "A critical review of epidemiologic studies of radiofrequency exposure and human cancers," *Environ Health Perspect*, 107, 155-168 (1999).

13 U.Bergqvist, "Review of epidemiological studies, pp. 147-170 in "*Mobile Communications Safety*", Edited by N.Kuster, Q.Balzano, and J.C.Lin. (London, 1997).

14 Independent Expert Group on Mobile Phones, *Mobile phones and health* (Oxford, 2000), pp. 1-160.

15 S.Milham, "Increased mortality in amateur radio operators due to lymphatic and hematopoietic malignancies," *Am J Epidemiol*, 127, 50-54 (1988).

16 T.Tynes, M.Hannevik, A.Andersen, A.I.Vistnes, and T.Haldorsen, "Incidence of breast cancer in Norwegian female radio and telegraph operators," *Cancer Causes Control*, 7, 197-204 (1996).

17 B.Armstrong, G.Theriault, P.Guenel, J.Deadman, M.Goldberg, and P.Heroux, "Association between exposure to pulsed electromagnetic fields and cancer in electric utility workers in Quebec, Canada, and France," *Am J Epidemiol*, 140, 805-820 (1994).

18 J.K.Grayson and T.J.Lyons, "Cancer incidence in United States Air Force aircrew, 1975-89," *Aviat Space Environ Med*, 67, 101-104 (1996).

19 T.L.Thomas, P.D.Stolley, A.Stemhagen, E.T.H.Fontham, M.L.Bleeker, P.A.Stewart, and R.N.Hoover, "Brain tumour mortality risk among men with electrical and electronic jobs: a case-control study," *J Natl Cancer Inst*, 79, 233-238 (1987).

20 C.Johansen, J.D.Boice, Jr., J.K.McLaughlin, and J.H.Olsen, "Cellular telephones and cancer - a nationwide cohort study in Denmark," *J Natl Cancer Inst*, 93, 203-207 (2001).

21 L.Hardell, Å.Näsman, A.Påhlson, A.Hallquist, and K.Hansson Mild, "Use of cellular telephones and the risk for brain tumours: a case-control study," *Int J Oncol*, 15, 113-116 (1999).

22 L.Hardell, A.Nasman, A.Pahlson, and A.Hallquist, "Case-control study on radiology work, medical x-ray investigations, and use of cellular telephones as risk factors for brain tumors," *MedGenMed*, May 4, 1-11 (2000).

23 J.E.Muscat, M.G.Malkin, S.Thompson, R.Shore, S.Stellman, D.McRee, A.I.Neugut, and E.L.Wynder, "Handheld cellular telephone use and risk of brain cancer," *JAMA*, 284, 3001-3007 (2000).

24 P.D.Inskip, R.E.Tarone, E.E.Hatch, T.C.Wilcosky, W.R.Shapiro, R.G.Selker, H.A.Fine, P.M.Black, J.S.Loeffler, and M.S.Linet, "Cellular-telephone use and brain tumors," *N Engl J Med*, 344, 79-86 (2001).

25 L.Hardell, A.Hallquist, K.Hansson Mild, M.Carlberg, A.Pahlson, and A.Lilja, "Cellular and cordless telephones and the risk for brain tumours," *Eur J Cancer Prev*, 11, 377-386 (2002).

26 J.E.Muscat, M.G.Malkin, R.E.Shore, S.Thompson, A.I.Neugut, S.D.Stellman, and J.Bruce, "Handheld cellular telephones and risk of acoustic neuroma," *Neurology*, 58, 1304-1306 (2002).

KEYWORD AND AUTHOR INDEX

Printed and bound by CPI Group (UK) Ltd, Croydon, CR0 4YY

15/09/2026

14962409-0003